Probability and Mathematical Statistics (Continued)

PURI, VILAPLANA, and WERTZ • New Perspectives in Theoretical and Applied Statistics
RANDLES and WOLFE • Introduction to the Theory of Nonparametric Statistics
RAO • Linear Statistical Inference and Its Applications, *Second Edition*
RAO • Real and Stochastic Analysis
RAO and SEDRANSK • W.G. Cochran's Impact on Statistics
RAO • Asymptotic Theory of Statistical Inference
ROBERTSON, WRIGHT and DYKSTRA • Order Restricted Statistical Inference
ROGERS and WILLIAMS • Diffusions, Markov Processes, and Martingales,Volume II: Îto Calculus
ROHATGI • An Introduction to Probability Theory and Mathematical Statistics
ROHATGI • Statistical Inference
ROSS • Stochastic Processes
RUBINSTEIN • Simulation and The Monte Carlo Method
RUZSA and SZEKELY • Algebraic Probability Theory
SCHEFFE • The Analysis of Variance
SEBER • Linear Regression Analysis
SEBER • Multivariate Observations
SEBER and WILD • Nonlinear Regression
SEN • Sequential Nonparametrics: Invariance Principles and Statistical Inference
SERFLING • Approximation Theorems of Mathematical Statistics
SHORACK and WELLNER • Empirical Processes with Applications to Statistics
STOYANOV • Counterexamples in Probability

Applied Probability and Statistics

ABRAHAM and LEDOLTER • Statistical Methods for Forecasting
AGRESTI • Analysis of Ordinal Categorical Data
AICKIN • Linear Statistical Analysis of Discrete Data
ANDERSON and LOYNES • The Teaching of Practical Statistics
ANDERSON, AUQUIER, HAUCK, OAKES, VANDAELE, and WEISBERG • Statistical Methods for Comparative Studies
ARTHANARI and DODGE • Mathematical Programming in Statistics
ASMUSSEN • Applied Probability and Queues
BAILEY • The Elements of Stochastic Processes with Applications to the Natural Sciences
BARNETT • Interpreting Multivariate Data
BARNETT and LEWIS • Outliers in Statistical Data, *Second Edition*
BARTHOLOMEW • Stochastic Models for Social Processes, *Third Edition*
BARTHOLOMEW and FORBES • Statistical Techniques for Manpower Planning
BATES and WATTS • Nonlinear Regression Analysis and Its Applications
BECK and ARNOLD • Parameter Estimation in Engineering and Science
BELSLEY, KUH, and WELSCH • Regression Diagnostics: Identifying Influential Data and Sources of Collinearity
BHAT • Elements of Applied Stochastic Processes, *Second Edition*
BLOOMFIELD • Fourier Analysis of Time Series: An Introduction
BOLLEN • Structural Equations with Latent Variables
BOX • R. A. Fisher, The Life of a Scientist
BOX and DRAPER • Empirical Model-Building and Response Surfaces
BOX and DRAPER • Evolutionary Operation: A Statistical Method for Process Improvement
BOX, HUNTER, and HUNTER • Statistics for Experimenters: An Introduction to Design, Data Analysis, and Model Building
BROWN and HOLLANDER • Statistics: A Biomedical Introduction
BUNKE and BUNKE • Statistical Inference in Linear Models, Volume I
CHAMBERS • Computational Methods for Data Analysis
CHATTERJEE and HADI • Sensitivity Analysis in Linear Regression
CHATTERJEE and PRICE • Regression Analysis by Example
CHOW • Econometric Analysis by Control Methods

Applied Probability and Statistics (Continued)

CLARKE and DISNEY • Probability and Random Processes: A First Course with Applications, *Second Edition*
COCHRAN • Sampling Techniques, *Third Edition*
COCHRAN and COX • Experimental Designs, *Second Edition*
CONOVER • Practical Nonparametric Statistics, *Second Edition*
CONOVER and IMAN • Introduction to Modern Business Statistics
CORNELL • Experiments with Mixtures: Designs, Models and The Analysis of Mixture Data
COX • Planning of Experiments
COX • A Handbook of Introductory Statistical Methods
DANIEL • Biostatistics: A Foundation for Analysis in the Health Sciences, *Fourth Edition*
DANIEL • Applications of Statistics to Industrial Experimentation
DANIEL and WOOD • Fitting Equations to Data: Computer Analysis of Multifactor Data, *Second Edition*
DAVID • Order Statistics, *Second Edition*
DAVISON • Multidimensional Scaling
DEGROOT, FIENBERG and KADANE • Statistics and the Law
DEMING • Sample Design in Business Research
DILLON and GOLDSTEIN • Multivariate Analysis: Methods and Applications
DODGE • Analysis of Experiments with Missing Data
DODGE and ROMIG • Sampling Inspection Tables, *Second Edition*
DOWDY and WEARDEN • Statistics for Research
DRAPER and SMITH • Applied Regression Analysis, *Second Edition*
DUNN • Basic Statistics: A Primer for the Biomedical Sciences, *Second Edition*
DUNN and CLARK • Applied Statistics: Analysis of Variance and Regression, *Second Edition*
ELANDT-JOHNSON and JOHNSON • Survival Models and Data Analysis
FLEISS • Statistical Methods for Rates and Proportions, *Second Edition*
FLEISS • The Design and Analysis of Clinical Experiments
FLURY • Common Principal Components and Related Multivariate Models
FOX • Linear Statistical Models and Related Methods
FRANKEN, KÖNIG, ARNDT, and SCHMIDT • Queues and Point Processes
GALLANT • Nonlinear Statistical Models
GIBBONS, OLKIN, and SOBEL • Selecting and Ordering Populations: A New Statistical Methodology
GNANADESIKAN • Methods for Statistical Data Analysis of Multivariate Observations
GREENBERG and WEBSTER • Advanced Econometrics: A Bridge to the Literature
GROSS and HARRIS • Fundamentals of Queueing Theory, *Second Edition*
GROVES, BIEMER, LYBERG, MASSEY, NICHOLLS, and WAKSBERG • Telephone Survey Methodology
GUPTA and PANCHAPAKESAN • Multiple Decision Procedures: Theory and Methodology of Selecting and Ranking Populations
GUTTMAN, WILKS, and HUNTER • Introductory Engineering Statistics, *Third Edition*
HAHN and SHAPIRO • Statistical Models in Engineering
HALD • Statistical Tables and Formulas
HALD • Statistical Theory with Engineering Applications
HAND • Discrimination and Classification
HEIBERGER • Computation for the Analysis of Designed Experiments
HOAGLIN, MOSTELLER and TUKEY • Exploring Data Tables, Trends and Shapes
HOAGLIN, MOSTELLER, and TUKEY • Understanding Robust and Exploratory Data Analysis
HOCHBERG and TAMHANE • Multiple Comparison Procedures
HOEL • Elementary Statistics, *Fourth Edition*
HOEL and JESSEN • Basic Statistics for Business and Economics, *Third Edition*
HOGG and KLUGMAN • Loss Distributions

Structural Equations
with Latent Variables

Structural Equations with Latent Variables

KENNETH A. BOLLEN

Department of Sociology
The University of North Carolina at Chapel Hill
Chapel Hill, North Carolina

A Wiley-Interscience Publication

JOHN WILEY & SONS

New York • Chichester • Brisbane • Toronto • Singapore

To Barbara
and to my parents

***Library of Congress Cataloging in Publication Data*:**

Bollen, Kenneth A.
Structural equations with latent variables/Kenneth A. Bollen.
p. cm.—(Wiley series in probability and mathematical statistics. Applied probability and statistics section, ISSN 0271-6356)
"A Wiley-Interscience publication."
Bibliography: p.
Includes index.
1. Social sciences—Statistical methods. 2. Latent variables.
I. Title. II. Series.
HA29.B732 1989
300′.29—dc 19 88-27272
ISBN 0-471-01171-1
CIP

Printed in the United States of America

10 9

Preface

Within the past decade the vocabulary of quantitative research in the social sciences has changed. Terms such as "LISREL," "covariance structures," "latent variables," "multiple indicators," and "path models" are commonplace. The structural equation models that lie behind these terms are a powerful generalization of earlier statistical approaches. They are changing researchers' perspectives on statistical modeling and building bridges between the way social scientists think substantively and the way they analyze data.

We can view these models in several ways. They are regression equations with less restrictive assumptions that allow measurement error in the explanatory as well as the dependent variables. They consist of factor analyses that permit direct and indirect effects between factors. They routinely include multiple indicators and latent variables. In brief, these models encompass and extend regression, econometric, and factor analysis procedures.

The book provides a comprehensive introduction to the general structural equation system, commonly known as the "LISREL model." One purpose of the book is to demonstrate the generality of this model. Rather than treating path analysis, recursive and nonrecursive models, classical econometrics, and confirmatory factor analysis as distinct and unique, I treat them as special cases of a common model. Another goal of the book is to emphasize the application of these techniques. Empirical examples appear throughout. To gain practice with the procedures, I encourage the reader to reestimate the examples, and then to devise and estimate new models. Several chapters contain some of the LISREL or EQS programs I used to obtain the results for the empirical examples. I have kept the examples as realistic as possible. This means that some of the initial specifications do *not* fit well. Through my experiences with students, colleagues, and in my own work, I frequently have found that the beginning model does not adequately describe the data. Respecification is often necessary. I note the difficulties this creates in proper interpretations of significance tests and the added importance of replication.

A final purpose is to emphasize the crucial role played by substantive expertise in most stages of the modeling process. Structural equation models are not very helpful if you have little idea about the subject matter. To begin the fitting process, the analysts must draw upon their knowledge to construct a multiequation system that specifies the relations between all latent variables, disturbances, and indicators. Furthermore they must turn to substantive information when respecifying models and when evaluating the final model. Empirical results can reveal that initial ideas are in error or they can suggest ways to modify a model, but they are given meaning only within the context of a substantively informed model.

Structural equation models can be presented in two ways. One is to start with the general model and then show its specializations to simpler models. The other is to begin with the simpler models and to build toward the general model. I have chosen the latter strategy. I start with the regression/econometric and factor analysis models and present them from the perspective of the general model. This has the advantage of gradually including new material while having types of models with which the reader is somewhat familiar. It also encourages viewing old techniques in a new light and shows the often unrealistic assumptions implicit in standard regression/econometric and factor analyses.

Specifically, I have organized the book as follows. Chapter 2 introduces several methodological tools. I present the model notation, covariances and covariance algebra, and a more detailed account of path analysis. Appendixes A and B at the end of the book provide reviews of matrix algebra and of asymptotic distribution theory. Chapter 3 addresses the issue of causality. Implicitly, the idea of causality pervades much of the structural equation writings. The meaning of causality is subject to much controversy. I raise some of the issues behind the controversy and present a structural equation perspective on the meaning of causality.

The regression/econometric models for observed variables are the subject of Chapter 4. Though many readers have experience with these, the covariance structure viewpoint will be new to many. The consequences of random measurement error in observed variable models is the topic of Chapter 5. The chapter shows why and when we should care about measurement error in observed variables.

Once we recognize that variables are measured with error, we need to consider the relation between the error-free variable and the observed variable. Chapter 6 is an examination of this relation. It introduces procedures for developing measures and explores the concepts of reliability and validity. Chapter 7 is on confirmatory factor analysis, which is used for estimating measurement models such as those in Chapter 6.

Finally, Chapters 8 and 9 provide the general structural equation model with latent variables. Chapter 8 emphasizes the "basics," whereas Chapter 9 treats more advanced topics such as arbitrary distribution estimators and the treatment of categorical observed variables.

The main motivation for writing this book arose from my experiences teaching at the Interuniversity Consortium for Political and Social Research (ICPSR) Summer Training Program in Methodology at the University of Michigan (1980–1988). I could not find a text suitable for graduate students and professionals with training in different disciplines. Nor could I find a comprehensive introduction to these procedures. I have written the book for social scientists, market researchers, applied statisticians, and other analysts who plan to use structural equation or LISREL models. I assume that readers have prior exposure and experience with matrix algebra and regression analysis. A background in factor analysis is helpful but not essential. Jöreskog and Sörbom's (1986) LISREL and Bentler's (1985) EQS are the two most popular structural equation software packages. I make frequent reference to them, but the ideas of the book extend beyond any specific program.

I have many people to thank for help in preparing this book. The Interuniversity Consortium for Political and Social Research (ICPSR) at the University of Michigan (Ann Arbor) has made it possible for me to teach these techniques for the last nine years in their Summer Program in Quantitative Methods. Bob Hoyer started me there. Hank Heitowit and the staff of ICPSR have continued to make it an ideal teaching environment. A number of the hundreds of graduate students, professors, and other professionals who attended the courses provided general and specific comments to improve the book. Gerhard Arminger (University of Wuppertal), Jan de Leeuw (University of Leiden), Raymond Horton (Lehigh University), Frederick Lorenz (Iowa State University), John Fox (York University), Robert Stine (University of Pennsylvania), Boone Turchi (University of North Carolina), and members of the Statistical and Mathematical Sociology Group at the University of North Carolina provided valuable comments on several of the chapters. Barbara Entwisle Bollen read several drafts of most chapters, and her feedback and ideas are reflected throughout the book. Without her encouragement, I do not know when or if I would have completed the book.

Brenda Le Blanc of Dartmouth College and Priscilla Preston and Jenny Neville of the University of North Carolina (Chapel Hill) performed expert typing above and beyond the call of duty. Stephen Birdsall, Jack Kasarda, and Larry Levine smoothed the path several times as I moved the manuscript from Dartmouth College to the University of North Carolina. I thank the Committee for Scholarly Publications, Artistic Exhibitions, and Performances at the University of North Carolina, who provided financial support needed to complete the book.

KENNETH A. BOLLEN

Chapel Hill, North Carolina

Contents

CHAPTER ONE

Introduction

Most researchers applying statistics think in terms of modeling the *individual observations*. In multiple regression or ANOVA (analysis of variance), for instance, we learn that the regression coefficients or the error variance estimates derive from the minimization of the sum of squared differences of the predicted and observed dependent variable for each case. Residual analyses display discrepancies between fitted and observed values for every member of the sample.

The methods of this book demand a reorientation. The procedures emphasize *covariances* rather than cases.[1] Instead of minimizing functions of observed and predicted individual values, we minimize the difference between the sample covariances and the covariances predicted by the model. The observed covariances minus the predicted covariances form the residuals. The fundamental hypothesis for these structural equation procedures is that the covariance matrix of the observed variables is a function of a set of parameters. If the model were correct and if we knew the parameters, the population covariance matrix would be exactly reproduced. Much of this book is about the equation that formalizes this fundamental hypothesis:

$$\Sigma = \Sigma(\boldsymbol{\theta}) \tag{1.1}$$

In (1.1), Σ (sigma) is the population covariance matrix of observed variables, $\boldsymbol{\theta}$ (theta) is a vector that contains the model parameters, and $\Sigma(\boldsymbol{\theta})$ is

[1]As is clear from several places in the book, individual cases that are outliers can severely affect covariances and estimates of parameters. Thus, with these techniques, researchers still need to check for outliers. In addition, in many cases (e.g., regression models) the minimizations based on individuals and minimizations based on the predicted and observed covariance matrices lead to the same parameter estimates.

the covariance matrix written as a function of $\boldsymbol{\theta}$. The simplicity of this equation is only surpassed by its generality. It provides a unified way of including many of the most widely used statistical techniques in the social sciences. Regression analysis, simultaneous equation systems, confirmatory factor analysis, canonical correlations, panel data analysis, ANOVA, analysis of covariance, and multiple indicator models are special cases of (1.1).

Let me illustrate. In a simple regression equation we have $y = \gamma x + \zeta$, where γ (gamma) is the regression coefficient, ζ (zeta) is the disturbance variable uncorrelated with x and the expected value of ζ, $E(\zeta)$, is zero. The y, x, and ζ are random variables. This model in terms of (1.1) is[2]

$$\begin{bmatrix} \mathrm{VAR}(y) & \\ \mathrm{COV}(x, y) & \mathrm{VAR}(x) \end{bmatrix} = \begin{bmatrix} \gamma^2 \mathrm{VAR}(x) + \mathrm{VAR}(\zeta) & \\ \gamma \mathrm{VAR}(x) & \mathrm{VAR}(x) \end{bmatrix} \tag{1.2}$$

where VAR( ) and COV( ) refer to the population variances and covariances of the elements in parentheses. In (1.2) the left-hand side is $\boldsymbol{\Sigma}$, and the right-hand side is $\boldsymbol{\Sigma}(\boldsymbol{\theta})$, with $\boldsymbol{\theta}$ containing γ, $\mathrm{VAR}(x)$, and $\mathrm{VAR}(\zeta)$ as parameters. The equation implies that each element on the left-hand side equals its corresponding element on the right-hand side. For example, $\mathrm{COV}(x, y) = \gamma \mathrm{VAR}(x)$ and $\mathrm{VAR}(y) = \gamma^2 \mathrm{VAR}(x) + \mathrm{VAR}(\zeta)$. I could modify this example to create a multiple regression by adding explanatory variables, or I could add equations and other variables to make it a simultaneous equations system such as that developed in classical econometrics. Both cases can be represented as special cases of equation (1.1), as I show in Chapter 4.

Instead of a regression model, consider two random variables, x_1 and x_2, that are indicators of a factor (or latent random variable) called ξ (xi). The dependence of the variables on the factor is $x_1 = \xi + \delta_1$ and $x_2 = \xi + \delta_2$, where δ_1 (delta) and δ_2 are random disturbance terms, uncorrelated with ξ and with each other, and $E(\delta_1) = E(\delta_2) = 0$. Equation (1.1) specializes to

$$\begin{bmatrix} \mathrm{VAR}(x_1) & \\ \mathrm{COV}(x_1, x_2) & \mathrm{VAR}(x_2) \end{bmatrix} = \begin{bmatrix} \phi + \mathrm{VAR}(\delta_1) & \\ \phi & \phi + \mathrm{VAR}(\delta_2) \end{bmatrix} \tag{1.3}$$

where ϕ (phi) is the variance of the latent factor ξ. Here $\boldsymbol{\theta}$ consists of three elements: ϕ, $\mathrm{VAR}(\delta_1)$, and $\mathrm{VAR}(\delta_2)$. The covariance matrix of the observed variables is a function of these three parameters. I could add more

[2] Given the symmetric nature of the covariance matrices only the lower half of these matrices is shown.

indicators and more latent factors, allow for coefficients ("factor loadings") relating the observed variables to the factors, and allow correlated disturbances creating an extremely general factor analysis model. As Chapter 7 demonstrates, this is a special case of the covariance structure equation (1.1).

Finally, a simple hybrid of the two preceding cases creates a simple system of equations. The first part is a regression equation of $y = \gamma\xi + \zeta$, where unlike the previous regression the independent random variable is unobserved. The last two equations are identical to the factor analysis example: $x_1 = \xi + \delta_1$ and $x_2 = \xi + \delta_2$. I assume that ζ, δ_1, and δ_2 are uncorrelated with ξ and with each other, and that each has an expected value of zero. The resulting structural equation system is a combination of factor analysis and regression-type models, but it is still a specialization of (1.1):

$$\begin{bmatrix} \mathrm{VAR}(y) & & \\ \mathrm{COV}(x_1, y) & \mathrm{VAR}(x_1) & \\ \mathrm{COV}(x_2, y) & \mathrm{COV}(x_2, x_1) & \mathrm{VAR}(x_2) \end{bmatrix}$$

$$= \begin{bmatrix} \gamma^2\phi + \mathrm{VAR}(\zeta) & & \\ \gamma\phi & \phi + \mathrm{VAR}(\delta_1) & \\ \gamma\phi & \phi & \phi + \mathrm{VAR}(\delta_2) \end{bmatrix} \tag{1.4}$$

These examples foreshadow the general nature of the models I treat. My emphasis is on systems of *linear equations*. By linear, I mean that the relations between all variables, latent and observed, can be represented in linear structural equations or they can be transformed to linear forms.[3] Structural equations that are nonlinear in the parameters are excluded. Nonlinear functions of parameters are, however, common in the *covariance* structure equation, $\boldsymbol{\Sigma} = \boldsymbol{\Sigma}(\boldsymbol{\theta})$. For instance, the last example had three linear structural equations: $y = \gamma\xi + \zeta$, $x_1 = \xi + \delta_1$, and $x_2 = \xi + \delta_2$. Each is linear in the variables and parameters. Yet the covariance structure (1.4) for this model shows that $\mathrm{COV}(x_1, y) = \gamma\phi$, which means that the $\mathrm{COV}(x_1, y)$ is a nonlinear function of γ and ϕ. Thus it is the structural equations linking the observed, latent, and disturbance variables that are linear, and not necessarily the covariance structure equations.

[3]Though an incredibly broad range of procedures falls under linear equations, those treated here are a special class of possible models of the general moment structure models (see Bentler, 1983).

The term "structural" stands for the assumption that the parameters are not just descriptive measures of association but rather that they reveal an invariant "causal" relation. I will have more to say about the meaning of "causality" with respect to these models in Chapter 3, but for now, let it suffice to say that the techniques do not "discover" causal relations. At best they show whether the causal assumptions embedded in a model match a sample of data. Also the models are for continuous latent and observed variables. The assumption of continuous observed variables is violated frequently in practice. In Chapter 9 I discuss the robustness of the standard procedures and the development of new ones for noncontinuous variables.

Structural equation models draw upon the rich traditions of several disciplines. I provide a brief description of their origins in the next section.

HISTORICAL BACKGROUND

Who invented general structural equation models? There is no simple answer to this question because many scholars have contributed to their development. The answer to this question is further complicated in that the models continue to unfold, becoming more general and more flexible. However, it is possible to outline various lines of research that have contributed to the evolution of these models.

My review is selective. More comprehensive discussions are available from the perspectives of sociology (Bielby and Hauser 1977), psychology (Bentler 1980; 1986), and economics (Goldberger 1972; Aigner et al. 1984). Two edited collections that represent the multidisciplinary origins of these techniques are the volumes by Goldberger and Duncan (1973) and Blalock ([1971] 1985). Other more recent collections are in Aigner and Goldberg (1977), Jöreskog and Wold (1982), the November 1982 issue of the *Journal of Marketing Research*, and the May–June 1983 issue of the *Journal of Econometrics*.

I begin by identifying three components present in today's general structural equation models: (1) path analysis, (2) the conceptual synthesis of latent variable and measurement models, and (3) general estimation procedures. By tracing the rise of each component, we gain a better idea about the origins of these procedures.

Let me consider path analysis first. The biometrician Sewall Wright (1918, 1921, 1934, 1960) is its inventor. Three aspects of path analysis are the path diagram, the equations relating correlations or covariances to parameters, and the decomposition of effects. The first aspect, the path diagram, is a pictorial representation of a system of simultaneous equations. It shows the relation between all variables, including disturbances and

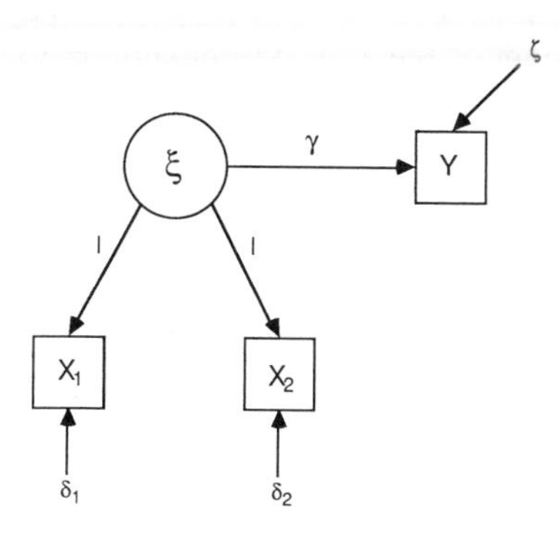

Figure 1.1 Path Diagram Example

errors. Figure 1.1 gives a path diagram for the last example of the previous section. It corresponds to the equations:

$$y = \gamma\xi + \zeta$$

$$x_1 = \xi + \delta_1$$

$$x_2 = \xi + \delta_2$$

where ζ, δ_1, and δ_2 are uncorrelated with each other and with ξ. Straight single-headed arrows represent one-way causal influences from the variable at the arrow base to the variable to which the arrow points. The implicit coefficients of one for the effects of ξ on x_1 and x_2 are made explicit in the diagram.

Using the path diagram, Wright proposed a set of rules for writing the equations that relate the correlations (or covariances) of variables to the model parameters; this constitutes the second aspect of path analysis. The equations are equivalent to covariance structure equations, an example of which appears in (1.4). He then proposed solving these equations for the unknown parameters, and substituting sample correlations or covariances for their population counterparts to obtain parameter estimates.

The third aspect of path analysis provides a means to distinguish direct, indirect, and total effects of one variable on another. The direct effects are those not mediated by any other variable; the indirect effects operate through at least one intervening variable, and the total effects is the sum of direct and all indirect effects.[4]

Wright's applications of path analysis proved amazing. His first article in 1918 was in contemporary terms a factor analysis in which he formulated

[4]I review path analysis more fully in Chapters 2 and 8.

and estimated a model of the size components of bone measurements. This was developed without knowledge of Spearman's (1904) work on factor analysis (Wright 1954, 15). Unobserved variables also appeared in some of his other applications. Goldberger (1972) credits Wright with pioneering the estimation of supply and demand equations with a treatment of identification and estimation more general than econometricians writing at the same time. His development of equations for covariances of variables in terms of model parameters is the same as that of (1.1), $\Sigma = \Sigma(\boldsymbol{\theta})$, except that he developed these equations from path diagrams rather than the matrix methods employed today.

With all these accomplishments it is surprising that social scientists and statisticians did not pay more attention to his work. As Bentler (1986) documents, psychometrics only flirted (e.g., Dunlap and Cureton 1930; Englehart 1936) with Wright's path analysis. Goldberger (1972) notes the neglect of econometricians and statisticians with a few exceptions (e.g., Fox 1958; Tukey 1954; Moran 1961; Dempster 1971). Wright's work also was overlooked in sociology until the 1960s. Partially in reaction to work by Simon (1954), Tukey (1954), and Turner and Stevens (1959), sociologists such as Blalock (1961, 1963, 1964), Boudon (1965), and Duncan (1966) saw the potential of path analysis and related "partial-correlation" techniques as a means to analyze nonexperimental data. Following these works, and particularly following Duncan's (1966) expository account, the late 1960s and early 1970s saw many applications of path analysis in the sociological journals. The rediscovery of path analysis in sociology diffused to political science and several other social science disciplines. Stimulated by work in sociology, Werts and Linn (1970) wrote an expository treatment of path analysis, but it was slow to catch on in psychology.

The next major boost to path analysis in the social sciences came when Jöreskog (1973), Keesing (1972), and Wiley (1973), who developed very general structural equation models, incorporated path diagrams and other features of path analysis into their presentations. Researchers know these techniques by the abbreviation of the JKW model (Bentler 1980), or more commonly as the LISREL model. The tremendous popularity of the LISREL model has facilitated the spread of path analysis. Path analysis has evolved over the years. Its present form has some elaboration in the symbols employed in path diagrams, has equations relating covariances to parameters that are derived with matrix operations rather than from "reading" the path diagram, and has a more refined and clearly defined decomposition of direct, indirect, and total effects (see, e.g., Duncan 1971; Alwin and Hauser 1975; Fox 1980; Graff and Schmidt 1982). But the contributions of Wright's work are still clear.

In addition to path analysis, the conceptual synthesis of latent variable and measurement models was essential to contemporary structural equation techniques. The factor analysis tradition spawned by Spearman (1904) emphasized the relation of latent factors to observed variables. The central concern was on what we now call the measurement model. The structural relations between latent variables other than their correlation (or lack of correlation) were not examined. In econometrics the focus was the structural relation between observed variables with an occasional reference to error-in-the-variable situations.

Wright's path analysis examples demonstrated that econometric-type models with variables measured with error could be identified and estimated. The conceptual synthesis of models containing structurally related latent variables and more elaborate measurement models was developed extensively in sociology during the 1960s and early 1970s. For instance, in 1963 Blalock argued that sociologists should use causal models containing both indicators and underlying variables to make inferences about the latent variables based on the covariances of the observed indicators. He suggested that observed variables can be causes or effects of latent variables or observed variables can directly affect each other. He contrasted this with the restrictive implicit assumptions of factor analysis where all indicators are viewed as effects of the latent variable. Duncan, Haller, and Portes (1968) developed a simultaneous equation model of peer influences on high school students' ambitions. The model included two latent variables reciprocally related, multiple indicators of the latent variables, and several background characteristics that directly affected the latent variables. Heise (1969) and others applied path analysis to separate the stability of latent variables from the reliability of measures.

This and related work in sociology during the 1960s and early 1970s demonstrated the potential of synthesizing econometric-type models with latent rather than observed variables and psychometric-type measurement models with indicators linked to latent variables. But their approach was by way of examples; they did not establish a general model that could be applied to any specific problems. It awaited the work of Jöreskog (1973), Keesing (1972), and Wiley (1973) for a practical general model to be proposed. Their models had two parts. The first was a latent variable model that was similar to the simultaneous equation model of econometrics except that all variables were latent ones. The second part was the measurement model that showed indicators as effects of the latent variables as in factor analyses. Matrix expressions for these models were presented so that they could apply to numerous individual problems. Jöreskog and Sörbom's LISREL programs were largely responsible for popularizing these structural equation models, as were the numerous publications and applications of Jöreskog (e.g., 1967, 1970, 1973, 1977, 1978) and his collaborators.

Bentler and Weeks (1980), McArdle and McDonald (1984), and others have proposed alternative representations of general structural equations. Though initially it seemed that these models were more general than the JKW model, most analysts now agree that both the new and old representations are capable of treating the range of linear models that typically occur in practice. I use what has come to be known as the "LISREL notation" throughout the book. To date, it is the most widely accepted representation. I will demonstrate ways to modify it to treat nonstandard applications in several of the chapters.

The last characteristic of the structural equation models are general estimation procedures. The early applications of path analysis by Wright and the sociologists influenced by his work used ad hoc estimation procedures to yield parameter estimates. There was little discussion of statistical inference and optimal ways of combining multiple estimates of a single parameter. Here work from econometrics and psychometrics proved indispensible. In econometrics the properties of estimators for structural equations with observed variables were well established (see, e.g., Goldberger 1964). In psychometrics the work of Lawley (1940), Anderson and Rubin (1956), and Jöreskog (1969) helped lay the foundations for hypothesis testing in factor analysis. Bock and Bargmann (1966) proposed an analysis of covariance structures to estimate the components of variance due to latent variables in multinormal observed variables. Jöreskog (1973) proposed a maximum likelihood estimator (based on the multinormality of the observed variables) for general structural equation models which is today the most widely used estimator. Jöreskog and Goldberger (1972) and Browne (1974, 1982, 1984) suggested generalized least squares (GLS) estimators that offer additional flexibility in the assumptions under which they apply. Browne (1982, 1984), for example, proposed estimators that assume arbitrary distributions or elliptical distributions for the observed variables. Bentler (1983) suggested estimators that treat higher-order product moments of the observed variables. He demonstrated that these moments can help identify model parameters that are not identified by the covariances and the gains in efficiency that may result. Muthén (1984, 1987), among others, has generalized these models to ordinal or limited observed variables.

Finally, this sketch of developments in structural equation models would not be complete without mentioning the computer software that has emerged. As I have already said, Jöreskog and Sörbom's LISREL software perhaps has been the single greatest factor leading to the spread of these techniques throughout the social sciences. It is now entering its seventh version. Bentler's (1985) EQS software has recently entered the field and also is likely to be widely used. McDonald (1980), Schoenberg (1987), and others have written programs with more limited circulations. The dual trend in software is that of providing more general and flexible models and programs that are more "user-friendly." Examples of the former are that LISREL VII and EQS allow a variety of estimators including Arbitrary distribution estimators and Muthén's (1987) LISCOMP allows ordinal or limited observed variables. At the same time there is a movement to allow programming through specifying equations (e.g., EQS and Jöreskog and Sörbom's SIMPLIS) rather than matrices.

CHAPTER TWO

Model Notation, Covariances, and Path Analysis

Readers of this book are likely to have diverse backgrounds in statistics. There is a need to establish some common knowledge. I assume that readers have prior exposure to matrix algebra. Appendix A at the end of the book provides a summary of basic matrix algebra for those wishing to review it. Appendix B gives an overview of asymptotic distribution theory which I use in several chapters. This chapter discusses three basic tools essential to understanding structural equation models. They are model notation, covariances, and path analysis.

MODEL NOTATION

Jöreskog (1973, 1977), Wiley (1973), and Keesling (1972) developed the notation on which I rely. Jöreskog and Sörbom's LISREL (LInear Structural RELationships) computer program popularized it, and many refer to it as the LISREL notation. I introduce the basic notation in this section and save the more specialized symbols for the later chapters where they are needed.

The full model consists of a *system of structural equations*. The equations contain random variables, structural parameters, and sometimes, nonrandom variables. The three types of random variables are latent, observed, and disturbance/error variables. The nonrandom variables are explanatory variables whose values remain the same in repeated random sampling (fixed or nonstochastic variables). These are less common than random explanatory variables.

The links between the variables are summarized in the *structural parameters*. The structural parameters are invariant constants that provide the "causal" relation between variables. The structural parameters may describe the causal link between unobserved variables, between observed variables, or between unobserved and observed variables. I further discuss the meaning of causality and the structural parameters in Chapter 3. The system of structural equations has two major subsystems: the latent variable model and the measurement model.

Latent Variable Model

Latent random variables represent unidimensional concepts in their purest form. Other terms for these are *unobserved* or *unmeasured variables* and *factors*. The observed variables or indicators of a latent variable contain random or systematic measurement errors, but the latent variable is free of these. Since all latent variables correspond to concepts, they are hypothetical variables. Concepts and latent variables, however, vary in their degree of abstractness. Intelligence, social class, power, and expectations are highly abstract latent variables that are central to many social science theories. Also important, but less abstract, are variables such as income, education, population size, and age. The latter type of latent variables are directly measurable, whereas the former are capable of being only indirectly measured. An example containing both types of latent variables is Emile Durkheim's hypothesis of the inverse relationship between social cohesion and suicide. Social cohesion refers to group solidarity, a fairly abstract latent variable. Suicide is directly observable. But this direct-indirect demarcation becomes blurred when one considers that some suicides are disguised or misclassified as some other form of death. Thus the measurement of suicide may not be as direct as it initially appears. I make no distinction between directly and indirectly observable latent variables for the latent variable models. Analytically they may be treated the same. Chapter 6 provides further discussion of the nature of latent variables, their measurement, and their scales.

The latent variable model encompasses the structural equations that summarize the relationships between latent variables. Sometimes this part of the model is called the "structural equation" or "causal model." I depart from this practice because it can be misleading. All equations in the model, both those for the latent variables and those for the measurement model, describe structural relationships. To apply structural to only the latent variable part of the full model suggests that the measurement model is not structural.

I use the relationship of political democracy to industrialization in developing countries to introduce the notation for latent variable models. International development researchers disagree about whether industrialization is positively associated with political democracy in Third World countries. The alternation between dictatorships and electoral regimes in some of these societies makes it difficult to discern whether any general association exists. Political democracy refers to the extent of political rights (e.g., fairness of elections) and political liberties (e.g., freedom of the press) in a country. Industrialization is the degree to which a society's economy is characterized by mechanized manufacturing processes. It is some of the consequences of industrialization (e.g., societal wealth, an educated population, advances in living standards) that are thought to enhance the chances of democracy. However, to keep the model simple, I do not include these intervening variables. Suppose that I have three latent random variables: political democracy in 1965 and 1960, and industrialization in 1960. I assume that political democracy in 1965 is a function of 1960 political democracy and industrialization. The 1960 industrialization level also affects the 1960 political democracy level. Nothing is said about the determinants of industrialization that lie outside of the model. Industrialization is an *exogenous* ("independent") latent variable and is symbolized as ξ_1 (xi). It is exogenous because its causes lie outside the model. The latent political democracy variables are *endogenous*; they are determined by variables within the model. Each endogenous latent variable is represented by η_i (eta). Political democracy in 1960 is represented as η_1 and 1965 democracy by η_2. The latent endogenous variables are only partially explained by the model. The unexplained component is represented by ζ_i (zeta) which is the random disturbance in the equation.

The terms exogenous and endogenous are model specific. It may be that an exogenous variable in one model is endogenous in another. Or, a variable shown as exogenous, in reality, may be influenced by a variable in the model. Regardless of these possibilities, the convention is to refer to variables as exogenous or endogenous based on their representation in a particular model.

The latent variable model for the current example is

$$\eta_1 = \gamma_{11}\xi_1 + \zeta_1 \tag{2.1}$$

$$\eta_2 = \beta_{21}\eta_1 + \gamma_{21}\xi_1 + \zeta_2 \tag{2.2}$$

The equations are *linear in the variables and linear in the parameters*. We can sometimes estimate equations that are nonlinear in the variables and linear in the parameters as in regression analysis. To date, however, practical and general means of implementing this for nonlinear functions of latent variables measured with error do not exist (see Chapter 9).

The random errors, ζ_1 and ζ_2, have expected values (means) of zero and are uncorrelated with the exogenous variable, industrialization (ξ_1). A constant is absent from the equations because the variables are deviated from their means.[1] This deviation form will simplify algebraic manipulations but does not affect the generality of the analysis. The β_{21} (beta) coefficient is the structural parameter that indicates the change in the expected value of η_2 after a one-unit increase in η_1 holding ξ_1 constant. The γ_{11} (gamma) and γ_{21} regression coefficients have analogous interpretations. The β_{21} coefficient is associated with the latent endogenous variable, whereas γ_{11} and γ_{21} are associated with the exogenous latent variable.

Equations (2.1) and (2.2) may be rewritten in matrix notation:

$$\begin{bmatrix} \eta_1 \\ \eta_2 \end{bmatrix} = \begin{bmatrix} 0 & 0 \\ \beta_{21} & 0 \end{bmatrix} \begin{bmatrix} \eta_1 \\ \eta_2 \end{bmatrix} + \begin{bmatrix} \gamma_{11} \\ \gamma_{21} \end{bmatrix} [\xi_1] + \begin{bmatrix} \zeta_1 \\ \zeta_2 \end{bmatrix} \tag{2.3}$$

which is more compactly written as

$$\boldsymbol{\eta} = \mathbf{B}\boldsymbol{\eta} + \boldsymbol{\Gamma}\boldsymbol{\xi} + \boldsymbol{\zeta} \tag{2.4}$$

The equation (2.4) is the general matrix representation of the structural equations for the latent variable model. Table 2.1 summarizes the notation for models relating latent variables, including each symbol's name, phonetic spelling, dimension, and definition.

Starting with the first variable, $\boldsymbol{\eta}$ is an $m \times 1$ vector of the latent endogenous random variables. In the industrialization–political democracy example m is equal to 2. The $\boldsymbol{\xi}$ vector is $n \times 1$, and it represents the n exogenous latent variables. Here, as in most cases, $\boldsymbol{\xi}$ is a vector of random variables. Occasionally, one or more of the ξ's are nonrandom. For the current example n is one since only industrialization (ξ_1) is exogenous. The errors in the equations or disturbances are represented by $\boldsymbol{\zeta}$, an $m \times 1$ vector. A ζ_i is associated with each η_i, with i running from 1 to m. The example has two ζ_i variables. The $\boldsymbol{\zeta}$ vector generally contains random variables.[2] As in a regression analysis the disturbance ζ_i includes those variables that influence η_i but are excluded from the η_i equation. We assume that these numerous omitted factors fused into ζ_i have $E(\zeta_i) = 0$

[1] In Chapters 4, 7, and 8 I will explain how to estimate models with a constant.

[2] One exception is if an equation is an identity (e.g., $\eta_1 = \eta_2 + \eta_3$) so that ζ_1 is zero and therefore constant.

Table 2.1 Notation for Latent (Unobserved) Variable Model

Structural Equation for Latent Variable Model

$$\boldsymbol{\eta} = \mathbf{B}\boldsymbol{\eta} + \boldsymbol{\Gamma}\boldsymbol{\xi} + \boldsymbol{\zeta}$$

Assumptions

$E(\boldsymbol{\eta}) = 0$
$E(\boldsymbol{\xi}) = 0$
$E(\boldsymbol{\zeta}) = 0$
$\boldsymbol{\zeta}$ uncorrelated with $\boldsymbol{\xi}$
$(\mathbf{I} - \mathbf{B})$ nonsingular

Symbol	Name	Phonetic Spelling	Dimension	Definition
			Variables	
$\boldsymbol{\eta}$	eta	*ā′ t ə* (or *ē′t ə*)	$m \times 1$	latent endogenous variables
$\boldsymbol{\xi}$	xi	*zī* (or ksē)	$n \times 1$	latent exogenous variables
$\boldsymbol{\zeta}$	zeta	zā′t ə (or zē′t ə)	$m \times 1$	latent errors in equations
			Coefficients	
$\mathbf{B}$	beta	*bā′t ə* (or bē′t ə)	$m \times m$	coefficient matrix for latent endogenous variables
$\boldsymbol{\Gamma}$	gamma	gam′ə	$m \times n$	coefficient matrix for latent exogenous variables
			Covariance Matrices	
$\boldsymbol{\Phi}$	phi	fī (or fē)	$n \times n$	$E(\boldsymbol{\xi}\boldsymbol{\xi}')$ (covariance matrix of $\boldsymbol{\xi}$)
$\boldsymbol{\Psi}$	psi	*sī* (or psē)	$m \times m$	$E(\boldsymbol{\zeta}\boldsymbol{\zeta}')$ (covariance matrix of $\boldsymbol{\zeta}$)

and are uncorrelated with the exogenous variables in $\boldsymbol{\xi}$. Otherwise, inconsistent coefficient estimators are likely.

We also assume that ζ_i is homoscedastic and nonautocorrelated. To clarify this assumption, suppose that I add an observation index to ζ_i so that ζ_{ik} refers to the value of ζ_i for the kth observation and ζ_{il} is ζ_i for the lth observation. The homoscedasticity assumption is that the VAR(ζ_i) is constant across cases [i.e., $E(\zeta_{ik}^2) = \mathrm{VAR}(\zeta_i)$ for all k]. The no autocorrelation assumption means that ζ_{ik} is uncorrelated with ζ_{il} for all k and l, where $k \neq l$ (i.e., $\mathrm{COV}(\zeta_{ik}, \zeta_{il}) = 0$ for $k \neq l$). Corrections for heteroscedastic or autocorrelated disturbances are well known for econometric-type models but hardly studied for the general structural equation model with latent variables. The homoscedasticity and no autocorrelation assumptions do *not* mean that the disturbances from two *different* equations need be uncorrelated nor that they need have the same variance. That is, $E(\zeta_{ik}^2) = \mathrm{VAR}(\zeta_i)$ is not the same as $E(\zeta_i^2) = E(\zeta_j^2)$, nor does $\mathrm{COV}(\zeta_{ik}, \zeta_{il}) = 0$ mean that $\mathrm{COV}(\zeta_i, \zeta_j) = 0$, where ζ_i and ζ_j are from separate equations.

The coefficient matrices are $\mathbf{B}$ and Γ. The $\mathbf{B}$ matrix is an $m \times m$ coefficient matrix for the latent endogenous variables. Its typical element is β_{ij} where i and j refer to row and column positions. The model assumes that $(\mathbf{I} - \mathbf{B})$ is nonsingular so that $(\mathbf{I} - \mathbf{B})^{-1}$ exists. This assumption enables (2.4) to be written in reduced form. Solving (2.4) algebraically so that only $\boldsymbol{\eta}$ appears on the left-hand side leads to the reduced form that I discuss in Chapter 4. The Γ matrix is the $m \times n$ coefficient matrix for the latent exogenous variables. Its elements are symbolized as γ_{ij}. For the industrialization–political democracy example,

$$\mathbf{B} = \begin{bmatrix} 0 & 0 \\ \beta_{21} & 0 \end{bmatrix}, \quad \boldsymbol{\eta} = \begin{bmatrix} \eta_1 \\ \eta_2 \end{bmatrix}, \quad \boldsymbol{\zeta} = \begin{bmatrix} \zeta_1 \\ \zeta_2 \end{bmatrix}, \quad \Gamma = \begin{bmatrix} \gamma_{11} \\ \gamma_{21} \end{bmatrix}, \quad \boldsymbol{\xi} = [\xi_1] \tag{2.5}$$

The main diagonal of $\mathbf{B}$ is always zero. This serves to remove η_i from the right-hand side of the ith equation for which it is the dependent variable. That is, we assume that a variable is not an immediate and instantaneous cause of itself. A zero in $\mathbf{B}$ also indicates the absence of an effect of one latent endogenous variable on another. That a zero appears in the (1, 2) position of $\mathbf{B}$ in (2.5) indicates that η_2 does not affect η_1. The Γ matrix in equation (2.5) is 2×1 since there are two endogenous latent variables and one exogenous latent variable. Since ξ_1 affects both η_1 and η_2, Γ contains no zero elements.

Two covariance matrices are part of the latent variable model in Table 2.1. A covariance matrix is an "unstandardized correlation matrix" with the variances of a variable down the main diagonal and the covariance (the product of the correlation between two variables times their standard deviations) of all pairs of variables in the off-diagonal.[3] The $n \times n$ covari-

[3] I treat covariances more fully later in this chapter.

ance matrix of the latent exogenous variables (or the ξ's) is $\mathbf{\Phi}$ (phi) with elements ϕ_{ij}. Like all covariance matrices, it is symmetric. If the variances of the ξ variables are equal to one, then $\mathbf{\Phi}$ is a correlation matrix. In the industrialization–political democracy example only one variable appears in $\boldsymbol{\xi}$, so $\mathbf{\Phi}$ is a scalar (i.e., ϕ_{11}) that equals the variance of ξ_1.

The $m \times m$ covariance matrix of the errors in the equations is $\mathbf{\Psi}$ (psi) with elements ψ_{ij}. Each element of the main diagonal of $\mathbf{\Psi}$ (ψ_{ii}) is the variance of the corresponding η_i variable that is unexplained by the explanatory variables included in the ith equation. In the current example $\mathbf{\Psi}$ is 2×2. The $(1,1)$ element is the variance of ζ_1, the $(2,2)$ element is the variance of ζ_2, and the off-diagonal elements—$(1,2)$ and $(2,1)$—are both equal to the covariance of ζ_1 with ζ_2. In this example I assume that the off-diagonal elements are zero.[4] As I show in Chapters 4 and 8, the covariance matrix for $\boldsymbol{\eta}$ is a function of $\mathbf{B}$, $\mathbf{\Gamma}$, $\mathbf{\Phi}$, and $\mathbf{\Psi}$. It does not have a special symbol.

Readers familiar with econometric texts will note the similarity between the structural equations for the latent variable model of Table 2.1 and the general representation of simultaneous equation systems (e.g., $\mathbf{By} + \mathbf{\Gamma x} = \mathbf{u}$, in Johnston 1984, 450). One difference is that both the endogenous and exogenous variables may be written on the left-hand side, leaving only $\boldsymbol{\zeta}$ on the right-hand side: $\mathbf{B}^*\boldsymbol{\eta} + \mathbf{\Gamma}^*\boldsymbol{\xi} = \boldsymbol{\zeta}$, with $\mathbf{B}^* = (\mathbf{I} - \mathbf{B})$ and $\mathbf{\Gamma}^* = -\mathbf{\Gamma}$. Most of the time some symbol other than $\boldsymbol{\zeta}$ represents the error in the equation (e.g., $\mathbf{u}$). These alternative representations matter little. Other differences are that in most econometric presentations $\mathbf{y}$ replaces $\boldsymbol{\eta}$ and $\mathbf{x}$ replaces $\boldsymbol{\xi}$. This difference is more than just a change in symbols. The classical econometric treatment assumes that the observed y and x variables are perfect measures of the latent η and ξ variables. Structural equations with latent variable models no longer have this assumption. In fact, the second major part of these models consists of structural equations linking the latent variables (the η's and ξ's) to the measured variables (the y's and x's). This part of the system is the measurement model.

Measurement Model

Like the latent variables, the observed variables have a variety of names, including manifest variables, measures, indicators, and proxies. I use these terms interchangeably. The latent variable model of industrialization and political democracy as described so far is exclusively in terms of unobserved variables. A test of this theory is only possible if I collect observable

[4]The ζ_1 and ζ_2 could be positively correlated since they influence the same latent variable separated only by five years. In other more elaborate models ψ_{12} might be a free parameter, but in this case it is not identified if freed. I return to the issue of identification in Chapter 4.

measures of these latent variables. One strategy is to use single indicators or proxy variables of political democracy and industrialization. Another option is to construct an index with two or more indicator variables for each of the concepts. The empirical analysis is of these observed indicators or indices, and researchers treat the results as tests of the relationships between the latent variables.

The underlying assumption of the preceding strategies is that the observed variables are perfectly correlated (or at least nearly so) with the latent variables that they measure. In most cases this is not true. Nearly all measures of abstract factors such as political democracy have far from perfect associations with the factor. The measurement model has structural equations that represent the link between the latent and observed variables, as an imperfect rather than a deterministic one. For this example I select three indicators of industrialization in 1960: gross national product (GNP) per capita (x_1), inanimate energy consumption per capita (x_2), and the percentage of the labor force in industry (x_3). For political democracy I have the same four indicators for 1960 and 1965: expert ratings of the freedom of the press (y_1 in 1960, y_5 in 1965), the freedom of political opposition (y_2 and y_6), the fairness of elections (y_3 and y_7), and the effectiveness of the elected legislature (y_4 and y_8). Thus each latent variable is measured with several observed variables.

Equations (2.6)–(2.7) provide a measurement model for these variables:

$$\begin{aligned} x_1 &= \lambda_1 \xi_1 + \delta_1 \\ x_2 &= \lambda_2 \xi_1 + \delta_2 \\ x_3 &= \lambda_3 \xi_1 + \delta_3 \end{aligned} \tag{2.6}$$

$$\begin{aligned} y_1 &= \lambda_4 \eta_1 + \epsilon_1, & y_5 &= \lambda_8 \eta_2 + \epsilon_5 \\ y_2 &= \lambda_5 \eta_1 + \epsilon_2, & y_6 &= \lambda_9 \eta_2 + \epsilon_6 \\ y_3 &= \lambda_6 \eta_1 + \epsilon_3, & y_7 &= \lambda_{10} \eta_2 + \epsilon_7 \\ y_4 &= \lambda_7 \eta_1 + \epsilon_4, & y_8 &= \lambda_{11} \eta_2 + \epsilon_8 \end{aligned} \tag{2.7}$$

As in the latent variable model, the variables in the measurement model are deviated from their means. The x_i variables ($i = 1, 2, 3$) stand for the three measures of ξ_1, industrialization, the y_1 to y_4 variables are measures of η_1, 1960 political democracy, and y_5 to y_8 measure η_2, 1965 democracy. Note that all the manifest variables depend on the latent variables. In some cases indicators may cause latent variables. This situation is discussed more fully in Chapters 3, 6, and 7.

The λ_i (lambda) coefficients are the magnitude of the expected change in the observed variable for a one unit change in the latent variable. These coefficients are regression coefficients for the effects of the latent variables on the observed variables. We must assign a scale to the latent variable to fully interpret the coefficients. Typically, analysts set a latent variable's scale equal to one of its indicators or standardize the variance of the latent variable to one. I will discuss this issue in more detail in chapters six and seven.

The δ_i (delta) and ϵ_i (epsilon) variables are the *errors of measurement* for x_i and y_i, respectively. They are disturbances that disrupt the relation between the latent and observed variables. The assumptions are that the errors of measurement have an expected value of zero, that they are uncorrelated with all ξ's, η's, and ζ's, and that δ_i and ϵ_j are uncorrelated for all i and j.

A correlation of δ_i and ϵ_j with any ξ's or η's can lead to inconsistent parameter estimators in a fashion analogous to a disturbance correlated with an explanatory variable in regression analysis. Sometimes in factor analysis δ_i and ϵ_j are called unique factors, and each δ_i and ϵ_j is divided into specific and nonspecific components. I will say more about this in Chapters 6 and 7, but till then I refer to the δ's and ϵ's as errors of measurement. Finally, we assume that each δ_i or ϵ_i is homoscedastic and nonautocorrelated across observations. This assumption parallels that made for the ζ_i's, the disturbances for the latent variable model.

Equations (2.6) and (2.7) may be more compactly written in matrix form as:

$$\mathbf{x} = \boldsymbol{\Lambda}_x \boldsymbol{\xi} + \boldsymbol{\delta} \tag{2.8}$$

$$\mathbf{y} = \boldsymbol{\Lambda}_y \boldsymbol{\eta} + \boldsymbol{\epsilon} \tag{2.9}$$

where

$$\mathbf{x} = \begin{bmatrix} x_1 \\ x_2 \\ x_3 \end{bmatrix}, \quad \boldsymbol{\Lambda}_x = \begin{bmatrix} \lambda_1 \\ \lambda_2 \\ \lambda_3 \end{bmatrix}, \quad \boldsymbol{\xi} = [\xi_1], \quad \boldsymbol{\delta} = \begin{bmatrix} \delta_1 \\ \delta_2 \\ \delta_3 \end{bmatrix} \tag{2.10a}$$

$$\mathbf{y} = \begin{bmatrix} y_1 \\ y_2 \\ y_3 \\ y_4 \\ y_5 \\ y_6 \\ y_7 \\ y_8 \end{bmatrix}, \quad \boldsymbol{\Lambda}_y = \begin{bmatrix} \lambda_4 & 0 \\ \lambda_5 & 0 \\ \lambda_6 & 0 \\ \lambda_7 & 0 \\ 0 & \lambda_8 \\ 0 & \lambda_9 \\ 0 & \lambda_{10} \\ 0 & \lambda_{11} \end{bmatrix}, \quad \boldsymbol{\eta} = \begin{bmatrix} \eta_1 \\ \eta_2 \end{bmatrix}, \quad \boldsymbol{\epsilon} = \begin{bmatrix} \epsilon_1 \\ \epsilon_2 \\ \epsilon_3 \\ \epsilon_4 \\ \epsilon_5 \\ \epsilon_6 \\ \epsilon_7 \\ \epsilon_8 \end{bmatrix} \tag{2.10b}$$

Equations (2.8) and (2.9) also are shown at the top of Table 2.2, which provides the notation for the measurement model. The random variables in $\mathbf{x}$ are indicators of the latent exogenous variables (the ξ's).[5] The random variables in $\mathbf{y}$ are indicators of the latent endogenous variables (the η's). In general, $\mathbf{x}$ is $q \times 1$ (where q is the number of indicators of $\boldsymbol{\xi}$) and $\mathbf{y}$ is $p \times 1$ (where p is the number of indicators of $\boldsymbol{\eta}$).

The Λ_y and Λ_x matrices contain the λ_i parameters, which are the structural coefficients linking the latent and manifest variables. The Λ_x matrix is $q \times n$ (where n is the number of ξ's) and Λ_y is $p \times m$ (where m is the number of η's). In confirmatory factor analysis or in models where $\mathbf{y} = \boldsymbol{\eta}$ or $\mathbf{x} = \boldsymbol{\xi}$, I double subscript λ (λ_{ij}). The i refers to x_i (y_i) and the j refers to the ξ_j (η_j) that influences x_i (y_i). When $\mathbf{x}$, $\boldsymbol{\xi}$, $\mathbf{y}$, and $\boldsymbol{\eta}$ are used in a model such as (2.10), the double subscripts can be confusing since each λ_{ij} can refer to two different parameters. Therefore I use a single subscript to consecutively number the λ's in these cases. The errors of measurement vector for $\mathbf{x}$ is $\boldsymbol{\delta}$, and $\boldsymbol{\delta}$ is $q \times 1$. The error vector for $\mathbf{y}$, $\boldsymbol{\epsilon}$, is $p \times 1$. Generally, $\boldsymbol{\delta}$ and $\boldsymbol{\epsilon}$ contain vectors of random variables.[6]

The last two matrices, Θ_δ and Θ_ϵ, are covariance matrices of the errors of measurement. The main diagonals contain the error variances associated with the indicators. The off-diagonal elements are the covariances of the errors of measurement for the different indicators. The Θ_δ matrix is $q \times q$ and has the error variances and their covariances for the x variables, and Θ_ϵ is a $p \times p$ matrix that contains the error variances and their covariances for the y variables. In this example I assume that the errors of measurement for the indicators of industrialization (x_1 to x_3) are uncorrelated so that Θ_δ is a diagonal matrix. This assumption is less defensible for Θ_ϵ because I have the same set of indicators at two points in time. It is likely that the error in measuring an indicator in 1960 is correlated with the error in measuring the same indicator in 1965. In addition the 1960 measures of y_2 and y_4 and the 1965 ones of y_6 and y_8 are from the same data source. The corresponding errors of measurement might be positively correlated due to systematic biases present in the source. Therefore the (4, 2), (5, 1), (6, 2), (7, 3), (8, 4), and (8, 6) off-diagonal elements of Θ_ϵ may be nonzero.

This example reveals some of the major features of structural equations with latent variables that are distinct from the standard regression approach. The models are more realistic in their allowance for measurement error in the observed variables. They allow random measurement error in $\boldsymbol{\epsilon}$

[5] The observed variables in $\mathbf{x}$ are not random when $\mathbf{x} = \boldsymbol{\xi}$ and $\mathbf{x}$ is fixed. See Chapter 4 for a discussion of fixed $\mathbf{x}$ variables.

[6] When any $\mathbf{y}$ or $\mathbf{x}$ has no measurement error the corresponding element in $\boldsymbol{\epsilon}$ or $\boldsymbol{\delta}$ is zero, a constant.

Table 2.2 Notation for Measurement Model

Structural Equations for Measurement Model

$$\mathbf{x} = \Lambda_x \boldsymbol{\xi} + \boldsymbol{\delta}$$
$$\mathbf{y} = \Lambda_y \boldsymbol{\eta} + \boldsymbol{\epsilon}$$

Assumptions

$E(\boldsymbol{\eta}) = 0$, $E(\boldsymbol{\xi}) = 0$, $E(\boldsymbol{\epsilon}) = 0$, and $E(\boldsymbol{\delta}) = 0$
$\boldsymbol{\epsilon}$ uncorrelated with $\boldsymbol{\eta}$, $\boldsymbol{\xi}$, and $\boldsymbol{\delta}$
$\boldsymbol{\delta}$ uncorrelated with $\boldsymbol{\xi}$, $\boldsymbol{\eta}$, and $\boldsymbol{\epsilon}$

Symbol	Name	Phonetic Spelling	Dimension	Definition
			Variables	
$\mathbf{y}$	—	—	$p \times 1$	observed indicators of $\boldsymbol{\eta}$
$\mathbf{x}$	—	—	$q \times 1$	observed indicators of $\boldsymbol{\xi}$
$\boldsymbol{\epsilon}$	epsilon	e*p*′ sə lo*n*′ (e*p*′ səlen)	$p \times 1$	measurement errors for $\mathbf{y}$
$\boldsymbol{\delta}$	delta	de*l*′ tə	$q \times 1$	measurement errors for $\mathbf{x}$
			Coefficients	
Λ_y	lambda *y*	la*m*′ də*y*	$p \times m$	coefficients relating $\mathbf{y}$ to $\boldsymbol{\eta}$
Λ_x	lambda *x*	lam′ də*x*	$q \times n$	coefficients relating $\mathbf{x}$ to $\boldsymbol{\xi}$
			Covariance Matrices	
$\boldsymbol{\Theta}_\epsilon$	theta-epsilon	th*ā*′ tə (th*ē*′ tə)-e*p*′ sə lon′	$p \times p$	$E(\boldsymbol{\epsilon\epsilon}')$ (covariance matrix of ϵ)
$\boldsymbol{\Theta}_\delta$	theta-delta	th*ā*′ tə (th*ē*′ tə)-de*l*′ tə	$q \times q$	$E(\boldsymbol{\delta\delta}')$ (covariance matrix of δ)

and $\boldsymbol{\delta}$, and systematic differences in scale are introduced with the λ coefficients. The error in measuring one variable can correlate with that of another. Multiple indicators can measure one latent variable. Furthermore researchers can analyze the relation between latent variables unobscured by measurement error. All of these features bring us closer to testing the hypotheses set forth in theories.

COVARIANCE

Covariance is a central concept for the models. In fact, another name for the general structural equation techniques is *analysis of covariance structures*. I review two aspects of covariances. One is covariance algebra which helps in deriving properties of the latent variable and measurement models. The other includes the factors that influence sample covariances which in turn can affect parameter estimates. I consider the covariance algebra first.

Covariance Algebra

Table 2.3 provides a summary of the definitions and common rules of covariance algebra. In Table 2.3 the $E(\cdot)$ refers to the expected value of the expression within the parentheses. The top half of Table 2.3 defines both the covariance and the variance. The capital X_1, X_2, and X_3 signify the original random variables rather than the mean deviation forms. When X_1 and X_2 have a positive linear association, the $\text{COV}(X_1, X_2)$ is positive. If they are inversely related the $\text{COV}(X_1, X_2)$ is negative, but it is zero if there is no linear association. Note that in the definition of covariance, I employ capital letters "COV" to signify *population* covariances. The population covariance matrix of the observed variables is Σ. I will discuss sample covariances shortly. The variance is the covariance of a variable with itself. Capital $\text{VAR}(X_1)$ represents the population variance of X_1. The main diagonal of Σ contains the variances of the observed variables.

Table 2.3 Definitions and Common Rules of Covariances

Definitions

$$\begin{aligned}\text{COV}(X_1, X_2) &= E[(X_1 - E(X_1))(X_2 - E(X_2))] \\ &= E(X_1 X_2) - E(X_1)E(X_2) \\ \text{VAR}(X_1) &= \text{COV}(X_1, X_1) \\ &= E[(X_1 - E(X_1))^2]\end{aligned}$$

Rules

c is a constant

X_1, X_2, X_3 are random variables

(1) $\text{COV}(c, X_1) = 0$

(2) $\text{COV}(cX_1, X_2) = c\,\text{COV}(X_1, X_2)$

(3) $\text{COV}(X_1 + X_2, X_3) = \text{COV}(X_1, X_3) + \text{COV}(X_2, X_3)$

Several examples help to illustrate the covariance rules. In the examples I assume that all disturbances have expected values of zero and that all random variables are deviated from their means. Unless stated otherwise, the lowercase x and y represent deviation forms of the original random variables X and Y. For the first example, suppose that you know the covariance of a latent variable, ξ_1, with an observed variable x_1. If you add a constant to x_1, the covariance is

$$\begin{aligned}\mathrm{COV}(\xi_1, x_1 + c) &= \mathrm{COV}(\xi_1, x_1) + \mathrm{COV}(\xi_1, c)\\ &= \mathrm{COV}(\xi_1, x_1)\end{aligned}$$

Thus the covariance of a latent variable and an observed variable is unchanged if a constant is added to the observed variable. This result is important given that we rarely have widely agreed-upon baselines (zero points) for measures of social science concepts. It shows that if the measure is changed by some constant value, this will not influence the covariance it has with the latent variable. The same example illustrates another point. Suppose that c is the mean of X_1 in its original form. Then $x_1 + c$ leads to X_1. The preceding example shows that the covariance of any random variable with another is the same regardless of whether the variables are in deviation or original form. If, however, the scale is changed to cx_1, this does change the covariance to $c\,\mathrm{COV}(\xi_1, x_1)$.

A second example addresses an issue in measurement. In psychometrics and other social science presentations it is often argued that two indicators, each positively related to the same concept, should have a positive covariance. Suppose that we have two indicators each related to the same ξ_1 such that $x_1 = \lambda_1\xi_1 + \delta_1$, $x_2 = \lambda_2\xi_1 + \delta_2$, $\mathrm{COV}(\xi_1, \delta_1) = \mathrm{COV}(\xi_1, \delta_2) = 0$, and λ_1 and $\lambda_2 > 0$. Must the $\mathrm{COV}(x_1, x_2)$ be positive?

$$\begin{aligned}\mathrm{COV}(x_1, x_2) &= \mathrm{COV}(\lambda_1\xi_1 + \delta_1, \lambda_2\xi_1 + \delta_2)\\ &= \lambda_1\lambda_2\phi_{11}\end{aligned}$$

For nonzero ϕ_{11}, the indicators x_1 and x_2 must have a positive covariance. (Since ϕ_{11} is a variance, it is positive for all nonconstant ξ_1.) As a second part, consider x_1 and x_2 as indicators of ξ_1 such that $\xi_1 = \lambda_1 x_1 + \lambda_2 x_2 + \delta_1$, $\mathrm{COV}(x_1, \delta_1) = \mathrm{COV}(x_2, \delta_1) = 0$, and λ_1 and $\lambda_2 > 0$. Must the $\mathrm{COV}(x_1, x_2) > 0$?

$$\mathrm{COV}(x_1, x_2) = ?$$

The x_1 and x_2 variables are exogenous. Their covariance is not determined within the model so that it can be positive, zero, or negative even if both x_1 and x_2 are positively related to ξ_1. An example is if the latent variable is exposure to discrimination (ξ_1) and the two indicators are race (x_1) and sex (x_2). Although x_1 and x_2 indicate exposure to discrimination, we would not expect them to have a positive correlation. Or, the latent variable could be social interaction (ξ_1) and the indicators time spent with friends (x_1) and time spent with family (x_2). These indicators might even have a negative covariance. This simple example shows that general statements about the necessity of indicators of the same concept to be positively associated require qualification.[7]

The covariance definitions and rules also apply to matrices and vectors. For instance, if $\mathbf{c}'$ is a vector of constants that conforms in multiplication with $\mathbf{x}$, then $\mathrm{COV}(\mathbf{x}, \mathbf{c}') = \mathbf{0}$. The $\mathrm{VAR}(\mathbf{x}) = \mathrm{COV}(\mathbf{x}, \mathbf{x}') = \boldsymbol{\Sigma}$, where all three symbols represent the population covariance matrix of $\mathbf{x}$. Typically, I use $\boldsymbol{\Sigma}$, sometimes subscripting it to refer to specific variables (e.g., $\boldsymbol{\Sigma}_{xy}$ = covariance matrix of $\mathbf{x}$ with $\mathbf{y}$).

Sample Covariances

So far I have limited the discussion to *population* covariances and variances. In practice, sample estimates of the variance and covariances are all that are available. The unbiased sample estimator of covariance is

$$\mathrm{cov}(X, Y) = \frac{\Sigma_{i=1}^{N}(X_i - \bar{X})(Y_i - \bar{Y})}{N - 1} \tag{2.11}$$

where $\mathrm{cov}(X, Y)$ represents the sample estimator of the covariance between the N random sample values of X and Y. The X_i and Y_i represent the values of X and Y for the ith observation. The $\bar{X}$ and $\bar{Y}$ are the sample means. In Appendix A I show that the sample covariance matrix is computed as

$$\mathbf{S} = \left(\frac{1}{N - 1}\right)\mathbf{Z}'\mathbf{Z} \tag{2.12}$$

where $\mathbf{Z}$ is a $N \times (p + q)$ matrix of deviation (from the means) scores for

[7] The implications of this for the internal consistency perspective on measurement is treated in Bollen (1984).

the $p + q$ observed variables. The **S** matrix is square and symmetric with the sample variances of the observed variables down its main diagonal and the sample covariances off the diagonal.

The sample covariance matrix is crucial to estimates of structural equation models. In LISREL, EQS, and other computer programs for the analysis of covariance structures, the sample covariance (or correlation) matrix often is the only data in the analysis. The parameter estimates depend on functions of the variances and covariances. For instance, in simple regression analysis the ordinary least squares (OLS) estimator for a regression coefficient is the ratio of the sample covariance of y and x to the variance of x [i.e., $\text{cov}(x, y)/\text{var}(x)$]. In more general structural equation systems the estimates result from far more complicated functions. For both the simple and the complex situation, factors that affect elements of the sample covariance matrix, **S**, have the potential to affect the parameter estimates.

Although, on average, the sample covariance matrix **S** equals the population covariance matrix Σ, some samples lead to an **S** closer to Σ than others. In addition to sampling fluctuations of **S**, other factors can affect its elements. One is a nonlinear relation between variables. The sample covariance or correlation are measures of *linear* association. Covariances like correlations can give a misleading impression of the association between two variables that have a curvilinear relation. In some substantive areas research is sufficiently developed to alert researchers to possible nonlinear links, and this knowledge can be incorporated into a model. For instance, demographic transition theory suggests that as mortality rates in countries decline, they follow a typical curvilinear pattern. Another example is that earnings initially increase with age, then eventually stabilize or even decline as individuals grow older. The more common situation is when the exact form of the relation is unknown. Scatterplots, partial plots (Belsley, Kuh, and Welsch 1980), comparisons of equation fits upon variable transformations, or related devices help to detect nonlinearity in observed variable equations such as regression models. Detection of nonlinearities in models with latent variables is a relatively underdeveloped area. McDonald (1967a, 1967b) suggests some procedures for exploring nonlinearities in factor analysis, but much more needs to be done to have procedures applicable to all types of structural equation models.

A second factor that can affect sample covariances and correlations are *outliers*. Outliers are observations with values that are distinct or distant from the bulk of the data. Outliers can have large or small effects on an analysis. Outliers that lead to substantial changes are *influential* observations. When influential cases are present the covariances provide a misleading summary of the association between most of the cases.

Table 2.4 U.S. Disposable Income Per Capita and Consumers' Expenditures Per Capita in Constant Dollars, 1922 to 1941

	Disposable Income (income)	Consumers' Expenditures (consum)
1922	433	394
1923	483	423
1924	479	437
1925	486	434
1926	494	447
1927	498	447
1928	511	466
1929	534	474
1930	478	439
1931	440	399
1932	372	350
1933	381	364
1934	419	392
1935	449	416
1936	511	463
1937	520	469
1938	477	444
1939	517	471
1940	548	494
1941	629	529

Source: Haavelmo (1953).

Detection of outliers can start with examining the univariate distributions of the observed variables. To illustrate this, I take data from Haavelmo (1953). His goal is to analyze the marginal propensity to consume as a function of income with aggregate U.S. data. That is, he seeks to estimate the proportion of U.S. disposable income spent for consumers' expenditures rather than "spent" for investment expenditures. Two variables in his analysis are U.S. disposable income per capita (income) and U.S. consumers' expenditures per capita (consum), both in constant dollars and for each year from 1922 to 1941. Table 2.4 lists these data.

A convenient means of summarizing sample univariate distributions are *stem-and-leaf* diagrams. The stem-and-leaf displays for consum and income are in Figure 2.1. To construct the display, each value of a variable is rounded to two digits. For instance, the 1930 income value of 478 is rounded to 48. Then all two-digit values are placed in ranked order for each variable. The first digit of each value is the stem, and it is indicated in the

Consumption

```
Stem-Leaf                  #
   5 3                     1
   4 55677779              8
   4 0223444               7
   3 5699                  4
   3
    ----+----+----+----+
Multiply Stem-Leaf by 100
```

Income

```
Stem-Leaf                  #
   6 3                     1
   5 5                     1
   5 011223                6
   4 5888899               7
   4 234                   3
   3 78                    2
    ----+----+----+----+
Multiply Stem-Leaf by 100
```

Figure 2.1 Stem-and-Leaf Displays for Haavelmo's (1953) U.S. Consumption and Income Data, 1922 to 1941

first column of the stem-and-leaf. The second digit is the leaf, and this is placed in the row that corresponds to its stem. At the base of the stem-and-leaf is the 10^p power that a value should be multiplied by to recover the proper decimal place for the original variable. Consider the top row for income in Figure 2.1. The stem is 6 and the leaf is 3, while the multiple is 10^2 which leads to 630 (the rounded value of 629 for 1941 income). The next to the bottom row represents 420, 430, and 440 which are the original values of 419 (in 1934), 433 (in 1922), and 440 (in 1931). The stem-and-leaf diagram is like a histogram turned on its side, but unlike the histogram, it allows the recovery of the original values with a two-digit degree of accuracy.[8]

The stem-and-leaf diagrams reveal that most income values are in the 450 to 500 range and most consumption values fall between 400 and 490.

[8]A brief introduction to stem-and-leaf displays and other tools of exploratory data analysis is in Leinhardt and Wasserman (1978). Also, if the range of values is large enough, the stem may have two or more digits so that the degree of accuracy would be greater than two digits.

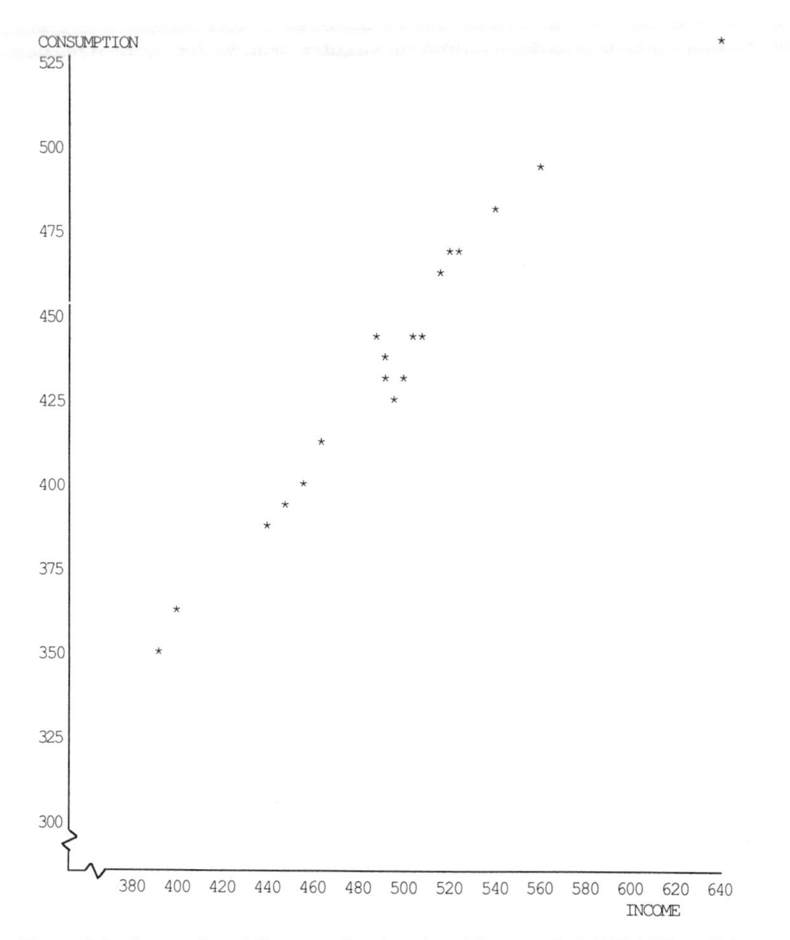

Figure 2.2 Scatterplot of Consumption (cons) and Income (inc) U.S. 1922 to 1941

The income value of 630 is the highest, and it is the most distant from the other values. For consumption the highest value, 530, is less distant from the other sample values. Both of these highest values are for 1941, which suggests that this observation bears special attention in subsequent analyses. A transformation of the variables might change the outlier status of this year, but to conform to Haavelmo's analysis, I stay with the original data.

With only two variables a scatterplot can aid the identification of outliers. Figure 2.2 is the scatterplot of consumption (consum) by income. This shows a very close linear association between the variables. The observation most distant from the bulk of the points in the upper-right quadrant is the one for 1941. If we imagine a straight line drawn through the points, this last case would fall slightly below the line. Whether this point would seriously affect parameter estimates can be assessed by doing analyses with and without the observation and by comparing estimates.

The sample covariance matrices of consumption and income with and without the 1941 data are

$$\mathbf{S} = \begin{bmatrix} 1889 & \\ 2504 & 3421 \end{bmatrix}, \qquad \mathbf{S}_{(i)} = \begin{bmatrix} 1505 & \\ 1863 & 2363 \end{bmatrix}$$

where $\mathbf{S}_{(i)}$ is the sample covariance matrix removing the ith observation (1941 in this case). When 1941 is dropped, there are substantial drops in the variances and covariance. Does this mean that the 1941 observation has a large effect on all estimates that are functions of the covariance matrix? Not necessarily, since it depends on how these elements combine to form an estimate. For instance, the correlation coefficient is the covariance of two variables divided by the product of their standard deviations. Dropping the 1941 case for consumption and income changes the correlation hardly at all ($r = 0.985$ vs. $r_{(i)} = 0.988$). If the object of study were the correlation, we would conclude that this outlier was not influential. But we cannot generalize from the effect on correlations to the effects on other estimates that are different functions of the elements of $\mathbf{S}$. In Chapter 4 I will examine the influence of this case on other more complex functions of $\mathbf{S}$.

The consumption and income example is atypical in that it contains only two variables. The standard application involves many more variables. Bivariate scattergrams often can identify deviant cases, but they do not always reveal multidimensional outliers (Daniel and Wood 1980, 50–53). Detection of multidimensional outliers is not a fully solved problem. In certain areas, such as single-equation regression analysis, several procedures are available.[9] In factor analysis or general systems of structural equations, the techniques are less developed. One general screening device, however, is to form a $N \times (p + q)$ matrix $\mathbf{Z}$ that contains all of the observed variables written as deviations from their mean values. Then define an $N \times N$ matrix $\mathbf{A} = \mathbf{Z}(\mathbf{Z}'\mathbf{Z})^{-1}\mathbf{Z}'$. The main diagonal of $\mathbf{A}$, called a_{ii}, has several useful interpretations. First, a_{ii} gives the "distance" of the ith case from the means for all of the variables. It has a range between zero and one such that the closer to one the more distant it is, while the closer to zero the nearer is the observation to the means. The $\Sigma_i^N a_{ii}$ equals $p + q$, the number of observed variables. This means that the average size of a_{ii} is $(p + q)/N$ so each a_{ii} can be compared to the average value as a means of judging its

[9]See, e.g., Belsley, Kuh, and Welsch (1980), Cook and Weisberg (1982), or Bollen and Jackman (1985).

magnitude. The relative size of a_{ii} also can be assessed by examining the univariate distribution of a_{ii} and noting any values that are much larger than the others.

I illustrate this procedure with the data in Table 2.5. The data come from a study that assesses the reliability and validity of human perceptions versus physical measures of cloud cover (Cermak and Bollen 1984). The first three columns of Table 2.5 contain the perception estimates of three judges of the percent of the visible sky containing clouds in each of 60 slides. The slides showed identical views of the sky at different times and days. The judgments range from 0% to 100% cloud cover.

I formed deviation scores for each variable and calculated $\mathbf{A}$ as $\mathbf{Z}(\mathbf{Z}'\mathbf{Z})^{-1}\mathbf{Z}'$. A stem-and-leaf display of the a_{ii}, the main diagonal of $\mathbf{A}$, is given in Figure 2.3. The sum of the a_{ii}'s is three, which corresponds to the three observed variables. The average value is 3/60, or 0.05, and all a_{ii} values lie in the 0 to 1 range. Two or three outliers are evident in the stem-and-leaf. The highest two a_{ii} have values of 0.301 and 0.307 and correspond to the 40th and 52nd observation. Both of these are over six times larger than the average a_{ii} value of 0.05. The third largest a_{ii} is 0.183, which is over three and a half times larger than the average and is observation 51. A weaker case could be made that the fourth highest value, 0.141, also is an outlier, but I concentrate on the first three.

I begin to assess the influence of the three outliers by calculating the covariance matrix with and without these three cases:

$$\mathbf{S} = \begin{bmatrix} 1301 & & \\ 1020 & 1463 & \\ 1237 & 1200 & 1404 \end{bmatrix}, \qquad \mathbf{S}_{(i)} = \begin{bmatrix} 1129 & & \\ 1170 & 1494 & \\ 1149 & 1313 & 1347 \end{bmatrix}$$

where $\mathbf{S}_{(i)}$ refers to the covariance matrix with the outliers removed and where i now refers to the set of observations 40, 51, and 52. In the consumption and income example all elements of the covariance matrix decreased when the outlier was removed. In this example some elements increase while others drop. For instance, the variance for the first cloud cover variable drops from 1301 to 1129, whereas the covariance for the first and second cloud cover variables increases from 1020 to 1170. The largest change in the correlation matrix for these variables is for the second cloud cover variable. The correlation of the first and second cloud cover variables shifts from 0.74 to 0.90 and that for the third and second variables increases from 0.84 to 0.93 when the outliers are removed. Thus these outliers influence the covariance and correlation matrices. I show in Chapter 7 that the outliers also affect a confirmatory factor analysis of these data.

Table 2.5 Three Estimates of Percent Cloud Cover for 60 Slides

OBS	COVER1	COVER2	COVER3
1	0	5	0
2	20	20	20
3	80	85	90
4	50	50	70
5	5	2	5
6	1	1	2
7	5	5	2
8	0	0	0
9	10	15	5
10	0	0	0
11	0	0	0
12	10	30	10
13	0	2	2
14	10	10	5
15	0	0	0
16	0	0	0
17	5	0	20
18	10	20	20
19	20	45	15
20	35	75	60
21	90	99	100
22	50	90	80
23	35	85	70
24	25	15	40
25	0	0	0
26	0	0	0
27	10	10	20
28	40	75	30
29	35	70	20
30	55	90	90
31	35	95	80
32	0	0	0
33	0	0	0
34	5	1	2
35	20	60	50
36	0	0	0
37	0	0	0
38	0	0	0
39	15	55	50
40	95	0	40
41	40	35	30
42	40	50	40
43	15	60	5

Table 2.5 *(Continued)*

OBS	COVER1	COVER2	COVER3
44	30	30	15
45	75	85	75
46	100	100	100
47	100	90	85
48	100	95	100
49	100	95	100
50	100	99	100
51	100	30	95
52	100	5	95
53	0	0	0
54	5	5	5
55	80	90	85
56	80	95	80
57	80	90	70
58	40	55	50
59	20	40	5
60	1	0	0

With influential cases like these, it is worthwhile determining why they are outliers. The cloud cover estimates for observations 40, 51, and 52 in Table 2.5 are drastically different from one another. For example, observation 52 has estimates of 100%, 5%, and 95%. The slides that correspond to the outliers reveal that the pictures were taken when the scene was obstructed by considerable haze. Under these conditions the judges could not easily estimate the cloud cover, and this led to the differences in estimates. The outliers indicate the inadequacy of this means of assessing cloud cover when hazy conditions are present and suggest that instructions for judging slides under these conditions should be developed.[10]

In general, outliers can increase, decrease, or have no effect on covariances and on the estimates based on these covariances. To determine which of these possibilities holds, researchers need to screen their data for maverick observations and to perform their analysis with and without these cases. A detailed examination of the outliers can suggest omitted variables, incorrect functional forms, clerical errors, or other neglected aspects of a study. In this way outliers can be the most valuable cases in an analysis.

[10] It also could be that the error variance for the cloud cover measures are associated with the degree of haziness, a possibility suggested to me by Ron Schoenberg.

Stem-Leaf	#
30 17	2
28	
26	
24	
22	
20	
18 3	1
16	
14 1	1
12	
10 56	2
8 2590	4
6 123588	6
4 445682799	9
2 002456	6
0 44613333456667778888888888888	29
----+----+----+----+----+	
Multiply Stem-Leaf by −0.01	

Figure 2.3 Stem-and-Leaf Display for a_{ii} of Cloud Cover Data

PATH ANALYSIS

As mentioned in Chapter 1, Sewall Wright's (1918, 1921) *path analysis* is a methodology for analyzing systems of structural equations. Contemporary applications emphasize three components of path analysis: (1) the path diagram, (2) decomposing covariances and correlations in terms of model parameters, and (3) the distinctions between direct, indirect, and total effects of one variable on another. I treat each of these in turn.

Path Diagrams

A *path diagram* is a pictorial representation of a system of simultaneous equations. One of the main advantages of a path diagram is that it presents a picture of the relationships that are assumed to hold. For many researchers this picture may represent the relationships more clearly than the equations. To understand path diagrams, it is necessary to define the symbols involved. Table 2.6 provides the primary symbols. The observed variables are enclosed in boxes. The unobserved or latent variables are circled, with the exception of the disturbance terms which are not enclosed. Straight single-headed arrows represent causal relations between the vari-

Table 2.6 Primary Symbols Used in Path Analysis

Symbol	Meaning
X_1	rectangular or square box signifies an observed or manifest variable
η_1	circle or ellipse signifies an unobserved or latent variable
ϵ_1	unenclosed variable signifies a disturbance term (error in either equation or measurement)
$\eta_1 \rightarrow Y_1$	straight arrow signifies assumption that variable at base of arrow "causes" variable at head of arrow
ξ_1, ξ_2	curved two-headed arrow signifies unanalyzed association between two variables
$\eta_1 \rightleftarrows \eta_2$	two straight single-headed arrows connecting two variables signifies feedback relation or reciprocal causation

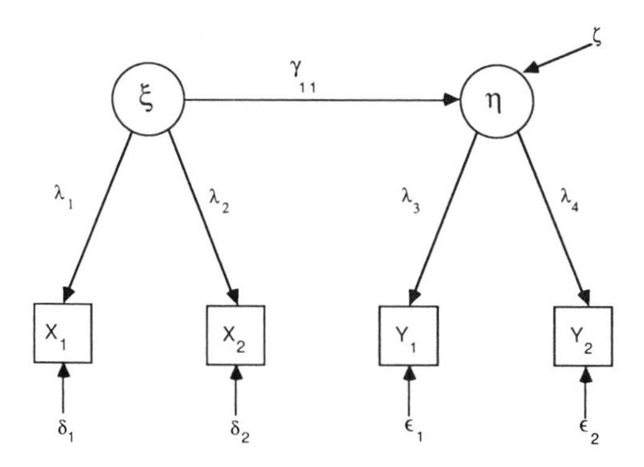

Figure 2.4 An Example of a Path Diagram

ables connected by the arrows. A curved two-headed arrow indicates an association between two variables. The variables may be associated for any of a number of reasons. The association may be due to both variables depending on some third variable(s), or the variables may have a causal relationship but this remains unspecified.

The path diagram in Figure 2.4 is equivalent to the following simultaneous system of equations:

$$\eta = \gamma_{11}\xi + \zeta$$

$$x_1 = \lambda_1\xi + \delta_1, \qquad x_2 = \lambda_2\xi + \delta_2$$

$$y_1 = \lambda_3\eta + \epsilon_1, \qquad y_2 = \lambda_4\eta + \epsilon_2$$

Assuming

$$\text{COV}(\xi, \delta_1) = 0 \quad \text{COV}(\xi, \delta_2) = 0 \quad \text{COV}(\xi, \epsilon_1) = 0 \quad \text{COV}(\xi, \epsilon_2) = 0$$

$$\text{COV}(\xi, \zeta) = 0 \quad \text{COV}(\eta, \epsilon_1) = 0 \quad \text{COV}(\eta, \epsilon_2) = 0 \quad \text{COV}(\delta_1, \delta_2) = 0$$

$$\text{COV}(\delta_1, \epsilon_1) = 0 \quad \text{COV}(\delta_1, \epsilon_2) = 0 \quad \text{COV}(\delta_2, \epsilon_1) = 0 \quad \text{COV}(\delta_2, \epsilon_2) = 0$$

$$\text{COV}(\epsilon_1, \epsilon_2) = 0 \quad \text{COV}(\delta_1, \zeta) = 0 \quad \text{COV}(\delta_2, \zeta) = 0 \quad \text{COV}(\epsilon_1, \zeta) = 0$$

$$\text{COV}(\epsilon_2, \zeta) = 0$$

I purposely have written all of the assumptions for disturbance terms since the same information is explicitly shown in the path diagram. In fact, all of the relationships are represented in the path diagram. For example, the fact that there is no arrow connecting δ_1 and δ_2 or ξ and ϵ_1 is equivalent to the assumption of a zero covariance between these variables. Thus path diagrams are another means of representing systems of equations.

Decomposition of Covariances and Correlations

Path analysis allows one to write the covariance or correlation between two variables as functions of the parameters of the model. One means of doing this is with covariance algebra.[11] To illustrate this, consider the simple model in Figure 2.5. It represents a single latent variable (ξ_1) that has four indicators (x_1 to x_4). All the errors of measurement (δ_i's) are uncorrelated, except for δ_2 and δ_3. The errors of measurement (δ_i's) are assumed to be uncorrelated with ξ_1 and $E(\delta_i) = 0$ for all i.

[11]An alternative more complicated means to decomposition is with the "first law of path analysis." See Kenney (1979) for a discussion of this.

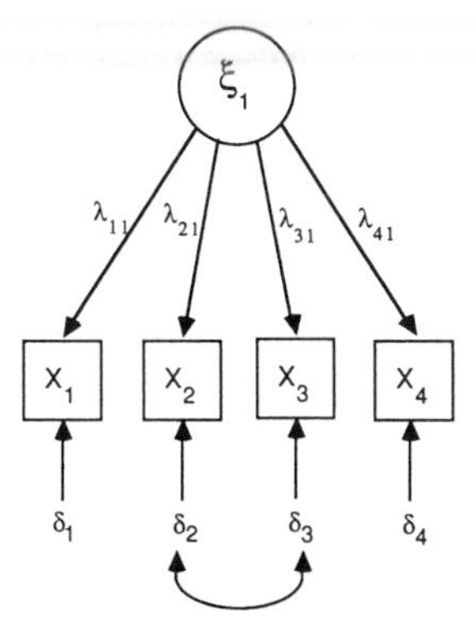

Figure 2.5 Path Diagram of a Single Latent Variable with Four Indicators

The decomposition of the $\text{COV}(x_1, x_4)$ is

$$\begin{aligned}\text{COV}(x_1, x_4) &= \text{COV}(\lambda_{11}\xi_1 + \delta_1, \lambda_{41}\xi_1 + \delta_4)\\ &= \lambda_{11}\lambda_{41}\phi_{11}\end{aligned}$$

The right-hand side of the top equation follows from the equations for x_1 and x_4 defined in the path diagram. This shows that the $\text{COV}(x_1, x_4)$ is a function of the effect of ξ_1 on x_1 and x_4 (i.e., λ_{11} and λ_{41}) and the variance of the latent variable ξ_1.

For complicated models this covariance algebra can be tedious. An alternative is to use matrix algebra to decompose covariances (or correlations) into the model parameters. As an example, consider the covariance matrix for the **x** variables or $\boldsymbol{\Sigma}$. The covariance matrix for **x** is the expected value of $\mathbf{xx}'$, where $\mathbf{x} = \boldsymbol{\Lambda}_x\boldsymbol{\xi} + \boldsymbol{\delta}$:

$$\begin{aligned}\mathbf{xx}' &= (\boldsymbol{\Lambda}_x\boldsymbol{\xi} + \boldsymbol{\delta})(\boldsymbol{\Lambda}_x\boldsymbol{\xi} + \boldsymbol{\delta})'\\ &= (\boldsymbol{\Lambda}_x\boldsymbol{\xi} + \boldsymbol{\delta})(\boldsymbol{\xi}'\boldsymbol{\Lambda}_x' + \boldsymbol{\delta}')\\ &= \boldsymbol{\Lambda}_x\boldsymbol{\xi}\boldsymbol{\xi}'\boldsymbol{\Lambda}_x' + \boldsymbol{\Lambda}_x\boldsymbol{\xi}\boldsymbol{\delta}' + \boldsymbol{\delta}\boldsymbol{\xi}'\boldsymbol{\Lambda}_x' + \boldsymbol{\delta}\boldsymbol{\delta}'\\ E(\mathbf{xx}') &= \boldsymbol{\Lambda}_x E(\boldsymbol{\xi}\boldsymbol{\xi}')\boldsymbol{\Lambda}_x' + \boldsymbol{\Lambda}_x E(\boldsymbol{\xi}\boldsymbol{\delta}') + E(\boldsymbol{\delta}\boldsymbol{\xi}')\boldsymbol{\Lambda}_x' + E(\boldsymbol{\delta}\boldsymbol{\delta}')\\ \boldsymbol{\Sigma} &= \boldsymbol{\Lambda}_x\boldsymbol{\Phi}\boldsymbol{\Lambda}_x' + \boldsymbol{\Theta}_\delta\end{aligned}$$

In this case $\boldsymbol{\Sigma}$, the covariance matrix of **x**, is decomposed in terms of the elements in $\boldsymbol{\Lambda}_x$, $\boldsymbol{\Phi}$, and $\boldsymbol{\Theta}_\delta$. As I show in Chapters 4, 7, and 8 the covariances for all observed variables can be decomposed into the model parameters in a similar fashion. These decompositions are important because they show that the parameters are related to the covariances and different parameter values lead to different covariances.

Total, Direct, and Indirect Effects

Path analysis distinguishes three types of effects: direct, indirect, and total effects. The direct effect is that influence of one variable on another that is unmediated by any other variables in a path model. The indirect effects of a variable are mediated by at least one intervening variable. The sum of the direct and indirect effects is the total effects:

$$\text{Total effects} = \text{Direct effect} + \text{Indirect effects}$$

The decomposition of effects always is with respect to a specific model. If the system of equations is altered by including or excluding variables, the estimates of total, direct, and indirect effects may change.

To make these types of effects more concrete, consider the model introduced in the model notation section relating 1960 industrialization to 1960 and 1965 political democracy in developing countries. The equations for this model are in (2.3), (2.10), (2.11), and in the assumptions in the discussion surrounding these equations.

The path model for these equations and assumptions is represented in Figure 2.6. Here ξ_1 is industrialization with indicators x_1 to x_3. The η_1 variable is 1960 political democracy with its four measures, y_1 to y_4, and η_2 is 1965 democracy measured by y_5 to y_8. An example of a direct effect is the effect of η_1 on η_2, that is, β_{21}. A one-unit change in η_1 leads to an expected direct change of β_{21} in η_2 net of ξ_1. There are no mediating variables between η_1 and η_2. The direct effect of ξ_1 on η_2 is γ_{21}, while λ_8 is the direct effect of η_2 on y_5.

To illustrate indirect effects, consider the influence of ξ_1 on η_2. The intervening variable in this case is η_1. A one-unit change in ξ_1 leads to an expected γ_{11} change in η_1. This γ_{11} change in η_1 leads to an expected β_{21} change in η_2. Thus the indirect effect of ξ_1 on η_2 is $\gamma_{11}\beta_{21}$. Following a similar procedure the indirect effect of η_1 on y_7 is $\beta_{21}\lambda_{10}$.

One variable's total effect on another is the sum of its direct effect and indirect effects. For instance, the total effect of ξ_1 on η_2 is

$$\begin{aligned}\text{Total effect} &= \text{Direct effect} + \text{Indirect effects}\\ &= \gamma_{21} + \gamma_{11}\beta_{21}\end{aligned}$$

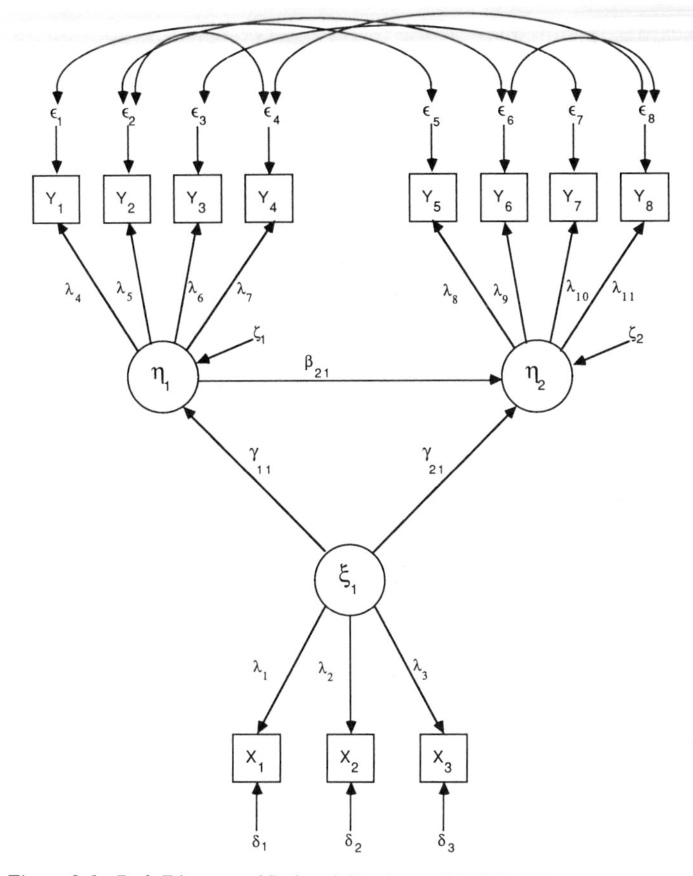

Figure 2.6 Path Diagram of Industrialization and Political Democracy Model

The total effect of ξ_1 on y_8 is

$$\begin{aligned}\text{Total effect} &= \text{Direct effect} + \quad \text{Indirect effects}\\ &= \quad 0 \quad +(\gamma_{21}\lambda_{11} + \gamma_{11}\beta_{21}\lambda_{11}).\end{aligned}$$

Since ξ_1 has no direct effect on y_8, the total effect comprises only indirect effects.

Considering each type of effect leads to a more complete understanding of the relation between variables than if these distinctions are not made. In the typical regression analysis the regression coefficient is an estimate of the direct effect of a variable. If we ignore the indirect effects that a variable may have through other variables, we may be grossly off in the assessment of its overall effect. For example, if in Figure 2.6 we claim that industrialization's (ξ_1) effect on 1965 political democracy (η_2) is γ_{21}, we would overlook a possibly large indirect effect of $\gamma_{11}\beta_{21}$. Similarly, if we claim that industrialization has no effect on 1965 political democracy based on a nonsignificant estimate of γ_{21}, we again would be in error if the indirect effect $\hat{\gamma}_{11}\hat{\beta}_{21}$ is significant. As I show in Chapter 8, matrix algebra provides an easier means of deriving these effects in all types of structural equation models.

To conclude this section, I should mention several misunderstandings about path analysis. One is the belief that path analysis is a methodology only appropriate when reciprocal or feedback relations are absent. This belief seems to stem from many of the early applications of path analysis in the social sciences which were restricted to one-way causation without any feedback. However, path diagrams, decompositions of covariances, or determination of total, direct, and indirect effects are not limited in this way. Indeed, as I mentioned in Chapter 1, Wright pioneered the estimation of nonrecursive models using path analysis.

A second belief is that path analysis deals exclusively with *standardized* regression coefficients. The standardized coefficients are defined as the usual unstandardized coefficient multiplied by the ratio of the standard deviation of the explanatory variable to the standard deviation of the variable it affects (see Chapter 4). To date, many analyses have reported standardized coefficients in path analyses. Path analysis, however, is not restricted to standardized coefficients. In fact, most of the examples I present are with unstandardized coefficients. A third incorrect belief about path analysis is that curvilinear relationships are not possible in path analysis. This too is not true. Just as with linear regression techniques, transformations of variables to capture curvilinear relationships may be used in path analysis. However, as I mentioned earlier, complications arise if latent variables have curvilinear relations with one another (see Chapter 9).

A final misunderstanding is the idea that path analysts are assuming that a path model that fits well proves causation. As I discuss in the next chapter, we can never prove causation with any technique. Rather, the purpose of path analysis is to determine if the *causal inferences* of a researcher are *consistent* with the data. If the path model does not fit the data, then revisions are needed, since one or more of the model assumptions are in error. If the path model is consistent with the data, this does not prove causation. Rather, it shows that our assumptions are not contradicted and may be valid. We only can say "may be valid" because other models and assumptions also may fit the data.

SUMMARY

This chapter provided three tools essential to subsequent chapters of the book. I repeatedly will apply the notation, covariances, and path analysis throughout the book. The next chapter examines the idea of causality that implicitly underlies the various structural equation techniques.

CHAPTER THREE

Causality and Causal Models

Most of this book is about statistics and data analysis: estimators, fitting functions, model fit, and identification. It is easy to forget the pervasive presence of the assumption of causality in structural equation models. This chapter is a reminder. It is a reminder of the causal assumptions made and the meaning of them. The purposes are to explore the nature of causality, the conditions of causation, and the limits of causal modeling.

NATURE OF CAUSALITY

Speculations about the nature of causality have a venerable history (see, e.g., Wallace 1972). No single definition of causality has emerged that is routinely employed in the sciences, humanities, and other fields where the term appears. Indeed, Nagel's conclusion of over two decades ago is true today: "... there is no uniquely correct explication of the term [cause]" (1965, 18).[1] Rather than presenting the many connotations of causality, I have a less ambitious goal. In this section I discuss the nature of causality as used in structural equation models.[2]

[1]One reaction to this ambiguity is to abandon the idea of causality as recommended by Bertrand Russell (1912–1913). He saw the concept as a "relic of a bygone age" that does more harm than good. His view was that the search for cause in the traditional sense should give way to functional relations represented in a series of differential equations. This, he claimed, was done in the more "advanced" sciences where causality was no longer discussed. Russell's views have been criticized on two fronts. First, it appears that even the sciences that Russell viewed as advanced did use cause or causality, contrary to his claim (Suppes 1970, 5–6). Second, Mackie (1974) and other philosophers of science see functional relations as an extension of the traditional concepts of causality, and not something fundamentally different.
[2]The pioneering work on causality in structural equation models by Wold (1956), Blalock (1964), Campbell and Stanley (1966), Simon (1954), and others has influenced my own thinking on causality as reflected in this section.

My starting point is a general definition. Consider one variable, say, y_1, which is *isolated* from all influences except from a second variable called x_1. If a change in y_1 accompanies a change in x_1, then x_1 is a *cause* of y_1. The definition of cause has three components: isolation, association, and the direction of influence. Association between two variables is not enough as is recognized in the well-known saying that correlation does not imply causation. Isolation comes first, then association. Even then we must establish that the association is due to x_1 influencing y_1 rather than the reverse.

Some argue that a variable can be a cause only if it can be subject to human manipulation (e.g., Holland 1986). According to this perspective, gender or race, for instance, cannot cause exposure to discrimination. This view leads to other counterintuitive ideas about causation: the moon does not cause the tides, tornadoes and hurricanes do not cause the destruction of property, and so on. I place no such restriction on what can be a cause. Isolation, association, and direction of influence are the three requirements for a cause. Human manipulation, such as occurs in an experiment, can be a tremendous aid toward creating isolation and establishing the direction of influence, but manipulation is neither a necessary nor sufficient condition of causality.

What makes it impossible to have absolute certainty that a variable is a cause is establishing that y_1 is isolated from all but x_1. Isolation is an unobtainable ideal. Isolation exists when y_1 and x_1 are in a "vacuum" that excludes all other influences. In reality traits that variables such as y_1 represent are part of a complex of characteristics of individuals, groups, or the other objects of study. The y_1 variable cannot occur in isolation since the units of analysis possess many characteristics besides x_1 on which they differ and some of these are likely to exert influence on y_1. Much of the debate about the causal status of a relation stems from doubts about whether the y_1 and x_1 association is due to these other factors. Without isolation of y_1, we can never be certain that x_1 causes y_1. Various experimental, quasi-experimental, and observational research designs attempt to approximate isolation through some form of control or randomization process. Regardless of the technique, the assumption of isolation remains a weak link in inferring cause and effect.

To gain a better understanding of the meaning of causality, I introduce an initially simple but increasingly elaborate example. I begin with a single observed variable y_1, whose causes are the object of investigation. A primitive first model for y_1 is

$$y_1 = \zeta_1 \tag{3.1}$$

Since ζ_1 is a disturbance variable that is made up of unnamed or unknown influences, equation (3.1) is not very informative. In essence we have no alleged cause other than a disturbance. Isolation and association are not issues. Normally, we assume that the disturbance influences y_1, instead of vice versa. We could view equation (3.1) as the representation of the extreme sceptic who believes that y_1 is a random variable incapable of being systematically related to other variables, so the best we can say is that it is a function of a disturbance. As an example suppose that y_1 is the number of motor vehicle fatalities (MVF) for each state in America in a given year. An argument to support (3.1) is that all MVFs are due to accidents and since accidents have unique and unpredictable origins, a random disturbance is the best representation.

A closer look at the disturbance term is worthwhile. The composition of ζ_1 in (3.1) is crucial to the possibility of explaining y_1. One case, as described earlier, is that it is an inherently stochastic variable with no identifiable components that can be separated and explicitly brought into the equation. If true, we have little hope of taking the explanation of y_1 any further. If ζ_1 is composed of one or more variables which we can bring into the model, the prognosis is much better.

The simplest composition is if ζ_1 consists of only one variable, say, x_1, with $\zeta_1 = f(x_1)$ and

$$y_1 = f(x_1) \tag{3.2}$$

As a further simplification I assume that $f(x_1) = \gamma_{11}x_1$ so that $f(x_1)$ is a simple linear function of x_1:

$$y_1 = \gamma_{11}x_1 \tag{3.3}$$

Equation (3.3) is a deterministic relation. For isolation to hold, no variable besides x_1 can be a cause of y_1. If the direction of influence from x_1 to y_1 is correct, the only way to change y_1 is by changing x_1. This follows since if y_1 could be changed without going through x_1, then equation (3.3) cannot be valid and y_1 is not isolated. The association is straightforward to assess, since for each unit shift in x_1, an exact γ_{11} shift in y_1 must occur.

Equation (3.3) corresponds to early historical accounts of causality in two ways. One is the tendency to treat bivariate relations: a single cause for a single effect. The other is the assumption that the cause-effect relation is deterministic. David Hume's ([1739] 1977) assumption of "constant conjunction" illustrates the latter: every time an object similar to the cause occurs, an event similar to the effect should follow. In (3.3) every difference in x_1 of one unit should result in exactly a γ_{11} difference in y_1. To place names on these variables, I maintain y_1 as motor vehicle fatalities (MVF), whereas x_1 might be the number of miles traveled by all motor vehicles in the state. The vehicle miles traveled increase the exposure of the population to accidents, and equation (3.3) shows that each unit increase in miles traveled leads to a γ_{11} increase in MVF.

Most readers are probably uncomfortable with equation (3.3). It represents a deterministic relation, and few would claim that this is reasonable. A single instance that does not conform to the equation is sufficient to falsify it. This is too rigid a requirement. A more reasonable standard is that x_1 has an *expected* influence of γ_{11} on y_1, but the actual values of y_1 are distributed around the predicted y_1. An alternative to equation (3.3) that includes this stochastic component is

$$y_1 = \gamma_{11}x_1 + \zeta_1 \tag{3.4}$$

Note that ζ_1 in (3.4) differs from ζ_1 in (3.1) in that in (3.4) ζ_1 is a disturbance that does not include x_1. To avoid excessive additional notation, I do not use new symbols to stand for the different disturbances, and I rely on the reader to keep in mind that as new variables are added to an equation the disturbance changes. I make the usual assumption that $E(\zeta_1)$ is zero.

Though adding a disturbance seems like a small change, it has serious implications for the concept of causality. For one thing the dual requirements of isolation and association are more difficult to evaluate than in the deterministic case. To establish x_1 as a cause of y_1, x_1 must be isolated from ζ_1. Since ζ_1 is an unobserved disturbance term, we cannot control it in any direct sense. Rather, we make assumptions about its behavior to create a *pseudo-isolation* condition. The most common assumption is that ζ_1 is uncorrelated with x_1. This is a standard assumption in regression analysis, and it enables us to assess the influence of x_1 on y_1 "isolated" from ζ_1. But the isolation is not perfect since for any observation, the y_1 to x_1 relation is disrupted by the disturbance. This is true whether the data come from an experiment where x_1 is a treatment applied with randomization to a subset of the observations or if the data are from nonexperimental sources.

Furthermore Hume's condition of constant conjunction, or any other definitions of causality that require the effect to be perfectly predictable from the cause, are no longer applicable. Instead, we move to a probabilistic view of causality (Suppes 1970). The predictability of y_1 now lies somewhere between equation (3.1), where all its variation is unexplainable, and equation (3.3), where all y_1's variation is accounted for by x_1.

Even (3.4) with x_1 and ζ_1 is a gross simplification if we expect ζ_1 to be uncorrelated with x_1. Many of the causes of y_1 besides x_1 are likely to be in the disturbance and to lead to a correlation of this disturbance with x_1.

Although an ideal randomized experiment might satisfy (3.4), it is more typical to need multiple explanatory variables for y_1:

$$
\begin{aligned}
y_1 &= \gamma_{11}x_1 + \gamma_{12}x_2 + \cdots + \gamma_{1q}x_q + \zeta_1 \\
y_1 &= \mathbf{\Gamma}_1\mathbf{x} + \zeta_1
\end{aligned}
\tag{3.5}
$$

where $\mathbf{\Gamma}_1$ is a $1 \times q$ row vector of $[\gamma_{11} \cdots \gamma_{1q}]$. We assume that $E(\zeta_1) = 0$ and the pseudo-isolation condition is that $\text{COV}(\mathbf{x}, \zeta_1) = \mathbf{0}$. The association is given by γ_{1j}, which is the expected change in y_1 for a one-unit change in x_j, when the remaining x's are held at constant values. Since we do not have control over ζ_1, we will not have an exact γ_{1j} change in y_1 each time x_j shifts one, and the other x's are unchanged. As was the case with a single explanatory variable, the disturbance term can disrupt the x_j and y_1 relation in any individual trial, even if it does not do so *on average*.

So far I have limited my remarks to observed variable models with y_1 and x's. The same points apply to all structural equation models. For instance, the typical implicit measurement model for an indicator x_1, and for the latent variable it measures (e.g., ξ_1), is

$$x_1 = \lambda_{11}\xi_1 \tag{3.6}$$

In essence this is a deterministic model like that in (3.3). A one-unit change in ξ_1 leads to an exact λ_{11} change in x_1. The λ_{11} typically is one so that x_1 and ξ_1 have the same scale. The major difference from (3.3) is that the explanatory variable is a "latent variable" in (3.6), though before both variables were observed variables. The deterministic relation in (3.6) blurs the distinction between latent and observed variables. Isolation holds if x_1 is removed from all influences save ξ_1. Even assuming that isolation and perfect association are present, a researcher must establish that causation runs from ξ_1 to x_1.

Most regression and classical econometric models implicitly assume a deterministic relation between measures and constructs. A more realistic model, however, would allow a disturbance or measurement error:

$$x_1 = \lambda_{11}\xi_1 + \delta_1 \tag{3.7}$$

with $E(\delta_1) = 0$. Pseudo-isolation for (3.7) is represented by the assumption that $\text{COV}(\xi_1, \delta_1) = 0$. We might even allow multiple latent variables to affect x_1:

$$
\begin{aligned}
x_1 &= \lambda_{11}\xi_1 + \lambda_{12}\xi_2 + \cdots + \lambda_{1n}\xi_n + \delta_1 \\
x_1 &= \mathbf{\Lambda}_1\boldsymbol{\xi} + \delta_1
\end{aligned}
\tag{3.8}
$$

The condition of pseudo-isolation is that $\text{COV}(\boldsymbol{\xi}, \delta_1) = \mathbf{0}$. Of course, I could write an analogous measurement model and assumptions in y's, η's, and ϵ's rather than the x's, ξ's, and δ's. Thus, to establish a structural or causal relation in measurement models, the same conditions must be satisfied as those for the previous models: isolation, association, and direction of causal relation. Here too we must settle for pseudo-isolation, because it is impossible to have ideal isolation. The same is true for the latent variable model.

In the next few sections I examine the three conditions of causality more closely.

ISOLATION

As I have just explained, we cannot isolate a dependent variable from all influences but a single explanatory variable, so it is impossible to make definitive statements about causes. We replace perfect isolation with pseudo-isolation by assuming that the disturbance (i.e., the composite of all omitted determinants) is uncorrelated with the exogenous variables of an equation.

To make the discussion more concrete, reconsider the observed variable model (3.5):

$$y_1 = \gamma_{11}x_1 + \gamma_{12}x_2 + \cdots + \gamma_{1q}x_q + \zeta_1$$

The disturbance ζ_1 is a random variable that is uncorrelated with x_1 to x_q. Typically, ζ_1 represents the net influence of the many variables that may exert minor influences on y_1 or that may not be known to the researcher. These omitted variables make up the disturbance term:

$$\zeta_1 = f(x_{q+1}, \ldots, x_\omega) \tag{3.9}$$

where $f(x_{q+1}, \ldots, x_\omega)$ is some function of x_{q+1} to x_ω. If we knew the function on the right-hand side of equation (3.9) and if we substituted it into (3.5), the relation of the x_i's ($i = 1, 2, \ldots, \omega$) to y_1 would be deterministic. We do not have $f(x_{q+1}, \ldots, x_\omega)$, so it is necessary to include ζ_1 without identifying its specific components. The assumption of pseudo-isolation is violated if

$$\text{COV}(\mathbf{x}, \zeta_1) = \text{COV}(\mathbf{x}, f(x_{q+1}, \ldots, x_w)) \neq \mathbf{0} \tag{3.10}$$

where $\mathbf{x}' = [x_1 \;\; x_2 \;\; \cdots \;\; x_q]$. When pseudo-isolation does not hold, the causal inferences regarding **x**'s impact on y_1 are jeopardized. If we assume that ζ_1 is an inherently stochastic term with no components, we still face the same threat to pseudo-isolation of COV(**x**, ζ_1) being nonzero.

A common situation that violates the pseudo-isolation condition is when a variable excluded from **x** that is part of ζ_1 leads to a correlation of ζ_1 and **x**. Such omitted factors can be intervening variables, can be common causes of y_1 and elements of **x**, or can have an ambiguous relation with **x** and y_1.

I first take up intervening variables. Consider the simple two-equation true model:

$$y_1 = \gamma_{11}x_1 + \zeta_1 \tag{3.11}$$

$$y_2 = \beta_{21}y_1 + \gamma_{21}x_1 + \zeta_2 \tag{3.12}$$

where $\text{COV}(\zeta_1, \zeta_2) = 0$, $\text{COV}(x_1, \zeta_1) = 0$, and $\text{COV}(x_1, \zeta_2) = 0$. The path diagram is in Figure 3.1. It is a *recursive model* where reciprocal or feedback causal relations are absent and the disturbances from each equation are uncorrelated (see Chapter 4). The disturbances ζ_1 and ζ_2 may contain a large number of omitted variables [as in (3.9)], but I assume that these composite terms of ζ_1 and ζ_2 are uncorrelated with x_1. The x_1 is an *exogenous variable*, which means that the variable is determined outside of the model. Its causes are not specified. The y_1 and y_2 variables are *endogenous*, since they are determined within the model. The variable y_1 is an intervening variable because part of x_1's influence on y_2 is through y_1. In the path analysis terminology of Chapter 2, γ_{21}, $\gamma_{11}\beta_{21}$, and $\gamma_{21} + \gamma_{11}\beta_{21}$ are, respectively, the direct, indirect, and total effects of x_1 on y_2.

Suppose that a researcher mistakenly omits the intervening y_1 variable from equation (3.12) and uses

$$y_2 = \gamma_{21}x_1 + \zeta_2^* \tag{3.13}$$

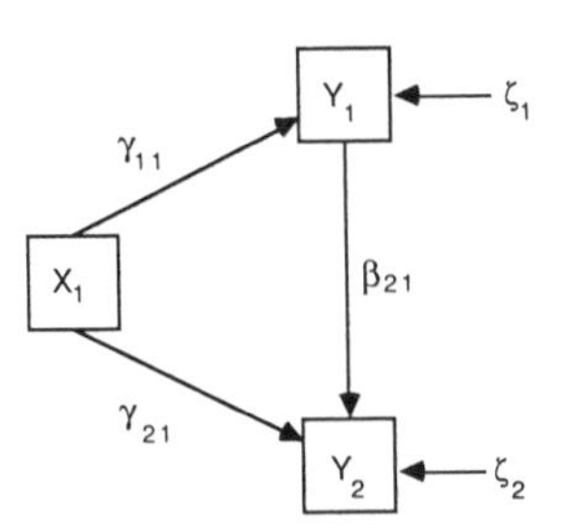

Figure 3.1 Path Diagram for Two-Equation Recursive Model

where $\zeta_2^* = \beta_{21} y_1 + \zeta_2$. The COV$(x_1, \zeta_2^*)$ is

$$\begin{aligned}\text{COV}(x_1, \zeta_2^*) &= \text{COV}(x_1, \beta_{21} y_1 + \zeta_2)\\ &= \text{COV}(x_1, \beta_{21}(\gamma_{11} x_1 + \zeta_1) + \zeta_2)\\ &= \beta_{21}\gamma_{11}\,\text{VAR}(x_1) \qquad (3.14)\end{aligned}$$

So long as VAR(x_1) is positive, $\beta_{21} \neq 0$, and $\gamma_{11} \neq 0$, the COV(x_1, ζ_2^*) is not zero, and pseudo-isolation is violated.

What are the consequences of omitting y_1 from (3.13)? The ordinary least squares (OLS) estimator, say $\hat{\gamma}_{21}^*$, is $[\text{cov}(y_2, x_1)]/\text{var}(x_1)$. Its probability limit (see Appendix B) is

$$\begin{aligned}\text{plim}(\hat{\gamma}_{21}^*) &= \text{plim}\left[\frac{\text{cov}(y_2, x_1)}{\text{var}(x_1)}\right]\\ &= \frac{\text{COV}(y_2, x_1)}{\text{VAR}(x_1)}\\ &= \gamma_{21} + \beta_{21}\gamma_{11}\\ &\equiv \gamma_{21}^* \qquad (3.15)\end{aligned}$$

So $\hat{\gamma}_{21}^*$ converges not to the direct effect, γ_{21}, but rather to the total effect, $\gamma_{21} + \beta_{21}\gamma_{11}$, of x_1 on y_2 which I represent as γ_{21}^*. Only when the indirect effect, $\beta_{21}\gamma_{11}$, is zero will the direct and total effects be equal. Otherwise, the total effect, γ_{21}^* $(= \gamma_{21} + \beta_{21}\gamma_{11})$ can be greater than or less than γ_{21}.

In practice, the omission of intervening variables pervades virtually all equations. For instance, an equation examining the relation of fertility to education omits intervening variables such as age at marriage, the cost of children, family size desires, contraceptive use, and the like. We may view the coefficient for education in the equation without these variables as the "total effect" of education in a hypothetical, more complete specification. The underspecified equation is less desirable than an equation that includes the intervening variables in that the understanding of the process is far more complete when the mediating factors are explicitly incorporated. However, often their omission from an equation is not excessively harmful so long as we remember that the coefficients for the included variables are net effects through the excluded intervening variables.

One potentially seriously misleading case is if the direct and indirect effects for a more fully specified model are of opposite signs but similar magnitudes. Then a left-out intervening variable can lead to the absolute value of a total effect considerably less than the absolute values of the direct or indirect effects of which it is comprised. This is one type of *suppressor relation*. McFatter (1979, 128) provides an interesting example. Consider Figure 3.1 and equations (3.11) and (3.12) as the true model. Suppose that y_2 is the number of errors made by assembly line workers, y_1 is the worker's boredom, and x_1 is his or her intelligence. Assume that the influence of intelligence (x_1) on boredom (y_1) is 0.707 ($= \gamma_{11}$), the influence of boredom (y_1) on number of errors (y_2) is the same ($\beta_{21} = 0.707$), and the effect of intelligence (x_1) on errors (y_2) is -0.50 ($= \gamma_{21}$). If boredom is omitted from the y_2 equation, intelligence's (x_1) coefficient in that equation is zero ($\gamma_{21}^* = 0$). The negative direct impact of intelligence on errors is suppressed unless boredom is controlled. Though this is the total effect of x_1 on y_2 in the original model, the zero γ_{21}^* may be misleading unless the researcher realizes the role of the boredom variable in the model. Suppressor relations also can occur in more complicated models with many explanatory variables.

One potential source of confusion is knowing whether a coefficient represents a direct, indirect, or total effect. For instance, if we know that we have omitted intervening factors between two variables, is the coefficient for the included explanatory variable a direct or total effect? Blalock (1964, 20) suggests that we restrict the terms intervening variables, direct, indirect, and total effects to those variables within a *specific* model. That is, a variable is intervening with respect to a particular explicit model. As work advances, researchers may introduce more intervening variables, but with regard to each model the direct and indirect effects are clear if we follow this advice.

Although most discussions of intervening variables concern observed variable regression models, the concept also is relevant in measurement models. Consider Figure 3.2. In part (a) of Figure 3.2, η_1 has a direct effect on y_1. Part (b) shows η_2 as an intervening variable between η_1 and y_1. The η_2 variable in part (b) has a direct effect on y_1, whereas η_1 has only an indirect effect. Suppose that η_1 is the inflationary expectations of a consumer and y_1 is the amount the consumer believes that the price of housing will increase in the next six months. The η_2 variable might be the consumer's inflation expectations for all durable goods (e.g., houses, cars, washing machines, dryers). If this is true, then overall expectations (η_1) only has its influence on y_1 through inflation expectations for durable goods (η_2). Future research might suggest more elaborate models so that η_2 no longer has a direct effect on y_1, but the direct and indirect effects of each variable will be clear with respect to any given model.

Intervening variables are one type of omitted variables that can lead to violations of the pseudo-isolation condition. Left-out common causes of the explanatory and dependent variables often pose a more serious threat. This

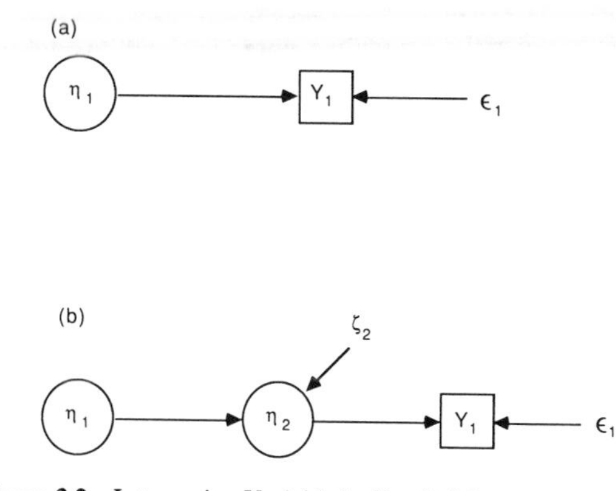

Figure 3.2 Intervening Variable in Simple Measurement Model

situation also can be explained by reference to Figure 3.1 and equations (3.11) and (3.12). Again consider (3.12) for y_2:

$$y_2 = \beta_{21} y_1 + \gamma_{21} x_1 + \zeta_2$$

It is easy to show that for recursive models such as (3.11) and (3.12) and their accompanying assumptions, the $\mathrm{COV}(y_1, \zeta_2)$ is zero. This means that ζ_1 is uncorrelated with y_1 and x_1 in the y_2 equation. Suppose that by mistake x_1 is omitted from the above equation, leading to

$$y_2 = \beta_{21} y_1 + \zeta_2^* \tag{3.16}$$

where $\zeta_2^* = \gamma_{21} x_1 + \zeta_2$. Equation (3.16) is an example of an equation with a common cause missing, since x_1 influences both y_2 and y_1. This violates the assumption that $\mathrm{COV}(y_1, \zeta_2^*)$ equals zero. In fact the $\mathrm{COV}(y_1, \zeta_2^*)$ equals $\gamma_{11}\gamma_{21}\,\mathrm{VAR}(x_1)$ so that pseudo-isolation does not hold except if γ_{11} or γ_{21} is zero. Furthermore the OLS estimator is $\hat{\beta}_{21}^* = [\mathrm{cov}(y_1/y_2)]/\mathrm{var}(y_1)$. This converges to

$$\begin{aligned}
\mathrm{plim}(\hat{\beta}_{21}^*) &= \frac{\mathrm{COV}(y_1, y_2)}{\mathrm{VAR}(y_1)} \\
&= \frac{\mathrm{COV}(y_1, \beta_{21} y_1 + \gamma_{21} x_1 + \zeta_2)}{\mathrm{VAR}(y_1)} \\
&= \beta_{21} + \gamma_{21} \frac{\mathrm{COV}(y_1, x_1)}{\mathrm{VAR}(y_1)} \\
&= \beta_{21} + \gamma_{21} b_{x_1 y_1} \\
&\equiv \beta_{21}^*
\end{aligned} \tag{3.17}$$

Table 3.1 Relation between the Absolute Values of β^*_{21} (= $\beta_{21} + \gamma_{21}b_{x_1y_1}$) and β_{21} when Omitted Common Cause

β_{21}	$\gamma_{21}b_{x_1y_1}$	$\lvert\beta^*_{21}\rvert$ vs. $\lvert\beta_{21}\rvert$	y_1 to y_2 Relation
$\neq 0$	0	=	no impact
0	$\neq 0$	>	totally spurious
> 0	> 0	>	partially spurious
< 0	< 0	>	partially spurious
> 0	< 0	<	suppressor
< 0	> 0	<	suppressor

So the true β_{21} is confounded with $\gamma_{21}b_{x_1y_1}$. The $b_{x_1y_1}$ coefficient comes from an *auxiliary regression* of x_1 on y_1. The auxiliary regression is not a structural equation, but it is a convenient means of summarizing the relation of the omitted x_1 variable and the included y_1 variable (see Theil 1957).

Table 3.1 gives the relation between β^*_{21} (= $\beta_{21} + \gamma_{21}b_{x_1y_1}$) and β_{21} under several different conditions. The relatively innocuous cases are when either x_1 has no effect on y_2 ($\gamma_{21} = 0$) or when x_1 and y_1 are uncorrelated ($b_{x_1y_1} = 0$), since then $\beta^*_{21} = \beta_{21}$ and we have an accurate measure of y_1's impact on y_2. Far more damaging is the instance where y_1 has no causal effect on y_2 ($\beta_{21} = 0$), but y_1 and y_2 both depend on x_1. Here β^*_{21} can take nonzero values, leading to a totally *spurious* relation between y_1 and y_2. The possibility of such spurious relations is the reason that the phrase "correlation is not causation" is appropriate. As an example, suppose that y_2 is the quality of a person's vision and y_1 is his or her proportion of gray scalp hairs. The variables correlate not because gray hair causes poor vision but because both are causally depended on age (x_1). On the other hand, we cannot automatically assume that all associations are spurious. This too should be demonstrated. For example, representatives of the tobacco industry sometimes argue that the correlation between smoking and cancer is spurious. One suggestion is that some people have a genetic predisposition to smoke and to get lung cancer. If such a factor is found, a stronger case for spuriousness could be made, but without it most remain skeptical of such a claim.

Rows three and four of Table 3.1 represent partially spurious cases where β_{21} and $\gamma_{21}b_{x_1y_1}$ have the same signs. This serves to inflate the absolute value of the coefficient β^*_{21} in the equation, which leaves out the common cause (x_1). As in the totally spurious situation we mistakenly attribute to y_1 part of the influence of x_1 on y_2.

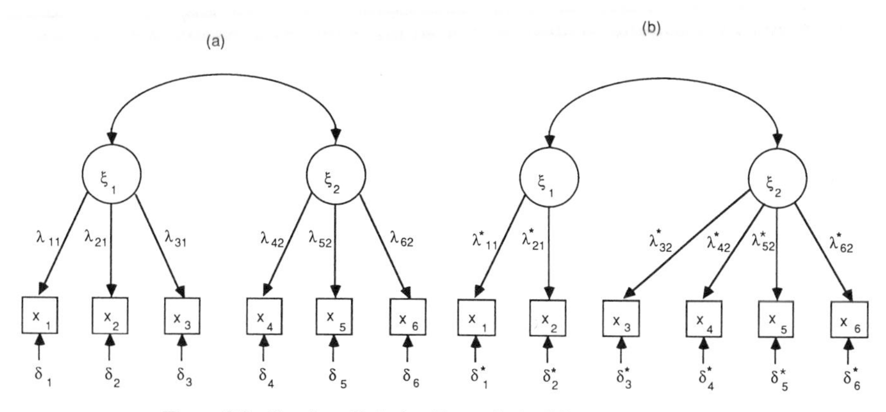

Figure 3.3 Spurious Relation Example for Measurement Model

Spurious relations can affect measurement models as well as observed variable ones. As an example, consider Figure 3.3. Suppose that ξ_1 refers to individual conservatism and ξ_2 refers to authoritarianism. The variables x_1 to x_3 are measures of conservatism while x_4 to x_6 are indicators of authoritarianism. A researcher may mistakenly treat x_3 as a measure of authoritarianism (ξ_2) and estimate model (b). The $\hat{\lambda}^*_{32}$ estimate may be the correct sign and statistically significant, but this is a spurious result of ξ_1's influence on x_3 and ξ_1's correlation with ξ_2. If we allowed ξ_1 to have a direct effect on x_3, then we would expect the new $\hat{\lambda}^*_{32}$ to be within sampling fluctuations of zero. The same spurious relations can occur in other measurement models when we misspecify the latent variables that affect an indicator and when the latent variables are correlated.

Another type of spurious relations in measurement models occurs when two observed variables are assumed to have a direct causal relation when in reality their association is due to their common dependence on a latent variable. For instance, two IQ tests are associated not because they have a causal relation but because both depend upon the latent variable of intelligence. Spurious relations in measurement models are probably just as prevalent as spurious relations in observed variable models. Indeed, the problem may be worse in measurement models given the lack of systematic attention to it.

Returning to Table 3.1 and the observed variable models (3.11) and (3.12) to which it refers, the last two rows give the conditions for suppressor relations. These occur when β_{21} and $\gamma_{21}b_{x_1y_1}$ have opposite direction effects and the common cause (x_1) is left out of the y_2 equation. Here the absolute value of the effect of y_1 on y_2 is underestimated or *suppressed*. Earlier I explained how the omission of intervening variables can lead to suppressor relations. This differs from the current situation in that before the suppressed effect was a measure of the total effect of the included variable. Now the suppressor is a common cause rather than an intervening variable. The β_{21}^* does not gauge the total effect of y_1 on y_2. Instead, it reflects the direct effect β_{21} combined with $\gamma_{21}b_{x_1y_1}$.

As an illustration, consider the earlier example where x_1 is intelligence, y_1 is boredom, and y_2 is number of errors made for an assembly line worker (see Figure 3.1). Suppose we remove x_1, the common cause of y_1 and y_2, from the y_2 equation. With $\gamma_{11} = 0.707$, $\gamma_{21} = -0.500$, and $\beta_{21} = 0.707$, we will have β_{21}^* $(= \beta_{21} + \gamma_{21}b_{x_1y_1})$ less than β_{21}, and boredom's influence on errors is suppressed.

As with spurious relations, suppressor effects have received the greatest attention in correlational and regression models, but the effect can occur in models with latent variables. Figure 3.4 illustrates one such case. Figure 3.4(*a*) is the true model, whereas (*b*) is the misspecified one where η_1 is not included as a determinant of y_3. If β_{21} and λ_{32} are positive and λ_{31} is negative, we may find the absolute value of $\hat{\lambda}_{32}^*$ suppressed relative to the absolute value of $\hat{\lambda}_{32}$ in the correct model. If $\hat{\lambda}_{32}^*$ is close enough to zero, we may mistakenly conclude that η_2 has no influence on y_3.

Many researchers suggest that a bivariate association between a cause and effect is a *necessary* condition to establish causality. The occurrence of suppressor relations casts doubt on this claim: no bivariate association can occur although a causal relation links two variables. The old saying that correlation does not prove causation should be complemented by the saying that *a lack of correlation does not disprove causation*. It is only when we isolate the cause and effect from all other influences that correlation is a necessary condition of causation. Thus, without isolation or without at least pseudo-isolation, correlation is neither a necessary nor a sufficient condition of causality.

Left-out intervening variables or left-out common causes are two ways in which the pseudo-isolation conditions can be violated. A third is when the omitted variable has an ambiguous relation to the explanatory variables. For instance, consider Figure 3.5 which has the equation

$$y_1 = \gamma_{11}x_1 + \gamma_{12}x_2 + \zeta_1 \tag{3.18}$$

where $\text{COV}(\zeta_1, x_i) = 0$ for $i = 1, 2$ and $E(\zeta_1) = 0$. If a researcher mistakenly leaves x_2 out of (3.18), we have

$$y_1 = \gamma_{11}x_1 + \zeta_1^* \tag{3.19}$$

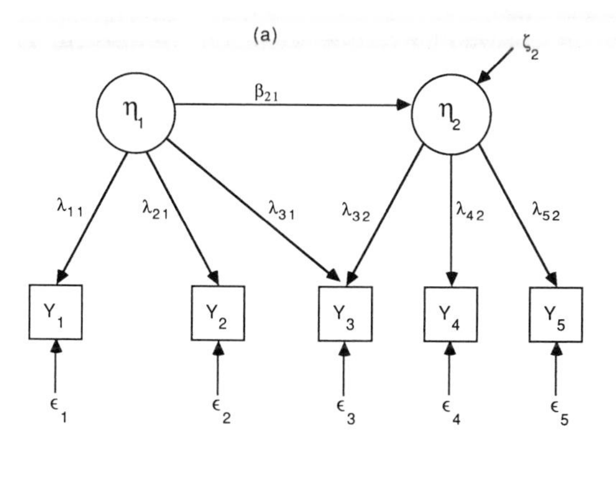

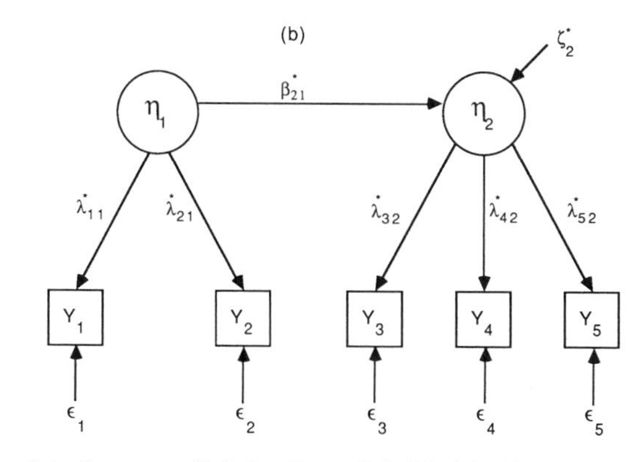

Figure 3.4 Suppressor Relation Example in Model with Latent Variables

where $\zeta_1^* = \gamma_{12}x_2 + \zeta_1$. The $\text{COV}(x_1, \zeta_1^*)$ is $\gamma_{12}\,\text{COV}(x_1, x_2)$ which is nonzero when γ_{12} and $\text{COV}(x_1, x_2)$ are not zero. The OLS estimator of $\hat{\gamma}_{11}^*$ converges to

$$\begin{aligned}
\text{plim}(\hat{\gamma}_{11}^*) &= \frac{\text{COV}(x_1, y_1)}{\text{VAR}(x_1)} \\
&= \gamma_{11} + \gamma_{12}\frac{\text{COV}(x_1, x_2)}{\text{VAR}(x_1)} \\
&= \gamma_{11} + \gamma_{12}b_{x_2x_1} \\
&\equiv \gamma_{11}^* \qquad (3.20)
\end{aligned}$$

Comparing γ_{11} to γ_{11}^*, we find results analogous to those in Table 3.1 where γ_{11} replaces β_{21} and $\gamma_{12}b_{x_2x_1}$ replaces $\gamma_{21}b_{x_1y_1}$. That is, γ_{11}^* $(= \gamma_{11} + \gamma_{12}b_{x_2x_1})$ can be higher, lower, or the same as γ_{11}. The major difference is that in the current model we do not know whether the omitted x_2 variable is an intervening variable, a common cause of x_1 and y_1, or if some other relation exists. All we know is that x_1 and x_2 are correlated, as indicated by the curved two-headed arrow connecting them in Figure 3.5. So a cost of not knowing the nature of the relation between x_1 and x_2 is that we can say less about the consequences of omitting x_2 from the y_1 equation.

For expository purposes I have kept the examples of omitted variables simple. Spurious, suppressor, and the other relations occur in more complex models with many explanatory variables and with or without latent variables. In general, omitted variables can lead to the violation of the pseudo-isolation condition of a zero correlation between the exogenous variables and the disturbance of an equation. This, in turn, can lead to inconsistent parameter estimators if the problem is ignored.

Omitted explanatory variables is the most obvious way in which pseudo-isolation is violated. Another way is when one or more of the explanatory variables contains random measurement error. For instance, suppose that the true model is

$$y_1 = \gamma_{11}\xi_1 + \gamma_{12}\xi_2 + \zeta_1 \tag{3.21}$$

where $E(\zeta_1) = 0$, $\text{COV}(\zeta_1, \xi_1) = 0$, and $\text{COV}(\zeta_1, \xi_2) = 0$. Also suppose that x_1 is a proxy for ξ_1 such that

$$x_1 = \xi_1 + \delta_1 \tag{3.22}$$

where $E(\delta_1) = 0$ and $\text{COV}(\xi_i, \delta_1) = 0$ for $i = 1, 2$. Combining (3.21) and

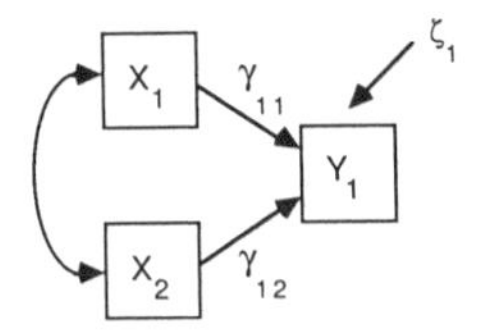

Figure 3.5 Path Diagram for Single Equation Model

(3.22), we have

$$\begin{aligned} y_1 &= \gamma_{11}(x_1 - \delta_1) + \gamma_{12}\xi_2 + \zeta_1 \\ &= \gamma_{11}x_1 + \gamma_{12}\xi_2 + \zeta_1 - \gamma_{11}\delta_1 \\ &= \gamma_{11}x_1 + \gamma_{12}\xi_2 + \zeta_1^* \end{aligned} \tag{3.23}$$

The last line is the equation with x_1 used in place of ξ_1. It is clear that the pseudo-isolation condition, $\text{COV}(x_1, \zeta_1^*) = 0$, is likely to fail, and the OLS estimators of (3.23) are inconsistent. To a large extent (3.23) is like the omitted variable situation where ξ_1 is the left-out variable that influences both x_1 and y_1. Chapter 5 gives a more comprehensive discussion of the consequences of measurement error.

Another less obvious violation of pseudo-isolation is when the "dependent" variable affects one or more of the "exogenous" variables. This might occur if the initial direction of influence is incorrect or if reciprocal causation exists. For instance, earlier I argued that vehicle miles traveled has a positive influence on motor vehicle fatalities (MVF). It also is possible that MVF reduces vehicle miles traveled by leading to fewer drivers. In this and other cases where the endogenous variables affect an "exogenous" one, the assumption that the disturbance is uncorrelated with x_1 is no longer defensible. The solution is to turn the former "exogenous" variable into an endogenous one so that it has a separate equation. We have estimators besides OLS to deal with such nonrecursive systems, provided it is an identified model for which the disturbances are uncorrelated with the truly exogenous variables. Chapter 4 treats such models.

A third less intuitive way in which a correlation is created between an explanatory variable and a disturbance can be shown by an example. Consider Figure 3.6. Part (*a*) is a recursive model with ζ_1 and ζ_2 uncorrelated with each other and with x_1, x_2, and x_3. Under these assumptions the $\text{COV}(y_1, \zeta_2)$ is zero, and OLS is a consistent estimator of the parameters for the y_2 equation. Part (*b*) of Figure 3.6 differs from (*a*) only in that $\text{COV}(\zeta_1, \zeta_2)$ is nonzero. This seemingly harmless change is sufficient to render $\text{COV}(y_1, \zeta_1) \neq 0$ and to make OLS an inconsistent estimator for the second equation. Alternative consistent estimators are available (see Chapter 4), but this example illustrates that having a nonzero covariance of disturbances is sufficient to create a correlation between a disturbance and an explanatory variable that is not present when the disturbances have no association.

A fourth way in which pseudo-isolation is violated is when a researcher specifies the wrong functional form for the relation between two variables. For instance, a linear relation between y_1 and x_1 may be used, but a curvilinear relation is correct. In many cases the variables may be trans-

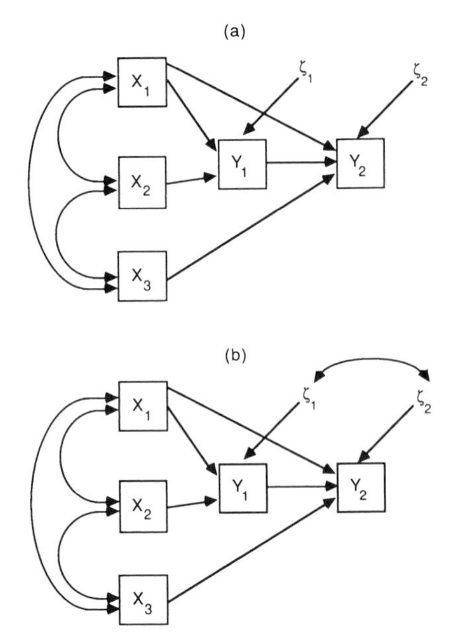

Figure 3.6 Path Diagrams for Two-Equation Model without (*a*) and with (*b*) Correlated Disturbances

formed so that the relations are linear in the transformed variables. If this is not done, nonzero correlations between the equation disturbance and the explanatory variables are likely, and pseudo-isolation fails.

Other situations also can break pseudo-isolation. For instance, the usual assumptions of a disturbance being uncorrelated with the explanatory variables may be violated if a lagged endogenous variable appears as an explanatory variable and the disturbances of that equation are autocorrelated (see, e.g., Johnston 1984, 360–371). A nonrandom subsample of the relevant population (see Heckman 1979) is another way to violate pseudo-isolation. Special procedures that take these problems into account are available, but if these difficulties are ignored, faulty causal inferences are likely.

In sum, many factors threaten the pseudo-isolation conditions necessary to establish a causal link between two variables. Omitted variables can inflate or deflate relations. Measurement errors, nonrandom sample selection, correlated disturbances, and other less obvious problems also can undermine pseudo-isolation. Though some research designs can lessen these potential problems, it is not possible to have certainty that two variables are totally isolated from other influences. Thus we should recognize the tentativeness of any claims for a causal relation while striving to eliminate as many threats to pseudo-isolation as is possible.

ASSOCIATION

In the preceding section I examined threats to isolation or pseudo-isolation. Now I take a closer look at association and the factors that make it difficult to establish. Unless otherwise stated, I assume that the condition of pseudo-isolation is satisfied and that all parameters are *identified*. The latter condition exists if it is possible to obtain unique values for each free parameter. I return to the issue of identification in several of the other chapters, but for now I assume that all parameters are identified.

When a putative cause and its effect are isolated from other influences, then the two variables should be associated. A bivariate association is neither a necessary nor sufficient condition for a causal relation. Rather, association *net of other influences* is necessary to establish causality. Of course, if an equation consists of a single explanatory variable and a disturbance that is uncorrelated with it, a bivariate association of the dependent and explanatory variable is all that is needed. For instance, most measurement models assume that an indicator is influenced by one latent variable and an error of measurement (i.e., $x_1 = \lambda_{11}\xi_1 + \delta_1$), $COV(\xi_1, \delta_1) = 0$, $E(\delta_1) = 0$) so that we expect a bivariate association of the measure and the latent variable. In the terminology of factor analysis, the factor complexity of the variable is one.

Does the common practice of assuming a bivariate relation between observed and latent variables make sense? In many cases it does not. To illustrate this, consider the following item taken from a widely used scale: "It's hardly fair to bring a child into the world with the way things look for the future." Respondents are asked to indicate their agreement or disagreement with this statement. The resulting responses form the observed variable. Many latent variables may underlie agreement or disagreement with the statement. One latent variable might be the respondent's fertility expectations or actual fertility experience. Those expecting to have children or those with children are likely to disagree with the statement. Belief in the likelihood of nuclear war or attitudes toward world population also could be latent variables driving the response, with those believing that war is imminent and that population is a problem more likely to agree with the statement than others. In fact, this item is from Srole's (1956) anomia scale. Anomia is a concept in sociology and social psychology that refers to an individual's feelings of isolation. Anomia as a latent variable may have a causal effect on the observed variable here, but it is likely that other latent variables also have an influence. The same is likely for many other indicators. In such cases, a bivariate association between a latent variable and its indicator is neither necessary nor sufficient for a causal relation between them. Nevertheless, the partial coefficient that corresponds to the association of the indicator and the latent variable, net of the other latent variables, should be nonzero if there is a causal relation.

Even under ideal conditions there are some complications to establishing this association. To illustrate, consider again (3.11) and (3.12), which I repeat here (also see Figure 3.1):

$$y_1 = \gamma_{11} x_1 + \zeta_1$$

$$y_2 = \beta_{21} y_1 + \gamma_{21} x_1 + \zeta_2$$

The γ_{11} coefficient gives the association of x_1 and y_1. The γ_{11} is a population parameter. We have $\hat{\gamma}_{11}$, a consistent estimator of γ_{11}, as the basis of making statements about γ_{11}. In any given sample the value of $\hat{\gamma}_{11}$ differs from γ_{11} because of sampling error. Usually we can estimate the probability that γ_{11} takes particular values. But these are probabilities not certainties, and mistakes in judging associations will occur. However, in practical terms we are usually willing to live with sampling error as long as we know its likely magnitude.

More problematic is when the standard errors or test statistics that are the basis of the tests of statistical significance are incorrect. One such case is if the disturbances, ζ_1, from the preceding y_1 equation are heteroscedastic or autocorrelated. Then OLS still provides a consistent estimator of γ_{11}, but the usual standard errors and test statistics for the coefficient estimator are not dependable. Thus we could make faulty inferences about the association of x_1 to y_1 because we have the wrong standard errors for $\hat{\gamma}_{11}$. Alternative estimators for regression equations that take into account heteroscedasticity or autocorrelation and provide suitable standard errors and test statistics are well-known (Johnson 1984). However, tests or corrections for heteroscedasticity or autocorrelated disturbances have received insufficient attention for models with latent variables. So for these models faulty inferences about association are possible.

Another complication in determining association is multicollinearity. Multicollinearity is the extent to which a linear dependence exists between an explanatory variable and the other explanatory variables in an equation. The simple bivariate correlation between explanatory variables is not sufficient for determining the extent of collinearity. The multiple correlation of each explanatory variable regressed on the other explanatory variables comes closer to measuring this dependence. A number of diagnostic techniques are available for collinearity in single-equation regression models (see, e.g., Belsley, Kuh, and Welsch 1980), but these are not discussed here.

Collinearity is a problem in nonexperimental research where many of the variables in an analysis are highly correlated. In experiments, researchers often can design explanatory variables so they are approximately (or exactly) uncorrelated. Collinearity generally increases the standard errors of the coefficients of the collinear variables (other things equal). The increased standard errors mean that we have greater uncertainty in the inferences that we make about the parameters. In intuitive terms it is difficult to estimate the unique effect of a variable if it always moves in conjunction with the other causal variables. The problem is discussed almost exclusively for regression models, but it also poses difficulties for measurement models (see Rindskopf 1984a). Figure 3.7 provides an example. It has two latent variables, morale (ξ_1) and sense of belonging (ξ_2). The seven indicators consist of three indicators of morale (x_1 to x_3), three indicators of belonging (x_5 to x_7), and one indicator (x_4) affected by both ξ_1 and ξ_2. The x_4 measure is really a measure of morale, but to illustrate collinearity problems, I allow ξ_2 to affect it. I collected data from x_1 to x_7 at a coeducational college. More details on the data are in Chapter 8. I use the female sample here. Morale (ξ_1) and sense of belonging (ξ_2) are highly correlated (0.9). When I estimated this model, all of the coefficient estimates represented in Figure 3.7 were significant except for the coefficients of ξ_1 and ξ_2 to x_4: $\hat{\lambda}_{41} = 1.61$ with a z-ratio of 1.5 and $\hat{\lambda}_{42} = -0.62$ with a z-ratio of -0.7. (The z-ratios are the ratios of the coefficients to their standard errors.) These estimates show the classic symptoms of collinearity. They have opposite signs, and the negative coefficient is counter to what I would predict. The standard errors are large, leading to nonsignificant results for these two coefficients. In addition the correlation between these parameter estimates is -0.977. Thus the collinearity between ξ_1 and ξ_2 makes it

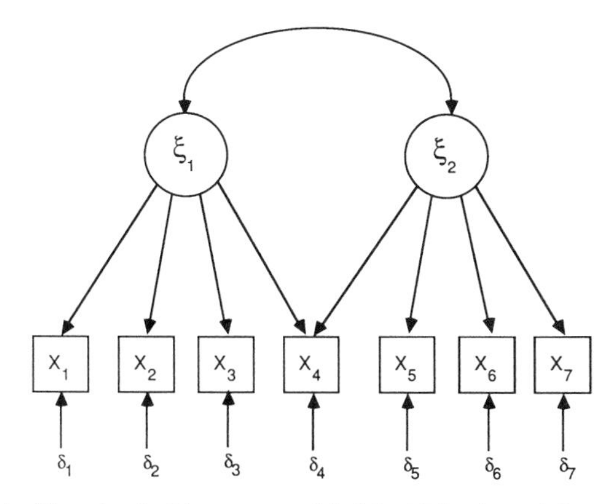

Figure 3.7 Multicollinearity in Measurement Model of Morale and Sense of Belonging for Females

extremely difficult to know if either or both variables have direct effects on x_4.

In the most extreme collinearity case one variable is an exact linear combination of the other variables. No unique estimate of the independent causal effect is then possible. For example, consider equation (3.24):

$$x_2 = \lambda_{21}\xi_1 + \lambda_{22}\xi_2 + \delta_2 \tag{3.24}$$

Suppose that $\xi_1 = \gamma_{12}\xi_2$ and that this relation is substituted in equation (3.24):

$$\begin{aligned} x_2 &= \gamma_{12}\lambda_{21}\xi_2 + \lambda_{22}\xi_2 + \delta_2 \\ &= \lambda^*\xi_2 + \delta_2 \end{aligned}$$

where

$$\lambda^* = \gamma_{12}\lambda_{21} + \lambda_{22} \tag{3.25}$$

Equation (3.25) shows that in essence only one distinct explanatory variable is needed—ξ_2. Although an unique estimate of λ^* may be obtained, unique estimates of λ_{21} and λ_{22} are not possible. Thus collinearity in measurement models has analogous consequences to those in regression problems.

Replication is an important check on whether an association is a sampling fluke. Although in theory the importance of replication is widely recognized, in practice replicative studies appear far too infrequently. The greatest attention and rewards seem to be for studies reporting novel findings or research. This emphasis is surely misplaced, since if an association is only found once or twice, this is not sufficient justification for considering it a causal relation. In the words of Popper (1959, 45), "We do not take even our own observations quite seriously, or accept them as scientific observations, until we have repeated and tested them." The degree of replication demanded depends on the development of research in an area. For example, in early stages merely finding a significant coefficient of the same sign relating two variables would satisfy a weak form of replication. Whereas in other areas replication may require that the parameters have the same value. The closer the results, the stronger is the replication.

Replication also provides a check on the pseudo-isolation assumption since if a model is correct, we expect it to hold in different samples and with different types of data. A relationship that is found in cross-sectional and time-series studies as well as in experimental and observational research is more likely to be causal than if it is only found in one of these settings. When results differ, this is a bad sign because it suggests misspecification but does not give the source of the error. Simon (1968, 1969) provides a good example of attempting to resolve contradictory cross-sectional and time-series results on the relation between income and fertility and income and suicide. Similar efforts in other areas would be helpful.

Replications has an added role because of the typical model-fitting process. Researchers often begin with parsimonious models. Usually these models do not fit the data, so they are modified. The end result of this sequence is that the reported significance levels for the final models are off to an unknown degree. Replication in an independent sample provides a check of whether the final models reflect capitalization on chance in one sample or a robust association.

Despite the advantages of replication, its limitation should be recognized. Spurious as well as causal relation can replicate. To use an old example, we may find a close relation between the number of fire trucks at a fire and the amount of damage. We may even be able to replicate this exactly with time-series and cross-sectional data. But even with the replication, it is clear that a control for the size of the fire would show the initial relation to be spurious.

In sum, association is the second condition for establishing causality. A causal variable should be associated with an effect after the two are pseudo-isolated. Sampling fluctuations are one threat to assessing such associations even when an equation is correct. Other problems are created when the standard errors and test statistics are biased due to uncorrected problems of heteroscedastic or autocorrelated disturbances or measurement errors. Replication of an association in an independent sample is one means to increase confidence that an association is robust.

DIRECTION OF CAUSATION

The plausibility of an association being causal rests on having the causal direction correct. In Hume's ([1739] 1977) highly influential treatment of causality, he required temporal priority as a condition of causality. That is, the alleged cause must precede the effect, and this is the primary means to ensure that the explanatory variable has causal primacy.

Hume's temporal priority raises questions for structural equation models. If temporal priority is required, then simultaneous relations are ruled out. Causes must come before effects, and simultaneous reciprocal causation—as in nonrecursive structural equation models—is not possible.[3]

[3]One criticism of temporal priority is that if we allow a time lag in cause-effect relations, then the influence of a variable must cross a gap in time before exerting its influence. Within this gap it is possible that an intervening variable enters to prevent the cause from having its effect.

Furthermore, if we accept that an interval between cause and effect must exist, the immediate question is how long is the interval? If we look for influence too early or too late, we can miss it. To illustrate, suppose that a researcher wants to analyze the causal relations between advertisements and sales for a product. He or she reasons that advertisements cause people to buy more of a product. The greater sales of the product cause more advertisements. It makes sense to expect some lag between the presentation of ads and the purchase of the product. Similarly, a delay between sales generating more ads is reasonable. Thus the temporal lag in cause-effect relations is believable in this and most examples. But what if only yearly data are available and we suspect that the causal lag is shorter than that? In cases where the observation period is longer than the temporal lag and two variables influence one another, the relation is sometimes approximated by two-way causal relations (Fisher 1970). That is, the analyst estimates models that allow reciprocal causation even though if shorter observation units (e.g., months, weeks, days) were available, the one-way nature of the causation and the temporal lag would be visible.

This assumes that the delay in response is known. Even with fine temporal gradations available, this would not solve the time lag problem. For this the investigator must draw on prior substantive or theoretical work. In most cases these sources are not specific enough to specify the time lags, and as a result empirical techniques are employed. In sum, we expect a time lag, no matter how small, between a cause and effect. But in practical terms the lag can be smaller than the observation interval. Under these circumstances the cause is placed in the same time period as the effect. If two variables influence each other, we allow a feedback relation.

Can a future event cause a present or past one? Sometimes it is suggested that this occurs when, for example, a future election causes campaigning activities or future taxes cause avoidance behavior in the present. In these and similar cases temporal priority is not violated. The future is not the cause; rather, it is the *expectation* of an election or of taxes that is the cause. And the expectation precedes the campaign or tax avoidance behavior.

Some researchers believe that if one variable is lagged behind a second variable and a significant coefficient occurs, then this helps to establish the causal priority of the lagged variable. Or, if a explanatory variable is measured at a time after the dependent variable, then critics charge that the estimated effect cannot represent causal influence.

The limitations of these practices can be illustrated in the bivariate case. Suppose that equations (3.26) and (3.27) represent the true relationships:

$$y_t = \gamma x_t + \zeta_t \tag{3.26}$$

$$x_t = \alpha x_{t-1} + \nu_t \tag{3.27}$$

where $COV(x_t, \zeta_t) = 0$, $COV(x_{t-1}, \zeta_t) = 0$, $COV(x_{t-1}, \nu_t) = 0$, and $COV(\zeta_{t-i}, \nu_{t-j}) = 0$ for all i and j. A researcher mistakenly believes that y is a cause of x. To test the causal priority of y, y is lagged behind:

$$x_t = \gamma^* y_{t-1} + \zeta_t^* \tag{3.28}$$

The researcher believes that a nonzero $\hat{\gamma}^*$ supports the causal priority of y but a zero coefficient does not.

The probability limit of the ordinary least squares estimator of γ^* is

$$\begin{aligned} \text{plim}(\hat{\gamma}^*) &= \frac{COV(x_t, y_{t-1})}{VAR(y_{t-1})} \\ &= \frac{COV(\alpha x_{t-1} + \nu_t, \gamma x_{t-1} + \zeta_{t-1})}{VAR(y_{t-1})} \\ &= \frac{\alpha\gamma\, VAR(x_{t-1})}{VAR(y_{t-1})} \end{aligned} \tag{3.29}$$

If neither γ nor α is zero, then (3.29) shows that a nonzero coefficient occurs even though the dependent variable is lagged behind the independent. Thus a simple lagging of a variable cannot establish causal priority.

In the second case the explanatory variable may only be available for a period after the dependent variable. This is represented in equation (3.30):

$$y_t = \gamma^* x_{t+1} + \zeta_t^* \tag{3.30}$$

The resulting coefficient here is

$$\begin{aligned} \text{plim}(\hat{\gamma}^*) &= \frac{COV(y_t, x_{t+1})}{VAR(x_{t+1})} \\ &= \frac{COV(\gamma x_t + \zeta_t, \alpha x_t + \nu_{t+1})}{VAR(x_{t+1})} \\ &= \frac{\alpha\gamma\, VAR(x_t)}{VAR(x_{t+1})} \\ &= \alpha\gamma, \qquad \text{when } VAR(x_t) = VAR(x_{t-1}) \end{aligned}$$

Notice that in this case the result only differs from the true causal parameter γ by α. If a rough estimate of α is available, then an estimate of γ may be obtained as well. Therefore leading the causal variable when the correct lag is not available may provide useful results. Of course this is clearly inferior to using the properly lagged variable. The basic lesson of these two examples (which generalize to the multivariate case) is that lagging or leading variables cannot establish causal priority.

A related idea views x as a cause of y, if we can predict y better by using past values of y and x than if we only use past values of y. For instance, consider the equation (3.31):

$$y_t = \sum_{i=1}^{k} \beta_i y_{t-i} + \sum_{j=1}^{l} \gamma_j x_{t-j} + \zeta_t \tag{3.31}$$

where ζ_t is not autocorrelated and it is uncorrelated with y_{t-i} and x_{t-j} for all i and $j > 0$. If we can reject the hypothesis that $\gamma_j = 0$ for all j, then x is a cause of y_t. If we do not reject H_0: $\gamma_j = 0$, then x is not a cause of y_t. We can write an analogous equation for x_t with its lagged values and y's lagged values as explanatory variables.

Procedures similar to this have arisen in several disciplines as a means to help assess causal direction or as tests of exogeneity. For instance, Campbell (1963) recommended cross-lagged correlations $r_{y_t x_{t-1}}$ and $r_{y_{t-1} x_t}$ to assess causal direction for panel data where y and x are available at two points of time for the same units of observation. The ambiguities of using correlations for this purpose were pointed out by sociologists Heise (1970) and Bohrnstedt (1969) who argued for a regression-based approach. Their recommendation was similar to (3.31), with the lags for y and x extended to one or at most a few periods of panel data. Finally, in econometrics Granger (1969), Sims (1972), and others proposed an approach similar to (3.31) for stationary time-series data.

Though these methods are appealing in their exploitation of time lags and lagged dependent variables, they do not obviate the need for the pseudo-isolation condition that the disturbance term ζ_t is uncorrelated with the lagged values of y and x. This assumption is automatically violated when ζ_t is autocorrelated. Omitted variables also are likely to undermine this assumption as long as only lagged values of y and x are included as potential causes of y_t. All of the previous discussion on omitted variables applies to these models. The equations can be expanded to include additional explanatory variables, but applications to date have rarely done so. Other comments on these approaches to causality are available in Duncan (1969), Zellner (1984), and Holland (1986), among others.

Determining the direction of causation between latent variables and indicators in measurement models can be troublesome. Following Blalock's (Namboodiri, Carter, and Blalock 1975, 569–572) suggestion, I distinguish

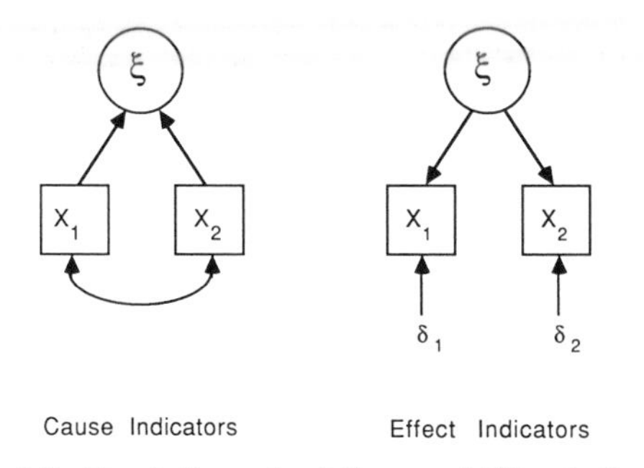

Figure 3.8 Simple Example of Cause and Effect Indicators

cause ("formative" or "induced") indicators from effect ("reflective") indicators (see Figure 3.8). Cause indicators are observed variables that are assumed to cause a latent variable. For effect indicators the latent variable causes the observed variable. Most researchers in the social sciences assume that indicators are effect indicators. Cause indicators are neglected despite their appropriateness in many instances.

Establishing causal priority is necessary to determine if an indicator is a cause or an effect of a latent variable. For instance, suppose that we use race and sex as indicators of exposure to discrimination. Since exposure to discrimination does not change a person's race or sex, these two variables are cause indicators. In research on stress exposure, analysts often use measures of significant, disruptive events experienced by respondents. These include marriage, divorce, unemployment, and job promotions. Clearly, these observed variables cause changes in stress exposure and are cause indicators.[4] Similar arguments demonstrate that the components of gross national product are cause indicators of the value of a nation's goods and services. On the other hand, many measures of attitudes are effect indicators. For instance, assume that the latent variable is self-esteem, and we ask the person to indicate his or her agreement with the statement: "I feel that I am as good as the next person." As a person's self-esteem increases we would expect the agreement with the statement to increase, though his or her response has little influence on the latent variable.

Temporal priority may be one means of establishing causal priority for measurement models. Thus we expect a state of anxiety to precede responses to questions designed to measure it. A change in education, income, and occupational prestige may temporally precede a change in socioeconomic status (SES) (see Hauser 1973). For all these examples it is difficult,

[4]One could make the argument that for a longer time lag, high stress levels contribute to the chances of divorce or removal from job. But these effects are unlikely in the short time frame usually assumed in measurement models.

however, to test temporal priority since the latent variable is unobserved. In this case we must rely heavily on "mental experiments" for determining causal priority. In such mental experiments a change in the latent variable is imagined. Then the researcher must decide if it is reasonable to expect a subsequent change in the observed variables. The reverse also should be done. That is, a change in the observed variables should be imagined and the researcher should decide if a change in the latent variable is reasonable. As an example consider education as an indicator of SES. If a person's SES increases, we do not necessarily expect education to increase. But if education increases, we expect some increase in SES. Thus education is better conceived of as a cause indicator of SES than as an effect.

It is theoretically possible that simultaneous reciprocal causation may exist between an indicator and a latent variable. This could occur where each may be reasonably thought of as a cause of the other and when the observation period exceeds the causal lag. For example, the latent variable of "financial health" of a company measured by stock prices may have such a relation. Greater financial health can cause higher stock prices and higher stock prices can increase financial health. Or, consider academic grade expectations as a latent variable and measured grade as the indicator. High grade expectations may influence measured grades and grades can influence expectations. I know of no empirical work that has tested possibilities like these, but it is clear that the estimation of such models could be difficult.[5]

Occasionally, empirical means can be designed to test for cause or effect indicators. The following example illustrates this. In some earlier work (Bollen 1982) I estimated a confirmatory factor analysis model of people's assessment of air quality. The data were responses to four questions on the color, clarity, odor, and overall quality of the air in a fixed location in Shenandoah National Park in Virginia. The hypothesized model was that all four measures were effect indicators of subjective air quality. A plausible alternative idea is that the "overall" measure is an effect indicator of the latent variable, subjective air quality, but the three other measures (color, clarity, and odor) are cause indicators of the latent variable.

Cermak (1983) designed an experiment to test these alternatives for the responses to the overall, color, and clarity questions. His reasoning was that if the second model is true, and the components of air quality form an assessment of air quality which in turn leads to the overall measure, then the time it takes for an individual to respond should be greater for the overall measure than for the other measures. If the first model is valid and all measures are dependent on the latent variable, then the response time

[5]One such difficulty is the identification problem discussed in Chapters 4, 7, and 8.

should be essentially the same.[6] Using slides of the identical locations and times for the original survey data, Cermak (1983) presented a random sequence of slides and a random order of air quality questions to subjects. He found no statistically significant differences in response time, thus lending support to the hypothesis that the measures are effect indicators. This is one example where research was designed to test the causal priority of indicators and latent variables. Analogous procedures might be employed in other areas.

Experiments also can help to establish the causal direction between two directly observed variables, say, x_1 and y_1. If the treatment, x_1, is randomized, applied, and y_1 changes, then it is implausible that y_1 causes x_1. Of course the condition of pseudo-isolation still needs to be met, but such experimental manipulation is excellent evidence of the causal priority.

In sum, establishing the direction of influence between two variables is necessary to establishing a causal relation. Knowing that one variable precedes another in time is probably the single most effective means of doing so, but as some of the examples in this section illustrate, this does not always work nor is it always clear that temporal priority is met (e.g., in models with latent variables and indicators). Sometimes experimental designs combined with nonexperimental ones can provide convincing evidence on causal direction, but even here the condition of pseudo-isolation must be met.

LIMITATIONS OF "CAUSAL" MODELING

I have described some of the limits of structural equation modeling throughout the chapter. Here I bring several points into sharper focus. The three subsections are (1) model-data versus model-reality consistency, (2) experimental and nonexperimental research designs, and (3) criticisms of structural equation modeling.

Model-Data versus Model-Reality Consistency

When evaluating a model, at least two broad standards are relevant. One is whether the model is consistent with the data. The other is whether the model is consistent with the "real world." Most applications of structural

[6]Some theories of perceptions might lead to different predictions of response time. So the experiment is conditional on Cermak's assumptions about perceptions.

equation techniques explicitly test the former and only implicitly treat the latter.

Model-data consistency is checked by comparing relations predicted by a model and its assumptions to those present in the data. In structural equation models some of the tests measure the discrepancy between the sample covariance matrix (**S**) and a model predicted covariance matrix ($\hat{\Sigma}$). To the extent that sampling fluctuations cannot explain their departure, the model and data are inconsistent. We also can check model-data consistency by comparing the magnitude, sign, and statistical significance of parameter estimates to those hypothesized in the model. In short, the model implies that the data should have certain characteristics that we can check.

Model-reality consistency is a more "slippery" issue. Here the question is whether the model mirrors real-world processes. For instance, does an econometric model of the U.S. economy really correspond to the behavior of the economy? Fully assessing model-reality consistency is not possible since it presupposes perfect knowledge of the "real" world with which to evaluate the model. In practice, we imperfectly evaluate model-reality consistency in several ways. One is comparing the predictions implied by a model to those observed in a context different from the data that supply the model parameter estimates. For instance, we might check the realism of an econometric model by contrasting its predictions of inflation rates to those observed in the future. If we are fortunate enough to be able to manipulate variables in the model, we can do so and see if the model correctly predicts the consequences. Or, we can examine the assumptions and relations embedded in a model and debate their validity based on other experiences or insights.

It is tempting to use model-data consistency as proof of model-reality consistency, but we could be misled by so doing. The problem lies in the asymmetric link between these two consistency checks. *If a model is consistent with reality, then the data should be consistent with the model. But, if the data are consistent with a model, this does not imply that the model corresponds to reality.*

Consider the first of these two statements. If we have a model that is valid, then the data are generated in accordance with it, and the empirical match between the model and data should be within the sampling error. The failure to find model-data consistency implies that the model does not capture the actual processes. So by detecting model-data inconsistencies, we reveal model-reality gaps. Herein lies the real power of causal modeling: we can reject models that are inconsistent with the data.

But we cannot forget the second statement: model-data consistency does not imply model-reality consistency. The problem is that the true model is only one of many that might match the data. When we find model-data consistency, we do not know whether we have a true model or one of the false ones. Herein lies the real weakness of causal modeling: model-data consistency is not sufficient for model-reality consistency.

For instance, as I mentioned previously, one test of model-data consistency is the extent to which a model predicted covariance matrix ($\hat{\Sigma}$) matches the sample covariance matrix (**S**). Even if these matrices perfectly match, the model can still be wrong. Indeed, for any exactly identified model, we have $\mathbf{S} = \hat{\Sigma}$. Furthermore, for any set of variables, many different models have the same model-data fit.

Figure 3.9 illustrates this for three observed variables y_1, y_2, and y_3. All of the models have the same fit to the data, even though they represent very different causal relations. For instance, in some models y_1 affects y_2, whereas in others the reverse is represented. In still others all three variables depend on a latent variable, and they have no causal influence on each other. If you consider that these are just a sample of the models where $\mathbf{S} = \hat{\Sigma}$ and that the possible models multiply when more than three variables are employed, you should see that model-data consistency alone cannot tell us model-reality consistency.

For an empirical illustration of this point, I took a covariance matrix of education (y_1), occupational prestige (y_2), and income (y_3) from Kluegel et al. (1977). The covariance matrix for 432 whites is

$$\mathbf{S} = \begin{bmatrix} 2.739 & & \\ 17.431 & 452.711 & \\ 1.448 & 13.656 & 4.831 \end{bmatrix}$$

I then estimated the three models that correspond to the top row of Figure 3.9. For the first model the coefficient matrix $\hat{\mathbf{B}}$ is

$$\begin{bmatrix} 0 & 0 & 0 \\ 0.495 & 0 & 0 \\ 0.336 & 0.126 & 0 \end{bmatrix}$$

These are standardized regression coefficients. I will talk more about standardized coefficients in Chapter 4, but in brief, they measure the expected change in the dependent variable in standard deviation units that accompanies a one standard deviation change in the explanatory variable, holding constant the other explanatory variables. Using all the parameter estimates from the first model, the original covariance matrix can be perfectly reproduced.

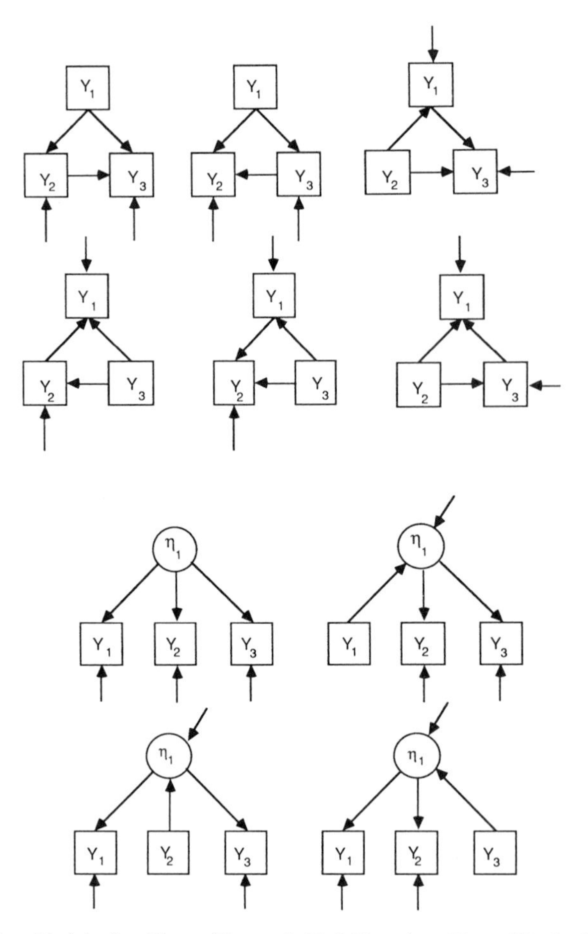

Figure 3.9 Ten Models for Three Observed Variables that Have "Perfect" Fit to the Covariance Matrix of Y_1 to Y_3

The $\hat{\mathbf{B}}$ coefficient matrices for the second and third models in the first row are

$$\begin{bmatrix} 0 & 0 & 0 \\ 0.450 & 0 & 0.113 \\ 0.398 & 0 & 0 \end{bmatrix} \qquad \begin{bmatrix} 0 & 0.495 & 0 \\ 0 & 0 & 0 \\ 0.336 & 0.126 & 0 \end{bmatrix}$$

Once again, the predicted covariance matrix ($\hat{\Sigma}$) equals the sample covariance matrix, **S**. The three models have identical fits, but they imply different causal relations. For instance, the third model has the unlikely relation of occupational prestige (y_2) causing a person's education (y_1). The second model shows income (y_3) causing occupational prestige (y_2). Since a job is acquired before earning income, this relation does not make sense.

Model fit alone cannot tell us which model is appropriate. Substantive knowledge plays a vital role in ruling out some of these possibilities. At the same time I recognize that knowledge in an area is often insufficiently developed to rule out false models. Indeed, researchers formulate models to help illuminate real-world relations.

A related problem is the failure to consider all plausible models with a good fit to the data. As Figure 3.9 illustrates, we can propose many models for even three variables. When we have more variables, the combination of possible relations multiplies tremendously. Recently, Glymour, Scheines, Spirtes, and Kelly (1987) proposed that the search for new models be automated using principles of artificial intelligence and microcomputers. Of course a considerable proportion of these computer-generated models will be substantively implausible, but the process could lead to a reasonable set of alternative models. It is too early to tell if such procedures will be effective.

Once we have a set of substantively reasonable models, the next step is to devise additional tests with new data that might help rule out some. For instance, if the direction of influence between two variables is ambiguous, we might be able to specify a model that allows reciprocal causation and then test the statistical significance of each effect. Or, we might devise an experiment that provides evidence on the direction of influences as was true for the air quality example I described earlier. We proceed by eliminating models, not proving them. This falsification perspective advocated by Karl Popper (1968) and others holds that a field advances by testing and comparing models to data and determining which models are the "fittest" to survive. The assumption is that the models not discarded by failing data tests are those closest to the true one.

Any model is an approximation to reality. A theory is an abstract set of ideas that links together concepts. A model is a formal representation of a theory. The theory at best approximates reality—and the model derived from it can do no better. Model building and modification is a process of successive approximation (Jeffreys 1983, 79). Due to the approximate nature of models and the impossibility of directly observing causality, all causal inferences must be regarded as tentative in the absolute sense, though subjectively we may have varying degrees of confidence in the relations being causal. Approximation is a characteristic of physical laws and models as well (Cartwright 1983), so the reader should not think that this is an issue unique to the social sciences or to structural equation models. As Glymour et al. (1987, 32–33) state: "In the natural sciences, nearly every exact, quantitative law ever proposed is known to be literally false. Kepler's laws are false, Ohm's law is false,..., and on and on. These theories are still used in physics and in chemistry and in engineering, even though they are known to be false. They are used because, although false, they are approximately correct. Approximation is the soul of science."

Model construction, then, is always done realizing that the model to some extent departs from actual conditions. The approximate nature of the model may take several forms. One is assuming that the structural equations are linear in the parameters. This is a reasonable starting point if we have no evidence that contradicts it. We sometimes are able to include curvilinear and nonlinear relations between variables by transforming variables leading to equations that are linear in the transformed variables. We also should be guided by the principle of parsimony. If we judge several models to fit the data and to approximate reality to the same degree, we should select the simplest model. We may find that a far more complicated model with many parameters has a slightly better data fit than a less elaborate one. We still may choose the simpler model if we believe that it is more likely to generalize to other samples than is the more complex one. At least we may be willing to suspend judgment until other data samples are scrutinized.

In sum, structural equation models face the same restrictions as other empirical methodologies. We can only reject a model—we can never prove a model to be valid. A good model-to-data fit does not mean that we have the true model. We need to examine other plausible specifications that fit; we need to explore various avenues to assess whether a model has a reasonable correspondence to reality.

Experimental and Nonexperimental Research

Two common misconceptions about structural equation models are (1) that these are statistical techniques only for nonexperimental (observational) data and (2) that true experiments solve the omitted variable problem that plagues structural equation analyses of nonexperimental data. The purpose of this section is to explain why both viewpoints are incorrect.

In the case of the first misconception, it is true that most applications of structural equation models use nonexperimental data and that most statistical analyses of experiments employ ANOVA or regression techniques. This apparent divergence is not so great when we realize that ANOVA and regression analysis are special cases of structural equation models (see Chapter 4). Both procedures are specialized in that they assume that the explanatory variables are measured without error. ANOVA is further restricted to categorical explanatory variables. Structural equation models proceed under less restrictive assumptions. Over 15 years ago Costner (1971), Miller (1971), Blalock (1971), and others suggested that analyses of experimental data could benefit from structural equations that allow measurement error, multiple indicators, and tests for confounding variables. Thus we see that the belief that structural equations apply only to nonexperimental data is incorrect on two counts. One is that the usual ANOVA and regression analyses of experiment are structural equation models, albeit specialized ones. The second is that the general structural equation techniques can allow more realistic analyses of experimental data that recognize the measurement errors and latent variables that are often present.

The second statement—that experiments solve the omitted variable problem—also is misleading. As explained in a previous section, omitted variables are an important way in which the conditions of pseudo-isolation are broken. Analysts have three general ways of controlling these extraneous factors: observation selection, statistical control, and randomization. In observation selection, analyses are restricted to a particular category of the potentially confounding variable. For example, in a study of the relation between a training program and job performance, we might limit the sample to females if we believe that gender can influence the relation. This type of control is available with either experimental or nonexperimental data. Its primary disadvantages are that we do not know whether the results hold for the unanalyzed groups and that we must determine in advance which variables to control.

The second major means of control is through statistical procedures. The task is to explicitly include in the analysis those characteristics suspected of influencing the dependent variable and associated with the other explanatory variables. Statistical control is available for both experimental and nonexperimental research. For the previous example of job training and performance, we could statistically control for gender by introducing a dummy variable to distinguish males from females and an interaction term of gender and job training. If age or seniority are possible confounders, these too could be statistically controlled. The main advantage of statistical control over control through selection of observations is that inferences can be made to a wider population. It shares a disadvantage with observation selection in that the researcher must know and have measures of the possible confounders. In practice, we are not aware of all the disruptive factors, or we do not have measures of them. This leads to the omitted variable problem I discussed earlier, and I refer the reader to the section on isolation for further details.

The ideal experimental research design utilizes a third and powerful means of control. It minimizes the effects of potentially confounding variables through randomization. The random assignment of subjects or cases to experimental and control conditions does not allow explicit control over potentially confounding factors, but it does enable us to establish probabilistic bounds on their influence. For instance, suppose we wish to examine the effectiveness of a new rehabilitation program for prisoners. Once we have selected a group for study, a random selection process determines whether a person is in the experimental or the control group. Although many factors can affect the prisoners' reactions to the experimental or traditional treatments, the method of random assignment on average prevents these systematic influences from being more prevalent in one group than in the other. One drawback is that the many variables whose influences are randomized are part of the disturbance. This increases the disturbance variance and makes it more difficult to detect an experimental effect than it would be if these extraneous variables were explicitly included in the equation.

In observation studies, control through randomization is unavailable. For instance, a nonexperimental study of prisoners might compare new and traditional programs in which the prisoners have not been randomly assigned to the treatments. Suppose that those in the new program are volunteers. Volunteers could have traits that affect the effectiveness of the program. They might be more receptive to change or more willing to cooperate, or they could differ from the nonvolunteers in several other ways that confound the determination of the new program effects. In such observational studies the researcher must use statistical controls and explicitly enter into a model the other variables suspected to cause the dependent variable. This can lead to elaborate models. For instance, Blalock (1979, 881) suggests that we may need "upwards of fifty variables" if we wish to disentangle many of the relationships of interest to sociologists. Econometric models of the U.S. economy often involve considerably more variables. The same may be needed in other social science problems.

Randomization is the reason that some believe that experimental designs solve the omitted variable problem. Though the advantages of randomization as a way to minimize the influence of some extraneous factors is clear, it does not always ensure that pseudo-isolation holds. That is, just like nonexperimental research, experimental research can be plagued by influential omitted variables. To illustrate this, suppose we have an experiment on the relation between frustration and aggression. We randomly assign individuals to experimental and control groups. To induce frustration, we administer electric shocks to subjects in the experimental group as they attempt to complete a task. We then have the experimental and control group respond to a scale gauging their feelings of aggression. With ANOVA, or equivalently with dummy variable regression procedures, we estimate

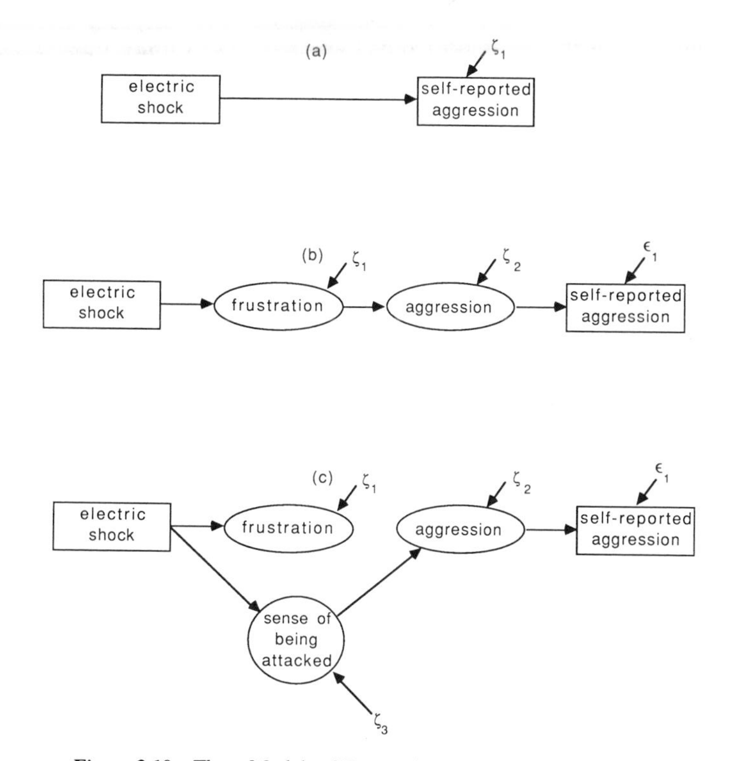

Figure 3.10 Three Models of Frustration-Aggression Experiment

whether a statistically and substantively significant difference exists between the groups.

Does a significant coefficient mean that frustration causes aggression? The answer is, not necessarily, and a structural equation model can help illuminate the reason. Figure 3.10(a) gives the path diagram of the two observed variables that the researcher analyzes. The first is a dummy variable showing whether the subject receives an electric shock or not. The second is the self-report of aggressiveness. This estimated model has an implicit, more complicated, underlying model. I show this in Figure 3.10(b). Here I explicitly include the hypothesized causal structure: electric shock causes frustration, frustration causes aggression, and aggression determines the self reported measure of aggression. The theory behind this experiment involves two latent variables, frustration and aggression, which are the omitted intervening variables between electric shock and the aggressiveness measures. Suppose that the influence of electric shocks on frustration is near zero or that the self-reported aggression measures have low reliability. This could lead to estimates of a near-zero association between the observed variables, even if the latent variables of frustration and aggression have a strong relationship. More generally, many analyses of experiments hypothesize a series of intervening variables through which a treatment leads to a change in an indicator, but rarely are the intervening variables included in an analysis. And randomization does not control for this type of omitted variable.

Another possible confounder occurs when the experimental manipulation changes variables other than the intended one. For instance, in Figure 3.10(c) I represent the electric shock as inducing frustration as intended but also a sense of being under attack. Perceiving this aggression may be the cause of aggressive feelings rather than these feelings being due to frustration. By only estimating the model in Figure 3.10(a), we would mistakenly attribute to frustration an effect due to a sense of being attacked. Randomization of the experimental treatment does not eliminate the problem. Employing a structural equation approach can help to investigate the plausibility of these alternative models. For instance, if researchers collect multiple indicators of self-reported frustration, of being under attack, and of aggression, they could estimate a general structural equation model with latent variables and test whether significant paths run from the experimental variable to the latent variables or whether any confounding effects are present. In any experiment we must be aware that variables other than the intended one may be influenced by the treatment and that these other variables could be responsible for the effects found.

Other types of omitted variables also may occur. When an experiment is relatively long, effects due to fatigue, decreased motivation, or boredom rather than due to the treatment variable may appear. Demand characteristics also can be confounders. These refer to the perceptions that subjects have of the experiment and that can affect their performance. If they perceive an experiment to measure intelligence, they may try to appear more intelligent, or if they see the goal as one of measuring cooperation, they may cooperate more so as to present themselves in a positive light. Conversely, if they perceive a treatment as intended to manipulate their behavior, they may take steps to avoid the predicted behavior in order to demonstrate their independence. The Hawthorne effect, where just the knowledge of being in an experiment affects subjects' performance, provides another type of an omitted variable not controlled by randomization.

Besides these subjects' effects, characteristics of the experimenter can confound an analysis. For instance, whether an experimenter is male or female, black or white, attractive or unattractive, might influence a subject's response to some types of experiments. The experimenter's degree of experience or personality characteristics (e.g., warmth or coldness, introvert or extrovert) can undermine some results. Even the expectations that an experimenter has about the outcomes of research can become an omitted variable that leads to spurious or suppressed relations. Rosenthal (1966) provides many examples of these and other experimenter effects.

Campbell and Stanley (1966) and Cook and Campbell (1976) provide a useful list of additional threats to valid causal inferences in experiments and quasiexperiments. Beyond these are the practical and ethical constraints. Ferber and Hirsch (1982) give a sobering assessment of the difficulties in large-scale social experiments such as the U.S. income maintenance ones. For instance, the social experiments cost many millions of dollars, require years of preparation, execution, and analysis, and often face ethical and legal constraints. In addition such experiments are temporary treatments designed to assess permanent policy changes. Researchers cannot know whether the behavior of subjects under these conditions will match those of a permanent program. Publicity about the experiment can influence the behavior of participants in the control or experimental group. And the list goes on. Lest anyone think that experiments in the physical sciences are immune from controversy, he or she can refer to Glymour et al. (1987, 28–29) for examples in physics and chemistry.

I raise these points not to suggest that we avoid experimental research. On the contrary, I believe that social science could greatly benefit by having more randomized experiments. Rather I hope these points will help to correct the neglect of the drawbacks of experimental research. Critics of nonexperimental research too often compare the *ideal* experiment to observational studies *in practice*. The most relevant comparison are experiments in practice to nonexperimental studies in practice. Such comparisons reveal strengths and limitations of each.

In sum, structural equation procedures are applicable to experimental as well as nonexperimental data. In fact, they could help to illuminate many of the problems that are untestable with the usual ANOVA and regression procedures. The randomization of treatment gives experimental designs a powerful advantage over nonexperimental designs. But we would be naive to believe that randomization removes all possible problems with omitted variables. Though the types of omitted variables may differ from those common in nonexperimental research, the consequences can be just as fatal to making correct causal inferences. Both nonexperimental and experimental research designs require the researcher to draw upon their substantive expertise so as to consider omitted variables that can violate pseudo-isolation. Clearly a powerful combination is to examine both experimental and nonexperimental evidence. The replication of results in each setting can increase confidence in the validity of the assumed relations.

Criticisms of Structural Equation Models

Though few have questioned the statistical theory that underlies structural equation models, some have criticized the typical application of the techniques. In the previous sections I have covered many of these comments. Here I call attention to three: the falsifiability of the models, the use of latent variables, and the distributional assumptions.

The first criticism suggests that it is not possible to disprove a structural equation model. For instance, Ling (1983) states: "... the researcher can never disconfirm a false causal assumption, regardless of the sample size or evidence, so long as the variables alleged to be causally related are correlated." It is difficult to understand the basis of this claim since in structural equation models we often find that two variables that are correlated have no relationship once other variables are controlled. Statistical significance tests for individual or groups of coefficient estimates are available.

Freedman (1986, 112) argues that "path analysis does not derive the causal theory from the data, or test any major part of it against the data." Most structural equation modelers would agree that they cannot "derive a theory from the data." In fact, an initial model precedes the data analysis. The justification for the comment that major parts of the model are not testable is unclear. As I discussed in the section on model-data consistency, we can test and reject structural equation (path) models. The tests of overall model fit presented in Chapters 4 and 7 are one advantage of these procedures. Thus the assertion that these models cannot be falsified has little basis. They cannot be proved valid, but models can be rejected.

A second criticism suggests that structural equation models are not believable because they sometimes incorporate latent variables. The assumption is that latent variables are only a product of the researcher's imagination and therefore have no scientific validity. If we accept this view and purge all scientific and statistical work of latent variables, we would eliminate much of contemporary science and statistics. Statistical models often include latent disturbances or latent probabilities. In physics the particles of quantum theory are unobservable in principle (Wallace 1972, 3). Biological species is an abstraction not always directly measurable. Symptoms often are the only way that medical scientists know that a disease is present. Even basic concepts such as time and space are less straightforwardly observed than previously thought prior to relativity theory. The list could go on. Indeed, latent variables are a part of, and necessary to, many scientific theories.

A related issue is the "naming problem" with latent variables (Cliff 1983). We may not correctly describe which latent variables are related to the indicators. This is a problem of validity: Is the named and defined latent variable truly related to the indicator, or is it some other latent variable linked to the measure? Unfortunately, this is a real problem in many applications. Often the concept that the latent variable represents is not clearly defined and not enough attention is given to tests of measurement validity. Chapters 6 and 7 treat these measurement issues in more detail.

A third criticism is the belief that the estimators of structural equation models have no value if the observed variables do not have a multinormal distribution. I will examine the issue in Chapters 4, 7, and 9, but for now it is worth noting that all the estimators I use are consistent estimators when the only assumption violated is that the observed variables are nonnormal. The tests of statistical significance are dependent on more restrictive conditions, but even here asymptotic distribution-free estimators are an option (see Chapter 9). Thus the assumptions about the distributions of the observed variables are no more restrictive, and even less restrictive, than the distributional assumptions routinely made for ANOVA or regression analysis.

SUMMARY

The concept of causality has many connotations. In this chapter I have provided a definition of causality oriented toward structural equation models. Isolation, association, and direction of causality are the three conditions used to establish a causal relation. Each condition is difficult to meet, but it is perhaps impossible to be certain that a cause and an effect are isolated from all other influences. We must regard all models as approximations to reality. The statistical tests can only disconfirm models; they can never prove a model or the causal relations within it. It will be helpful to remember this as we examine the empirical examples in the book as well as when we apply these procedures in research. Finally, we should realize that the problems of demonstrating isolation, association, and direction of causation are age-old issues. This chapter has merely described them as they are manifested in structural equation models.

CHAPTER FOUR

Structural Equation Models with Observed Variables

Regression-based models are common in the social sciences. They may consist of a single equation oriented toward explaining one endogenous (dependent) variable or multiequation models with a number of endogenous variables and reciprocal relationships. These "classical econometric models," as they are sometimes called, all have in common the assumption that the endogenous and exogenous variables are directly observed with no measurement error. If measurement error is allowed, it is assumed to occur only in endogenous variables that do not serve as explanatory variables in any equation. Thus, with few exceptions, the observed $\mathbf{y}$ and $\mathbf{x}$ are assumed to equal the corresponding $\boldsymbol{\eta}$ and $\boldsymbol{\xi}$.

In this chapter I examine structural equation models with observed variables. I do this for two reasons. First, they are the most common structural equation models. In fact, many readers will have some experience with estimating regression or path analysis models of this type. Second, these models are a special case of the more general structural equation procedures with latent variables that are discussed in later chapters. The major topics of this chapter—model specification, the implied covariance matrix, identification, and estimation—will recur for the other models.

MODEL SPECIFICATION

Equation (4.1) is a general representation of structural equations with observed variables (classical econometric model):

$$\mathbf{y} = \mathbf{B}\mathbf{y} + \boldsymbol{\Gamma}\mathbf{x} + \boldsymbol{\zeta} \tag{4.1}$$

where

$\mathbf{B} = m \times m$ coefficient matrix

$\Gamma = m \times n$ coefficient matrix

$\mathbf{y} = p \times 1$ vector of endogenous variables

$\mathbf{x} = q \times 1$ vector of exogenous variables

$\boldsymbol{\zeta} = p \times 1$ vector of errors in the equations

Since the disturbances $\boldsymbol{\zeta}$ represent random errors in the relationships between the y's and x's these models are sometimes referred to as *errors in the equations* models. The standard assumption is that the errors ($\boldsymbol{\zeta}$) are uncorrelated with $\mathbf{x}$.

The implicit measurement model for structural equations with observed variables is

$$\begin{aligned} \mathbf{y} &= \boldsymbol{\eta} \\ \mathbf{x} &= \boldsymbol{\xi} \end{aligned} \tag{4.2}$$

where

$\mathbf{y} = p \times 1$ vector of manifest (observed) variables

$\mathbf{x} = q \times 1$ vector of manifest (observed) variables

Simply put, $\mathbf{x}$ and $\mathbf{y}$ are assumed to exactly represent the latent $\boldsymbol{\xi}$ and $\boldsymbol{\eta}$ and only one indicator is used for each latent variable. The number of y variables equals the number of η variables ($p = m$) and the number of x variables equals the number of ξ variables ($q = n$).

The two major types of structural equations with observed variables are recursive and nonrecursive ones. Recursive models are systems of equations that contain no reciprocal causation or feedback loops. When this is true, it is possible to write $\mathbf{B}$ as a lower triangular matrix. In addition the covariance matrix of the errors in equations ($\boldsymbol{\Psi}$) is diagonal.[1] This means that the

[1]If $\boldsymbol{\Psi}$ is not diagonal but the other conditions for recursive models are met, the model is sometimes called *partially recursive*. Occasionally, the term recursive is used to refer to any model with $\mathbf{B}$ lower triangular regardless of whether $\boldsymbol{\Psi}$ is diagonal. I do not follow this practice.

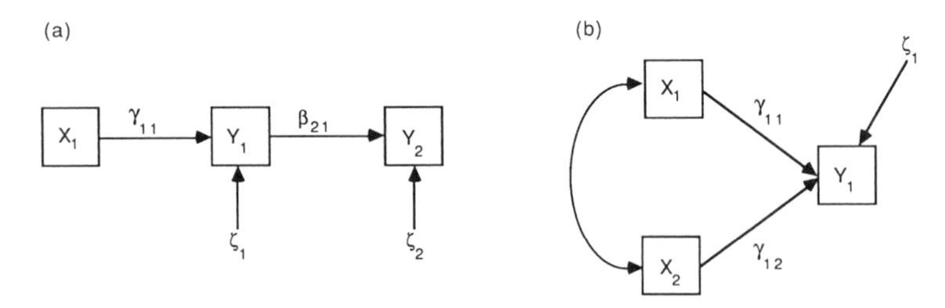

Figure 4.1 Two Examples of Recursive Structural Equation Models

disturbances for one equation are uncorrelated with the disturbances of the other equations. For example, if y_1 causes y_2, y_2 cannot affect y_1, either directly or through a chain of other variables. Furthermore the disturbance, ζ_1, for the y_1 equation is uncorrelated with ζ_2 for the y_2 equation. Figure 4.1 provides two hypothetical examples of recursive models.

The path diagram for an empirical example is in Figure 4.2. The data are from a study of union sentiment among southern nonunion textile workers (McDonald and Clelland 1984). The variables are years in textile mill (x_1), age (x_2), deference (or submissiveness) to managers (y_1), support for labor activism (y_2), and sentiment toward unions (y_3).[2] In addition to formulating the relation between workers' attitudes, the model separately assesses the influence of seniority and age. It specifies that age (x_2) affects deference (y_1) and attitude toward activism (y_2), but not union sentiment (y_3). Seniority (x_1) affects only union sentiment (y_3). The causal ordering among the endogenous variables hypothesized by McDonald and Clelland is that deference (y_1) influences attitudes toward activism (y_2) and unions (y_3) and that activism affects union sentiment. The disturbances (ζ's) are uncorrelated across equations. The matrix equation for the model is

$$\begin{bmatrix} y_1 \\ y_2 \\ y_3 \end{bmatrix} = \begin{bmatrix} 0 & 0 & 0 \\ \beta_{21} & 0 & 0 \\ \beta_{31} & \beta_{32} & 0 \end{bmatrix} \begin{bmatrix} y_1 \\ y_2 \\ y_3 \end{bmatrix} + \begin{bmatrix} 0 & \gamma_{12} \\ 0 & \gamma_{22} \\ \gamma_{31} & 0 \end{bmatrix} \begin{bmatrix} x_1 \\ x_2 \end{bmatrix} + \begin{bmatrix} \zeta_1 \\ \zeta_2 \\ \zeta_3 \end{bmatrix} \tag{4.3}$$

[2]This is a simplified version of the model presented in McDonald and Clelland (1984). I use the logarithm of years in mill because of the skewness and extreme values in the unlogged variable. The other variables did not require transformations.

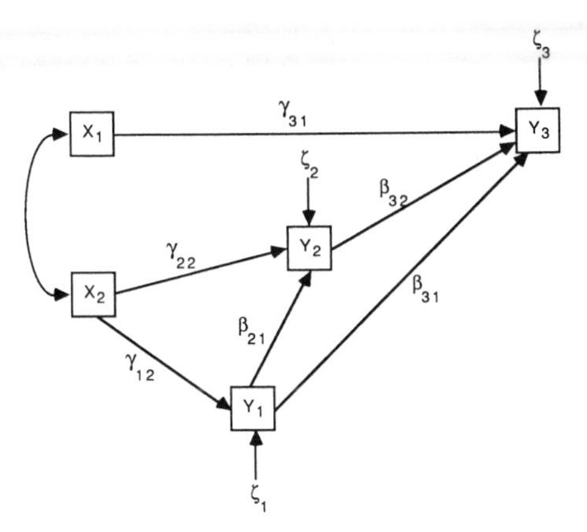

Figure 4.2 Union Sentiment Model for Southern Textile Workers

Since **B** is lower triangular and $\boldsymbol{\Psi}$ is diagonal, the union sentiment model is recursive.

The second major type of structural equation with observed variables is nonrecursive. Nonrecursive models contain reciprocal causation, feedback loops, or they have correlated disturbances. Unlike recursive models, **B** is not lower triangular, or the $\boldsymbol{\Psi}$ matrix is not diagonal. Two examples are in Figure 4.3.

I use the relation between objective measures of socioeconomic status and an individual's perception of their status as an empirical example. There are five variables: income (x_1), occupational prestige (x_2), subjective income (y_1), subjective occupational prestige (y_2), and subjective overall

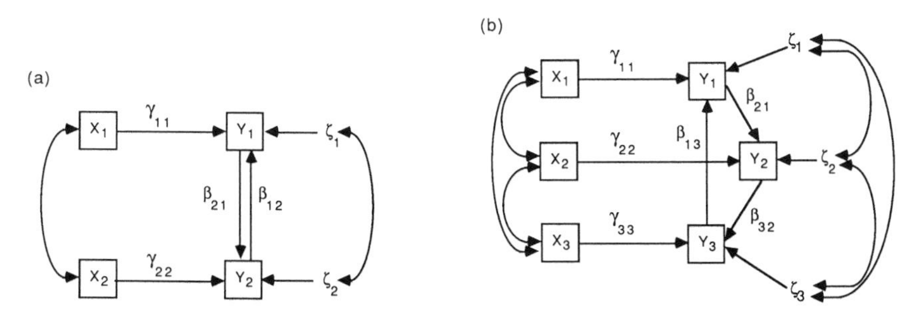

Figure 4.3 Two Examples of Nonrecursive Structural Equation Models

social status (y_3). The first two variables (x_1 and x_2) are measures of actual income and occupational prestige. The y_1 and y_2 indicators are an individual's perceived income and perceived occupational prestige, respectively. These need not be the same as x_1 and x_2 since individuals' actual incomes and occupations may vary from their assessment of their social standing on these two variables. The last variable, y_3, is the overall assessment individuals make of their social status. Like y_1 and y_2, this is a subjective measure.

One plausible model for these variables is that actual income (x_1) directly affects subjective income (y_1), actual occupational prestige (x_2) directly affects subjective occupational prestige (y_2), and that subjective income (y_1) and subjective occupational prestige (y_2) directly influence each other as well as subjective overall socioeconomic status (y_3).

These ideas are embodied in the path diagram in Figure 4.4. The **B**, **Γ**, **Ψ**, and **Φ** for this model are

$$\mathbf{B} = \begin{bmatrix} 0 & \beta_{12} & 0 \\ \beta_{21} & 0 & 0 \\ \beta_{31} & \beta_{32} & 0 \end{bmatrix}, \qquad \mathbf{\Gamma} = \begin{bmatrix} \gamma_{11} & 0 \\ 0 & \gamma_{22} \\ 0 & 0 \end{bmatrix} \tag{4.4}$$

$$\mathbf{\Psi} = \begin{bmatrix} \psi_{11} & & \\ \psi_{12} & \psi_{22} & \\ \psi_{13} & \psi_{23} & \psi_{33} \end{bmatrix}, \qquad \mathbf{\Phi} = \begin{bmatrix} \phi_{11} & \\ \phi_{21} & \phi_{22} \end{bmatrix} \tag{4.5}$$

Unlike a recursive model, **B** cannot be written as a lower triangular matrix, and **Ψ** is not diagonal.

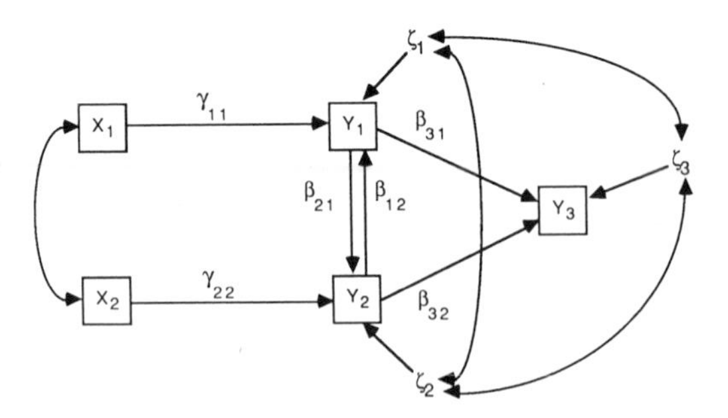

Figure 4.4 Example of Nonrecursive Model of Objective and Subjective Status

I will return to the union sentiment and the objective and perceived status examples throughout the chapter.

IMPLIED COVARIANCE MATRIX

As stated in Chapter 1, the basic hypothesis of the general structural equation model is

$$\boldsymbol{\Sigma} = \boldsymbol{\Sigma}(\boldsymbol{\theta}) \tag{4.6}$$

where $\boldsymbol{\Sigma}$ is the population covariance matrix of $\mathbf{y}$ and $\mathbf{x}$ and $\boldsymbol{\Sigma}(\boldsymbol{\theta})$ is the covariance matrix written as a function of the free model parameters in $\boldsymbol{\theta}$. Equation (4.6) implies that each element of the covariance matrix is a function of one or more model parameters. I derive the specialization of (4.6) for observed variable models in this section. The relation of $\boldsymbol{\Sigma}$ to $\boldsymbol{\Sigma}(\boldsymbol{\theta})$ is basic to an understanding of identification, estimation, and assessments of model fit.

I assemble $\boldsymbol{\Sigma}(\boldsymbol{\theta})$ in three pieces: (1) the covariance matrix of $\mathbf{y}$, (2) the covariance matrix of $\mathbf{x}$ with $\mathbf{y}$, and (3) the covariance matrix of $\mathbf{x}$. Consider first $\boldsymbol{\Sigma}_{yy}(\boldsymbol{\theta})$, the implied covariance matrix of $\mathbf{y}$:

$$\begin{aligned}
\boldsymbol{\Sigma}_{yy}(\boldsymbol{\theta}) &= E(\mathbf{y}\mathbf{y}') \\
&= E\left[(\mathbf{I}-\mathbf{B})^{-1}(\boldsymbol{\Gamma}\mathbf{x}+\boldsymbol{\zeta})\left((\mathbf{I}-\mathbf{B})^{-1}(\boldsymbol{\Gamma}\mathbf{x}+\boldsymbol{\zeta})\right)'\right] \\
&= E\left[(\mathbf{I}-\mathbf{B})^{-1}(\boldsymbol{\Gamma}\mathbf{x}+\boldsymbol{\zeta})(\mathbf{x}'\boldsymbol{\Gamma}'+\boldsymbol{\zeta}')(\mathbf{I}-\mathbf{B})^{-1'}\right] \\
&= (\mathbf{I}-\mathbf{B})^{-1}\left(E(\boldsymbol{\Gamma}\mathbf{x}\mathbf{x}'\boldsymbol{\Gamma}')+E(\boldsymbol{\Gamma}\mathbf{x}\boldsymbol{\zeta}')+E(\boldsymbol{\zeta}\mathbf{x}'\boldsymbol{\Gamma}')+E(\boldsymbol{\zeta}\boldsymbol{\zeta}')\right)(\mathbf{I}-\mathbf{B})^{-1'} \\
&= (\mathbf{I}-\mathbf{B})^{-1}(\boldsymbol{\Gamma}\boldsymbol{\Phi}\boldsymbol{\Gamma}'+\boldsymbol{\Psi})(\mathbf{I}-\mathbf{B})^{-1'}
\end{aligned} \tag{4.7}$$

where

$$\boldsymbol{\Phi} = \text{covariance matrix of } \mathbf{x}$$
$$\boldsymbol{\Psi} = \text{covariance matrix of } \boldsymbol{\zeta}$$

The implied covariance matrix of **x**, $\boldsymbol{\Sigma}_{xx}(\boldsymbol{\theta})$ is equal to $\boldsymbol{\Phi}$, or

$$\begin{aligned}\boldsymbol{\Sigma}_{xx}(\boldsymbol{\theta}) &= E(\mathbf{x}\mathbf{x}') \\ &= \boldsymbol{\Phi}\end{aligned} \tag{4.8}$$

The final part of the implied covariance matrix is $\boldsymbol{\Sigma}_{xy}(\boldsymbol{\theta})$, the implied covariance of **x** with **y**:

$$\begin{aligned}\boldsymbol{\Sigma}_{xy}(\boldsymbol{\theta}) &= E(\mathbf{x}\mathbf{y}') \\ &= E\left[\mathbf{x}\left((\mathbf{I}-\mathbf{B})^{-1}(\boldsymbol{\Gamma}\mathbf{x}+\boldsymbol{\zeta})\right)'\right] \\ &= \boldsymbol{\Phi}\boldsymbol{\Gamma}'(\mathbf{I}-\mathbf{B})^{-1'}\end{aligned} \tag{4.9}$$

After assembling equations (4.7) to (4.9), the implied covariance matrix of **y** and **x** is

$$\boldsymbol{\Sigma}(\boldsymbol{\theta}) = \begin{bmatrix} (\mathbf{I}-\mathbf{B})^{-1}(\boldsymbol{\Gamma}\boldsymbol{\Phi}\boldsymbol{\Gamma}'+\boldsymbol{\Psi})(\mathbf{I}-\mathbf{B})^{-1'} & (\mathbf{I}-\mathbf{B})^{-1}\boldsymbol{\Gamma}\boldsymbol{\Phi} \\ \boldsymbol{\Phi}\boldsymbol{\Gamma}'(\mathbf{I}-\mathbf{B})^{-1'} & \boldsymbol{\Phi} \end{bmatrix} \tag{4.10}$$

I illustrate (4.10) with a hypothetical example that is a causal chain model $x_1 \rightarrow y_1 \rightarrow y_2$ [see Figure 4.1(a) for path diagram]:

$$y_1 = \gamma_{11}x_1 + \zeta_1 \tag{4.11}$$

$$y_2 = \beta_{21}y_1 + \zeta_2 \tag{4.12}$$

where $\text{COV}(\zeta_1, x_1)$, $\text{COV}(\zeta_1, \zeta_2)$, and $\text{COV}(x_1, \zeta_2)$ are zero. The matrices for this model are

$$\mathbf{B} = \begin{bmatrix} 0 & 0 \\ \beta_{21} & 0 \end{bmatrix}, \qquad \boldsymbol{\Gamma} = \begin{bmatrix} \gamma_{11} \\ 0 \end{bmatrix} \tag{4.13}$$

$$\boldsymbol{\Psi} = \begin{bmatrix} \psi_{11} & 0 \\ 0 & \psi_{22} \end{bmatrix}, \qquad \boldsymbol{\Phi} = [\phi_{11}] \tag{4.14}$$

Substituting (4.13) and (4.14) into (4.10) and using (4.6) leads to

$$\boldsymbol{\Sigma} = \boldsymbol{\Sigma}(\boldsymbol{\theta}) \tag{4.15}$$

$$\begin{bmatrix} \text{VAR}(y_1) & & \\ \text{COV}(y_2, y_1) & \text{VAR}(y_2) & \\ \text{COV}(x_1, y_1) & \text{COV}(x_1, y_2) & \text{VAR}(x_1) \end{bmatrix} = \begin{bmatrix} \gamma_{11}^2\phi_{11} + \psi_{11} & & \\ \beta_{21}(\gamma_{11}^2\phi_{11} + \psi_{11}) & \beta_{21}^2(\gamma_{11}^2\phi_{11} + \psi_{11}) + \psi_{22} & \\ \gamma_{11}\phi_{11} & \beta_{21}\gamma_{11}\phi_{11} & \phi_{11} \end{bmatrix}$$

The left-hand side of (4.15) is the population covariance matrix of y_1, y_2, and x_1, whereas the right-hand side represents each variance and covariance in terms of the unknown model parameters. Only the bottom half of each matrix is shown. The elements above the main diagonal are the same as the ones below it. In general, a covariance matrix for $p + q$ variables has $(1/2)(p + q)(p + q + 1)$ nonredundant elements. For this example, there are six ($= (1/2)(3)(4)$) elements, which implies six equations. For example, (4.15) shows that $\text{VAR}(y_1) = \gamma_{11}^2\phi_{11} + \psi_{11}$ and $\text{COV}(x_1, y_1) = \gamma_{11}\phi_{11}$. The six equations make clear the dependence of the population variances and covariances on the model parameters.

The same steps could be followed to show the relation of $\boldsymbol{\Sigma}$ to $\boldsymbol{\Sigma}(\boldsymbol{\theta})$ for the union sentiment and the objective and perceived social status examples as well as for any other structural equations with observed variables. Substitute the $\mathbf{B}$, $\boldsymbol{\Gamma}$, $\boldsymbol{\Psi}$, and $\boldsymbol{\Phi}$ of the model into (4.10). The union sentiment and the social status models, each with five variables, have 15 ($= (1/2)(5)(6)$) nonredundant elements in their covariance matrices, which implies 15 equations for each model. In practice, $(\mathbf{I} - \mathbf{B})^{-1}$, which is required for (4.10), is tedious to compute when $\mathbf{B}$ is large.

If researchers are mostly interested in relatively few elements of the covariance matrix, covariance algebra provides an alternative derivation. The analyst simply uses algebraic manipulations of the structural equations (if needed) and the rules of covariance algebra reviewed in chapter two. For instance, the structural equations for the perceived income (y_1) equation and the perceived occupational prestige equation (y_2) are [see Figure 4.4 or equation (4.4)]

$$y_1 = \beta_{12}y_2 + \gamma_{11}x_1 + \zeta_1 \tag{4.16}$$

$$y_2 = \beta_{21}y_1 + \gamma_{22}x_2 + \zeta_2 \tag{4.17}$$

Write the equations so that they are in *reduced-form*. That is, the only variables on the right-hand side should be exogenous variables. The reduced-form of (4.16) is

$$y_1 = (1 - \beta_{12}\beta_{21})^{-1}(\gamma_{11}x_1 + \beta_{12}\gamma_{22}x_2 + \zeta_1 + \beta_{12}\zeta_2) \tag{4.18}$$

[Equation (4.18) follows upon substitution of the right-hand side of (4.17) for y_2 in (4.16) and solving for y_1.]

The covariance of objective income (x_1) and perceived income (y_1) is

$$\begin{aligned} \mathrm{COV}(x_1, y_1) &= \mathrm{COV}\big(x_1, (1 - \beta_{12}\beta_{21})^{-1}(\gamma_{11}x_1 + \beta_{12}\gamma_{22}x_2 + \zeta_1 + \beta_{12}\zeta_2)\big) \\ &= (1 - \beta_{12}\beta_{21})^{-1}(\gamma_{11}\phi_{11} + \beta_{12}\gamma_{22}\phi_{12}) \end{aligned} \tag{4.19}$$

A similar series of steps leads to any of the other variance or covariance equations.

In summary, once a model is specified, the variances and covariances are functions of the model parameters. Attempts to establish that unique values can be found for these parameters introduces the issue of identification.

IDENTIFICATION

Identification is a topic relevant to all structural equation models. In this section I first provide some general comments on the issue. I follow this with specific comments and rules for the identification of structural equations with observed variables.

Investigations of identification begin with one or more equations relating known and unknown parameters. By "known" parameters, I do *not* mean that the exact values of the parameters are known. Rather, I mean parameters that are known to be identified. Generally, these parameters are population characteristics of the distribution of the observed variables such as their variances and covariances for which consistent sample estimators are readily available and for which identification is typically not an issue. The "unknown" parameters are the parameters whose identification status is not known. The researcher must establish whether unique values exist for these. The unknown parameters are from the structural equation model. Identification is demonstrated by showing that the unknown parameters are functions only of the identified parameters *and* that these functions lead to unique solutions. If this can be done, the unknown parameters are *identi-*

fied; otherwise, one or more parameters are *unidentified*. Thus the goal is to solve for the unknown parameters in terms of the known-to-be-identified parameters.

I illustrate this with a simple example. Suppose that VAR(y) is the identified parameter, θ_1 and θ_2 are the unknown parameters, and the equation relating these is VAR(y) = $\theta_1 + \theta_2$. The identification issue is whether unique values of θ_1 and θ_2 follow from this equation. Clearly, with two unknowns in one equation, θ_1 and θ_2, are not identified. For any given value of VAR(y), an infinite set of values of θ_1 and θ_2 satisfy the equation VAR(y) = $\theta_1 + \theta_2$. However, adding a second equation of $\theta_1 = \theta_2$ would ensure identification with each parameter equal to VAR(y)/2. Since VAR(y)/2 is a unique value for any given value of VAR(y), θ_1 and θ_2 are identified parameters.

The same general principles hold for more complicated structural equation models. The known-to-be-identified parameters are the elements of $\boldsymbol{\Sigma}$, the population covariance matrix of the observed variables. The parameters whose identification status is unknown are in $\boldsymbol{\theta}$, where $\boldsymbol{\theta}$ contains the t free and (nonredundant) constrained parameters of $\mathbf{B}$, $\boldsymbol{\Gamma}$, $\boldsymbol{\Phi}$, and $\boldsymbol{\Psi}$. The equation relating $\boldsymbol{\Sigma}$ to $\boldsymbol{\theta}$ is the covariance structure hypothesis, $\boldsymbol{\Sigma} = \boldsymbol{\Sigma}(\boldsymbol{\theta})$, which I presented in the last section. *If an unknown parameter in* $\boldsymbol{\theta}$ *can be written as a function of one or more elements of* $\boldsymbol{\Sigma}$, *that parameter is identified.*[3] *If all unknown parameters in* $\boldsymbol{\theta}$ *are identified, then the model is identified.*

An alternative definition of identification begins by considering two $t \times 1$ vectors $\boldsymbol{\theta}_1$ and $\boldsymbol{\theta}_2$, each of which contains specific values for the unknown parameters in $\boldsymbol{\theta}$. I can form the implied covariance matrices, $\boldsymbol{\Sigma}(\boldsymbol{\theta}_1)$ and $\boldsymbol{\Sigma}(\boldsymbol{\theta}_2)$, for each of these solution vectors. If the model is identified, all $\boldsymbol{\theta}_1$ and $\boldsymbol{\theta}_2$ solutions where $\boldsymbol{\Sigma}(\boldsymbol{\theta}_1) = \boldsymbol{\Sigma}(\boldsymbol{\theta}_2)$ must have $\boldsymbol{\theta}_1 = \boldsymbol{\theta}_2$. If a pair of vectors $\boldsymbol{\theta}_1$ and $\boldsymbol{\theta}_2$ exists such that $\boldsymbol{\Sigma}(\boldsymbol{\theta}_1) = \boldsymbol{\Sigma}(\boldsymbol{\theta}_2)$ and $\boldsymbol{\theta}_1 \neq \boldsymbol{\theta}_2$, then $\boldsymbol{\theta}$ is not identified.

Overidentification of a parameter refers to an excess of identifying information. For instance, suppose that for the unknown parameter ϕ_{11}, we have ϕ_{11} = VAR(x_1) and ϕ_{11} = COV(x_1, y_1), where VAR(x_1) and COV(x_1, y_1) are identified. Either of these equalities alone would establish

[3] I assume that the function of $\boldsymbol{\Sigma}$ leads to a unique solution. In most cases this is true, but it is possible to have more than one solution. For instance, suppose that $\boldsymbol{\theta}_1$ equals the square root of VAR(y_2). The solutions are the positive and negative square roots of VAR(y_2). Other information often rules out all but one solution. In this example, if $\boldsymbol{\theta}_1$ is a variance parameter, a negative value is not permissible. Usually, if an unknown parameter is a function of the identified parameters, multiple solutions are uncommon.

the identification of ϕ_{11}. Taken together, ϕ_{11} is overidentified. One possible source of confusion is thinking that because ϕ_{11} has two equations, it assumes two different values. This is not true. In the population the $\mathrm{VAR}(x_1)$ and $\mathrm{COV}(x_1, y_1)$ have the same value when the model is correct so that ϕ_{11} takes only one value.

A model is overidentified when each parameter is identified and at least one parameter is overidentified. A model is exactly identified when each parameter is identified but none is overidentified. A similar distinction is made for equations. In practice, the term identified models (or equations) refers to exactly identified and overidentified models (or equations). An underidentified model (or equation) has at least one parameter that cannot be identified. Generally, the identified parameters in an underidentified model can be consistently estimated, though this is not true for the unidentified parameters. Sometimes functions of unidentified parameters are identified. In the earlier example with the single equation of $\mathrm{VAR}(y) = \theta_1 + \theta_2$, the *sum* of $\theta_1 + \theta_2$ is identified, though the individual parameters $\theta_1 + \theta_2$ are not. Such identified *functions* of unidentified parameters often can be consistently estimated, though this is not true for the individual unidentified parameters.

As the preceding discussion makes clear, identification is made possible from information about the distribution of the observed variables y and x. If the variables have a multinormal distribution, then the parameters that characterize the distribution of the observed variables are the population means ($\boldsymbol{\mu}$) and the population covariance matrix ($\boldsymbol{\Sigma}$). These are first- and second-order moments of the distribution. The third and higher-order moments are either zero or functions of the lower-order moments. They are redundant and do not provide any information that can aid in model identification. For variables that are *not* multinormally distributed, higher-order moments of the distribution may help identify parameters. I do not consider this topic here (see Bentler 1983). Of the first- and second-order moments, I have shown that the second-order moment, $\boldsymbol{\Sigma}$, is at the core of the covariance structure hypothesis, $\boldsymbol{\Sigma} = \boldsymbol{\Sigma}(\boldsymbol{\theta})$. And I have defined the identification of $\boldsymbol{\theta}$ in terms of the elements of $\boldsymbol{\Sigma}$. For now, I ignore the first-order moment $\boldsymbol{\mu}$ since all variables are in deviation form and $\boldsymbol{\mu}$ is not structured as a function of $\boldsymbol{\theta}$. Later, when I estimate the regression constants of equations, I will make use of $\boldsymbol{\mu}$. So for my purposes, the population covariance matrix of the observed variables, $\boldsymbol{\Sigma}$, contains the known-to-be-identified parameters from which we must demonstrate that the unknown parameters in $\boldsymbol{\theta}$ can be identified.

Identification is not a problem of too few cases. It is defined in terms of the population parameters. The *population* covariance matrix is the source of identified information. The *parameters* refer to population, not to sample values. So no matter how big the sample, an unidentified parameter remains unidentified.

As should be evident, model identification in structural equations with observed variables is not possible without placing restrictions on model parameters. A researcher may specify that all the elements in Γ, Φ, Ψ, and $\mathbf{B}$ are free in an attempt to "let the data tell us" how these variables are related. Unfortunately, this asks more from the data than it can offer since such a model is underidentified and the true values of the parameters cannot be distinguished from an infinite number of false ones. For the data to tell us relationships, we must restrict the parameters before asking the data to speak. The most common restrictions set some elements of $\mathbf{B}$, Γ, Φ, or Ψ to zero or some other constant. Others impose equality or inequality constraints on parameters.

Although it is not obvious, I have already imposed two restrictions necessary for identification. Both appear in equation (4.1), which is the structural equation $\mathbf{y} = \mathbf{B}\mathbf{y} + \Gamma\mathbf{x} + \zeta$. First, the main diagonal of $\mathbf{B}$ is fixed at zero. If this were not done, each endogenous variable would be shown as having a direct effect on itself. I follow the standard convention of setting the diagonal of $\mathbf{B}$ to zero so that the dependent variable of each equation appears on the left-hand side with an implicit coefficient of one. This is sometimes referred to as the *normalization* convention, and without it, the structural equation models would be underidentified.

A second identification convention that is often taken for granted is that the coefficient matrix for ζ in equation (4.1) always is an identity matrix. This means that each zeta appears in only one equation with a coefficient of one. The ζ vector contains variables that unlike $\mathbf{y}$ and $\mathbf{x}$ cannot be directly observed. These latent variables must have a scale to make them interpretable. Two ways to do this are by setting the latent variable's variance to a constant (e.g., to one) or by scaling it to one of the observed variables by setting its coefficient to one. I will have more to say about the scaling conventions in Chapters 6 and 7, but suffice it to say that the ζ's are typically given a scale by setting their coefficients in their respective equations to one rather than fixing their variance to some constant value.

These two identification constraints are so routinely used that many analysts are not aware that they are imposing them when they estimate a regression or econometric model. In most cases these conventions are not enough to identify multiequation models, and other information must be brought to the specification.

An illustration of an identified model is a macroeconomic example from Haavelmo (1953). Some of his data were examined in Chapter 2. Haavelmo's interest is the marginal propensity to consume. This is the part of disposable income that goes to the purchase of consumption goods. He treats aggregate disposable income (y_1) as equal to investment expenditures (x_1) plus consumption expenditures (y_2). Consumption expenditures (y_2) are a function of disposable income (y_1) plus a disturbance term. The two equation system is

$$y_1 = y_2 + x_1 \tag{4.20}$$

$$y_2 = \beta_{21} y_1 + \zeta_2 \tag{4.21}$$

where the COV(ζ_2, x_1) is zero. The first equation is an identity. The **B**, Γ, $\mathbf{\Psi}$, and $\mathbf{\Phi}$ are

$$\mathbf{B} = \begin{bmatrix} 0 & 1 \\ \beta_{21} & 0 \end{bmatrix}, \qquad \Gamma = \begin{bmatrix} 1 \\ 0 \end{bmatrix} \tag{4.22}$$

$$\mathbf{\Psi} = \begin{bmatrix} 0 & 0 \\ 0 & \psi_{22} \end{bmatrix}, \qquad \mathbf{\Phi} = [\phi_{11}] \tag{4.23}$$

Later in this chapter I will estimate this model with Haavelmo's data, but for now, I examine its identification. To do so, substitute (4.22) and (4.23) into (4.10). This leads to

$$\Sigma(\theta) = \begin{bmatrix} (1-\beta_{21})^{-2}(\phi_{11}+\psi_{22}) & & \\ (1-\beta_{21})^{-2}(\beta_{21}\phi_{11}+\psi_{22}) & (1-\beta_{21})^{-2}(\beta_{21}^2\phi_{11}+\psi_{22}) & \\ (1-\beta_{21})^{-1}\phi_{11} & (1-\beta_{21})^{-1}\phi_{11}\beta_{21} & \phi_{11} \end{bmatrix} \tag{4.24}$$

When a model includes an identity [see eq. (4.20)], the Σ and $\Sigma(\theta)$ matrices are singular. Virtually all the discussion of identification and estimation assumes that these matrices are nonsingular. A simple solution is to eliminate variables that are linear combinations of other variables and to use a subset of linearly independent variables to form Σ and $\Sigma(\theta)$. In (4.24) I do this by eliminating y_2 and the second row and column of $\Sigma(\theta)$ which correspond to y_2. The result is a 2×2 $\Sigma(\theta)$ and Σ. The covariance structure hypothesis of $\Sigma = \Sigma(\theta)$ has three unknown parameters ($\theta' = [\beta_{21}\ \psi_{22}\ \phi_{11}]'$). The model is identified if we can write each element of θ as a function only of the elements of Σ.

Consider the three equations taken from $\boldsymbol{\Sigma} = \boldsymbol{\Sigma}(\boldsymbol{\theta})$:

$$\begin{aligned} \operatorname{VAR}(x_1) &= \phi_{11} \\ \operatorname{COV}(x_1, y_1) &= (1 - \beta_{21})^{-1}\phi_{11} \\ \operatorname{VAR}(y_1) &= (1 - \beta_{21})^{-2}(\phi_{11} + \psi_{22}) \end{aligned} \tag{4.25}$$

The solution for ϕ_{11} is obvious, and algebraic manipulations lead to

$$\begin{aligned} \phi_{11} &= \operatorname{VAR}(x_1) \\ \beta_{21} &= 1 - \frac{\operatorname{VAR}(x_1)}{\operatorname{COV}(x_1, y_1)} \\ \psi_{22} &= \left[\frac{\operatorname{VAR}(x_1)}{\operatorname{COV}(x_1, y_1)}\right]^2 \operatorname{VAR}(y_1) - \operatorname{VAR}(x_1) \end{aligned} \tag{4.26}$$

Equation (4.26) demonstrates that Haavelmo's two-equation model is identified.

Although the algebraic manipulations of the $\boldsymbol{\Sigma} = \boldsymbol{\Sigma}(\boldsymbol{\theta})$ equation can help establish identification, it loses its practicality as the model increases in complexity. Consider the union sentiment or the socioeconomic status examples with five observed variables and 15 equations. Attempting to solve for the unknown parameters through algebraic means would be error-prone and time-consuming. Fortunately, other procedures are available to assess identification.

t-Rule

The easiest test to apply is a *necessary but not sufficient* condition of identification.[4] This necessary condition is perfectly general and can be applied to all models discussed in this chapter as well as to all the models treated later. The t-rule for identification is that the number of nonredundant elements in the covariance matrix of the observed variables must be greater than or equal to the number of unknown parameters in $\boldsymbol{\theta}$:

$$t \leq \left(\tfrac{1}{2}\right)(p + q)(p + q + 1) \tag{4.27}$$

[4] In my discussion of identification I ignore inequality constraints on parameters that may help in identification. As I mentioned earlier, I assume that $\boldsymbol{\Sigma}$ and $\boldsymbol{\Sigma}(\boldsymbol{\theta})$ are nonsingular.

where $p + q$ is the number of observed variables and t is the number of free parameters in $\boldsymbol{\theta}$. The right-hand side of (4.27) is the number of nonredundant elements in $\boldsymbol{\Sigma}$. Each of these variances or covariances is known to be identified, and equation (4.10) shows that each is a function of one or more of the t-elements in $\boldsymbol{\theta}$. This leads to $(\frac{1}{2})(p+q)(p+q+1)$ equations in t unknowns. If the number of unknowns exceeds the number of equations, the identification of $\boldsymbol{\theta}$ is not possible. For the socio-economic status example, equations (4.4) and (4.5) have 15 unknowns, and $(\frac{1}{2})(p+q)(p+q+1)$ is 15 so that the t-rule is met. The union sentiment example [see (4.3)] also satisfies the t-rule.

This necessary condition of identification is extremely useful since it allows us to quickly discover underidentified models. Its major limitation is that meeting the necessary condition does not guarantee identification. Fortunately, additional rules of identification have been devised for structural equations with observed variables that take particular forms.

Null B Rule

In a multiequation model where no endogenous variable affects any other endogenous variable, the **B** matrix is zero. An example is the two-equation model:

$$\begin{aligned} y_1 &= \gamma_{11}x_1 + \gamma_{12}x_2 + \zeta_1 \\ y_2 &= \gamma_{21}x_1 + \gamma_{23}x_3 + \zeta_2 \\ \text{COV}(x_i, \zeta_j) &= 0, \qquad \text{for } i = 1,2,3,\ j = 1,2 \end{aligned} \tag{4.28}$$

The **B** matrix is zero since y_1 does not affect y_2, nor does y_2 affect y_1.

To establish the identification of any model where **B** is zero, I show that the unknown parameters in $\boldsymbol{\Gamma}$, $\boldsymbol{\Phi}$, and $\boldsymbol{\Psi}$ are functions of the identified parameters of $\boldsymbol{\Sigma}$. Substituting $\mathbf{B} = \mathbf{0}$ into equation (4.10) and partitioning $\boldsymbol{\Sigma}$ into four parts leads to

$$\boldsymbol{\Sigma} = \boldsymbol{\Sigma}(\boldsymbol{\theta}) \tag{4.29}$$

$$\begin{bmatrix} \boldsymbol{\Sigma}_{yy} & \boldsymbol{\Sigma}_{yx} \\ \boldsymbol{\Sigma}_{xy} & \boldsymbol{\Sigma}_{xx} \end{bmatrix} = \begin{bmatrix} (\boldsymbol{\Gamma\Phi\Gamma}' + \boldsymbol{\Psi}) & \boldsymbol{\Gamma\Phi} \\ \boldsymbol{\Phi\Gamma}' & \boldsymbol{\Phi} \end{bmatrix} \tag{4.30}$$

The lower-right quadrant of (4.30) reveals that $\boldsymbol{\Phi} = \boldsymbol{\Sigma}_{xx}$ so that $\boldsymbol{\Phi}$ is identified. Using the lower-left quadrant results in

$$\begin{aligned} \mathbf{\Phi\Gamma}' &= \mathbf{\Sigma}_{xy} \\ \mathbf{\Sigma}_{xx}\mathbf{\Gamma}' &= \mathbf{\Sigma}_{xy} \\ \mathbf{\Gamma}' &= \mathbf{\Sigma}_{xx}^{-1}\mathbf{\Sigma}_{xy} \end{aligned} \tag{4.31}$$

The second step of (4.31) follows from substituting $\mathbf{\Sigma}_{xx}$ in for $\mathbf{\Phi}$, and the last occurs by premultiplying both sides by $\mathbf{\Sigma}_{xx}^{-1}$, where $\mathbf{\Sigma}_{xx}$ must be nonsingular. The bottom line confirms that $\mathbf{\Gamma}$ is a function of known-to-be-identified covariance matrices, and is itself identified. Finally, the upper-left quadrant of (4.30) solved for $\mathbf{\Psi}$ creates

$$\begin{aligned} \mathbf{\Psi} &= \mathbf{\Sigma}_{yy} - \mathbf{\Gamma\Phi\Gamma}' \\ &= \mathbf{\Sigma}_{yy} - \mathbf{\Sigma}_{yx}\mathbf{\Sigma}_{xx}^{-1}\mathbf{\Sigma}_{xx}\mathbf{\Sigma}_{xx}^{-1}\mathbf{\Sigma}_{xy} \\ &= \mathbf{\Sigma}_{yy} - \mathbf{\Sigma}_{yx}\mathbf{\Sigma}_{xx}^{-1}\mathbf{\Sigma}_{xy} \end{aligned} \tag{4.32}$$

Thus, when $\mathbf{B} = \mathbf{0}$, $\mathbf{\Phi}$, $\mathbf{\Gamma}$,and $\mathbf{\Psi}$ can each be written as functions of the identified covariance matrices of the observed variables and are therefore identified. I refer to this identification condition as the *Null* **B** *Rule*. If the disturbances from one equation are uncorrelated with those from the other equations in a system (i.e., $\mathbf{\Psi}$ is diagonal), then these equations may be treated as separate, or "unrelated." If $\mathbf{\Psi}$ is not diagonal so that the disturbances from at least two equations are correlated, then such a model is sometimes called "seemingly unrelated regressions" (see Kmenta 1986). In either case the unknown parameters in $\mathbf{\Gamma}$, $\mathbf{\Psi}$, and $\mathbf{\Phi}$ are identified.

The Null **B** Rule is a *sufficient* condition for identification of a model. This means that if **B** is null, the unknown parameters are identified. However, the Null **B** Rule is not necessary for identification.

Recursive Rule

Like the Null **B** Rule, the Recursive Rule is a sufficient condition for model identification, but it is not a necessary one. Unlike the Null **B** Rule, the Recursive Rule does not require $\mathbf{B} = \mathbf{0}$. For the recursive rule to apply, the **B** matrix must be triangular, and the $\mathbf{\Psi}$ matrix must be diagonal. A more exact description of the condition for **B** is that the matrix can be written as a lower triangular matrix. This qualification is required since in some models the subscripting or ordering of the y variables can mask a lower triangular **B**. If both conditions hold, then the model is identified.

The union sentiment model provides an example. The **B** is lower triangular, and $\boldsymbol{\Psi}$ is diagonal [see (4.3)]. Its three equations are

$$y_1 = \gamma_{12} x_2 + \zeta_1 \tag{4.33}$$

$$y_2 = \beta_{21} y_1 + \gamma_{22} x_2 + \zeta_2 \tag{4.34}$$

$$y_3 = \beta_{31} y_1 + \beta_{32} y_2 + \gamma_{31} x_1 + \zeta_3 \tag{4.35}$$

A property of all recursive models is that for a given equation, the disturbance term is uncorrelated with the explanatory variables. This is not surprising for exogenous explanatory variables since we assume that ζ is uncorrelated with **x**. For instance, in the union sentiment equations, the $\mathrm{COV}(x_2, \zeta_2)$, $\mathrm{COV}(x_2, \zeta_1)$, and $\mathrm{COV}(x_1, \zeta_3)$ are zero. However, the zero covariances of the disturbance and the endogenous variables, which are explanatory variables in an equation, is less obvious. To illustrate how this occurs, consider (4.34). The $\mathrm{COV}(\zeta_2, y_1)$ is

$$\begin{aligned} \mathrm{COV}(\zeta_2, y_1) &= COV(\zeta_2, \gamma_{12} x_2 + \zeta_1) \\ &= 0 \end{aligned} \tag{4.36}$$

So ζ_2 is uncorrelated with y_1 and x_2, the two explanatory variables of the second equation. Similarly, for the third equation $\mathrm{COV}(\zeta_3, y_1)$ and $\mathrm{COV}(\zeta_3, y_2)$ are zero. The zero for $\mathrm{COV}(\zeta_3, y_1)$ follows since y_1 is a function of x_2 and ζ_1, and both are uncorrelated with ζ_3. A zero for $\mathrm{COV}(\zeta_3, y_2)$ occurs because y_2 is a function of x_2, ζ_1, and ζ_2, all of which are uncorrelated with ζ_3. In general for the ith equation in any recursive model, ζ_i is uncorrelated with the endogenous variables which are explanatory variables in that equation. This is due to these endogenous variables being functions of the exogenous variables and the disturbances from other equations, both of which are uncorrelated with ζ_i.

I will make use of this property in demonstrating that **B**, $\boldsymbol{\Gamma}$, $\boldsymbol{\Phi}$, and $\boldsymbol{\Psi}$ in recursive models are identified. Intuitively, if all the explanatory variables of an equation are uncorrelated with the disturbance, then it is like a standard regression equation and such equations are identified. I establish the identification of these parameters more rigorously shortly. I do this by treating one equation at a time and using a slightly different notation. The ith equation of a recursive model is

$$y_i = [\boldsymbol{\beta}_i' \mid \boldsymbol{\gamma}_i'] \mathbf{z}_i + \zeta_i \tag{4.37}$$

where y_i is the dependent variable and ζ_i is the disturbance term for the ith equation. The $\boldsymbol{\beta}_i'$ vector is the ith row of $\mathbf{B}$, removing all zero values and leaving only free parameters; $\boldsymbol{\gamma}_i'$ is similarly defined for $\boldsymbol{\Gamma}$; and $\mathbf{z}_i$ is the subset of variables in $\mathbf{y}$ and $\mathbf{x}$ which have direct effects on y_i. In other words, the column vector $\mathbf{z}_i$ contains only those variables that belong in the y_i equation, and $[\boldsymbol{\beta}_i' \mid \boldsymbol{\gamma}_i']$ contains their coefficients. As an example, in the union sentiment model equation (4.35), $\mathbf{z}_3'$ is $[y_1\ y_2\ x_1]'$, $\boldsymbol{\beta}_3'$ is $[\beta_{31}\ \beta_{32}]$, and $\boldsymbol{\gamma}_3'$ is $[\gamma_{31}]$.

Post multiplying both sides of (4.37) by $\mathbf{z}_i'$ and taking expectations leads to

$$\boldsymbol{\sigma}_{y_i z_i}' = [\boldsymbol{\beta}_i' \mid \boldsymbol{\gamma}_i']\boldsymbol{\Sigma}_{z_i z_i} + \boldsymbol{\sigma}_{\zeta_i z_i}' \tag{4.38}$$

where $\boldsymbol{\sigma}_{y_i z_i}'$ is the covariance row vector of y_i with its explanatory variables, $\boldsymbol{\Sigma}_{z_i z_i}$ is the nonsingular covariance matrix of $\mathbf{z}_i$, and $\boldsymbol{\sigma}_{\zeta_i z_i}'$ is the covariance vector of ζ_i with the explanatory variables for the ith equation. As shown earlier, ζ_i is uncorrelated with the explanatory variables for the ith equation, so $\boldsymbol{\sigma}_{\zeta_i z_i}'$ is zero. Dropping the last term of (4.38) and solving for $[\boldsymbol{\beta}_i' \mid \boldsymbol{\gamma}_i']$ results in[5]

$$[\boldsymbol{\beta}_i' \mid \boldsymbol{\gamma}_i'] = \boldsymbol{\sigma}_{y_i z_i}'\boldsymbol{\Sigma}_{z_i z_i}^{-1} \tag{4.39}$$

The covariances of the observed variables $\boldsymbol{\sigma}_{y_i z_i}'$ and $\boldsymbol{\Sigma}_{z_i z_i}^{-1}$ are identified, and since $[\boldsymbol{\beta}_i' \mid \boldsymbol{\gamma}_i']$ is a function of identified parameters, it too is identified.

To establish the identification of ψ_{ii}, I postmultiply both sides of (4.37) by y_i' and take expectations. After simplification this is

$$\mathrm{VAR}(y_i) = [\boldsymbol{\beta}_i' \mid \boldsymbol{\gamma}_i']\boldsymbol{\Sigma}_{z_i z_i}\begin{bmatrix} \boldsymbol{\beta}_i \\ \hline \boldsymbol{\gamma}_i \end{bmatrix} + \psi_{ii} \tag{4.40}$$

Solving (4.40) for ψ_{ii} and substituting (4.39) in for $[\boldsymbol{\beta}_i' \mid \boldsymbol{\gamma}_i']$,

$$\psi_{ii} = \mathrm{VAR}(y_i) - \boldsymbol{\sigma}_{y_i z_i}'\boldsymbol{\Sigma}_{z_i z_i}^{-1}\boldsymbol{\sigma}_{z_i y_i} \tag{4.41}$$

The ψ_{ii} is a function of identified variances and covariances, and so it is identified.

These results suffice to show that $\mathbf{B}$, $\boldsymbol{\Gamma}$, $\boldsymbol{\Phi}$, and $\boldsymbol{\Psi}$ in recursive models are identified. Equation (4.39) proves that for the ith equation, the ith row of free parameters in $\mathbf{B}$ and $\boldsymbol{\Gamma}$ are identified, and (4.41) shows the same for ψ_{ii}.

[5]Note that the transpose of (4.39) is the usual OLS formula for regression coefficients if the sample counterparts to the population covariance matrices are used. It should come as no surprise that OLS is a consistent estimator of the coefficients in recursive models.

Since these relations hold for all the equations, **B**, Γ, and Ψ are identified. Furthermore, since $\Phi = \Sigma_{xx}$ [see (4.10)] for all the models of this chapter, it too is identified. Thus all parameters in recursive models are identified.

Since the recursive rule is a sufficient condition for model identification, it is possible for **B** to be nontriangular or for Ψ not to be diagonal, and for the model still to be identified. This leads to the next identification rules.

Rank and Order Conditions

Except for the *t*-rule, the other identification rules placed restrictions on either **B** or Ψ. Nonrecursive models fail to satisfy these restrictions and must have their identification established in some other way. In this section I present the order and rank conditions that apply to many nonrecursive systems. Like the Null **B** Rule and Recursive Rule, the rank and order conditions of identification are for models that assume no measurement error and that assume that all exogenous variables (**x**) are uncorrelated with the errors in the equations (ζ). They differ from these prior rules in several ways, however. First, the rank and order conditions apply where **B** assumes any form as long as $(\mathbf{I} - \mathbf{B})$ is nonsingular. Second, they help to determine the identification of one equation at a time. For the Null **B** and Recursive rules, the identification of the whole model is established if the conditions are met. To show identification of a model with the rank and order conditions, each equation must meet the conditions.

A third difference, which is more subtle, is that the rank and order conditions assume that Ψ contains no restrictions. That is, no element of Ψ is constrained to a fixed value (e.g., zero) or to any other constraint. This has the advantage that if all equations of a model meet the rank and order conditions we know that all elements of Ψ are identified and may be estimated. The disadvantage is that if we have knowledge that certain elements of Ψ should be restricted, this knowledge is not utilized. In fact, it is possible that constraints on Ψ may help identify one or more parameters that would not otherwise be identified as was true for recursive models.

The order condition is the easiest to apply. If the only type of restriction in an equation is excluding variables, then the order condition may be stated as follows: *A necessary condition for an equation to be identified is that the number of variables excluded from the equation be at least* $p - 1$. To understand the basis of this rule, consider a single equation from a model

$$y_i = [\boldsymbol{\beta}_i' \mid \boldsymbol{\gamma}_i']\mathbf{z}_i + \zeta_i \tag{4.42}$$

Equation (4.42) has the same notation as that of (4.37) from the section on recursive models. The major difference is that I have all of the **y** and **x** variables in $\mathbf{z}_i$ except y_i. The $\boldsymbol{\beta}_i'$ vector equals the ith row of $\mathbf{B}$, excluding the normalization coefficient of zero and is $1 \times (p-1)$. The $\boldsymbol{\gamma}_i'$ is $1 \times q$ and equals the ith row of $\boldsymbol{\Gamma}$.

Post-multiply both sides of (4.42) by $\mathbf{x}'$ and take expectations:

$$\boldsymbol{\sigma}_{y_i x}' = [\boldsymbol{\beta}_i' \mid \boldsymbol{\gamma}_i'] \boldsymbol{\Sigma}_{z_i x} \tag{4.43}$$

If $\boldsymbol{\beta}_i'$ and $\boldsymbol{\gamma}_i'$ are functions only of the identified covariance elements in $\boldsymbol{\sigma}_{y_i x}'$ and $\boldsymbol{\Sigma}_{z_i x}$, they are identified. A necessary but not sufficient condition for this is that the number of equations implied by (4.43) be at least equal to the number of unknown, free parameters in $[\boldsymbol{\beta}_i' \mid \boldsymbol{\gamma}_i']$. The number of equations is the number of elements in $\boldsymbol{\sigma}_{y_i x}'$ which is q. This follows since q covariances with y_i result from the q variables in $\mathbf{x}$. The number of unknowns in $\boldsymbol{\beta}_i'$ is $(p-1)$, and in $\boldsymbol{\gamma}_i'$ it is q, since no restrictions are placed on these vectors. Clearly, with q equations in $(p-1)+q$ unknowns, $\boldsymbol{\beta}_i'$ and $\boldsymbol{\gamma}_i'$ cannot be identified. The $(p-1)+q$ unknowns must be reduced to q. A necessary condition for identification is that $(p-1)$ restrictions be placed on the elements of $\boldsymbol{\beta}_i'$ and $\boldsymbol{\gamma}_i'$. The most common restriction is setting some elements to zero so that the corresponding variable is omitted from the equation. If excluded variables are the only type of restrictions, then $(p-1)$ variables must be excluded from the ith equation to make possible identification. This is the *order condition*.

I use the objective and subjective socioeconomic status example as an illustration. The matrix equation for this model is

$$\begin{bmatrix} y_1 \\ y_2 \\ y_3 \end{bmatrix} = \begin{bmatrix} 0 & \beta_{12} & 0 \\ \beta_{21} & 0 & 0 \\ \beta_{31} & \beta_{32} & 0 \end{bmatrix} \begin{bmatrix} y_1 \\ y_2 \\ y_3 \end{bmatrix} + \begin{bmatrix} \gamma_{11} & 0 \\ 0 & \gamma_{22} \\ 0 & 0 \end{bmatrix} \begin{bmatrix} x_1 \\ x_2 \end{bmatrix} + \begin{bmatrix} \zeta_1 \\ \zeta_2 \\ \zeta_3 \end{bmatrix}$$

Suppose that for the first equation β_{13} and γ_{12} were not restricted to zero:

$$y_1 = \beta_{12} y_2 + \beta_{13} y_3 + \gamma_{11} x_1 + \gamma_{12} x_2 + \zeta_1 \tag{4.44}$$

This is in the form of (4.42). Next, multiply both sides by the exogenous variables and take expectations:

$$\begin{aligned} \mathrm{COV}(y_1, x_1) = {} & \beta_{12}\,\mathrm{COV}(y_2, x_1) + \beta_{13}\,\mathrm{COV}(y_3, x_1) + \gamma_{11}\,\mathrm{VAR}(x_1) \\ & + \gamma_{12}\,\mathrm{COV}(x_2, x_1) \end{aligned} \tag{4.45}$$

$$\begin{aligned} \mathrm{COV}(y_1, x_2) = {} & \beta_{12}\,\mathrm{COV}(y_2, x_2) + \beta_{13}\,\mathrm{COV}(y_3, x_2) + \gamma_{11}\,\mathrm{COV}(x_1, x_2) \\ & + \gamma_{12}\,\mathrm{VAR}(x_2) \end{aligned} \tag{4.46}$$

The result is two equations in four unknowns ($\beta_{12}, \beta_{13}, \gamma_{11}, \gamma_{12}$), and it is clear that a unique solution for these parameters is not possible. The order condition requires ($p - 1$) or 2 exclusions from the first equation. The original specification of $\beta_{13} = 0$ and $\gamma_{12} = 0$ satisfies this requirement so that the original first equation meets this necessary condition of identification.

One way to check the order condition for all equations in a model is to form a matrix, say, $\mathbf{C}$, which is $[(\mathbf{I} - \mathbf{B}) \mid -\mathbf{\Gamma}]$. Then for each row, count the number of zero elements. If a row has ($p - 1$) or more zeros, it meets the order condition. For the objective and subjective socioeconomic status example, $\mathbf{C}$ is

$$\mathbf{C} = \begin{bmatrix} 1 & -\beta_{12} & 0 & -\gamma_{11} & 0 \\ -\beta_{21} & 1 & 0 & 0 & -\gamma_{22} \\ -\beta_{31} & -\beta_{32} & 1 & 0 & 0 \end{bmatrix} \tag{4.47}$$

Each row of $\mathbf{C}$ has ($p - 1$) or 2 exclusions, so each equation satisfies the order condition.

The order condition is an easy way to rule out underidentified equations for nonrecursive models with $\mathbf{\Psi}$ free. As a necessary but not sufficient condition, passing the order condition does not guarantee identification. Underidentification can occur if it is possible to produce a new equation with the same form but different parameters from the old equation, by using a linear combination of the other equations in a model. One such instance occurs when two or more equations have identical restrictions.

The two following equations provide an illustration:

$$y_1 = \beta_{12} y_2 + \gamma_{11} x_1 + \gamma_{12} x_2 + \zeta_1 \tag{4.48}$$

$$y_2 = \beta_{21} y_1 + \gamma_{22} x_2 + \zeta_2 \tag{4.49}$$

Suppose that we exclude x_2 from both equations ($\gamma_{12} = \gamma_{22} = 0$). The $\mathbf{C}$ matrix is

$$\mathbf{C} = \begin{bmatrix} 1 & -\beta_{12} & -\gamma_{11} & 0 \\ -\beta_{21} & 1 & 0 & 0 \end{bmatrix} \tag{4.50}$$

The order condition for both equations is met since each equation has 1 ($= p - 1$) exclusion. Multiply the second row of $\mathbf{C}$ by a constant called a

and add it to the first row:

$$\begin{bmatrix} 1 - \beta_{21}a & -\beta_{12} + a & -\gamma_{11} & 0 \\ -\beta_{21} & 1 & 0 & 0 \end{bmatrix} \tag{4.51}$$

Next, divide each element of the first row by $(1 - a\beta_{21})$:

$$\mathbf{C}^* = \begin{bmatrix} 1 & -\beta_{12}^* & -\gamma_{11}^* & 0 \\ -\beta_{21} & 1 & 0 & 0 \end{bmatrix} \tag{4.52}$$

where

$$-\beta_{12}^* = \left(\frac{-\beta_{12} + a}{1 - a\beta_{21}} \right) \quad \text{and} \quad -\gamma_{11}^* = \frac{-\gamma_{11}}{(1 - a\beta_{21})}.$$

The first equation represented by the first row of $\mathbf{C}^*$ has the same form and excluded variables as that shown in the first row of $\mathbf{C}$ in (4.50). An infinite set of β_{12}^* and γ_{11}^* values can be produced that do not equal the true β_{12} and γ_{11}. So β_{12} and γ_{11} are not identified, even though the order condition for the first equation is satisfied.

I apply a similar procedure to the second equation. I multiply the first row of $\mathbf{C}$ in (4.50) by a and add it to the second row:

$$\mathbf{C}^* = \begin{bmatrix} 1 & -\beta_{12} & -\gamma_{11} & 0 \\ -\beta_{21} + a & 1 - a\beta_{12} & -a\gamma_{11} & 0 \end{bmatrix} \tag{4.53}$$

This new second row of $\mathbf{C}^*$ has a different form than the second row of $\mathbf{C}$. The (2, 3) position of $\mathbf{C}^*$ is $-a\gamma_{11}$, whereas the corresponding element of $\mathbf{C}$ is zero. Provided $-\gamma_{11}$ or a is not zero, no value of a leads to the same form as the second row of $\mathbf{C}$. Thus a new second equation indistinguishable from the old one cannot be produced from a linear combination of the first equation. The second equation is identified.

Although in this two equation system it is easy to determine whether new "imposter" equations are possible, in more complex multiequation systems the above procedure is more difficult to apply. The rank condition of identification accomplishes the same purpose with less effort. The rank rule starts with $\mathbf{C}$ (i.e., $[(\mathbf{I} - \mathbf{B}) \mid -\boldsymbol{\Gamma}]$). To check the identification of the ith equation, delete all columns of $\mathbf{C}$ that do not have zeros in the ith row of $\mathbf{C}$. Use the remaining columns to form a new matrix, $\mathbf{C}_i$. *A necessary and sufficient condition for the identification of the ith equation is that the rank of* $\mathbf{C}_i$ *equals* $(p - 1)$. This is the rank condition. As an illustration, consider the two-equation model in equations (4.48) and (4.49). The $\mathbf{C}$ matrix is in (4.50). I examine the identification of the first equation. The only zero in the first row is in the fourth column, so the first three columns are deleted and $\mathbf{C}_1$ is

$$\mathbf{C}_1 = \begin{bmatrix} 0 \\ 0 \end{bmatrix} \tag{4.54}$$

The rank of a matrix or vector is the number of independent rows and columns. With both elements of $\mathbf{C}_1$ zero, its rank is zero. With a rank less than one, the first equation is not identified as was shown earlier. For the second equation of $\mathbf{C}$ in (4.50), $\mathbf{C}_2$ is

$$\mathbf{C}_2 = \begin{bmatrix} -\gamma_{11} & 0 \\ 0 & 0 \end{bmatrix} \tag{4.55}$$

Except when γ_{11} is zero, the rank $(\mathbf{C}_2)$ is one, which satisfies the rank condition, and the second equation is identified.

The rank condition also can be checked for the objective and subjective status example. The $\mathbf{C}$ for this model is in (4.47). For the first equation, $\mathbf{C}_1$ is

$$\mathbf{C}_1 = \begin{bmatrix} 0 & 0 \\ 0 & -\gamma_{22} \\ 1 & 0 \end{bmatrix}$$

Unless $\gamma_{22} = 0$, $\mathbf{C}_1$ has two independent rows and columns, so its rank is two which satisfies the rank condition. Similarly, $\mathbf{C}_2$ and $\mathbf{C}_3$ also have a rank of two, and thus all equations are identified.

Equation systems with identity relations are treated the same, except that the equation that represents the identity does not require rank and order conditions since it is identified by definition. I illustrate this with Haavelmo's marginal propensity to consume model [see (4.20) to (4.23)]. The $\mathbf{C}$ for his model is

$$\mathbf{C} = \begin{bmatrix} 1 & -1 & -1 \\ -\beta_{21} & 1 & 0 \end{bmatrix}$$

The first row is the identity equation, which is identified. The $\mathbf{C}_2$ matrix is

$$\mathbf{C}_2 = \begin{bmatrix} -1 \\ 0 \end{bmatrix}$$

The rank is one, which satisfies the rank condition, so the equation is identified.

Exclusion restrictions are the most common type of restriction placed on elements of **B** and Γ. These represent the absence of a variable from an equation by setting its corresponding coefficient in **B** or Γ to zero. I limited my discussion of the rank and order conditions to exclusion restrictions, but other types are possible. For instance, suppose that the influence of two exogenous variables in an equation are unknown but are assumed to be equal. Or a linear function of two or more parameters might equal a known constant. Inequality or nonlinear constraints also are possible. I do not treat these topics here but refer the reader to Fisher (1966) and Johnston (1984).

Also remember that the rank and order condition assume that no restrictions are placed on $\boldsymbol{\Psi}$. If $\boldsymbol{\Psi}$ is restricted, it is still possible for the model to be identified, even if it fails the rank and order conditions. For instance, if I apply the order condition to the union sentiment example [see (4.3)], the last equation fails. Yet this equation is identified. The reason is that the union sentiment model is recursive so that $\boldsymbol{\Psi}$ is restricted. As shown in the section on recursive models, these restrictions and the triangular **B** enable the model to be identified. Thus, when restrictions are placed on $\boldsymbol{\Psi}$, the order rule is no longer necessary, and the rank rule is no longer necessary and sufficient for identification.

The rank and order conditions establish the identification status of equations. If each equation meets the rank rule, then the model as a whole is identified. Both these conditions have the advantage that they do not require a special form for the **B** matrix. They assume no restrictions for the $\boldsymbol{\Psi}$ matrix. This may be a disadvantage if additional restrictions on **B**, $\boldsymbol{\Psi}$, or the Γ matrix are appropriate.

Summary of Identification Rules

Table 4.1 summarizes the identification rules discussed in this section. The t-rule, the Null **B** Rule, and Recursive Rule are conditions for the identification of the model as a whole. The first is only a necessary condition, but the second and third are sufficient conditions. The t-rule is the most general rule and applies to all of the models I have discussed in this chapter. The Null **B** Rule is appropriate whenever $\mathbf{B} = \mathbf{0}$, regardless of the form of $\boldsymbol{\Psi}$. The recursive rule is only appropriate for models with triangular **B** matrices and diagonal $\boldsymbol{\Psi}$ matrices.

Finally, the rank and order conditions establish the identification status of equations. If each equation meets the rank rule, then the model as a whole is identified. Both conditions allow any nonsingular $(\mathbf{I} - \mathbf{B})$ matrix

Table 4.1 Identification Rules for Structural Equations with Observed Variables Assuming No Measurement Error ($\mathbf{y} = \mathbf{B}\mathbf{y} + \boldsymbol{\Gamma}\mathbf{x} + \boldsymbol{\zeta}$)

Identification Rule	Evaluates	Requirements	Necessary Condition	Sufficient Condition
t-Rule	model	$t \leq (\frac{1}{2})(p+q)(p+q+1)$	yes	no
Null $\mathbf{B}$ Rule	model	$\mathbf{B} = \mathbf{0}$	no	yes
Recursive Rule	model	$\mathbf{B}$ triangular $\boldsymbol{\Psi}$ diagonal	no	yes
Order Condition	equation	restrictions $\geq p - 1$ $\boldsymbol{\Psi}$ free	yes[a]	no
Rank Condition	equation	rank $(\mathbf{C}_i) = p - 1$ $\boldsymbol{\Psi}$ free	yes[a]	yes[a]

[a] This characterization of the rank and order conditions assumes that all elements in $\boldsymbol{\Psi}$ are free.

and assume no restrictions for the $\boldsymbol{\Psi}$ matrix. Although these rules cover most structural equation models with observed variables, they are not comprehensive. Identification rules for block-recursive and some other models are available (see, e.g., Fox, 1984, 247–251; Bekker and Pollock 1986).

ESTIMATION

The estimation procedures derive from the relation of the covariance matrix of the observed variables to the structural parameters. Earlier in this chapter I showed that $\boldsymbol{\Sigma}(\boldsymbol{\theta})$ is

$$\boldsymbol{\Sigma}(\boldsymbol{\theta}) = \begin{bmatrix} (\mathbf{I}-\mathbf{B})^{-1}(\boldsymbol{\Gamma}\boldsymbol{\Phi}\boldsymbol{\Gamma}' + \boldsymbol{\Psi})(\mathbf{I}-\mathbf{B})^{-1'} & (\mathbf{I}-\mathbf{B})^{-1}\boldsymbol{\Gamma}\boldsymbol{\Phi} \\ \boldsymbol{\Phi}\boldsymbol{\Gamma}'(\mathbf{I}-\mathbf{B})^{-1'} & \boldsymbol{\Phi} \end{bmatrix} \tag{4.56}$$

If the structural equation model is correct and the population parameters are known, then $\boldsymbol{\Sigma}$ will equal $\boldsymbol{\Sigma}(\boldsymbol{\theta})$. For instance, consider the simple structural equation:

$$y_1 = x_1 + \zeta_1 \tag{4.57}$$

In (4.57) I have set γ_{11} equal to 1. The population covariance matrix for y_1

and x_1 is

$$\Sigma = \begin{bmatrix} \text{VAR}(y_1) & \text{COV}(y_1, x_1) \\ \text{COV}(x_1, y_1) & \text{VAR}(x_1) \end{bmatrix} \tag{4.58}$$

The Σ matrix in terms of the structural parameters is

$$\Sigma(\theta) = \begin{bmatrix} \phi_{11} + \psi_{11} & \phi_{11} \\ \phi_{11} & \phi_{11} \end{bmatrix} \tag{4.59}$$

Assuming that the model is correct and that the population parameters are known, each element in (4.58) should equal the corresponding element in (4.59). The ϕ_{11} parameter is overidentified since it equals $\text{VAR}(x_1)$ and $\text{COV}(x_1, y_1)$.

In practice, we do not know either the population covariances and variances or the parameters. The task is to form sample estimates of the unknown parameters based on sample estimates of the covariance matrix. The sample covariance matrix, **S**, for y_1 and x_1 is

$$\mathbf{S} = \begin{bmatrix} \text{var}(y_1) & \text{cov}(y_1, x_1) \\ \text{cov}(x_1, y_1) & \text{var}(x_1) \end{bmatrix} \tag{4.60}$$

Once we select values for ϕ_{11} and ψ_{11} (represented as $\hat{\phi}_{11}$ and $\hat{\psi}_{11}$), the implied covariance matrix, $\hat{\Sigma}$, can be formed by substituting $\hat{\phi}_{11}$ and $\hat{\psi}_{11}$ into (4.59):

$$\hat{\Sigma} = \begin{bmatrix} \hat{\phi}_{11} + \hat{\psi}_{11} & \hat{\phi}_{11} \\ \hat{\phi}_{11} & \hat{\phi}_{11} \end{bmatrix} \tag{4.61}$$

The $\hat{\Sigma}$ stands for the implied covariance matrix $\Sigma(\theta)$, with $\hat{\theta}$ replacing θ [i.e., $\hat{\Sigma} = \Sigma(\hat{\theta})$]. We choose the values for $\hat{\phi}_{11}$ and $\hat{\psi}_{11}$ such that $\hat{\Sigma}$ is as close to **S** as possible.

To illustrate this process, suppose that **S** is

$$\mathbf{S} = \begin{bmatrix} 10 & 6 \\ 6 & 4 \end{bmatrix} \tag{4.62}$$

Next we choose $\hat{\phi}_{11}$ equal to 7 and $\hat{\psi}_{11}$ equal to 3. This makes (4.61):

$$\hat{\Sigma} = \begin{bmatrix} 10 & 7 \\ 7 & 7 \end{bmatrix} \tag{4.63}$$

The residual matrix $(\mathbf{S} - \hat{\boldsymbol{\Sigma}})$ indicates how "close" $\hat{\boldsymbol{\Sigma}}$ is to $\mathbf{S}$:

$$(\mathbf{S} - \hat{\boldsymbol{\Sigma}}) = \begin{bmatrix} 0 & -1 \\ -1 & -3 \end{bmatrix} \tag{4.64}$$

Although this set of estimates leads to a perfect match to the sample $\text{var}(y_1)$, the fit is less adequate for $\text{cov}(x_1, y_1)$ and the $\text{var}(x_1)$. The $\hat{\boldsymbol{\Sigma}}$ overpredicts these elements.

Consider a new set of values with $\hat{\phi}_{11} = 5$ and $\hat{\psi}_{11} = 5$. The $\hat{\boldsymbol{\Sigma}}$ is now

$$\hat{\boldsymbol{\Sigma}} = \begin{bmatrix} 10 & 5 \\ 5 & 5 \end{bmatrix} \tag{4.65}$$

The residual matrix $(\mathbf{S} - \hat{\boldsymbol{\Sigma}})$ is

$$(\mathbf{S} - \hat{\boldsymbol{\Sigma}}) = \begin{bmatrix} 0 & 1 \\ 1 & -1 \end{bmatrix} \tag{4.66}$$

Although these values do not lead to a perfect match of $\hat{\boldsymbol{\Sigma}}$ to $\mathbf{S}$, the second set of values seem to lead to a better fit than the first.

The same process, but with more complexities, occurs for the general structural equation model with observed variables. The unknown parameters in $\mathbf{B}$, $\boldsymbol{\Gamma}$, $\boldsymbol{\Phi}$, and $\boldsymbol{\Psi}$ are estimated so that the implied covariance matrix, $\hat{\boldsymbol{\Sigma}}$, is close to the sample covariance matrix $\mathbf{S}$. To know when our estimates are as "close" as possible, we must define "close"—that is, we require a function that is to be minimized. Many different fitting functions for the task are possible. The fitting functions $F(\mathbf{S}, \boldsymbol{\Sigma}(\boldsymbol{\theta}))$ are based on $\mathbf{S}$, the sample covariance matrix, and $\boldsymbol{\Sigma}(\boldsymbol{\theta})$, the implied covariance matrix of structural parameters. If estimates of $\boldsymbol{\theta}$ are substituted in $\boldsymbol{\Sigma}(\boldsymbol{\theta})$, this leads to the implied covariance matrix, $\hat{\boldsymbol{\Sigma}}$. The value of the fitting function for $\hat{\boldsymbol{\theta}}$ is $F(\mathbf{S}, \hat{\boldsymbol{\Sigma}})$. In the previous section I took $(\mathbf{S} - \hat{\boldsymbol{\Sigma}})$ as an indicator of deviation in fit, and I attempted to reduce the discrepancy between each element in $\hat{\boldsymbol{\Sigma}}$ and that in $\mathbf{S}$. The fitting functions I will present here have the following properties: (1) $F(\mathbf{S}, \boldsymbol{\Sigma}(\boldsymbol{\theta}))$ is a scalar, (2) $F(\mathbf{S}, \boldsymbol{\Sigma}(\boldsymbol{\theta})) \geq 0$, (3) $F(\mathbf{S}, \boldsymbol{\Sigma}(\boldsymbol{\theta})) = 0$ if and only if $\boldsymbol{\Sigma}(\boldsymbol{\theta}) = \mathbf{S}$, and (4) $F(\mathbf{S}, \boldsymbol{\Sigma}(\boldsymbol{\theta}))$ is continuous in $\mathbf{S}$ and $\boldsymbol{\Sigma}(\boldsymbol{\theta})$. According to Browne (1984, 66), minimizing fitting functions that satisfy these conditions leads to consistent estimators of $\boldsymbol{\theta}$. I present three such functions: maximum likelihood (ML), unweighted least squares (ULS), and generalized least squares (GLS). The use of these fitting methods extends to all the models in the book.

Maximum Likelihood (ML)

To date, the most widely used fitting function for general structural equation models is the maximum likelihood (ML) function. The fitting function that is minimized is

$$F_{\mathrm{ML}} = \log|\boldsymbol{\Sigma}(\boldsymbol{\theta})| + \operatorname{tr}\left(\mathbf{S}\boldsymbol{\Sigma}^{-1}(\boldsymbol{\theta})\right) - \log|\mathbf{S}| - (p+q) \tag{4.67}$$

Generally, we assume that $\boldsymbol{\Sigma}(\boldsymbol{\theta})$ and $\mathbf{S}$ are positive-definite which means that they are nonsingular. Otherwise, it would be possible for the undefined log of zero to appear in F_{ML}. Appendixes 4A and 4B provide the derivation of (4.67) which is based on the assumption that $\mathbf{y}$ and $\mathbf{x}$ have a multinormal distribution or that $\mathbf{S}$ has a Wishart distribution.

To verify that F_{ML} is zero when $\hat{\boldsymbol{\Sigma}} = \mathbf{S}$, substitute $\hat{\boldsymbol{\Sigma}}$ for $\boldsymbol{\Sigma}(\boldsymbol{\theta})$ and $\hat{\boldsymbol{\Sigma}} = \mathbf{S}$ in (4.67). In this case

$$F_{\mathrm{ML}} = \log|\mathbf{S}| + \operatorname{tr}(\mathbf{I}) - \log|\mathbf{S}| - (p+q)$$

where $\operatorname{tr}(\mathbf{I}) = p + q$, and F_{ML} is zero. Thus, when we have a model that perfectly predicts the values of the sample covariance matrix, a perfect fit is indicated by a zero.

To further demonstrate the operation of this function, I return to the structural equation, $y_1 = x_1 + \zeta_1$. The $\mathbf{S}$ and $\boldsymbol{\Sigma}(\boldsymbol{\theta})$ are in equations (4.60) and (4.59). After substituting $\hat{\boldsymbol{\Sigma}}$ for $\boldsymbol{\Sigma}(\boldsymbol{\theta})$, F_{ML} is

$$\begin{aligned} F_{\mathrm{ML}} = {} & \log\left(\hat{\psi}_{11}\hat{\phi}_{11}\right) + \hat{\psi}_{11}^{-1}\left(\operatorname{var}(y_1) - 2\operatorname{cov}(y_1, x_1) + \operatorname{var}(x_1)\right) \\ & + \hat{\phi}_{11}^{-1}\operatorname{var}(x_1) - \log\left[\operatorname{var}(y_1)\operatorname{var}(x_1) - \left(\operatorname{cov}(y_1, x_1)\right)^2\right] - 2 \end{aligned} \tag{4.68}$$

A necessary condition for the minimization of F_{ML} is that $\hat{\phi}_{11}$ and $\hat{\psi}_{11}$ be chosen so that the partial derivatives of F_{ML} with respect to $\hat{\phi}_{11}$ and $\hat{\psi}_{11}$ are zero. The partial derivatives are

$$\frac{\partial F_{\mathrm{ML}}}{\partial \hat{\phi}_{11}} = \hat{\phi}_{11}^{-1} - \hat{\phi}_{11}^{-2}\operatorname{var}(x_1) \tag{4.69}$$

$$\frac{\partial F_{\mathrm{ML}}}{\partial \hat{\psi}_{11}} = \hat{\psi}_{11}^{-1} - \hat{\psi}_{11}^{-2}\left(\operatorname{var}(y_1) - 2\operatorname{cov}(y_1, x_1) + \operatorname{var}(x_1)\right) \tag{4.70}$$

Setting (4.69) and (4.70) to zero and solving for $\hat{\phi}_{11}$ and $\hat{\psi}_{11}$ leads to

$$\hat{\phi}_{11} = \text{var}(x_1) \tag{4.71}$$

$$\hat{\psi}_{11} = \text{var}(y_1) - 2\,\text{cov}(y_1, x_1) + \text{var}(x_1) \tag{4.72}$$

A sufficient condition for these values to minimize F_{ML} is that the matrix formed by taking the second partial derivatives of the fitting function with respect to $\hat{\phi}_{11}$ and $\hat{\psi}_{11}$ be positive-definite. This matrix is

$$\begin{bmatrix} -\hat{\phi}_{11}^{-2} + 2\hat{\phi}_{11}^{-3}\,\text{var}(x_1) & 0 \\ 0 & -\hat{\psi}_{11}^{-2} + 2\hat{\psi}_{11}^{-3}(\text{var}(y_1) - 2\,\text{cov}(y_1, x_1) + \text{var}(x_1)) \end{bmatrix} \tag{4.73}$$

Setting $\hat{\phi}_{11}$ and $\hat{\psi}_{11}$ to the values in (4.71) and (4.72) and simplifying shows that (4.73) is positive-definite for positive values of $\hat{\phi}_{11}$ and $\hat{\psi}_{11}$. Thus the solution for the parameters in (4.71) and (4.72) do minimize F_{ML}. Suppose that we have sample values of $\text{var}(y_1) = 10$, $\text{cov}(x_1, y_1) = 5$, and $\text{var}(x_1) = 8$. The estimates for $\hat{\phi}_{11}$ and $\hat{\psi}_{11}$ would both equal 8.

The example illustrates a case where explicit solutions for the structural parameters which minimize F_{ML} exist. In general, F_{ML} is a more complicated nonlinear function of the structural parameters, and explicit solutions are not always found. Instead, an iterative numerical procedure is necessary to find the free and equality constrained unknowns in $\mathbf{B}$, $\mathbf{\Gamma}$, $\mathbf{\Phi}$, and $\mathbf{\Psi}$ that minimize F_{ML}. Several numerical procedures are available. I describe and illustrate one such procedure in Appendix 4C.

ML estimators have several important properties. The properties are asymptotic so that they hold in large samples (see Appendix B). First, although they may be biased in small samples, ML estimators are asymptotically unbiased. Second, the ML estimator is consistent ($\text{plim}\,\hat{\theta} = \theta$, and $\hat{\theta}$ is ML estimator and θ is population parameter). Third, they are asymptotically efficient so that among consistent estimators, none has a smaller asymptotic variance. Furthermore the distribution of the estimator approximates a normal distribution as sample size increases (i.e., they are asymptotically normally distributed). This last property suggests that if you know the standard error of the estimated parameters, then the ratio of the estimated parameter to its standard error should approximate a Z-distribution for large samples.

In Appendix 4B I show that the asymptotic covariance matrix for the ML estimator of $\boldsymbol{\theta}$ is

$$\left(\frac{2}{(N-1)}\right)\left\{E\left[\frac{\partial^2 F_{\mathrm{ML}}}{\partial\boldsymbol{\theta}\,\partial\boldsymbol{\theta}'}\right]\right\}^{-1}$$

When $\hat{\boldsymbol{\theta}}$ is substituted for $\boldsymbol{\theta}$, we have an estimated asymptotic covariance matrix with the estimated asymptotic variances of $\hat{\boldsymbol{\theta}}$ down the main diagonal and the estimated covariances in the off-diagonal elements. Thus tests of statistical significance for $\hat{\boldsymbol{\theta}}$ are possible. In the $y_1 = x_1 + \zeta_1$ example the matrix of second partial derivatives of F_{ML} with respect to $\hat{\phi}_{11}$ and $\hat{\psi}_{11}$ is in equation (4.73). Taking the expected value and inverting this matrix, and then multiplying it by $2/(N-1)$ gives the asymptotic covariance matrix with the variances of $\hat{\phi}_{11}$ and $\hat{\psi}_{11}$ down the main diagonal. Using the sample variances, covariance, $\hat{\phi}_{11}$ $(= 8)$, and $\hat{\psi}_{11}$ $(= 8)$ from above and an N of 201 leads to estimated asymptotic standard errors of 0.8 for both $\hat{\phi}_{11}$ and $\hat{\psi}_{11}$. Each estimate is highly statistically significant.

Another important characteristic is that with few exceptions F_{ML} is *scale invariant* and *scale free* (see Swaminathan and Algina 1978). These properties have to do with the consequences of changing the measurement units of one or more of the observed variables (e.g., dollars to cents or a scale range from 0–10 to 0–100). A fitting function, say, $F(\mathbf{S}, \boldsymbol{\Sigma}(\boldsymbol{\theta}))$, is scale invariant if $F(\mathbf{S}, \boldsymbol{\Sigma}(\boldsymbol{\theta})) = F[\mathbf{DSD}, \mathbf{D}\boldsymbol{\Sigma}(\boldsymbol{\theta})\mathbf{D}]$, where $\mathbf{D}$ is a diagonal, nonsingular matrix with positive elements down its diagonal. For instance, if the main diagonal of $\mathbf{D}$ consists of the inverses of the standard deviations for the observed variables, then $\mathbf{DSD}$ is a correlation matrix. The scale invariance property implies that the value of the fit function is the same when $\mathbf{DSD}$ is substituted for $\mathbf{S}$ and $\mathbf{D}\boldsymbol{\Sigma}(\boldsymbol{\theta})\mathbf{D}$ is substituted for $\boldsymbol{\Sigma}(\boldsymbol{\theta})$. Making these substitutions into $F_{\mathrm{ML}}(\mathbf{S}, \boldsymbol{\Sigma}(\boldsymbol{\theta}))$ of (4.67) shows that it is scale invariant. Therefore the values of the fit function are the same for the correlation and the covariance matrices, or more generally, they are the same for any change of scale.

A related concept is *scale freeness*. This has to do with maintaining an equivalency between the structural parameters and estimates in a model with the original variables and those in a model with linearly transformed variables. To understand this property, consider the structural equation with the original variables:

$$\mathbf{y} = \mathbf{B}\mathbf{y} + \boldsymbol{\Gamma}\mathbf{x} + \boldsymbol{\zeta} \tag{4.74}$$

Suppose that $\mathbf{y}$ and $\mathbf{x}$ are rescaled so that $\mathbf{y}^* = \mathbf{D}_y\mathbf{y}$ and $\mathbf{x}^* = \mathbf{D}_x\mathbf{x}$, where

$\mathbf{D}_y$ and $\mathbf{D}_x$ are the diagonal matrices containing the scaling factors for $\mathbf{y}$ and $\mathbf{x}$, respectively. Now the structural equation is

$$\mathbf{y}^* = \mathbf{B}^*\mathbf{y}^* + \mathbf{\Gamma}^*\mathbf{x}^* + \boldsymbol{\zeta}^* \tag{4.75}$$

For equation (4.74) in the original scale and for (4.75) in the transformed scale to be equivalent, the original one is first premultiplied by $\mathbf{D}_y$:

$$\mathbf{y}^* = \mathbf{D}_y\mathbf{y} = \mathbf{D}_y\mathbf{B}\mathbf{y} + \mathbf{D}_y\mathbf{\Gamma}\mathbf{x} + \mathbf{D}_y\boldsymbol{\zeta} \tag{4.76}$$

The left-hand side of (4.76) matches the left-hand side of (4.75). To match the right-hand sides, the three terms in (4.75) must equal the corresponding terms of (4.76). For instance, $\mathbf{B}^*\mathbf{y}^*$ must equal $\mathbf{D}_y\mathbf{B}\mathbf{y}$. For this to hold, $\mathbf{B}^*$ must equal $\mathbf{D}_y\mathbf{B}\mathbf{D}_y^{-1}$. Similarly, we must have $\mathbf{\Gamma}^* = \mathbf{D}_y\mathbf{\Gamma}\mathbf{D}_x^{-1}$, $\boldsymbol{\zeta}^* = \mathbf{D}_y\boldsymbol{\zeta}$, $\mathbf{\Phi}^* = \mathbf{D}_x\mathbf{\Phi}\mathbf{D}_x$, and $\mathbf{\Psi}^* = \mathbf{D}_y\mathbf{\Psi}\mathbf{D}_y$. Parameters and their estimators that have these properties are scale free. This means that we can easily move from the values for the transformed and untransformed data, by knowing the diagonal scaling matrices. In general, F_{ML} is scale free. Thus, when the scale of one or more variables is changed, the ML estimates from the models with the transformed and untransformed variables generally have a simple relationship. The exceptions are mostly cases where parameter elements are constrained to nonzero constant values or are subject to equality or inequality restrictions that run counter to the new scalings (Browne 1982, 75–77; Swaminathan and Algina 1978). The equation $y_1 = x_1 + \zeta_1$ provides an example. If γ_{11} is kept at one and the correlation matrix is analyzed rather than the covariance matrix, the parameters and estimates are not scale free. Examples such as this are the exception rather than the rule.

A final important aspect of the F_{ML} estimator is that it provides a test of overall model fit for overidentified models. The asymptotic distribution of $(N - 1)F_{ML}$ is a χ^2 distribution with $\frac{1}{2}(p + q)(p + q + 1) - t$ degrees of freedom, where t is the number of free parameters and F_{ML} is the value of the fitting function evaluated at the final estimates. The null hypothesis of the chi-square test is H_0: $\Sigma = \Sigma(\boldsymbol{\theta})$. This implies that the overidentifying restrictions for the model are correct. Rejection of H_0 suggests that at least one restriction is in error so that $\Sigma \neq \Sigma(\boldsymbol{\theta})$. I will discuss the chi-square test in more detail in Chapter 7 but, for now, realize that its justification depends on having a sufficiently large sample, on the multinormality of the observed variables, and on the validity of $\Sigma = \Sigma(\boldsymbol{\theta})$.

The $y_1 = x_1 + \zeta_1$ example provides an illustration of an overidentified model. The covariance structure $\Sigma = \Sigma(\boldsymbol{\theta})$ implies that $\phi_{11} = \text{COV}(x_1, y_1)$ and $\phi_{11} = \text{VAR}(x_1)$. From this, the overidentifying restriction is that $\text{COV}(x_1, y_1) = \text{VAR}(x_1)$. If the model is correct, this relation holds in the population but need not be exact in any sample. The chi-square test determines the probability that the overidentification restriction is true given $\mathbf{S}$. Using the previous values of $\hat{\phi}_{11} = 8$, $\hat{\psi} = 8$, $\text{var}(y_1) = 10$, $\text{cov}(x_1, y_1) = 5$, and $\text{var}(x_1) = 8$, F_{ML} is 0.152. With $N = 201$, the chi-square estimate is 30.4, with one $(= (\frac{1}{2})(2)(3) - 2)$ degree of freedom. The chi-square estimate is highly statistically significant ($p < 0.001$), making it unlikely that the model is valid.

Unweighted Least Squares (ULS)

The ULS fitting function is

$$F_{\text{ULS}} = (\tfrac{1}{2})\text{tr}\left[(\mathbf{S} - \boldsymbol{\Sigma}(\boldsymbol{\theta}))^2\right] \tag{4.77}$$

Although it may not be obvious, F_{ULS} minimizes one-half the sum of squares of each element in the residual matrix $(\mathbf{S} - \boldsymbol{\Sigma}(\boldsymbol{\theta}))$. An analogy to ordinary least squares regression is evident. In OLS the sum of squares of the residual term is minimized. The error is the discrepancy between the observed dependent variable and the one predicted by the model. With F_{ULS} we minimize the sum of squares of each element in the residual matrix $(\mathbf{S} - \boldsymbol{\Sigma}(\boldsymbol{\theta}))$. The residual matrix in this case consists of the differences between the sample variances and covariances and the corresponding ones predicted by the model.

To clarify the operation of this function, consider the structural equation $y_1 = x_1 + \zeta_1$. The $\mathbf{S}$ and $\boldsymbol{\Sigma}(\boldsymbol{\theta})$ for this model are in equations (4.60) and (4.59). Substituting $\hat{\boldsymbol{\Sigma}}$ for $\boldsymbol{\Sigma}(\boldsymbol{\theta})$, F_{ULS} is

$$F_{\text{ULS}} = \tfrac{1}{2}\Big(\big(\text{var}(y_1) - \hat{\phi}_{11} - \hat{\psi}_{11}\big)^2 + 2\big(\text{cov}(y_1, x_1) - \hat{\phi}_{11}\big)^2 + \big(\text{var}(x_1) - \hat{\phi}_{11}\big)^2\Big) \tag{4.78}$$

As (4.78) makes clear, estimates for $\hat{\phi}_{11}$ and $\hat{\psi}_{11}$ are selected to reduce the squared discrepancies between $\text{var}(y_1)$ and $(\hat{\phi}_{11} + \hat{\psi}_{11})$, $\text{cov}(y_1, x_1)$ and $\hat{\phi}_{11}$, and the $\text{var}(x_1)$ and $\hat{\phi}_{11}$.

A necessary condition for the minimization of F_{ULS} is that $\hat{\phi}_{11}$ and $\hat{\psi}_{11}$ be chosen so that the partial derivatives of the fitting function with respect to $\hat{\phi}_{11}$ and $\hat{\psi}_{11}$ are zero. The partial derivatives are

$$\frac{\partial F_{\text{ULS}}}{\partial \hat{\phi}_{11}} = -\text{var}(y_1) - 2\,\text{cov}(y_1, x_1) - \text{var}(x_1) + 4\hat{\phi}_{11} + \hat{\psi}_{11} \tag{4.79}$$

$$\frac{\partial F_{\text{ULS}}}{\partial \hat{\psi}_{11}} = -\text{var}(y_1) + \hat{\phi}_{11} + \hat{\psi}_{11} \tag{4.80}$$

Setting (4.79) and (4.80) to zero and solving the two equations for $\hat{\phi}_{11}$ and $\hat{\psi}_{11}$ in terms of the covariances and variances of y_1 and x_1 results in

$$\hat{\phi}_{11} = \frac{\operatorname{var}(x_1) + 2\operatorname{cov}(y_1, x_1)}{3} \tag{4.81}$$

$$\hat{\psi}_{11} = \operatorname{var}(y_1) - \frac{\operatorname{var}(x_1) + 2\operatorname{cov}(y_1, x_1)}{3} \tag{4.82}$$

The matrix formed by taking the second partial derivatives of the fitting function with respect to $\hat{\phi}_{11}$ and $\hat{\psi}_{11}$ is positive-definite, so setting $\hat{\phi}_{11}$ equal to (4.81) and $\hat{\psi}_{11}$ equal to (4.82) minimizes F_{ULS}.

As (4.81) makes clear, the F_{ULS} fitting function gives greater weight to the $\operatorname{cov}(y_1, x_1)$ than to the $\operatorname{var}(x_1)$ in estimating $\hat{\phi}_{11}$. This results because the off-diagonal elements appear twice in the trace of the square of $(\mathbf{S} - \hat{\boldsymbol{\Sigma}})$, whereas the main diagonal elements appear only once [see (4.78)].

With the same sample covariance matrix as before, $\operatorname{var}(y_1) = 10$, $\operatorname{cov}(x_1, y_1) = 5$, and the $\operatorname{var}(x_1) = 8$, the estimates are $\hat{\phi}_{11} = 6$ and $\hat{\psi}_{11} = 4$. Remember that for this model, $(y_1 = x_1 + \zeta_1)$ both the $\operatorname{VAR}(x_1)$ and the $\operatorname{COV}(x_1, y_1)$ should equal ϕ_{11} in the population. The ϕ_{11} is overidentified and is estimated by combining the $\operatorname{var}(x_1)$ and the $\operatorname{cov}(x_1, y_1)$ in a manner that gives greater weight to $\operatorname{cov}(x_1, y_1)$. The ULS $\hat{\phi}_{11}$ of 6 is one unit away from the $\operatorname{cov}(x_1, y_1)$ of 5 and two units away from the $\operatorname{var}(x_1)$ of 8. Both $\hat{\phi}_{11}$ and $\hat{\psi}_{11}$ for ULS are lower than the ML estimates ($\hat{\phi}_{11} = \hat{\psi}_{11} = 8$). Since different fitting functions are minimized, differences are expected, though often the ULS and ML estimates are very close. In more complex models simple ULS solutions for the unknown parameters are unavailable. As with the ML method, iterative numerical techniques are required (see Appendix 4C).

Given its simplicity, F_{ULS} has the advantage of being intuitively pleasing. As with all the fitting functions presented in this book, it leads to a consistent estimator of $\boldsymbol{\theta}$, and this is without the assumption that the observed variables have a particular distribution as long as $\boldsymbol{\theta}$ is identified. Browne (1982) suggests ways to calculate tests of statistical significance for the estimates resulting from ULS. In short, ULS is an easy to understand fitting function that leads to a consistent estimator of $\boldsymbol{\theta}$ for which tests of statistical significance can be developed.

There are disadvantages. First, ULS does not lead to the asymptotically most efficient estimator of $\boldsymbol{\theta}$. The ML estimator has greater efficiency. Second, F_{ULS} is not scale invariant, nor is it scale free. I described these properties in the subsection on F_{ML}. The values of F_{ULS} differ when correlation instead of covariance matrices are analyzed, or more generally, it can differ with any change of scale. The lack of scale freeness means that a relatively simple relation does not exist between the parameters and estimates when different scalings of the variables are used. I illustrate these properties when I consider the empirical examples later in this chapter. Finally, a test of overidentification is not as readily available for F_{ULS} (but see Browne 1984, 1982) as it is for F_{ML}.

Generalized Least Squares (GLS)

F_{ULS} minimizes squared deviations between the observed elements of $\mathbf{S}$ and the corresponding predicted elements of $\boldsymbol{\Sigma}(\boldsymbol{\theta})$ in a fashion analogous to OLS regression. The main difference is that OLS treats the observed and predicted y's for *individual observations*, whereas observed and predicted covariances are the focus of F_{ULS}. A problem with F_{ULS} is that it implicitly weights all elements of $(\mathbf{S} - \boldsymbol{\Sigma}(\boldsymbol{\theta}))$ as if they have the same variances and covariances with other elements. This is similar to the inappropriate application of OLS when the disturbances from a regression equation are heteroscedastic or autocorrelated. The solution in regression analysis is to employ generalized least squares (GLS) which weights observations to correct for the unequal variances or nonzero covariances of the disturbances. Building on this analogy, it would seem reasonable to apply a GLS fitting function that weights the elements of $(\mathbf{S} - \boldsymbol{\Sigma}(\boldsymbol{\theta}))$ according to their variances and covariances with other elements.

A general form of the GLS fitting function is

$$F_{GLS} = \left(\tfrac{1}{2}\right)\operatorname{tr}\left(\left\{[\mathbf{S} - \boldsymbol{\Sigma}(\boldsymbol{\theta})]\mathbf{W}^{-1}\right\}^2\right) \tag{4.83}$$

where $\mathbf{W}^{-1}$ is a weight matrix for the residual matrix. The weight matrix $\mathbf{W}^{-1}$ is either a random matrix that converges in probability to a positive-definite matrix as $N \to \infty$, or it is a positive-definite matrix of constants. F_{ULS} is a special case of F_{GLS}, where $\mathbf{W}^{-1} = \mathbf{I}$. The $\hat{\boldsymbol{\theta}}$ from F_{GLS} with any $\mathbf{W}^{-1}$ that satisfies the preceding condition does have some desirable characteristics. Like F_{ML} and F_{ULS}, $\hat{\boldsymbol{\theta}}$ from F_{GLS} is a consistent estimator of $\boldsymbol{\theta}$. Also the asymptotic distribution of $\hat{\boldsymbol{\theta}}$ is multinormal with a known asymptotic covariance matrix so that tests of statistical significance are possible (see Browne 1982; 1984). Not all choices of $\mathbf{W}^{-1}$ lead to efficient estimators, however, and the asymptotic covariance matrix can be quite complicated. The F_{GLS} with $\mathbf{W}^{-1} = \mathbf{I}$ (i.e., F_{ULS}) is an example.

Two assumptions about the elements of $\mathbf{S}$ lead to a simple condition for selecting the "correct" weighting matrix $\mathbf{W}^{-1}$ and to optimal properties for the GLS $\hat{\boldsymbol{\theta}}$. The assumptions are (1) $E(s_{ij}) = \sigma_{ij}$ and (2) the asymptotic distribution of the elements of $\mathbf{S}$ is multinormal with means of σ_{ij} and asymptotic covariances of s_{ij} and s_{gh} equal to $N^{-1}(\sigma_{ig}\sigma_{jh} + \sigma_{ih}\sigma_{jg})$. The first assumption simply requires that the $E(s_{ij})$ exists and that s_{ij} is the unbiased estimator of σ_{ij}. The multinormal asymptotic distribution of the elements of $\mathbf{S}$ is quite general. A sufficient condition for it to hold is that the observations are independent and identically distributed and that the fourth-order moments of $\mathbf{x}$ and $\mathbf{y}$ exist. The key assumption is that $\text{ACOV}(s_{ij}, s_{gh}) = N^{-1}(\sigma_{ig}\sigma_{jh} + \sigma_{ih}\sigma_{jg})$. This is satisfied if $\mathbf{x}$ and $\mathbf{y}$ are multinormally distributed, but it also is true for other distributions without excessive kurtosis (Browne 1974). I will say more about kurtosis and estimators with good properties with nonnormal data in Chapter 9, but for now, note that these assumptions are somewhat less restrictive than those made to derive F_{ML}.

If the assumptions are satisfied, then $\mathbf{W}^{-1}$ should be chosen so that $\text{plim}\, \mathbf{W}^{-1} = c\boldsymbol{\Sigma}^{-1}$, where c is any constant (typically $c = 1$). The GLS $\hat{\boldsymbol{\theta}}$ with $\mathbf{W}^{-1}$ so selected has many of the same desirable asymptotic properties as that of $\hat{\boldsymbol{\theta}}$ from F_{ML}. That is, the GLS estimator has an asymptotic multinormal distribution, and it is asymptotically efficient. Furthermore, the asymptotic covariance matrix of $\hat{\boldsymbol{\theta}}$ from F_{GLS} takes a much simpler form than is true when $\text{plim}\, \mathbf{W}^{-1} \neq c\boldsymbol{\Sigma}^{-1}$. The asymptotic covariance matrix of $\hat{\boldsymbol{\theta}}$ is $(2/(N-1))$ times the inverse of the expected value of the information matrix: $(2/(N-1))[E(\partial^2 F_{\text{GLS}}/\partial\boldsymbol{\theta}\,\partial\boldsymbol{\theta}')]^{-1}$ (Jöreskog 1981). Estimated asymptotic covariances and standard errors are available by substituting $\hat{\boldsymbol{\theta}}$ for $\boldsymbol{\theta}$.

Although many $\mathbf{W}^{-1}$ are consistent estimators of $\boldsymbol{\Sigma}^{-1}$, the most common choice is $\mathbf{W}^{-1} = \mathbf{S}^{-1}$:

$$
\begin{aligned}
F_{\text{GLS}} &= \left(\tfrac{1}{2}\right)\text{tr}\left(\left\{[\mathbf{S} - \boldsymbol{\Sigma}(\boldsymbol{\theta})]\mathbf{S}^{-1}\right\}^2\right) \\
&= \left(\tfrac{1}{2}\right)\text{tr}\left\{\left[\mathbf{I} - \boldsymbol{\Sigma}(\boldsymbol{\theta})\mathbf{S}^{-1}\right]^2\right\}
\end{aligned}
\tag{4.84}
$$

This F_{GLS} is found in both LISREL and EQS. Henceforth I use F_{GLS} to refer to the GLS fitting function with $\mathbf{W}^{-1} = \mathbf{S}^{-1}$ unless otherwise noted.

Another choice of $\mathbf{W}^{-1}$ where $\text{plim}\, \mathbf{W}^{-1} = \boldsymbol{\Sigma}^{-1}$ is $\mathbf{W}^{-1} = \boldsymbol{\Sigma}^{-1}(\hat{\boldsymbol{\theta}})$, where $\hat{\boldsymbol{\theta}}$ minimizes F_{ML}. Applying $F_{\text{GLS}}(\mathbf{W}^{-1} = \boldsymbol{\Sigma}^{-1}(\hat{\boldsymbol{\theta}}))$ leads to the same GLS $\hat{\boldsymbol{\theta}}$ as found with F_{ML} (Lee and Jennrich 1969). In this sense F_{ML} is a member of F_{GLS} as given in (4.83). As a consequence, the asymptotic properties of the ML $\hat{\boldsymbol{\theta}}$ do not depend on the multinormality of $\mathbf{y}$ and $\mathbf{x}$ which I used to derive F_{ML} (see Appendixes 4A and 4B). Rather, the asymptotic normality and asymptotic efficiency follow under the less restrictive assumptions that justified the GLS estimator with $\text{plim}\, \mathbf{W}^{-1} = \boldsymbol{\Sigma}^{-1}$.

Unlike F_{ULS} but like F_{ML}, F_{GLS} is scale invariant and scale free. These desirable properties were described in the F_{ML} section, and they hold here as well. An added benefit of F_{GLS} is that $(N-1)F_{\text{GLS}}$ evaluated at the final estimates has an asymptotic chi-square distribution when the model is correct. The df are $(1/2)(p+q)(p+q+1)-t$. Thus, like $(N-1)F_{\text{ML}}$, $(N-1)F_{\text{GLS}}$ approximates a chi-square variate in large samples provided that $\operatorname{plim} \mathbf{W}^{-1} = \boldsymbol{\Sigma}^{-1}$. If the model is valid, $(N-1)F_{\text{ML}}$ and $(N-1)F_{\text{GLS}}$ are asymptotically equivalent so that in large samples these estimated chi-squares should be close.

Consider again the example of $y_1 = x_1 + \zeta_1$ with $\operatorname{var}(y_1) = 10$, $\operatorname{cov}(x_1, y_1) = 5$, and $\operatorname{var}(x_1) = 8$. The GLS estimators are $\hat{\phi}_{11} = 6.03$ and $\hat{\psi}_{11} = 6.03$. These are lower than the ML estimates of 8 for both parameters and also different from the ULS estimates of 6 and 4, respectively. With an N of 201 the chi-square estimate $(N-1)F_{\text{GLS}}$ is 24.7 with 1 df which is a bit lower than the 30.4 estimate for the ML solution. Both chi-square estimates are highly significant suggesting that H_0: $\boldsymbol{\Sigma} = \boldsymbol{\Sigma}(\boldsymbol{\theta})$ should be rejected.

Though F_{GLS} has desirable properties, it too has its limits. If the distribution of the observed variables has very "fat" or "thin" tails, the asymptotic covariances of s_{ij} and s_{gh} can deviate from $N^{-1}(\sigma_{ig}\sigma_{jh} + \sigma_{ih}\sigma_{jg})$ so that this assumption is violated. If so, the usual asymptotic standard errors and chi-square tests need no longer be accurate for significance testing. Studies of robustness of these estimators are being developed (e.g., Satorra and Bentler 1986). Another consideration is that even when the assumptions for $\mathbf{S}$ are met, the properties of the estimators are asymptotic. Very little is known about the small sample behavior of $\hat{\boldsymbol{\theta}}$ from F_{GLS}, but it appears that it has a bias toward zero in small samples.

Other Estimators

Those familiar with regression and econometric procedures are familiar with other estimators for structural equations with observed variables. Ordinary least squares (OLS) is appropriate for recursive systems. For nonrecursive models, two-stage least squares (2SLS) is the most common estimation procedure, but others such as three-stage least squares (3SLS), full-information maximum likelihood (FIML) are well-known. Although traditional econometric procedures and the ones covered in this chapter are distinct in several ways, there are some points in common. For instance, Jöreskog (1973) shows that the $\mathbf{F}_{\text{ML}}$ is equivalent to FIML estimation for the observed variable models covered here. Also, for recursive models, the OLS estimator is the maximum likelihood estimator, provided the ζ's are normally distributed. For further information on OLS, 2SLS, 3SLS, and related procedures, see Johnston (1984), Wonnacott and Wonnacott (1979), or Fox (1984).

Empirical Examples

I illustrate the ML, ULS, and GLS estimators with the socioeconomic status, union sentiment, and marginal propensity to consume examples. The path diagram for the socioeconomic status model is in Figure 4.4. It presents the relation between three self-perception measures of socioeconomic status [perceived income (y_1), perceived occupational prestige (y_2), and perceived overall status (y_3)] and two objective ones [income (x_1) and occupational prestige (x_2)]. The matrix representation of the path diagram is in (4.30). The rank and order conditions establish the model's identification. Since the number of unknowns in $\boldsymbol{\theta}$ equals the number of nonredundant covariance elements in $\mathbf{S}$, the model is exactly identified.

I utilize data provided in Kluegel et al. (1977)[6] and restrict my analysis to the covariance matrix for the sample of whites ($N = 432$). The covariance matrix for y_1, y_2, y_3, x_1, and x_2 is

$$\mathbf{S} = \begin{bmatrix} 0.449 & & & & \\ 0.166 & 0.410 & & & \\ 0.226 & 0.173 & 0.393 & & \\ 0.564 & 0.259 & 0.382 & 4.831 & \\ 2.366 & 3.840 & 3.082 & 13.656 & 452.711 \end{bmatrix}$$

Appendix 4D lists the LISREL and EQS programs to estimate the parameters with the ML fitting function.

Table 4.2 reports the estimates for the above model resulting from the ML, ULS, and GLS fitting functions. All sets of estimates are identical. This is because the model is exactly identified and $\hat{\boldsymbol{\Sigma}}$ must equal $\mathbf{S}$ for exactly identified models. To produce the same $\hat{\boldsymbol{\Sigma}}$, the ML, ULS, and GLS estimates must be the same.

Do the parameter estimates make sense? At a minimum the directions of effects seem reasonable. The $\hat{\gamma}_{11}$ and $\hat{\gamma}_{22}$ show that actual income (x_1) has a positive effect on an individual's perception of their ranking in income (y_1) and that occupational prestige (x_2) positively affects subjective prestige (y_2).[7] Similarly, $\hat{\beta}_{31}$ and $\hat{\beta}_{32}$ show that the greater one's perceived income

[6]The sample of respondents is from the 1969 Gary Area Project of the Institute for Social Research at Indiana University.

[7]The 0.007 estimate of γ_{22} is a result of the larger-scale values for actual occupational prestige compared to the subjective prestige variable. The former (x_2) has a mean of 36.7 and a standard deviation of 21.3, and the latter (y_2) has a mean of 1.5 and a standard deviation of 0.64.

Table 4.2 The ML, ULS, and GLS Estimates for Subjective and Objective Social Status Model (N = 432)

Parameter	ML, ULS, and GLS Estimate (standard error)
β_{21}	0.29 (0.12)
β_{12}	0.25 (0.18)
β_{31}	0.43 (0.14)
β_{32}	0.54 (0.19)
γ_{11}	0.10 (0.02)
γ_{22}	0.007 (0.001)
ϕ_{11}	4.83 (0.33)
ϕ_{21}	13.66 (2.35)
ϕ_{22}	452.71 (30.84)
ψ_{11}	0.34 (0.03)
ψ_{21}	−0.06 (0.08)
ψ_{22}	0.33 (0.02)
ψ_{31}	−0.03 (0.05)
ψ_{32}	−0.10 (0.06)
ψ_{33}	0.29 (0.04)
$R^2_{y_1}$	0.24
$R^2_{y_2}$	0.20
$R^2_{y_3}$	0.26
Coefficient of Determination	0.32

(y_1) and perceived occupational status (y_2), the higher one tends to perceive their overall social status (y_3). Finally, $\hat{\beta}_{21}$, and $\hat{\beta}_{12}$ indicate a positive reciprocal relation for subjective income (y_1) and subjective occupational prestige (y_2). An estimated asymptotic standard error is available for the ML and GLS estimates. These are identical for both estimators and are in parentheses below the coefficients. The ratio of the ML and GLS estimate to the asymptotic standard error approximates a Z-test statistic for the null hypothesis that the coefficient is zero. An estimate roughly twice its standard error is statistically significant at an alpha of 0.05 for a two-tail test. All estimates, except $\hat{\beta}_{12}$, $\hat{\psi}_{21}$, $\hat{\psi}_{31}$, and $\hat{\psi}_{32}$, meet this criterion. Although none of the off-diagonal elements of $\hat{\Psi}$ are statistically significant, it is surprising to find that they are negative. If there were any nonzero covariances between equation disturbances, I would expect them to be positive. The estimate of subjective occupational prestige's (y_2) influence on subjective income (y_1) is less than one and a half times its standard error. This raises doubts about the validity of the hypothesis that subjective occupation (y_2) affects subjective income (y_1).

Beneath the coefficient estimates are the $R^2_{y_i}$'s and the coefficients of determination. The $R^2_{y_i}$'s are calculated as[8]

$$R^2_{y_i} = 1 - \frac{\hat{\psi}_{ii}}{\hat{\sigma}^2_{y_i}} \tag{4.85}$$

Where $\hat{\sigma}^2_{y_i}$ is the predicted variance of y_i from $\hat{\Sigma}$. For proper solutions $\hat{\psi}_{ii}$ is nonnegative so that $R^2_{y_i}$ has a maximum of one. The closer to one, the better is the fit. The $R^2_{y_i}$'s from Table 4.2 are moderate in size compared to the squared multiple correlation coefficients typical in cross-sectional survey research.

The coefficient of determination is

$$\text{coefficient of determination} = 1 - \frac{|\hat{\Psi}|}{|\hat{\Sigma}_{yy}|} \tag{4.86}$$

where $\hat{\Sigma}_{yy}$ is the covariance matrix of y taken from $\hat{\Sigma}$. The $|\hat{\Psi}|$ is the determinant of the estimate of the covariance matrix for the equation errors. This shows the joint effect of the model variables on the endogenous variables. It varies between zero and one.

[8]In LISREL VI the R^2 calculation uses the sample variance of y_i in the denominator instead of $\hat{\sigma}^2_{y_i}$ in (4.85). In many but not all cases these will be the same.

The coefficient of determination has some properties of which the researcher should be aware. One is that it can be sensitive to small numbers in the main diagonal of $\hat{\Psi}$. For instance, suppose that

$$\hat{\Psi} = \begin{bmatrix} 0 & & \\ 0 & 10 & \\ 0 & 0 & 10 \end{bmatrix} \tag{4.87}$$

This might occur if equation one is an identity relation. The $|\hat{\Psi}|$ is zero and the coefficient of determination is one even with equations two and three having possibly large unexplained variance terms. In a less extreme case one equation might have a relatively small $\hat{\psi}_{ii}$ term, while the other equations have large unexplained variances. The coefficient of determination could be misleadingly small because of the one low $\hat{\psi}_{ii}$.

A second characteristic to keep in mind is illustrated with the following example:

$$\hat{\Psi} = \begin{bmatrix} 0.5 & 0 \\ 0 & 0.5 \end{bmatrix}, \qquad \hat{\Sigma}_{yy} = \begin{bmatrix} 1 & 0.5 \\ 0.5 & 1 \end{bmatrix} \tag{4.88}$$

$$\hat{\Psi} = \begin{bmatrix} 0.5 & 0 & 0 \\ 0 & 0.5 & 0 \\ 0 & 0 & 0.5 \end{bmatrix}, \quad \hat{\Sigma}_{yy} = \begin{bmatrix} 1 & 0.5 & 0.5 \\ 0.5 & 1 & 0.5 \\ 0.5 & 0.5 & 1 \end{bmatrix} \tag{4.89}$$

In (4.88) the R^2's for the first and second equations are 0.5. The R^2's for all three equations in (4.89) also are 0.5. Yet, the coefficient of determinations differ, being 0.67 for the two-equation model and 0.75 for the three-equation model. So, even if a model has identical R^2's for each equation, it is possible for the coefficient of determination to be greater for the model with more equations. The coefficients of determination for the example is 0.32 and moderate in size. This value is higher than any of the R^2's for the individual equations so it is not a good gauge of the "typical" R^2 in the model.

In sum, based on the sign and significance of the coefficient estimates, the $R^2_{y_i}$'s, and the coefficients of determination, the model seems consistent with the data. The major exceptions are the nonsignificance of $\hat{\beta}_{12}$ and the off-diagonal elements of $\hat{\Psi}$. Since the model is exactly identified the chi-square test of overidentification cannot help in the assessment of model fit.

The second example analyzes data from McDonald and Clelland (1984) on union sentiment (see Figure 4.2). The sample consists of 173 southern textile workers and the five variables are deference (or submission) to managers (y_1), support of labor activism (y_2), sentiment toward unions (y_3), years in mill (x_1), and age (x_2). The matrix equation for the model is

$$\begin{bmatrix} y_1 \\ y_2 \\ y_3 \end{bmatrix} = \begin{bmatrix} 0 & 0 & 0 \\ \beta_{21} & 0 & 0 \\ \beta_{31} & \beta_{32} & 0 \end{bmatrix} \begin{bmatrix} y_1 \\ y_2 \\ y_3 \end{bmatrix} + \begin{bmatrix} 0 & \gamma_{12} \\ 0 & \gamma_{22} \\ \gamma_{31} & 0 \end{bmatrix} \begin{bmatrix} x_1 \\ x_2 \end{bmatrix} + \begin{bmatrix} \zeta_1 \\ \zeta_2 \\ \zeta_3 \end{bmatrix}$$

As shown in the last section, this model is identified based on the recursive rule. The covariance matrix is

$$\mathbf{S} = \begin{bmatrix} 14.610 & & & & \\ -5.250 & 11.017 & & & \\ -8.057 & 11.087 & 31.971 & & \\ -0.482 & 0.677 & 1.559 & 1.021 & \\ -18.857 & 17.861 & 28.250 & 7.139 & 215.662 \end{bmatrix}$$

The order of the variables in $\mathbf{S}$ from top to bottom is y_1, y_2, y_3, x_1, and x_2. These data were screened for outliers with the procedures I described in Chapter 2. I found several outliers but their influence was minor.

Table 4.3 reports the ULS, GLS, and ML estimates for the model. The negative coefficient for age in the deference (y_1) equation suggests that submissive attitudes are less for older than for younger workers. Age (x_2) has a positive effect ($\hat{\gamma}_{22}$) and deference a negative one ($\hat{\beta}_{21}$) on activism (y_2). Union sentiment (y_3) is inversely related to deference (see $\hat{\beta}_{31}$) and positively related to activism (see $\hat{\beta}_{32}$) and to years in the mill (see $\hat{\gamma}_{31}$). Only about 10% of the variance in deference ($R^2_{y_1}$) is explained by age while about 23% of the variance in activism ($R^2_{y_2}$) is accounted for by age and deference. The highest R^2 is about 0.4 for union sentiment with years in the mill, deference, and activism the explanatory variables. All GLS and ML estimates are at least twice their standard errors. The GLS and ML estimates and standard errors are very close. The ULS estimates are nearly the same except for $\hat{\gamma}_{31}$, $\hat{\beta}_{31}$, $\hat{\psi}_{33}$, and $\hat{\phi}_{11}$. This may be due to the lack of either scale invariance or scale freeness for ULS and the large variance for age compared to other variables in the analysis.

To illustrate the scale dependency of ULS, I form $\mathbf{y}^* = \mathbf{D}_y\mathbf{y}$ and $\mathbf{x}^* = \mathbf{D}_x\mathbf{x}$, where

$$\mathbf{D}_y = \begin{bmatrix} 1 & 0 & 0 \\ 0 & 1 & 0 \\ 0 & 0 & 1 \end{bmatrix}, \qquad \mathbf{D}_x = \begin{bmatrix} 1 & 0 \\ 0 & 0.1 \end{bmatrix} \tag{4.90}$$

Table 4.3 The ULS, GLS, and ML Estimates for Union Sentiment Model ($N = 173$)

Parameter	ULS Estimate	GLS Estimate (standard error)	ML Estimate (standard error)
γ_{12}	−0.087	−0.088	−0.087
		(0.019)	(0.019)
γ_{22}	0.058	0.058	0.058
		(0.016)	(0.016)
γ_{31}	1.332	0.861	0.860
		(0.342)	(0.340)
β_{21}	−0.284	−0.288	−0.285
		(0.062)	(0.062)
β_{31}	−0.192	−0.214	−0.218
		(0.098)	(0.097)
β_{32}	0.846	0.853	0.850
		(0.112)	(0.112)
ψ_{11}	12.959	12.904	12.961
		(1.395)	(1.398)
ψ_{22}	8.475	8.444	8.489
		(0.913)	(0.915)
ψ_{33}	18.838	19.343	19.454
		(2.092)	(2.098)
ϕ_{11}	0.774	1.014	1.021
		(0.110)	(0.110)
ϕ_{21}	7.139	7.139	7.139
		(1.256)	(1.256)
ϕ_{22}	215.663	214.843	215.662
		(23.23)	(23.255)
$R^2_{y_1}$	0.113	0.113	0.113
$R^2_{y_2}$	0.229	0.233	0.229
$R^2_{y_3}$	0.410	0.392	0.390
Coefficient of Determination	0.229	0.206	0.205
$\chi^2(\text{df} = 3)$	—	1.25	1.26

The impact of this is that $\mathbf{y}^*$ and x_1^* keep the same scales as $\mathbf{y}$ and x_1, respectively, but x_2^* (age) is multipled by 0.1 to reduce the magnitude of its variance. The ULS fitting function applied to this new covariance matrix with the scale of x_2 reduced does not lead to scale free estimates. For instance, the $\hat{\Gamma}$ corresponding to the original variables and $\hat{\Gamma}^*$ for the rescaled ones are

$$\hat{\Gamma} = \begin{bmatrix} 0 & -0.087 \\ 0 & 0.058 \\ 1.332 & 0 \end{bmatrix}, \quad \hat{\Gamma}^* = \begin{bmatrix} 0 & -0.839 \\ 0 & 0.604 \\ 1.096 & 0 \end{bmatrix} \tag{4.91}$$

If these were scale free estimates, then $\hat{\Gamma}^* = \mathbf{D}_y \hat{\Gamma} \mathbf{D}_x^{-1}$. As is easily verified, this relation does not hold. Note also that $\hat{\gamma}_{31}$ and $\hat{\gamma}_{31}^*$ are not equal, even though I changed neither y_3 nor x_1's scales.

The ML counterparts to (4.91) are

$$\hat{\Gamma} = \begin{bmatrix} 0 & -0.087 \\ 0 & 0.058 \\ 0.860 & 0 \end{bmatrix}, \quad \hat{\Gamma}^* = \begin{bmatrix} 0 & -0.874 \\ 0 & 0.579 \\ 0.860 & 0 \end{bmatrix} \tag{4.92}$$

Within the margins of rounding error $\hat{\Gamma}^* = \mathbf{D}_y \hat{\Gamma} \mathbf{D}_x^{-1}$ which is expected since the ML estimates are scale free. Furthermore the value for the ULS fit function evaluated at its final estimates differs for the original and for the rescaled variables, though they are the same for the ML fit function. This example illustrates the disadvantages of the scale dependency of ULS.

A chi-square estimate is readily available for the GLS and ML solutions. Both chi-square estimates are about 1.3 with 3 df so that the H_0: $\Sigma = \Sigma(\boldsymbol{\theta})$ cannot be rejected, and the implied covariance matrix, $\hat{\Sigma}$, does an excellent job of reproducing the sample covariance matrix, $\mathbf{S}$.

The final empirical example is Haavelmo's model of the marginal propensity to consume. The model is

$$\begin{bmatrix} y_1 \\ y_2 \end{bmatrix} = \begin{bmatrix} 0 & 1 \\ \beta_{21} & 0 \end{bmatrix} \begin{bmatrix} y_1 \\ y_2 \end{bmatrix} + \begin{bmatrix} 1 \\ 0 \end{bmatrix} [x_1] + \begin{bmatrix} 0 \\ \zeta_2 \end{bmatrix}$$

where y_1 is per capita disposable income, y_2 is per capita consumption expenditures, and x_1 is per capita investment expenditures. The sample covariance matrix is

$$\mathbf{S} = \begin{bmatrix} 3421 & & \\ 2504 & 1889 & \\ 916 & 616 & 301 \end{bmatrix}$$

The $\mathbf{S}$ is singular since $y_1 = y_2 + x_1$ and a singular $\mathbf{S}$ is a problem for F_{ML} and F_{GLS} which involve $\log|\mathbf{S}|$ and $\mathbf{S}^{-1}$, respectively. A simple solution is to remove one of the variables involved in the singularity (e.g., y_1) and show (with the coefficient matrices and error vector) that the removed variable is an exact linear function of the other variables (see also Jöreskog and Sorbom 1986b, III.42–III.54). The ML estimates for the second equation are

$$y_2 = \underset{(0.035)}{0.672}\, y_1 + \zeta_2$$

This means that for every dollar change in per capita disposable income, on average about 67% goes toward consumption. Before leaving this example, recall that in Chapter 2 I found that 1941 was an outlier for these data. To see if it influences the model fit, I reestimated the equation omitting this case:

$$y_2 = \underset{(0.038)}{0.715}\, y_1 + \zeta_2$$

Dropping the 1941 observation leads to a higher estimate of the marginal propensity to consume.

In summary, this section presents and illustrates three fitting functions that are consistent estimators of $\boldsymbol{\theta}$. The ML and GLS ones are asymptotically efficient when the assumption of multinormality holds or when the distribution of the variables have normal kurtosis, whereas the ULS generally is inefficient. The estimates from all fitting functions were generally close for the examples, though this was not true for all parameters.

FURTHER TOPICS

In this section I discuss several other topics that arise when utilizing observed variable models. These are standardized and unstandardized coefficients, alternative assumptions for $\mathbf{x}$, interaction terms, and equations with intercepts.

Standardized and Unstandardized Coefficients

The $\hat{\mathbf{B}}$ and $\hat{\boldsymbol{\Gamma}}$ from ML, ULS, and GLS estimators reported in Tables 4.2 and 4.3 contain unstandardized coefficients that depend upon the units in which the variables are scaled. If two explanatory variables in an equation have the same units then a comparison of their unstandardized coefficients provides an idea of their relative influence. For instance, consider the equation

$$Y_1 = \alpha_1 + \gamma_{11} X_1 + \gamma_{12} X_2 + \zeta_1 \tag{4.93}$$

where Y_1 is job satisfaction, X_1 is yearly income, and X_2 is yearly salary bonus. Suppose X_1 and X_2 are measured in hundreds of dollars. Since income and bonus have the same units, we can compare γ_{11} and γ_{12} to see which has the greatest effect on job satisfaction for the identical increase in dollars.

Often explanatory variables have different scales. This makes the assessment of relative direct influences difficult. Suppose that in (4.93) X_2 is years of education instead of salary bonus. A comparison of γ_{11} and γ_{12} is not very informative since X_1 is in dollars while X_2 is in years. Or, for the union sentiment equation in Table 4.3, it is difficult to compare the effect of activism and years in mill since their units differ. To lessen this difficulty, analysts adjust the unstandardized coefficients to make them "dimensionless." That is, analysts modify the coefficients so that they are in similar units. Economists frequently use the *elasticities* of variables defined as the expected percentage change in the dependent variable for a 1% change in an explanatory variable net of the other explanatory variables. Elasticity at a point is $(\partial Y/\partial X)(X/Y)$. For linear models such as those treated here, $(\partial Y/\partial X)$ equals the unstandardized coefficient associated with X. You also must choose a point (X/Y) at which to evaluate elasticity. The most common choice is at the sample means $\bar{X}/\bar{Y}$. For equation (4.93), the estimates for the elasticities at the means for X_1 and X_2 are $\hat{\gamma}_{11}(\bar{X}_1/\bar{Y}_1)$ and $\hat{\gamma}_{12}(\bar{X}_2/\bar{Y}_1)$. Comparing these, you can estimate which 1% change in the independent variables leads to a larger percent change in the dependent variable. Since both changes are in percentage units, it is easier to compare the effects, than when they are in their original units. With the exception of (4.93), the examples I have introduced do not have elasticities at the mean since the variables are in deviation form with means of zero. Even with raw scores, the elasticities may not be meaningful. Elasticities require that X and Y have an unambiguous zero point and equal intervals between adjacent points (i.e., ratio level measures).

When this is not true, sociologists, psychologists, and political scientists frequently use another dimensionless gauge of relative influence called the *standardized* regression coefficient. The standardized coefficient for $\hat{\beta}_{ij}$ is $\hat{\beta}_{ij}(\hat{\sigma}_{Y_j}/\hat{\sigma}_{Y_i})$ that for $\hat{\gamma}_{ij}$ is $\hat{\gamma}_{ij}(\hat{\sigma}_{X_j}/\hat{\sigma}_{Y_i})$, where $\hat{\sigma}_{Y_j}$, $\hat{\sigma}_{Y_i}$, and $\hat{\sigma}_{X_i}$ are the model predicted standard deviations of Y_j, Y_i, and X_j. The standardized coefficient shows the mean response in standard deviation units of the dependent variable for a one standard deviation change in an explanatory variable, holding constant the other variables in a model. Thus you compare the shift in standard deviation units of the dependent variable that accompanies shifts of one standard deviation in the explanatory variables as a means to assess relative effects.

There is an interesting similarity in the elasticity at the mean and the standardized regression coefficients. They differ only in the ratio that multiples the unstandardized coefficient. For elasticity, it is $(\bar{X}/\bar{Y})$, but for the standardized coefficient, it is $(\hat{\sigma}_X/\hat{\sigma}_Y)$. We can interpret the elasticity at the mean as the shift in Y that accompanies a change in X from 0 to $\bar{X}$, as a proportion of $\bar{Y}$. Whereas the standardized regression coefficient is the shift in Y that accompanies a change in X from 0 to a value of $\hat{\sigma}_X$, as a proportion of $\hat{\sigma}_Y$. From this perspective, the standardized regression coefficient is a special case of elasticity evaluated at the point, where X is set to the value of $\hat{\sigma}_X$ and Y is set to $\hat{\sigma}_Y$.

Several cautionary points need to be noted. First, the assessment of relative impact via standardized coefficients need not be the same as that for elasticities at the means. Neither is a definitive measure of the *importance* of a variable. The researcher must decide which, if either, measure is appropriate to the problem being studied. Second, the unstandardized coefficient could be multiplied by $(\hat{\sigma}_X/\hat{\sigma}_Y)$ or by (s_X/s_Y), where s_X and s_Y are the square roots of sample estimates of the variance of X and Y. In exactly identified models (e.g., the typical regression equation) there is no difference. However, in some overidentified cases the constraints imposed may be such that the model implied standard deviations do not equal the sample standard deviations. Also in some overidentified models the closeness of the implied standard deviation to the sample standard deviation depends on the fitting function that is optimized. For consistency, I will always use the ratio of model implied standard deviations.

Another source of confusion is the testing of statistical significance for standardized coefficients or elasticities. A common practice with standardized coefficients is to perform a multiple regression analysis based on the correlation matrix of variables. This results in standardized regression coefficients, but the standard error associated with the coefficient is not correct when $\mathbf{x}$ is random. The problem is that the standardized coefficient is a function of the unstandardized coefficient as well as the ratio of $\hat{\sigma}_X/\hat{\sigma}_Y$. The ratio $\hat{\sigma}_X/\hat{\sigma}_Y$ is a random variable that takes different values from sample to sample. Its distribution helps determine the distribution of the standardized coefficient and this contribution is not taken into account by the usual formulas for the standard errors (Bentler and Lee 1983; Browne 1982). The same point applies to estimates of elasticity at the mean (Valentine, 1980).

An additional point is the hazard of comparing standardized coefficients or elasticities for the same variable across different groups. For instance, a researcher may want to determine if education has the same effect on income for males and females. They may find differences in standardized coefficients only because the sample implied standard deviations differ for males and females even though the unstandardized coefficients are identical. A similar point holds for elasticities; $\bar{X}/\bar{Y}$ can differ in different groups and thus give the appearance that the relation between variables is distinct even when the $\hat{\beta}$'s (or $\hat{\gamma}$'s) are identical. In general, comparisons of a variable's influence in different groups should be made with unstandardized coefficients. I will return to this point in Chapter 8.

As an illustration of standardized coefficients, consider the activism (y_2) equation in Table 4.3. The unstandardized ML coefficients for the effects of age (x_2) and deference (y_1) on activism are 0.058 and -0.285. Age is measured in years, but deference is a composite index so it is difficult to judge the relative impact of these variables. The standardized coefficients are 0.26 for age and -0.33 for deference. This indicates a larger average change in standard deviation units of activism for a one standard deviation difference in deference than for a one standard deviation difference in age net of the other variables.

Alternative Assumptions for x

The derivation of F_{ML} assumes that $(\mathbf{y}', \mathbf{x}')_i'$ is independently sampled from a multinormal distribution where $i = 1, 2, \ldots, N$. In many applications the multinormality assumption, particularly for $\mathbf{x}$, is not even approximately true. For instance, $\mathbf{x}$ may contain dummy variables, interaction, or other product terms of the exogenous variables, and these have nonnormal distributions.

What are the consequences for the F_{ML} estimator of violating the multinormality assumption for $\mathbf{x}$? Surprisingly, even under less restrictive conditions the estimator has desirable properties (see also F_{GLS} section). First, the consistency of the estimators from F_{ML} (or F_{GLS} and F_{ULS}) does not depend on the multinormality of $\mathbf{x}$ (Browne 1982). So as N grows large these estimators eventually converge on the parameters. Of course this says nothing about the validity of the usual tests of statistical significance for the model and coefficients. But, if we can assume that $\mathbf{x}$ and $\boldsymbol{\zeta}$ are independent, that $\boldsymbol{\zeta}$ has a multinormal distribution with a mean of $\mathbf{0}$ and covariance matrix $\boldsymbol{\Psi}$, and that the distribution of $\mathbf{x}$ does not depend on the parameters $\mathbf{B}$, $\boldsymbol{\Gamma}$, and $\boldsymbol{\Psi}$, then the F_{ML} fitting function leads to the same ML estimators as does full-information maximum likelihood (FIML). This follows since under the above assumptions, maximizing the resulting likelihood function for $\mathbf{B}$, $\mathbf{\Gamma}$, and $\mathbf{\Psi}$ leads to the F_{ML} estimator regardless of $\mathbf{x}$'s distribution (Jöreskog 1973, 94; Johnston 1984, 281–285).[9] Thus F_{ML} can be justified under the above assumptions when $\mathbf{x}$ is not multinormal, and we can apply the usual tests of statistical significance.

In this case we still consider $\mathbf{y}$ and $\mathbf{x}$ as random variables sampled independently but instead of both variables having a multinormal distribution, the distribution of $\mathbf{y}$ *conditional on* $\mathbf{x}$ is multivariate normal. When $\mathbf{x}$ is independent of $\boldsymbol{\zeta}$, the above properties hold conditional on *any* values of $\mathbf{x}$ so that the conditional distribution assumption is not as limited as it might appear.

Another assumption often made in classical approaches to regression is that $\mathbf{x}$ is fixed. Fixed $\mathbf{x}$ or nonstochastic exogenous variables means that the set of values for $\mathbf{x}$ remain unchanged from one sample to another. Unlike before, $\mathbf{x}$ is a set of constants, not random variables. However, as before, $\mathbf{y}$ is random. Since $\mathbf{x}$ is the same for each sample, there is no need to assume a population distribution for $\mathbf{x}$. Maintaining the assumption that $\boldsymbol{\zeta}$ has a multinormal distribution with a mean of $\mathbf{0}$ and a covariance matrix of $\mathbf{\Psi}$, the F_{ML} estimators are FIML (Johnston, 1984, 490–492) when $\mathbf{x}$ is fixed. We can view this as a special case of assuming the multinormality of $\mathbf{y}$ conditional on $\mathbf{x}$, but here the distribution of $\mathbf{x}$ is degenerate since it takes the same values in repeated samples.

Typically, the F_{ML} estimator of $\mathbf{\Phi}$ is the unbiased sample covariance matrix for $\mathbf{x}$, $\mathbf{S}_{xx}$. The sample covariances and variances have $N - 1$ in the denominator. When $\mathbf{x}$ is fixed this correction for "sample bias" is not needed. There is no sampling variability in $\mathbf{x}$. The "covariances" of $\mathbf{x}$ stay the same. Therefore the elements in $\mathbf{S}_{xx}$ should be divided by N rather than $N - 1$. A simple adjustment is to multiply the $\mathbf{S}_{xx}$ by $(N - 1)/N$. For large samples, this leads to only minor changes.[10]

In sum, the assumption that $\mathbf{y}$ and $\mathbf{x}$ are sampled independently from a multinormal distribution can be replaced with either of two alternatives that lead to the same ML estimators, standard errors, and tests of signifi-

[9]Generally, the estimator for $\mathbf{\Phi}$ will be the sample covariance matrix of $\mathbf{x}$. In particular, this is true when $\mathbf{x}$ is multinormal.

[10]LISREL VI does not make this adjustment when the "fixed $\mathbf{x}$" option is chosen. Also there is the potential for confusion about the degrees of freedom for the chi-square test of a model when $\mathbf{x}$ is fixed. Normally, the df $= (1/2)(p + q)(p + q + 1) - t$. However, with fixed $\mathbf{x}$ the elements in $\mathbf{\Phi}$ are no longer free parameters that count toward t, and the $(1/2)(q)(q + 1)$ nonredundant elements in $\mathbf{S}_{xx}$ are no longer elements that should be counted in the total number of nonredundant elements in $\mathbf{S}$. If t does *not* include any elements from $\mathbf{\Phi}$, then the df $= (1/2)(p)(p + 2q + 1) - t$; if the elements of $\mathbf{\Phi}$ are counted as if they are free, the usual df formula will give the correct df.

cance for $\mathbf{B}$, $\mathbf{\Gamma}$, and $\mathbf{\Psi}$ as before. The first alternative assumes that $\mathbf{x}$ is a random variable distributed independently of $\boldsymbol{\zeta}$, whereas the second assumes that $\mathbf{x}$ is constant in repeated samples. Both alternatives assume that $\boldsymbol{\zeta}$ is multinormal with a covariance matrix of $\mathbf{\Psi}$. Though these options are appealing, they are not always appropriate. For instance, in most nonexperimental research, $\mathbf{x}$ is random rather than fixed and varies as new units are sampled. Or, if $\mathbf{x}$ contains lagged endogenous variables (lagged values of y), we cannot assume $\mathbf{x}$ is fixed. But more important, $\mathbf{x}$ and $\boldsymbol{\zeta}$ are no longer independently distributed, though they can be contemporaneously uncorrelated (Johnston 1984, 360–361). Also, when we know that some $\mathbf{x}$ variables contain random measurement error (e.g., $\mathbf{x} = \boldsymbol{\xi} + \boldsymbol{\zeta}$) or that one or more of the $\mathbf{x}$ variables are influenced by at least one of the $\mathbf{y}$ variables, then $\mathbf{x}$ is a random not fixed vector (or at least one variable in $\mathbf{x}$ is random). Furthermore random measurement error or a reciprocal relation between variables in $\mathbf{y}$ and $\mathbf{x}$ violates the assumption of independence (and uncorrelatedness) between $\mathbf{x}$ and $\boldsymbol{\zeta}$ and thereby undermines the consistency of the F_{ML} and the other estimators. In the case of reciprocal $y - x$ relations, the x variables involved should be treated as endogenous (y's), leaving only those truly exogenous variables in $\mathbf{x}$. In Chapter 5 I discuss the consequence of random measurement error for variables in $\mathbf{x}$.

Another possibility I have not discussed is when $\mathbf{y}$ conditional on $\mathbf{x}$ (or with fixed $\mathbf{x}$) does not have a multinormal distribution. Although the parameter estimators are still consistent, their asymptotic covariance matrix may not be accurate. Correct formulas for the asymptotic covariance matrix under nonnormality are given in Browne (1984) and Arminger and Schoenberg (1987). Alternatively, in Chapter 9, I present other estimators that do not require multinormality.

Interaction Terms

Sometimes the influence of a combination of variables is greater than the sum of their linear effects. For instance, in the union sentiment example long seniority, together with older ages, may lessen deference more than the separate linear effects of these variables. We can incorporate interaction effects in observed variable structural equations in a way analogous to that done in regression analysis: define a new variable that is a combination of two or more variables and include it as an explanatory variable.

I illustrate this with the deference (Y_1) equation for the union sentiment model. Suppose that the interaction term is the product of seniority (X_1) and age (X_2) so that $X_3 = X_1 X_2$. The equation with the linear and interaction terms is

$$Y_1 = \alpha_1 + \gamma_{11}X_1 + \gamma_{12}X_2 + \gamma_{13}X_3 + \zeta_1 \tag{4.94}$$

For the most part the analysis proceeds in the usual way in which X_3 becomes an exogenous variable added to $\mathbf{S}$ and $\mathbf{\Phi}$, with its coefficient an element of $\mathbf{\Gamma}$. However, two complications arise. One is in the interpretation of the coefficients of the variables in the interaction. Usually in a model like (4.94) without the interaction term (i.e., without $\gamma_{13}X_3$), we can say that for a one-unit difference in X_1, we expect an average difference of γ_{11} in Y_1 net of X_2. With an interaction involving X_1, we need to modify this. For a one-unit difference in X_1, we expect an average difference of $\gamma_{11} + \gamma_{13}X_2$. The influence of X_1 depends on the values of γ_{13} and X_2. An analogous interpretation is needed for X_2's influence (see Stolzenberg 1979).

A second complication is that I cannot assume that the observed variables in (4.94) have a multinormal distribution even if Y_1, X_1, and X_2 do. The problem is that the product of two normally distributed variables (i.e., X_3) is not normally distributed. However, if I assume either that $\mathbf{x}$ is independent of ζ, or that $\mathbf{x}$ is fixed, and that ζ has a multinormal distribution, then the usual properties for the F_{ML} estimators hold as described in the last section. The F_{ML} estimates and standard errors for the coefficient estimates in equation (4.94) are as follows:

$$\begin{aligned}\hat{Y}_1 = \underset{(1.345)}{16.031} + \underset{(0.926)}{0.792}X_1 - \underset{(0.033)}{0.076}X_2 - \underset{(0.019)}{0.013}X_3\end{aligned}$$

To test the presence of an interaction effect, we need to examine $\hat{\gamma}_{13}$ and its statistical significance. It has a negative sign as predicted, but the estimate is smaller than its standard error and is not statistically significant. It appears that there is not an interaction effect of seniority (X_1) and age (X_2) on deference (Y_1). Note that squared terms can be treated in a way analogous to the interaction terms.

Intercepts

As in the last subsection, an equation intercept is sometimes required. In most regression and econometric software the intercept is automatically provided. In this subsection I explain how to obtain intercepts with LISREL VI since it is not part of the standard output. The procedure has three steps: (1) analyze the moment matrix, (2) add a new X variable that consists of only ones, and (3) use the "fixed $\mathbf{x}$" option.

Consider the three equations for the union sentiment model with intercept terms represented by α_i:

$$\begin{bmatrix} Y_1 \\ Y_2 \\ Y_3 \end{bmatrix} = \begin{bmatrix} 0 & 0 & 0 \\ \beta_{21} & 0 & 0 \\ \beta_{31} & \beta_{32} & 0 \end{bmatrix} \begin{bmatrix} Y_1 \\ Y_2 \\ Y_3 \end{bmatrix} + \begin{bmatrix} 0 & \gamma_{12} & \alpha_1 \\ 0 & \gamma_{22} & \alpha_2 \\ \gamma_{31} & 0 & \alpha_3 \end{bmatrix} \begin{bmatrix} X_1 \\ X_2 \\ 1 \end{bmatrix} + \begin{bmatrix} \zeta_1 \\ \zeta_2 \\ \zeta_3 \end{bmatrix} \tag{4.95}$$

Equation (4.95) is the same as that for the original model except that the last column of $\boldsymbol{\Gamma}$ now contains the intercept terms and the last element of $\mathbf{x}$ is the constant one. The moment matrix consists of the sum of the products for each pair of variables in their raw form divided by N [e.g., $(\Sigma X_1 X_2)/N$]. The last row in the moment matrix contains the means of all the observed variables, and the last element is one which is the mean of the constant term. Finally, treat the $\mathbf{x}$ matrix as fixed. The α_i estimates for (4.81) and their standard errors based on the ML fitting function are

$$\alpha_1 = \underset{(0.871)}{16.764}$$

$$\alpha_2 = \underset{(1.257)}{9.134}$$

$$\alpha_3 = \underset{(1.930)}{8.694}$$

The other coefficients are unchanged by the inclusion of intercepts. Jöreskog and Sörbom (1986, V.14–V.17) describe an alternative procedure which can be employed if the $\mathbf{x}$ variables (except for the constant) are treated as random.

Analysis of Variance (ANOVA) and Related Techniques

It is well-known that ANOVA is a special case of regression analysis where the explanatory variables consist of dummy variables. For instance, consider

$$Y_1 = \alpha + \gamma_{11} X_1 + \gamma_{12} X_2 + \gamma_{13} X_3 + \zeta \tag{4.96}$$

where X_1 and X_2 are dummy variables and X_3 equals $X_1^* X_2$, an interaction term. This regression formulation corresponds to an ANOVA model with two "factors" (X_1 and X_2) and an interaction term (X_3). Similarly, I could add continuous variables to the model to represent the analysis of covariance models or add more equations for other Y's to represent multivariate analysis of variance (MANOVA) models (Pedhazur 1982). Since single- and multiequation regression models are part of the family of observed variable structural equations, it follows that ANOVA, analysis of covariance, and MANOVA are special cases of the models of this chapter (see Kenny 1979, 184–205). Tests for differences in means, slopes, and interactions are possible by comparing the fit of equations that constrain the differences to zero to the fit of equations where this constraint is not in force. Chapter 7 presents measures of fit and tests of statistical significance for such comparisons. If the variances of the disturbances differs across categories of the dummy variables, then the multiple group analysis I describe in Chapter 8 is an option.

SUMMARY

This chapter reviewed structural equations with observed variables. The area is familiar to readers experienced with regression analysis, though the presentation was in the context of the general structural equation model. Many of the ideas introduced here reappear in later chapters. For instance, the relation of the covariance matrix of the observed variables to the model parameters is fundamental to all the models I consider. The estimation procedures have general application. I explored measures of model fit that are closely related to those common in regression analysis and a chi-square test of overall model fit. In Chapter 7 I present many more fit measures that are appropriate for all the structural equation models, including those of this chapter. The next chapter discusses the consequences of estimating observed variable models when one or more of the variables contain measurement error.

APPENDIX 4A DERIVATION OF F_{ML} (y and x MULTINORMAL)

The ML fitting function derives from the multinormality of **y** and **x**.[1] There are two alternative ways of showing this. One works from the multinormal probability distribution function directly, and the second begins with the Wishart distribution of the sample covariance matrix (**S**). In this appendix I show the first derivation, and in Appendix 4B I treat the second one.

[1]The section on Alternative Assumptions for **x** in this chapter shows that the F_{ML} estimator can be justified on grounds besides multinormality.

I start by reviewing ML estimation. ML estimation begins with a random sample of N independently and identically distributed observations for a random variable Z. The probability density function of each Z_i (for $i = 1, 2, \ldots, N$) is $f(Z_i; \theta)$, where θ is a fixed parameter that helps to determine the probability density of the Z's. When each Z_i is independently drawn from the others, their joint probability density function is

$$f(Z_1, Z_2, \ldots, Z_N; \theta) = f(Z_1; \theta) f(Z_2; \theta) \ldots f(Z_N; \theta) \quad (4A.1)$$

The joint density is the product of the marginal densities of Z_i since $Z_1, Z_2, \ldots, Z_N$ are independent.

Once we observe a specific set of values for $Z_1, Z_2, \ldots, Z_N$ in a sample, we can write a function:

$$L(\theta; Z_1, Z_2, \ldots, Z_N) = L(\theta; Z_1) L(\theta; Z_2) \ldots L(\theta; Z_N) \quad (4A.2)$$

where $L(\theta; Z_i)$ is the value of $f(Z_i; \theta)$ when Z_i is at its sample value. Equation (4A.2) is the *likelihood function*, abbreviated as $L(\theta)$ or just L. Although (4A.1) and (4A.2) look similar, they have important differences. In (4A.1) θ is a fixed parameter, and the Z_i's are random variables. In (4A.2) the Z_i's are fixed values in a particular sample and the magnitude of $L(\theta)$ is a function of θ. The θ is a fixed parameter in the population, but since it is unknown, we can try different trial values of it to see how $L(\theta)$ changes. Define $\hat{\theta}$ as an estimator of θ. In maximum likelihood estimation we want the $\hat{\theta}$ that leads to the greatest probability (or likelihood) of generating the given sample values of Z_i. That is, we choose the $\hat{\theta}$ that maximizes (4A.2) for the sample of values $Z_1, Z_2, \ldots, Z_N$.

To find the maximum of $L(\theta)$, it usually is convenient to maximize the logarithm (log) of $L(\theta)$. This does not change the value of θ since the log of a number is a monotonic function of that number. To find θ, take the derivative of $\log L(\theta)$ with respect to θ, that is, $[d \log L(\theta)]/d\theta$, set this quantity to zero, and solve for θ. If this value is substituted in for θ in the second derivative of $\log L(\theta)$ with respect to θ and the resulting value is negative, this is sufficient to show that the value of θ maximizes $\log L(\theta)$.

In sum, in ML estimation we assume that a random variable has a known probability density function. For a given sample of observations of the random variable, we can write a likelihood function, $L(\theta)$, that is a function of θ. Generally, we can simplify the search for the ML estimator by using $\log L(\theta)$. Choosing a value $\hat{\theta}$ that maximizes $\log L(\theta)$ leads to the ML estimator for the sample.

In deriving F_{ML}, the set of N independent observations are of the multinormal random variables $\mathbf{y}$ and $\mathbf{x}$. If we combine $\mathbf{y}$ and $\mathbf{x}$ into a single

$(p+q)\times 1$ vector $\mathbf{z}$, where $\mathbf{z}$ consists of deviation scores, its probability density function is

$$f(\mathbf{z};\Sigma) = (2\pi)^{-(p+q)/2}|\Sigma|^{-1/2}\exp\left[\left(-\tfrac{1}{2}\right)\mathbf{z}'\Sigma^{-1}\mathbf{z}\right] \qquad (4A.3)$$

For a random sample of N independent observations of $\mathbf{z}$, the joint density is

$$f(\mathbf{z}_1,\mathbf{z}_2,\ldots,\mathbf{z}_N;\Sigma) = f(\mathbf{z}_1;\Sigma)f(\mathbf{z}_2;\Sigma)\ldots f(\mathbf{z}_N;\Sigma) \qquad (4A.4)$$

Once we observe a given sample, the likelihood function is

$$L(\boldsymbol{\theta}) = (2\pi)^{-N(p+q)/2}|\Sigma(\boldsymbol{\theta})|^{-N/2}\exp\left[-\tfrac{1}{2}\sum_{i=1}^{N}\mathbf{z}_i'\Sigma^{-1}(\boldsymbol{\theta})\mathbf{z}_i\right] \qquad (4A.5)$$

I substitute $\Sigma(\boldsymbol{\theta})$ for Σ based on the covariance structure hypothesis that $\Sigma = \Sigma(\boldsymbol{\theta})$ and to make explicit the role of $\boldsymbol{\theta}$ in the likelihood function.

The log of the likelihood function is

$$\log L(\boldsymbol{\theta}) = \frac{-N(p+q)}{2}\log(2\pi) - \left(\frac{N}{2}\right)\log|\Sigma(\boldsymbol{\theta})| - \left(\frac{1}{2}\right)\sum_{i=1}^{N}\mathbf{z}_i'\Sigma^{-1}(\boldsymbol{\theta})\mathbf{z}_i \qquad (4A.6)$$

Rewrite the last term in (4A.6) as

$$\begin{aligned}
-\left(\frac{1}{2}\right)\sum_{i=1}^{N}\mathbf{z}_i'\Sigma^{-1}(\boldsymbol{\theta})\mathbf{z}_i &= -\left(\frac{1}{2}\right)\sum_{i=1}^{N}\operatorname{tr}\left[\mathbf{z}_i'\Sigma^{-1}(\boldsymbol{\theta})\mathbf{z}_i\right] \\
&= -\left(\frac{N}{2}\right)\sum_{i=1}^{N}\operatorname{tr}\left[\mathbf{N}^{-1}\mathbf{z}_i\mathbf{z}_i'\Sigma^{-1}(\boldsymbol{\theta})\right] \\
&= -\left(\frac{N}{2}\right)\operatorname{tr}\left[\mathbf{S}^*\Sigma^{-1}(\boldsymbol{\theta})\right] \qquad (4A.7)
\end{aligned}$$

where $\mathbf{S}^*$ is the sample ML estimator of the covariance matrix which employs N rather than $(N-1)$ in the denominator. [The first step of (4A.7) follows since a scalar equals its trace, while the second uses the property that tr(ABC) = tr(CAB) for conforming matrices.] Equation (4A.7) allows us to rewrite the $\log L(\boldsymbol{\theta})$ as

$$\begin{aligned}\log L(\boldsymbol{\theta}) &= \text{constant} - \left(\frac{N}{2}\right)\log|\boldsymbol{\Sigma}(\boldsymbol{\theta})| - \left(\frac{N}{2}\right)\operatorname{tr}\left[\mathbf{S}^*\boldsymbol{\Sigma}^{-1}(\boldsymbol{\theta})\right] \\ &= \text{constant} - \left(\frac{N}{2}\right)\left\{\log|\boldsymbol{\Sigma}(\boldsymbol{\theta})| + \operatorname{tr}\left[\mathbf{S}^*\boldsymbol{\Sigma}^{-1}(\boldsymbol{\theta})\right]\right\} \end{aligned} \quad (4A.8)$$

Compare (4A.8) to F_{ML}:

$$F_{ML} = \log|\boldsymbol{\Sigma}(\boldsymbol{\theta})| + \operatorname{tr}\left[\mathbf{S}\boldsymbol{\Sigma}^{-1}(\boldsymbol{\theta})\right] - \log|\mathbf{S}| - (p+q) \quad (4A.9)$$

Equations (4A.9) and (4A.8) differ in several ways that are largely inconsequential for the estimation of $\boldsymbol{\theta}$. The "constant" term in (4A.8) has no influence on the choice of $\hat{\boldsymbol{\theta}}$, so its absence from (4A.9) has no impact. Similarly, the inclusion of $(-\log|\mathbf{S}| - (p+q))$ in (4A.9) does not affect the $\hat{\boldsymbol{\theta}}$ choice because, for a given sample, $\mathbf{S}$ and $(p+q)$ are constant. The only effect of the $(-N/2)$ term present in (4A.8) but not in (4A.9) is to lead us to minimize rather than maximize (4A.9) because of the change from a minus to a plus sign.

The final difference is the unbiased sample covariance matrix $\mathbf{S}$ is in (4A.9), while the ML estimator $\mathbf{S}^*$ is in (4A.8). Since $\mathbf{S}^* = [(N-1)/N]\mathbf{S}$ these matrices will be essentially equal in large samples. Thus, except for the mostly inconsequently difference in $\mathbf{S}^*$ and $\mathbf{S}$, the F_{ML} and ML expression lead to the same estimator $\hat{\boldsymbol{\theta}}$. Appendix 4C discusses and provides an illustration of how to determine the $\hat{\boldsymbol{\theta}}$ value that minimizes the F_{ML} fitting function.

In Appendix 4B I show that the Wishart Distribution for $\mathbf{S}$ leads to a ML estimator that employs $\mathbf{S}$ rather than $\mathbf{S}^*$.

APPENDIX 4B DERIVATION OF F_{ML} (S WISHART DISTRIBUTION)

If $\mathbf{y}$ and $\mathbf{x}$ have a multinormal distribution, then the unbiased sample covariance matrix, $\mathbf{S}$, has a *Wishart Distribution* (Anderson 1958, 154–159). For $\mathbf{S}$ in a given sample, the $\log L(\boldsymbol{\theta})$ is

$$\log L(\boldsymbol{\theta}) = \log\left[\frac{|\mathbf{S}|^{(N^*-(p+q)-1)/2}\exp\left\{-(N^*/2)\operatorname{tr}\left[\mathbf{S}\boldsymbol{\Sigma}^{-1}(\boldsymbol{\theta})\right]\right\}(N^*/2)^{(p+q)N^*/2}}{\pi^{(p+q)((p+q)-1)/4}|\boldsymbol{\Sigma}(\boldsymbol{\theta})|^{N^*/2}\prod_{i=1}^{p+q}\Gamma[(1/2)(N^*+1-i)]}\right] \quad (4B.1)$$

In (4B.1) $N^* = N - 1$ and the gamma function $(\Gamma[(1/2)(N^* + 1 - i)]$ is a constant for a given sample size N and number of variables $(p + q)$. The $\prod_{i=1}^{p+q}$ symbol is the standard one to represent the product of the function that follows it, with i running from 1 to $(p + q)$.

When the logs are taken and the expression simplified, (4B.1) becomes

$$\log L(\boldsymbol{\theta}) = -\left(\frac{N^*}{2}\right)\left\{\log|\boldsymbol{\Sigma}(\boldsymbol{\theta})| + \operatorname{tr}\left[\mathbf{S}\boldsymbol{\Sigma}^{-1}(\boldsymbol{\theta})\right]\right\} + \text{constant} \quad (4\text{B}.2)$$

In (4B.2) the constant term contains all the remaining terms from (4B.1) that do not change once the sample is given. Comparing (4B.2) to F_{ML} [see (4A.9)], we see that the value of $\hat{\boldsymbol{\theta}}$ that *maximizes* $\log L(\boldsymbol{\theta})$ will minimize F_{ML} and that both functions contain $\mathbf{S}$.

The asymptotic covariance matrix of the ML estimator of $\boldsymbol{\theta}$ is

$$\operatorname{ACOV}(\hat{\boldsymbol{\theta}}) = \left\{-E\left[\frac{\partial^2 \log L(\boldsymbol{\theta})}{\partial\boldsymbol{\theta}\,\partial\boldsymbol{\theta}'}\right]\right\}^{-1} \quad (4\text{B}.3)$$

The matrix within the braces is the *information matrix*. The inverse of the information matrix is a $t \times t$ matrix with the asymptotic variances of the ML estimator $\hat{\boldsymbol{\theta}}$ down its main diagonal and the asymptotic covariances in the off-diagonal positions.

What is the relation between the $\operatorname{ACOV}(\hat{\boldsymbol{\theta}})$ in (4B.3) for the $\log L(\boldsymbol{\theta})$ and that for F_{ML}? This is an important question since it is F_{ML} not $\log L(\boldsymbol{\theta})$ that we use for the fitting function. From (4B.2) we can see that the partial derivatives of $\log L(\boldsymbol{\theta})$ with respect to $\boldsymbol{\theta}$ are zero for the constant term. Equation (4A.9) reveals that for F_{ML} the partials for the last two terms are zero. The relation between these second-order partials is

$$\frac{-\partial^2 \log L(\boldsymbol{\theta})}{\partial\boldsymbol{\theta}\,\partial\boldsymbol{\theta}'} = \left(\frac{N^*}{2}\right)\frac{\partial^2 F_{\mathrm{ML}}}{\partial\boldsymbol{\theta}\,\partial\boldsymbol{\theta}'} \quad (4\text{B}.4)$$

Combining (4B.4) and (4B.3),

$$\operatorname{ACOV}(\hat{\boldsymbol{\theta}}) = \left(\frac{2}{N^*}\right)\left\{E\left[\frac{\partial^2 F_{\mathrm{ML}}}{\partial\boldsymbol{\theta}\,\partial\boldsymbol{\theta}'}\right]\right\}^{-1} \quad (4\text{B}.5)$$

Evaluating (4B.5) by substituting $\hat{\boldsymbol{\theta}}$ for $\boldsymbol{\theta}$ leads to an estimated asymptotic covariance matrix for $\hat{\boldsymbol{\theta}}$ from F_{ML}. An illustration of this matrix was given in the ML section in this chapter.

APPENDIX 4C NUMERICAL SOLUTIONS TO MINIMIZE FITTING FUNCTIONS

In this section I provide an overview of numerical methods to minimize the fitting functions for covariance structure models. My goal is to illustrate the iterative process for a simple model. For further discussions of numerical minimization in nonlinear models, see Bard (1974), Goldfeld and Quandt (1972), Kennedy and Gentle (1980), and Judge et al. (1980, ch. 17).

A necessary condition for finding the minimum of any function, say, $f(\boldsymbol{\theta})$, is to set the partial derivative of $f(\boldsymbol{\theta})$ with respect to each θ_i to zero and to solve for θ_i. A sufficient condition for these values to minimize $f(\boldsymbol{\theta})$ is for the matrix of second partial derivatives, $\partial^2 f(\boldsymbol{\theta})/\partial\boldsymbol{\theta}\,\partial\boldsymbol{\theta}'$, to be positive-definite at these values.

Consider the first-order partials. If $\boldsymbol{\theta}$ is $t \times 1$, then

$$\frac{\partial f(\boldsymbol{\theta})}{\partial \theta_i} = 0, \qquad \text{for } i = 1, 2, \ldots, t \tag{4C.1}$$

After taking partial derivatives, (4C.1) shows that there are as many equations as there are θ_i's in $\boldsymbol{\theta}$. In some cases simple algebraic solutions for the unknown θ_i's are derivable from the t equations of (4C.1). For instance, in linear multiple regression where the fitting function $f(\boldsymbol{\theta})$ is the sum of squared residuals and $\boldsymbol{\theta}$ contains the unknown regression parameters, (4C.1) results in t equations, linear in θ_i. Explicit solutions for the regression parameters are well-known (e.g., Fox 1984, 33–35). In the general structural equation model where $f(\boldsymbol{\theta})$ is the F_{ML}, F_{GLS}, or F_{ULS} fitting functions, (4C.1) results in t equations that are typically nonlinear in the parameters. Explicit solutions for the parameters usually are not accessible. In these cases minimization with numerical methods are necessary.

The numerical methods begin with an objective function to be minimized.[1] In the chapter I defined three objective functions, F_{ML}, F_{GLS}, and F_{ULS}. The goal is to develop a sequence of values for $\boldsymbol{\theta}$ such that the last vector in the sequence minimizes one of these objective functions. The first trial values for $\boldsymbol{\theta}$ are symbolized as $\boldsymbol{\theta}^{(1)}$, the second as $\boldsymbol{\theta}^{(2)}$, and so on, with the last ones as $\boldsymbol{\theta}^{(l)}$. The sequence of trial values for $\boldsymbol{\theta}$ is $\boldsymbol{\theta}^{(1)}, \boldsymbol{\theta}^{(2)}, \ldots, \boldsymbol{\theta}^{(l)}$.

The three key issues in minimization are (1) the selection of initial or starting values, $\boldsymbol{\theta}^{(1)}$, (2) the rules for moving from one step in the sequence

[1]The strategy to maximize $f(\boldsymbol{\theta})$ is the same as that to minimize, except for the change in directions. Minimization instead of maximization of fitting functions such as F_{ML} is called for since F_{ML} is related to the negative of the likelihood function. See Appendix 4A.

to the next (from $\theta^{(i)}$ to $\theta^{(i+1)}$), and, (3) when to stop the iterations. The next three subsections treat these problems.

Starting Values

The starting or initial values in $\theta^{(1)}$ affect the numerical minimization in several ways. One is that the initial values determine the number of iterations needed to reach the final solution. Starting values close to the final ones usually reduce the iterations required. Also starting values far from the final ones increase the probability of finding local minima for the fitting function rather than the global one or of not finding a convergent solution.

There are several strategies for selecting initial values. One is to use a noniterative procedure to estimate the model parameters and to use these estimates. Jöreskog and Sörbom's (1986) LISREL VI program provides an instrumental variable technique for this purpose. I describe their procedure in Chapter 9. In some situations this automatic provision of starting values is not an option. For instance, in LISREL IV and prior versions and in the 1985 and 1986 versions of Bentler's EQS, automatic starting values are not available. Also in a few cases the automatic starting values do not lead to convergence, so user-provided values are required.

OLS estimates of many of the structural equation models of this chapter provide a useful beginning point. Indeed, in the case of the usual multiple regression or recursive models with normally distributed disturbances, the OLS parameter estimates are identical to the estimates from F_{ML}. If OLS estimates are not readily available, researchers may take initial values from prior research with a similar model and variables.

Alternatively, the researcher can use rough guidelines to select starting values. Table 4C.1 presents some possibilities. The left column lists the parameter to be estimated, and the two last columns give the starting values and the value of a. In the table a is a constant, but it refers to a different constant for each parameter. Researchers must decide, based on their substantive knowledge, whether a should be positive or negative for β_{ij}, γ_{ij}, or ψ_{ij}. The far right column gives rough rules for selecting the magnitude of a. These depend on the researcher's best subjective estimate of the strength of the relation, fit, or correlation. The guidelines for β_{ij} and γ_{ij} originate with the connection between standardized and unstandardized coefficients. For instance, the standardized β_{ij} is β_{ij} (s.d. of y_j)/(s.d. of y_i), where s.d. abbreviates "standard deviation." If the standardized coefficient equals a, then the unstandardized coefficient equals a (s.d. of y_i/s.d. of y_j). The value of a corresponds to the expected magnitude of a standardized coefficient. Of course, when dealing with standardized variables, the ratio of

Table 4C.1 Guidelines for Starting Values in Structural Equations with Observed Variables

Parameters	Starting Value	Value of a
$\beta_{ij}(i \neq j)$	$a\left(\frac{\text{s.d. of } y_i}{\text{s.d. of } y_j}\right)$	$\lvert a\rvert = 0.9$ "strong" $\lvert a\rvert = 0.4$ "moderate" $\lvert a\rvert = 0.2$ "weak"
$\gamma_{ij}(i \neq j)$	$a\left(\frac{\text{s.d. of } y_i}{\text{s.d. of } x_j}\right)$	$\lvert a\rvert = 0.9$ "strong" $\lvert a\rvert = 0.4$ "moderate" $\lvert a\rvert = 0.2$ "weak"
ψ_{ii}	$a \operatorname{var}(y_i)$ $0 \leq a \leq 1$	$a = 0.2$ "strong" fit $a = 0.4$ "moderate" fit $a = 0.9$ "weak" fit
$\psi_{ij}(i \neq j)$	$a(\psi_{ii}\psi_{jj})^{1/2}$ $-1 < a < 1$	$\lvert a\rvert = 0.9$ "strong" correlation $\lvert a\rvert = 0.4$ "moderate" correlation $\lvert a\rvert = 0.2$ "weak" correlation
$\mathbf{\Phi}$	sample covariance matrix for $\mathbf{x}$	

standard deviations is one, and the starting value should just be a. The condition for the initial values of ψ_{ii} restricts a to prevent the error variance from being negative or greater than the var(y_i). Restrictions of a for ψ_{ij} stop it from creating a correlation of ζ_i with ζ_j which exceeds plus or minus one. Thus with these guidelines researchers should be able to select reasonable starting values, when these are not otherwise available.

Steps in the Sequence

With starting values in place, the question is how to move from the initial $\boldsymbol{\theta}^{(1)}$ to the next trial values $\boldsymbol{\theta}^{(2)}$, or more generally, how do we move from $\boldsymbol{\theta}^{(i)}$ to $\boldsymbol{\theta}^{(i+1)}$? The most basic criterion is that as we progress through the sequence $\theta^{(1)}$, $\boldsymbol{\theta}^{(2)}, \ldots, \boldsymbol{\theta}^{(l)}$ the values of the fitting function should decline, with the fitting function at its minimum for $\boldsymbol{\theta}^{(l)}$. Ideally, for each step, $F(\theta^{(i+1)})$ is less than $F(\boldsymbol{\theta}^{(i)})$, where $F(\cdot)$ represents the value of the fitting function at the trial values. However, even good fitting functions do not always decrease monotonically.

Multiple parameters complicate the minimization process. Yet I can illustrate many of the basic issues with a one-parameter model. Consider

$$y_1 = x_1 + \zeta_1 \tag{4C.2}$$

where $\phi_{11} = 1$, $\text{COV}(x_1, \zeta_1) = 0$, $E(\zeta_1) = 0$, and y_1 and x_1 are in deviation form. This example is similar to one I presented in Chapter 4 when introducing F_{ML}, F_{GLS}, and F_{ULS}, except that in (4C.2) I also assume that the ϕ_{11}, the variance of the explanatory variable, equals one.

The covariance structure equation in this situation is

$$\begin{array}{ccc} \boldsymbol{\Sigma} & = & \boldsymbol{\Sigma}(\boldsymbol{\theta}) \\ \begin{bmatrix} \text{VAR}(y_1) & \text{COV}(x_1, y_1) \\ \text{COV}(x_1, y_1) & \text{VAR}(x_1) \end{bmatrix} & = & \begin{bmatrix} 1 + \psi_{11} & 1 \\ 1 & 1 \end{bmatrix} \end{array} \tag{4C.3}$$

The only unknown in $\boldsymbol{\theta}$ is ψ_{11}. To simplify the notation, I drop the subscripts from ψ_{11} and refer to it as ψ. The F_{ML} fitting function is

$$F_{\text{ML}} = \log|\boldsymbol{\Sigma}(\boldsymbol{\theta})| + \text{tr}\left[\mathbf{S}\boldsymbol{\Sigma}^{-1}(\boldsymbol{\theta})\right] - \log|\mathbf{S}| - (p + q) \tag{4C.4}$$

Suppose the sample covariance matrix is

$$\mathbf{S} = \begin{bmatrix} 2 & 1 \\ 1 & 1 \end{bmatrix} \tag{4C.5}$$

Substituting $\hat{\boldsymbol{\Sigma}}$ for $\boldsymbol{\Sigma}(\boldsymbol{\theta})$ in (4C.3) and $\mathbf{S}$ from (4C.5) into F_{ML} of (4C.4) results in

$$\begin{aligned} F_{\text{ML}} = {} & \log\left|\begin{bmatrix} 1 + \hat{\psi} & 1 \\ 1 & 1 \end{bmatrix}\right| + \text{tr}\left(\begin{bmatrix} 2 & 1 \\ 1 & 1 \end{bmatrix}\begin{bmatrix} 1 + \hat{\psi} & 1 \\ 1 & 1 \end{bmatrix}^{-1}\right) \\ & - \log\left|\begin{bmatrix} 2 & 1 \\ 1 & 1 \end{bmatrix}\right| - 2 \end{aligned} \tag{4C.6}$$

which after simplification is

$$F_{\text{ML}} = \log \hat{\psi} + \hat{\psi}^{-1} - 1 \tag{4C.7}$$

Given the simplicity of this example, we can see that setting $\hat{\psi}$ to one leads to the minimum of $F_{\text{ML}} = 0$. But, to illustrate the behavior of F_{ML}, I try different values for $\hat{\psi}$.

The first two columns of Table 4C.2 show the values of $\hat{\psi}$ and F_{ML} when $\hat{\psi}$ varies between 0.1 and 2.0 by increments of 0.1. Figure 4C.1 plots the relation between F_{ML} (ordinate axis) and $\hat{\psi}$ (abscissa axis) when $\hat{\psi}$ ranges from 0.1 to 10.0. It is obvious from Table 4C.2 and Figure 4C.1 that F_{ML} is

Table 4C.2 Values for $\hat{\psi}$, F_{ML}, and $\partial F_{\mathrm{ML}} / \partial \hat{\psi}$ When $\hat{\psi}$ Ranges from 0.1 to 2.0 by 0.1 Increments

$\hat{\psi}$	F_{ML}	$\partial F_{\mathrm{ML}}/\partial \hat{\psi}$
0.1	6.697	−90.000
0.2	2.391	−20.000
0.3	1.129	−7.778
0.4	0.584	−3.750
0.5	0.307	−2.000
0.6	0.156	−1.111
0.7	0.072	−0.612
0.8	0.027	−0.313
0.9	0.006	−0.123
1.0	0.000	0.000
1.1	0.004	0.083
1.2	0.016	0.139
1.3	0.032	0.178
1.4	0.051	0.204
1.5	0.072	0.222
1.6	0.095	0.234
1.7	0.119	0.242
1.8	0.143	0.247
1.9	0.168	0.249
2.0	0.193	0.250

at its minimum of zero when $\hat{\psi}$ is one. But suppose we did not know this and started at some other initial value, say, $\hat{\psi}^{(1)}$. What should $\hat{\psi}^{(2)}$ be? Table 4C.2 shows that if $\hat{\psi}^{(1)}$ is less than one, we should increase the value of $\hat{\psi}$ to reduce F_{ML} (provided we do not increase it too much). When $\hat{\psi}^{(1)}$ is larger than one, F_{ML} decreases if $\hat{\psi}^{(2)}$ is less than $\hat{\psi}^{(1)}$ (provided we do not decrease it too much). If we draw a line tangent to the curve for F_{ML} at any point $\hat{\psi}^{(1)}$ in Figure 4C.1, the slope of the line is negative when $\hat{\psi}^{(1)}$ is less than one, positive when $\hat{\psi}^{(1)}$ is greater than one, and zero at the minimum. This suggests that $\partial F_{\mathrm{ML}}/\partial \hat{\psi}$, which gives the slopes of these tangent lines, can provide the direction to move so that F_{ML} decreases. The $\partial F_{\mathrm{ML}}/\partial \hat{\psi}$ is the *gradient* of the fitting function. In general, a negative gradient suggests that the trial value for a parameter should be increased, whereas a positive gradient means that it should be decreased. The last column in Table 4C.2 provides the gradient for different values of $\hat{\psi}^{(i)}$. The gradient is negative when $0 < \hat{\psi}^{(i)} < 1$; it is positive when $\hat{\psi}^{(i)} > 1$; and it is zero when $\hat{\psi}^{(i)}$ equals one. The direction in which to change $\hat{\psi}^{(i)}$ is $-\partial F_{\mathrm{ML}}/\partial \hat{\psi}$.

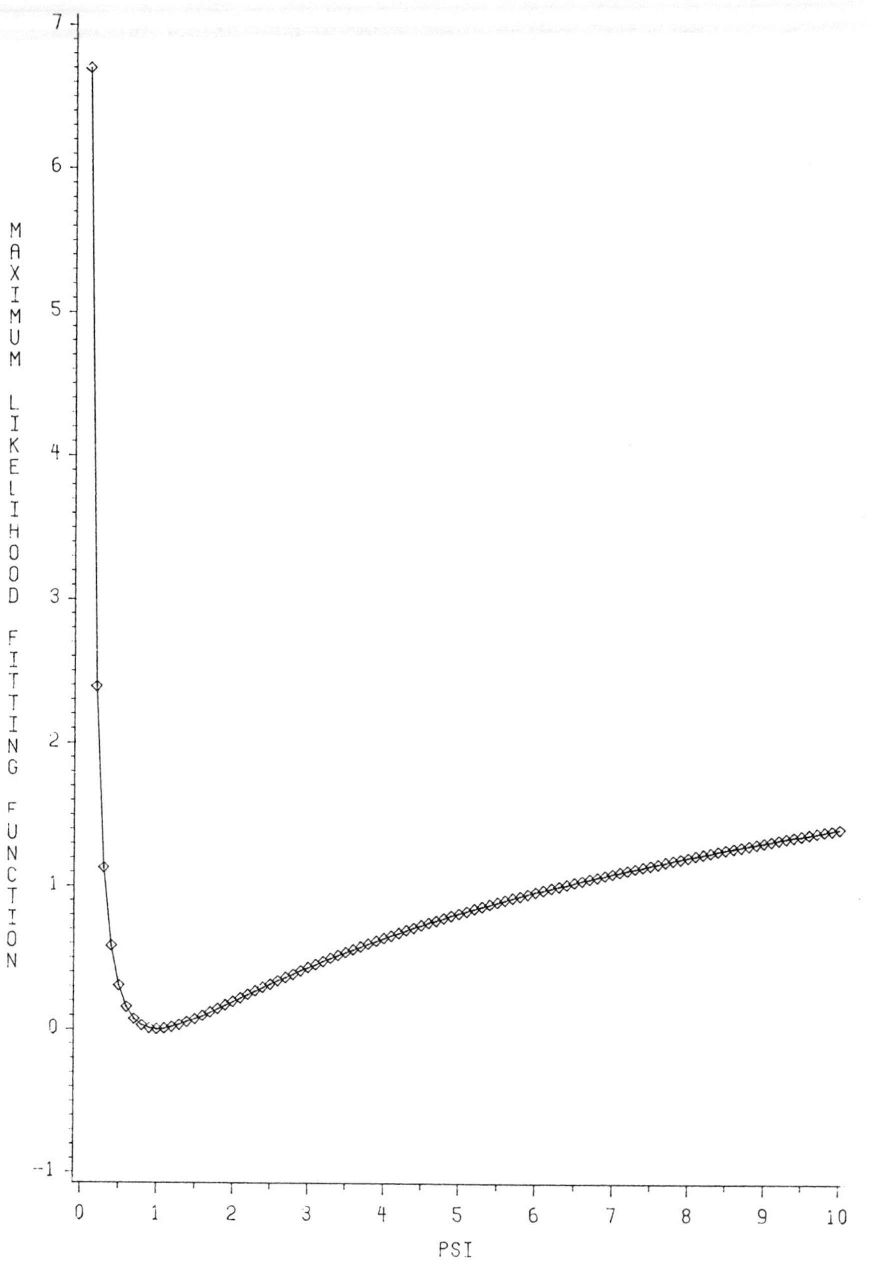
MAXIMUM LIKELIHOOD FITTING FUNCTION
7
6
5
4
3
2
1
0
-1
0
1
2
3
4
5
6
7
8
9
10
PSI

Figure 4C.1

The gradient plays a role in most numerical minimization procedures. For a vector $\hat{\boldsymbol{\theta}}$ of estimates for unknown parameters, the $\hat{\boldsymbol{\theta}}^{(i+1)}$ trial values for many of these procedures is

$$\hat{\boldsymbol{\theta}}^{(i+1)} = \hat{\boldsymbol{\theta}}^{(i)} - \mathbf{C}^{(i)}\mathbf{g}^{(i)} \tag{4C.8}$$

where $\mathbf{g}^{(i)}$ is the gradient vector $\partial F_{\text{ML}}/\partial\hat{\boldsymbol{\theta}}$ at $\hat{\boldsymbol{\theta}}^{(i)}$, and $\mathbf{C}^{(i)}$ is a positive-definite matrix. Many minimization methods are distinguished by their choice of $\mathbf{C}^{(i)}$. The simplest choice is $\mathbf{C}^{(i)}$ as an identity matrix. This leads to the *method of steepest descent*. An analyst can adjust the "step length" by multiplying $\mathbf{C}^{(i)}$ by a constant (e.g., $\frac{1}{2}$) to alter the resulting change in $\hat{\boldsymbol{\theta}}^{(i+1)}$. The main disadvantages of the steepest descend method are that it is very slow and it is not very sensitive to the different shapes that F_{ML} may take.

Another choice for $\mathbf{C}^{(i)}$ is the inverse of the second partial derivatives (i.e., inverse Hessian matrix) of F_{ML} with respect to $\boldsymbol{\theta}$ or $[\partial^2 F_{\text{ML}}/\partial\hat{\boldsymbol{\theta}}\,\partial\hat{\boldsymbol{\theta}}']^{-1}$ at $\hat{\boldsymbol{\theta}}^{(i)}$. This choice of $\mathbf{C}^{(i)}$ is based on a Taylor series expansion of F_{ML} of $\hat{\boldsymbol{\theta}}$ around $\hat{\boldsymbol{\theta}}^{(i)}$. The main disadvantage of this Newton-Raphson algorithm is that the analytic first and second partial derivatives of F_{ML} with respect to $\hat{\boldsymbol{\theta}}$ are needed. Also its computational needs are heavy since the inverse of the second partials is calculated in each step. In general, $\hat{\boldsymbol{\theta}}^{(i+1)}$ is determined by

$$\hat{\boldsymbol{\theta}}^{(i+1)} = \hat{\boldsymbol{\theta}}^{(i)} - \left[\frac{\partial^2 F_{\text{ML}}}{\partial\hat{\boldsymbol{\theta}}\,\partial\hat{\boldsymbol{\theta}}'}\right]^{-1}\left[\frac{\partial F_{\text{ML}}}{\partial\hat{\boldsymbol{\theta}}}\right] \tag{4C.9}$$

I illustrate the Newton-Raphson algorithm for the previous F_{ML} example. The analytic first derivative of F_{ML} is

$$\frac{dF_{\text{ML}}}{d\hat{\psi}} = \hat{\psi}^{-1}(1 - \hat{\psi}^{-1}) \tag{4C.10}$$

The second derivative is

$$\frac{d^2F_{\text{ML}}}{d\hat{\psi}^2} = \frac{d\left[\hat{\psi}^{-1}(1 - \hat{\psi}^{-1})\right]}{d\hat{\psi}^2}$$

$$= \frac{2 - \hat{\psi}}{\hat{\psi}^3} \tag{4C.11}$$

The sequence from $\hat{\psi}^{(i)}$ to $\hat{\psi}^{(i+1)}$ is

$$\hat{\psi}^{(i+1)} = \hat{\psi}^{(i)} - \left[\frac{2 - \hat{\psi}^{(i)}}{\left(\hat{\psi}^{(i)}\right)^3}\right]^{-1} \left[\left(\hat{\psi}^{(i)}\right)^{-1}\left(1 - \left(\hat{\psi}^{(i)}\right)^{-1}\right)\right]$$

$$= \hat{\psi}^{(i)} - \frac{\hat{\psi}^{(i)}\left(\hat{\psi}^{(i)} - 1\right)}{2 - \hat{\psi}^{(i)}}$$

With an initial $\hat{\psi}^{(1)}$ of 0.5, $\hat{\psi}^{(2)}$ is

$$\hat{\psi}^{(2)} = 0.5 + 0.167 = 0.667$$

Repeating this process leads to a sequence of $\psi^{(i)}$ as follows:

Iteration i =	**1**	**2**	**3**	**4**	**5**	**6**
$\hat{\psi}^{(i)} =$	0.500	0.667	0.833	0.952	0.996	1.000

Thus, the algorithm converges to the minimum in relatively few trials.

Jöreskog and Sörbom's LISREL program uses a modification of Fletcher and Powell's minimization procedure (see Gruvaeus and Jöreskog 1970; Jöreskog 1977, 272). Unlike the Newton-Raphson method, the Fletcher-Powell one does not require the inverse of the analytic second partials in each iteration. Instead, this matrix is built up through adjustments after each iteration. Bentler's (1985, 55–56) EQS uses a modified Gauss-Newton method of minimization. Though the minimization procedures are more complicated than those shown here, the methods of this section illustrate the basic process.

Stopping Criteria

The final issue in numerical minimization is when to stop the iterations. Several criteria are possible. For instance, if the differences in the fit function from one iteration to the next differ by less than some very small value, we could stop. Or, if there is little difference in the parameter estimates from one iteration to the next, we may say that the method has converged. It also is possible to set limits on the number of iterations or on the computer time to end the iterations. Typically, when the iterations cease due to an excessive number of iterations or excessive CPU time, the solution has not converged and the estimates should not be used.

Jöreskog and Sörbom's (1986) LISREL VI program stops the iterations at 250 if the sequence does not converge before then. Bentler's (1985) EQS has a default cutoff of 30 iterations for F_{ML}, F_{GLS}, and F_{ULS}, but it can be set as high as 500. Also in EQS a default convergent criteria is met when the average difference of parameter estimates for two iterations is 0.001 or less.

It is possible for several local minima of a fitting function to exist, but Jöreskog and Sörbom (1986, I.31) suggest that in practice this is rare. One possible but not definitive check for local minima is to rerun an analysis with starting values relatively far from the first run and then to determine if convergence to the same estimates occurs.

APPENDIX 4D ILLUSTRATIONS OF LISREL AND EQS PROGRAMS

Jöreskog and Sörbom's (1986) LISREL and Bentler's (1985) EQS are the two most widely used software packages for estimating covariance structure models. The purpose of this appendix is to show how both programs can be used for the objective-subjective socioeconomic status example in the chapter. This is not a substitute for the LISREL or EQS manuals. The manuals contain detailed instructions with complete descriptions of options as well as many empirical examples and interpretations of output. The LISREL program is distributed by Scientific Software Inc. (P.O. Box 536, Mooreville, IN 46158-0536) and is an "add-on" procedure for SPSSX. The EQS program is distributed by BMDP Statistical Software, Inc. (1440 Sepulveda Boulevard, Los Angeles, CA 90025). Of course procedures (e.g., JCL) to access these programs will depend on the user's institution. Alternatively, microcomputer versions of the programs are available from the above addresses.

LISREL Program

I begin with a general LISREL programming strategy, which has five steps:

1. *Draw a path diagram of the model (optional step).* The starting point for structural equations is a model. The path diagram often is a convenient way for researchers to represent their models. Figure 4.4 gives the path diagram for the objective-subjective socioeconomic status example.
2. *Write the equations for the model.* If you have drawn a path diagram, you can "read" it to determine the equations. List an equation for

each endogenous variable, while noting any covariances between the disturbances. Represent the equations in matrix notation. For the SES example,

$$\begin{bmatrix} y_1 \\ y_2 \\ y_3 \end{bmatrix} = \begin{bmatrix} 0 & \beta_{12} & 0 \\ \beta_{21} & 0 & 0 \\ \beta_{31} & \beta_{32} & 0 \end{bmatrix} \begin{bmatrix} y_1 \\ y_2 \\ y_3 \end{bmatrix} + \begin{bmatrix} \gamma_{11} & 0 \\ 0 & \gamma_{22} \\ 0 & 0 \end{bmatrix} \begin{bmatrix} x_1 \\ x_2 \end{bmatrix} + \begin{bmatrix} \zeta_1 \\ \zeta_2 \\ \zeta_3 \end{bmatrix}$$

where $\boldsymbol{\Psi}$ is a free matrix.

3. *Write the eight-parameter matrices for the model.* The LISREL model has eight-parameter matrices: $\mathbf{B}$(BE), Γ(GA), Φ(PH), Ψ(PS), Λ_x(LX), Λ_y(LY), Θ_δ(TD), and Θ_ϵ(TE). The LISREL program abbreviations for each of these is listed in parentheses. With structural equations with observed variables, Λ_x(LX) and Λ_y(LY) are identity matrices and Θ_δ(TD) and Θ_ϵ(TE) are zero matrices. This leaves only four matrices to describe: $\mathbf{B}$(BE), Γ(GA), Φ(PH), and Ψ(PS). Equations (4.4) and (4.5) in the chapter list these matrices for the SES example.
4. *Declare whether each parameter matrix initially should be free or fixed.* If most elements of a matrix are free parameters, use the MODEL statement (which follows) or defaults to declare it free. Otherwise, set it as fixed. For the SES example LX, LY, TD, and TE have all fixed elements as described in step 3. Most of the elements in BE and GA are fixed. The PS are PH matrices are free. In the program that follows I show how these declarations are made.
5. *Modify the individual parameters within matrices that still need adjustment.* After step 4 all elements in a matrix are either fixed or free. It is necessary to adjust each matrix so that the status of each parameter is correct. Individual parameters can be freed with an FR card, fixed with an FI card, or constrained to equal other parameters with an EQ card. For instance, γ_{11}(GA 1 1) and γ_{22}(GA 2 2) need to be freed in the GA matrix for the SES example. This is illustrated next.

This listing is of the LISREL VI program to obtain the ML estimates for the objective-subjective SES example of this chapter. The numbers in parentheses at the far right are not part of the commands but refer to the explanations listed at the end of the program. Generally, the same problem can be programmed in several ways so that this program is not the only way to obtain the ML estimates.

```
TITLE OBJEC.-SUBJ. SES ML
INPUT PROGRAM                                              (1)
```

```
NUMERIC DUMMY
END FILE
END INPUT PROGRAM
USERPROC NAME = LISREL
OBJECTIVE-SUBJECTIVE STATUS ML                                    (2)
DA NI = 5 NO = 432 MA = CM                                        (3)
KM                                                                (4)
*
1.000
 .292   1.000
 .282    .184   1.000
 .166    .383    .386   1.000
 .231    .277    .431    .537   1.000
SD                                                                (5)
*
21.277  2.198   0.640   0.670   0.627                             (6)
LA
*
'OCC' 'INC' 'SUBOCC' 'SUBINC' 'SUBGEN'                            (7)
SE
4 3 5 2 1/
MO NY = 3 NX = 2 NE = 3 NK = 2 LY = ID LX = ID BE = FU GA = FI TE = FI TD = FI   (8)
FR GA 1 1 GA 2 2 BE 1 2 BE 2 1 BE 3 1 BE 3 2                      (9)
OU TV SE SS TO                                                    (10)
END USER
```

(1) Cards for use with SPSSX. The first six cards and the last one are instructions to enable the user to access LISREL when it is part of SPSSX and when using matrix input. The instructions will differ when raw data is the input (see SPSSX documentation). Freestanding versions of LISREL do not require these cards.

(2) Title card. This is the name you give to the program. The title can be up to 80 columns wide. It can be lengthened to more than 80 characters if you place a "C" as the last character on the first card and continue it on a second card immediately following it.

(3) Data card. NI is the number of variables in the input matrix. NO is the number of observations. MA is the type of matrix to be analyzed, where CM is the covariance matrix. Since MA = CM is the program default, this keyword could be omitted.

(4) Input correlation matrix. The KM card informs the program that a correlation matrix follows. The * in column one on the new line after KM means that you are reading in the matrix elements in free format

—that is, each element of the matrix is separated by a blank. The default option is to read only the lower half of the correlation matrix, row-wise.

(5) Standard deviation. The SD and * cards tell the program that you are reading in the standard deviations on the next card(s) and that they are in free format. The order of the standard deviations must correspond to the order of the variables in the input matrix. Standard deviations must be supplied along with the correlation matrix when you intend to analyze the covariance matrix.

(6) Labels. The labels are read in the same manner as the standard deviations, corresponding to the order of the variables in the input correlation matrix. Each label is enclosed by single quotes.

(7) Selection card. The selection card reorders the variables in the input correlation matrix into the order of entry that the LISREL program assumes, the endogenous variables first (in this case, $y_1 - y_3$), followed by the exogenous variables ($x_1 - x_2$). The numbers on the selection card correspond to the original position of each variable in the input correlation matrix. (You may also reorder the variables by listing the variable names in order of model entry.) The selection card also can be used to read a subset of the variables from the input matrix. When using the card to subset, the card must be terminated with a "/". The "/" could have been omitted in this setup. If the select lines are omitted entirely, the program assumes that all variables in the matrix are to be included in the analysis and that the variables are already in the proper order.

(8) Model card. This line specifies the model. Here the user explicitly or implicitly (by defaults) declares the number of y, x, η, and ξ variables and the form and mode of the eight model matrices. The most common matrix forms are full (FU), zero (ZE), identity (ID), and diagonal (DI). The mode is fixed (FI) or free (FR). The FI sets all elements of the named matrix to fixed values. Unless otherwise specified, the fixed values are zeroes. The FR mode lets all elements of the matrix be estimated. MO must appear in the first two columns of this card. NY and NX indicate the number of y and x observed variables in the model. NE and NK are the number of endogenous and exogenous latent variables. Since I assume no measurement error in $\mathbf{y}$ or $\mathbf{x}$, LY and LX are fixed to identity matrices, and TE and TD are fixed to zero. BE = FU and GA = FI change the default specifications of the $\mathbf{B}$ and Γ matrices. The $\mathbf{B}$ is specified as a full matrix with every element fixed to 0. GA = FI fixes all elements of Γ to 0. The default for Ψ(PS) and Φ(PH) are symmetric and free. The

LISREL manual provides a complete listing of the defaults, modes, and special models available with the MO card.

(9) Fixed (FI) and free (FR) cards. Although the model card specifies whether an entire matrix is fixed or free, most models require that a subset of matrix elements be freed or fixed for estimation. FI and FR cards fix or free individual matrix elements. For example, the model parameters card specifies Γ as full and fixed. The FR card frees Γ elements (1, 1) and (2, 2) for estimation. [It also frees several elements of **B**(BE).] If you wish to fix a matrix element to a value other than 0, you must supply that value on a separate ST or VA card. FI and FR cards can be up to 80 columns wide. The command can be continued on the next line by placing a "C" as the last symbol on the line.

(10) Output card. This line requests specific LISREL output options. OU must appear in the first two columns of the card. TV and SE request that "t-values" (actually *z*-values) and asymptotic standard errors for the parameter estimates be printed. SS requests the LISREL-standardized solution. TO prints the output with 80 character records. Other output options are available. The OU card also tells the program which estimation procedure should be used. The default, which is used here, is the maximum likelihood (ML) estimator.

EQS Program

I recommend the following strategy for EQS programming:

1. *Draw a path diagram (optional step).* See description under LISREL programming.
2. *Write the equations for the model.* For EQS, latent endogenous (η) and exogenous variables (ξ) are represented by F, the error in equations (ζ) by D, and the unique factors by E. The symbol for the observed variables is V in place of y and x. The equations are not in matrix notation but in scalar form. For the SES example, three equations would be listed, and the nonzero covariances between D_i and D_j noted.
3. *Note all parameters to be estimated.* All nonfixed coefficients for the equations from the last step are marked with $*$ to signify that they are to be estimated. Also you should list all variances and covariances between exogenous variables or disturbances that are to be estimated. The SES example has

$$y_1 = \beta_{12}^* y_2 + \gamma_{11}^* x_1 + \zeta_1$$
$$y_2 = \beta_{21}^* y_1 + \gamma_{22}^* x_2 + \zeta_2$$
$$y_3 = \beta_{31}^* y_1 + \gamma_{32}^* y_2 + \zeta_3$$

where the variances and covariances of x_1 and x_2 are free parameters, as are the variances and covariances of ζ_1 to ζ_3. A listing for the objective-subjective SES example follows. The explanation listed at the end of the program are numbered according to the numbers appearing at the right-hand margin.

```
/TITLE                                                        (1)
  OBJECTIVE-SUBJECTIVE SOCIOECONOMIC STATUS ML
/SPEC                                                         (2)
  CAS = 432; VAR = 5; ME = ML; MA = COV;
/LAB
  V1 = OCC; V2 = INC; V3 = SUBOCC; V4 = SUBINC;
  V5 = SUBGEN;
/EQU                                                          (4)
  V3 = .01*V1 + .2*V4 + D3; V4 = .2*V2 + .2*V3 + D4;
    V5 = .4*V4 + .4*V3 + D5;
/VAR                                                          (5)
  D3 TO D5 = .3*; V1 = 453*; V2 = 4.8*;
/COV                                                          (6)
  V1,V2 = 13.7*; D3,D4 = 0*; D3,D5 = 0*; D4,D5 = 0*;
/MAT                                                          (7)
  1.000
   .292    1.000
   .282     .184    1.000
   .166     .383     .386    1.000
   .231     .277     .431     .537    1.000
/STA                                                          (8)
  21.277   2.198    .640     .670     .627
/END                                                          (9)
```

(1) Title section. Each section begins with a "/" and a keyword that identifies the section to follow. The next line gives the title to the program. One or more lines can be listed.

(2) Specification section. This section gives the number of cases (CAS = 432), the number of input variables (VAR = 5), the method of estimation (ME = ML, where ML is maximum likelihood), and the matrix to be analyzed (MA = COV, where COV is covariance matrix). Since MA = COV; is the program default, this statement could have been omitted.

(3) Label section. Labels for latent or observed variables can be provided. A V followed by a number refers to the number of the observed variable. The order of numbers is determined by the order of the variables in the input matrix.

(4) Equation section. This section lists one equation for each endogenous variable. The variables with a direct effect on the endogenous variable and a disturbance term are listed to the right of the equal sign. A starting value for the parameter estimate precedes the variable on the right-hand side of an equation. An * indicates that the value for the parameter is to be estimated. For instance, in the SES model, subjective occupation prestige (V3) depends on "objective" occupation prestige (V1), subjective income (V4), and a disturbance (D3). The coefficients of V1 and V4 are to be estimated with starting values of .01 and .2, respectively. Each of the remaining two endogenous variables has a similarly constructed equation. The starting values are derived from the researcher's best guesses about the influences of one variable on another. It is important to consider the units of measurement or variances of variables when selecting a starting value (see Appendix 4C). LISREL VI automatically provides a starting values. Future versions of EQS will have this feature.

(5) Variance section. This section specifies the fixed or free values for the variances of all exogenous variables and disturbances. Endogenous variables should not be listed in the variance section. The SES example requires three variances for the three disturbance terms D3 to D5 and two for the exogenous variables V1 and V2. I have used the TO convention to refer to a range of consecutively numbered variables (D3 to D5). The * indicates free parameters.

(6) Covariance section. Here the fixed and free covariances among the exogenous variables or among the disturbances are specified. The convention is to list the two variables separated by a comma, followed by an equal sign and the value for the covariance. In the SES model the covariances of D3 and D4, D3 and D5, and D4 and D5 are free parameters, with starting values of 0. The covariance of V1 and V2 is free, with a starting value of 13.7 (the sample covariance of these two variables).

(7) Matrix input section. The matrix is either a correlation or covariance matrix. The lower half of a matrix with elements separated by blanks is expected unless different instructions were given in the /SPEC section. Note that this section does not end with a semicolon.

(8) Standard deviation section. The program requires the standard deviations of the observed variables to transform the correlation matrix into a covariance matrix. There should be as many standard deviations as variables in the correlation matrix. Note that this section does not end with a semicolon.

(9) Signifies end of program.

CHAPTER FIVE

The Consequences of Measurement Error

Structural equation models with observed variables was the topic of the preceding chapter. There I assumed that each variable was a perfect measure of its corresponding latent variable. This assumption is common in empirical research, though most often it is implicit rather than explicit. We need only look at our own research to know that the assumption is often incorrect. Whether we analyze attitudes, behaviors, dollars, or counts, the measures often contain both random and nonrandom errors.

In this chapter I examine the consequences of measurement error. The chapter is cumulative in that results derived in one section are important to the understanding of succeeding sections. I begin with the consequences for the mean and variance of one variable. Then I examine the covariance, correlation, and the simple regression between two imperfectly measured variables. I follow this with the topics of measurement error in multiple regression and in multiequation systems. In each section I draw a comparison of the results with and without measurement error, thus allowing an assessment of the differences that the error makes.

UNIVARIATE CONSEQUENCES

Researchers routinely calculate the mean and variance of a variable. In this section I derive the consequences of employing these statistics for an observed variable as proxies for the mean and variance of the true variable. To illustrate the univariate consequences of measurement error, I begin with one observed variable, X_1:

$$X_1 = \nu_1 + \lambda_{11}\xi_1 + \delta_1 \tag{5.1}$$

where ν_1 (upsilon) is an intercept, λ_{11} is a scaling constant, $E(\delta_1) = 0$, and $\text{COV}(\xi_1, \delta_1) = 0$. I do *not* use deviation scores for X_1 and ξ_1. As an example ξ_1 could be the achievement motive of salespeople and X_1 could be a scale to gauge it. The λ_{11} is the slope relating the measure of achievement to the true achievement level, and δ_1 represents the random error in measured achievement which is orthogonal to true achievement and $E(\delta_1) = 0$.

Represent the population mean of ξ_1 by κ_1 and the population mean of X_1 by μ_{x_1}. The relation between the mean of X_1 and the mean of ξ_1 is

$$\begin{aligned} E(X_1) &= E(\nu_1 + \lambda_{11}\xi_1 + \delta_1) \\ \mu_{x_1} &= \nu_1 + \lambda_{11}\kappa_1 \end{aligned} \tag{5.2}$$

Expression (5.2) shows that the relation between μ_{x_1} and κ_1 depends on the values of ν_1 and λ_{11}. A brief digression on the scaling of ξ_1 and the scaling constants ν_1 and λ_{11} is necessary to take this comparison forward. Virtually all latent variables have ambiguous scales. There are no inherent units of anxiety, expectations, stability, satisfaction, or many other constructs. Latent variables are given scales through agreements on the optimal units in which to measure them. For instance, to measure physical length, we could assign ξ_1 a metric scale. The metric system is not the only scale for length, but it is one that emerges from scientific consensus.[1] To give ξ_1 the metric scale, we must know the units of measurement for X_1 and then select appropriate values for ν_1 and λ_{11}. If we wish ξ_1 to be in centimeters, and have X_1 in centimeters, we can set ν_1 to zero and λ_{11} to one. Alternatively, suppose that we keep X_1 in centimeters but we want ξ_1 in inches. Now λ_{11} should be 2.54 (1 in. = 2.54 cm), but ν_1 remains at zero.

When agreement about the measurement unit for ξ_1 is absent, then the scale choice is largely arbitrary. With only one variable as in (5.1) a convenient choice is to give ξ_1 the same scale as X_1 in the sense that ν_1 is zero and λ_{11} is one. The result is $X_1 = \xi_1 + \delta_1$ so that a one-unit difference in ξ_1 corresponds to an expected difference of one unit in X_1. Under this choice of ν_1 and λ_{11}, equation (5.2) reveals that the mean of the observed variable is the same as the mean of the latent variable.

Two indicators of ξ_1 complicate the situation. Consider equations (5.3) and (5.4):

$$X_1 = \nu_1 + \lambda_{11}\xi_1 + \delta_1 \tag{5.3}$$

$$X_2 = \nu_2 + \lambda_{21}\xi_1 + \delta_2 \tag{5.4}$$

[1] The prevalence of the English system of measurement in the United States shows that the consensus on the scale for length is not complete.

Though we can select X_1 to scale ξ_1 (i.e., $\nu_1 = 0$, $\lambda_{11} = 1$), the equation for X_2 need not have $\nu_2 = 0$ or $\lambda_{21} = 1$. Under this scaling of ξ_1, μ_{x_1} equals κ_1, but the relation for X_2 is $\mu_{x_2} = \nu_2 + \lambda_{21}\kappa_1$.

As an illustration, suppose that ξ_1 is the achievement motivation of employees. We have the company's records of hours worked per week for X_1 and self-reported hours worked per week for X_2. We could scale ξ_1 in hours by setting ν_1 to zero and λ_{11} to one. Then the mean of X_1 would equal the mean of ξ_1. However, for this choice of scaling, the mean of self-reported hours (μ_{x_2}) could give a misleading idea of κ_1. For instance, self-reports of hours may be overstated by a constant amount ($\nu_2 > 0$), or each one-hour increase in ξ_1 may lead to more than one hour in self-reported time ($\lambda_{21} > 1$) on average. So the choice of scale for ξ_1 affects the relation between the mean of ξ_1 and the means of observed variables.

In sum, when we have a single measure of a latent variable, it is convenient to assign ξ_1 the same scale as X_1. The mean of X_1 then equals the mean of ξ_1. With a second measure of ξ_1, say, X_2, the mean X_2 can depart from the mean of ξ_1.

Up to this point I have couched the discussion in terms of the population, but in practice random samples of cases are the basis of estimates. The sample mean of the observed variable X_1 is represented as $\bar{X}_1$, and it is generally an unbiased estimator of μ_{x_1}. From (5.2) it is clear that the expected value of the sample mean $\bar{X}_1$ is $\nu_1 + \lambda_{11}\kappa_1$. Thus, when $\nu_1 = 0$ and $\lambda_{11} = 1$, $\bar{X}_1$, is an unbiased estimator of κ_1.

The variance of the observed variable compared to the variance of the latent variable is also of interest. Using equation (5.1) as the basic model and applying the definition of the variance as the covariance of a variable with itself, the variance of X_1 is

$$\mathrm{VAR}(X_1) = \lambda_{11}^2\phi_{11} + \mathrm{VAR}(\delta_1) \tag{5.5}$$

where ϕ_{11} is the variance of ξ_1. Notice that the constant ν_1 does not affect the variance of X_1. Without further restrictions on λ_{11}, the $\mathrm{VAR}(X_1)$ can be greater than, less than, or even equal to ϕ_{11}. However, the most common case is when we scale ξ_1 to X_1 so that λ_{11} is one. Looking at (5.5), we see that with a λ_{11} of one, $\mathrm{VAR}(X_1)$ equals ϕ_{11} only if the $\mathrm{VAR}(\delta_1)$ is zero. But this is a trivial case since it signifies an absence of random measurement error. Otherwise, the $\mathrm{VAR}(X_1)$ is greater than ϕ_{11}.

The discussion comparing $\mathrm{VAR}(X_1)$ to ϕ_{11} has been exclusively in terms of population variances. Turning our attention to the usual sample estimator [represented as $\mathrm{var}(X_1)$], we have

$$\mathrm{var}(X_1) = \frac{\Sigma_{i=1}^{N}(X_{1i} - \bar{X}_1)^2}{N - 1} \tag{5.6}$$

It is well known that the expected value of this sample variance is VAR(X_1). Since I have shown that the VAR(X_1) equals $\lambda_{11}^2\phi_{11}$ + VAR(δ_1), it follows that var(X_1) generally is a biased estimator that overestimates ϕ_{11} when λ_{11} is one.

As an illustration, consider an achievement motive scale employed by Bagozzi (1980, 258) in a study of performance and satisfaction for 122 industrial salesmen. One of his achievement scales consists of self-reported measures of the degree to which a salesperson values work. Its variance is 3.8, and the variance of δ_1 is about 2.5.[2] With the true achievement motive scaled to the observed measure (i.e., $\lambda_{11} = 1$), the variance of the measure (= 3.8) greatly exceeds the variance of the latent variable (= 1.3).

BIVARIATE AND SIMPLE REGRESSION CONSEQUENCES

In this section I examine the association of two variables when either or both measures contain error. My focus is on the effects of measurement error on the covariance, correlation, and the unstandardized and standardized regression coefficients. The strategy is the same as in the last section where I compare the results from the ideal case without error to those in the more typical case with error. The true model governing the variables is

$$\begin{aligned} x &= \lambda_1 \xi + \delta \\ y &= \lambda_2 \eta + \epsilon \\ \eta &= \gamma \xi + \zeta \end{aligned} \tag{5.7}$$

To simplify the notation, I return to deviation scores for x, y, η, and ξ. And I do not use subscripts for the variables nor for γ since their meaning is clear without them. In (5.7) λ_1, λ_2, and γ are nonzero constants, and the $E(\delta)$, $E(\epsilon)$, and $E(\zeta)$ are zero. The δ and ϵ variables represent errors of measurement for x and y, respectively. They are uncorrelated with each other and with η, ξ, and ζ. The error in the equation, ζ, also is uncorrelated with ξ. The first two equations of (5.7) describe the measurement model of this simple system. With only one indicator per latent variable, we can scale the latent variables by setting λ_1 and λ_2 to one.

Our goal is to know γ, but we only observe x and y. In practice, y is regressed on x to estimate γ. The differences between the true model with

[2] I estimated the variance of δ_1 by using a two-factor model of achievement and satisfaction, with two indicators of each latent variable and the covariance matrix derived from Bagozzi (1980, 259). Confirmatory factor analysis is explained in Chapter 7.

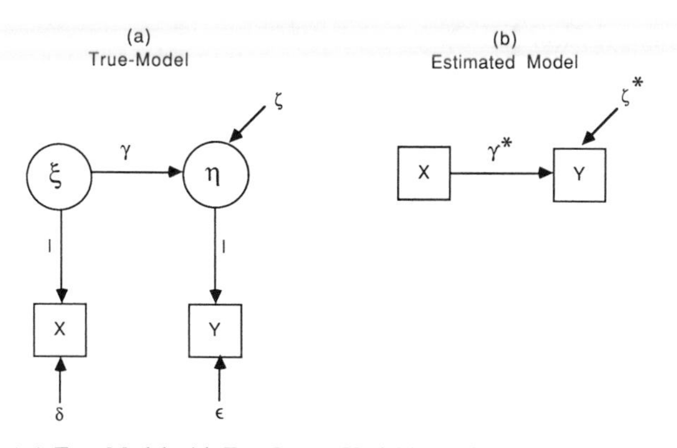

Figure 5.1 (*a*) A True Model with Two Latent Variables and (*b*) Estimated Model with Only Observed Variables

latent variables and the estimated one with observed variables are depicted in Figure 5.1. Figure 5.1(*a*) is the true model contained in equation (5.7) above. Figure 5.1(*b*) represents a model in which measurement error is ignored or assumed to be negligible. The γ and ζ in the observed variable model are superscripted with asterisks to distinguish them from the corresponding γ and ζ in the true model. In general, γ^* will not equal γ, and ζ^* will not equal ζ. The following paragraphs demonstrate this conclusion.

To compare γ^* to γ, a useful place to start is with the covariance between ξ and η and the covariance of x and y. The covariance of the latent variables is

$$\begin{aligned} \text{COV}(\xi, \eta) &= \text{COV}(\xi, \gamma\xi + \zeta) \\ &= \gamma\phi \end{aligned} \tag{5.8}$$

The covariance of the observed variables is

$$\begin{aligned} \text{COV}(x, y) &= \text{COV}(\xi + \delta, \eta + \epsilon) \\ &= \gamma\phi \end{aligned} \tag{5.9}$$

The covariance of the observed variables (x, y) equals the covariance of the latent variables (ξ, η).

These covariances are useful in comparing the regression coefficients γ^* to γ. As is easily shown, the $\text{COV}(\xi, \eta)$ divided by ϕ equals γ. Using the analogous expression for γ^*,

$$\begin{aligned}\gamma^* &= \frac{\mathrm{COV}(x, y)}{\mathrm{VAR}(x)} \\ &= \gamma\left[\frac{\phi}{\mathrm{VAR}(x)}\right] \end{aligned} \tag{5.10}$$

The quantity within brackets on the right-hand side of (5.10) equals one only when error variance in x is absent. Otherwise, it varies between zero and one and is the squared correlation between the latent and observed variables. Sometimes this is called ρ_{xx}, the reliability coefficient of x (see Chapter 6). Without perfect reliability, measurement error means that the absolute magnitude of $\gamma^* < \gamma$; that is, the coefficient for the observed variables is attenuated relative to that for the true latent variables.

To illustrate, consider the influence of achievement motive on job satisfaction. The data are from Bagozzi (1980), and I described the achievement measure (x) in the last section. The job satisfaction scale (y) gauges a salesperson's satisfaction with promotion opportunities, pay, general work situation, and other dimensions of the job. If job satisfaction (y) is regressed on the achievement measure (x), the estimated slope is 0.36. The reliability of the achievement measure is about 0.34. Assuming that the latent variables are given the same scale as the observed variables, the slope relating the latent job satisfaction variable (η) to the latent achievement motive variable (ξ) is 1.06. This nearly threefold increase (0.36 vs. 1.06) in the slope shows the serious impact that measurement error can have if not taken into account. These results are not affected by the reliability of the job satisfaction variable. In the case where λ_1 and λ_2 are not one (or not equal), the relation between γ and γ^* is $\gamma^* = \gamma(\lambda_2/\lambda_1)\rho_{xx}$. The attenuation of γ^* need not follow since (λ_2/λ_1) might exceed one. However, with a single indicator per latent variable this would be an unusual situation.

To continue the assessment of the differences between the true and estimated models shown in Figure 5.1, I now turn to the correlation of x and y (ρ_{xy}) compared with the correlation of ξ and η ($\rho_{\xi\eta}$). I first derive an expression for the squared correlation of ξ and η:

$$\begin{aligned}\rho_{\xi\eta}^2 &= \frac{[\mathrm{COV}(\xi, \eta)]^2}{\phi \mathrm{VAR}(\eta)} \\ &= \frac{\gamma^2\phi}{\mathrm{VAR}(\eta)} \end{aligned} \tag{5.11}$$

I use this result in examining the squared correlation of the observed variables x and y:[3]

$$\begin{aligned}
\rho_{xy}^2 &= \frac{[\mathrm{COV}(x, y)]^2}{\mathrm{VAR}(x)\mathrm{VAR}(y)} \\
&= \frac{\gamma^2\phi^2}{\mathrm{VAR}(x)\mathrm{VAR}(y)} \\
&= \rho_{xx}\frac{\gamma^2\phi}{\mathrm{VAR}(y)}\frac{\mathrm{VAR}(\eta)}{\mathrm{VAR}(\eta)} \\
&= \rho_{xx}\rho_{yy}\rho_{\xi\eta}^2 \qquad (5.12)
\end{aligned}$$

Equations (5.12) show that the squared correlation of x and y (ρ_{xy}^2) equals the product of the reliabilities of x and y ($\rho_{xx}\rho_{yy}$) times the squared correlation of the latent variables ($\rho_{\xi\eta}^2$). Since reliability coefficients range between 0 and 1, ρ_{xy} will be less than or equal to $\rho_{\xi\eta}$. Equation (5.12) also provides the results for the standardized regression coefficient, since for the case of a simple regression it is equivalent to the correlation of the explanatory and dependent variables. It shows that the standardized regression coefficient for the observed variables cannot exceed that for the latent variables.

Equation (5.12) contains still further information. Solving it for $\rho_{\xi\eta}$ leads to

$$\rho_{\xi\eta} = \frac{\rho_{xy}}{[\rho_{xx}\rho_{yy}]^{1/2}} \qquad (5.13)$$

Equation (5.13) is the classical formula for the correction for attenuation of correlation coefficients due to random measurement error, a formula well known in psychology and sociology. If the reliabilities of x and y are known, the correlation between the latent variables can be determined from (5.13) knowing the correlation of x and y. For instance, the correlation of the job satisfaction scale (y) with achievement motivation (x) is 0.20. The reliability of the achievement measure is 0.34, and the reliability for job satisfaction is 0.66. Applying formula (5.13), the correlation of the latent satisfaction and latent achievement variables is 0.42, a value that is more than twice that for the observed variables.

[3] I use a simplification in this derivation suggested by Jerome Katrichis.

Thus far, in deriving the consequences of measurement error in the bivariate context, I have dealt with the population covariances, regression coefficients, and correlations.[4] The sample estimator of the population covariance is

$$\operatorname{cov}(x, y) = \frac{\sum_{i=1}^{N}(x_i y_i)}{N - 1} \tag{5.14}$$

The sample covariance is well known to be an unbiased and consistent estimator of the population covariance $\operatorname{COV}(x, y)$. In (5.9) I showed that $\operatorname{COV}(x, y)$ equals $\gamma\phi$. The sample $\operatorname{cov}(x, y)$ is an unbiased and consistent estimator of the covariance of the latent variables.

Ordinary least squares (OLS) is the most frequently used estimator of regression coefficients. In the case of simple regression, the OLS estimator $\hat{\gamma}^*$, when y is regressed on x, is

$$\hat{\gamma}^* = \frac{\operatorname{cov}(y, x)}{\operatorname{var}(x)} \tag{5.15}$$

Using the asymptotic theory reviewed in Appendix B, we can determine if $\hat{\gamma}^*$ is a consistent estimator. If $\hat{\gamma}^*$ is a consistent estimator of γ, the probability limits of $\hat{\gamma}^*$ as N approaches infinity [i.e., $\operatorname{plim}(\hat{\gamma}^*)$] should converge to γ. Instead, we find

$$\begin{aligned}\operatorname{plim}(\hat{\gamma}^*) &= \operatorname{plim}\left[\frac{\operatorname{cov}(y, x)}{\operatorname{var}(x)}\right] \\ &= \frac{\operatorname{COV}(y, x)}{\operatorname{VAR}(x)} \\ &= \gamma\rho_{xx}\end{aligned} \tag{5.16}$$

Note that $\operatorname{plim}(\hat{\gamma}^*)$ equals the same quantity as that in equation (5.10). Thus the earlier results for the relation of γ^* to γ apply here. Generally, $\hat{\gamma}^*$ is an inconsistent estimator of γ.

[4] In the achievement-satisfaction example I have ignored sampling fluctuations.

Finally, the sample estimator of the bivariate squared correlation is

$$\frac{[\mathrm{cov}(y, x)]^2}{\mathrm{var}(y)\mathrm{var}(x)} \tag{5.17}$$

The $\mathrm{plim}(\hat{\rho}^2_{xy})$ is

$$\begin{aligned}\mathrm{plim}(\hat{\rho}^2_{xy}) &= \frac{\mathrm{plim}[\mathrm{cov}(y, x)]^2}{\mathrm{plim}[\mathrm{var}(y)]\mathrm{plim}[\mathrm{var}(x)]} \\ &= \frac{[\mathrm{COV}(x, y)]^2}{\mathrm{VAR}(x)\mathrm{VAR}(y)} \\ &= \rho_{xy}\rho_{yy}\rho^2_{\xi\eta}\end{aligned} \tag{5.18}$$

The last step in equation (5.18) follows from substituting the results of (5.12). Equation (5.18) indicates that unless there is no random measurement error in x and y, $\hat{\rho}^2_{xy}$ is an inconsistent estimator of $\rho^2_{\xi\eta}$.

Let me summarize the major findings of this section. The covariance of two observed variables equals the covariance of their corresponding latent variables (when $\lambda_1 = \lambda_2 = 1$). The covariance of x and y divided by the variance of x leads to the regression slope, γ^*, for y regressed on x. The absolute value of the slope γ^* is less than the absolute value of γ whenever x contains random error (i.e., $\gamma^* = \rho_{xx}\gamma$). This relation is not affected by random error in y. The squared correlation between two measures is attenuated relative to the latent variables whenever the reliability of x or y is less than one (i.e., $\rho^2_{xy} = \rho_{xx}\rho_{yy}\rho^2_{\xi\eta}$). In short, we are more likely to be wrong than right if we assume that the relation between two observed variables is the same as that between the latent variables. Since the latent variables are of primary interest, the potential for misleading findings is great if measurement error is ignored. The relations become more complicated if λ_1 and λ_2 are not equal to one.

CONSEQUENCES IN MULTIPLE REGRESSION

The last section examined the consequences of errors of measurement for simple regression. I now take up multiple regression in which the evaluation of the consequences of measurement error becomes more complex. My strategy for assessing the impact of error remains the same as in prior sections. I compare the true models with latent variables to the estimated ones with observed variables and contrast the regression coefficients and multiple correlation coefficient from each.

I begin with a formal statement of the models. The true model relating the latent concepts is

$$\eta = \Gamma\boldsymbol{\xi} + \zeta \tag{5.19}$$

where η represents a single latent endogenous variable, $\boldsymbol{\xi}$ is an $n \times 1$ vector of exogenous variables, Γ is a $1 \times n$ vector of coefficients, and ζ is the disturbance term which is uncorrelated with $\boldsymbol{\xi}$ and $E(\zeta) = \mathbf{0}$. The estimated model with observed variables is

$$y = \Gamma^*\mathbf{x} + \zeta^* \tag{5.20}$$

where the coefficient matrix (Γ^*) and the disturbance term (ζ^*) have asterisks to distinguish them from the Γ and ζ of (5.19).

The relation of $\mathbf{x}$ to $\boldsymbol{\xi}$ and y to η is

$$\begin{aligned} \mathbf{x} &= \boldsymbol{\xi} + \boldsymbol{\delta} \\ y &= \eta + \epsilon \end{aligned} \tag{5.21}$$

The errors of measurement ($\boldsymbol{\delta}, \epsilon$) are uncorrelated with ξ, η, ζ, and with each other. In (5.21) $\mathbf{x}$, $\boldsymbol{\xi}$, and $\boldsymbol{\delta}$ are $q \times 1$, and y, η, and ϵ are scalars. To simplify the discussion, I assume one, and only one, observed variable x for each latent variable ξ. Moreover I assume that each ξ influences only one x. Only one y variable is used for η. Furthermore the observed variables have the same scale as the latent variables (i.e., $\Lambda_{\mathbf{x}} = \mathbf{I}$ and $\Lambda_y = 1$). In the typical regression application our interest is in Γ, which provides the impact of the explanatory variables on the dependent variable. However, if we ignore the error in y and $\mathbf{x}$ we end up with Γ^*. What is the relation of Γ^* to Γ? To begin to answer this question, I show that Γ is obtainable from the product of covariance matrices of the latent variables. Then I develop the analogous expression with observed variables and its relation to Γ. Consider first the latent variable model of equation (5.19). The covariance matrix of the exogenous variables, ξ, is $\boldsymbol{\Phi}$. The covariance matrix of $\boldsymbol{\xi}$ and η, symbolized as $\boldsymbol{\Sigma}_{\xi\eta'}$ is:

$$\begin{aligned} \boldsymbol{\Sigma}_{\xi\eta'} &= \mathrm{COV}(\xi, \eta') \\ &= \mathrm{COV}(\xi, \xi'\Gamma' + \zeta') \\ &= \boldsymbol{\Phi}\Gamma' \end{aligned} \tag{5.22}$$

Multiplying $\mathbf{\Phi}^{-1}$ times $\mathbf{\Sigma}_{\xi\eta'}$ leads to the structural parameters in $\mathbf{\Gamma}'$:

$$\begin{aligned}\mathbf{\Phi}^{-1}\mathbf{\Sigma}_{\xi\eta'} &= \mathbf{\Phi}^{-1}\mathbf{\Phi}\mathbf{\Gamma}' \\ &= \mathbf{\Gamma}' \end{aligned} \tag{5.23}$$

We cannot compute $\mathbf{\Phi}^{-1}\mathbf{\Sigma}_{\xi\eta'}$ since we do not have $\boldsymbol{\xi}$ and η. Instead of $\mathbf{\Phi}^{-1}\mathbf{\Sigma}_{\xi\eta'}$ we utilize its counterpart for the observed variables $\mathbf{\Gamma}^{*\prime} = \mathbf{\Sigma}_{xx'}^{-1}\mathbf{\Sigma}_{xy'}$, where $\mathbf{\Sigma}_{xx'}^{-1}$, is the inverse of the covariance matrix of $\mathbf{x}$ and $\mathbf{\Sigma}_{xy'}$ is the covariance matrix of $\mathbf{x}$ and y. Only if it leads to $\mathbf{\Gamma}'$ will $\mathbf{\Sigma}_{xx'}^{-1}\mathbf{\Sigma}_{xy'}$ be appropriate. To see what this product is, consider first $\mathbf{\Sigma}_{xy'}$:

$$\begin{aligned}\mathbf{\Sigma}_{xy'} &= \text{COV}(\mathbf{x}, y') \\ &= \text{COV}(\mathbf{x}, \eta' + \epsilon') \\ &= \text{COV}(\mathbf{x}, \eta') \\ &= \text{COV}(\mathbf{x}, \boldsymbol{\xi}'\mathbf{\Gamma}' + \zeta') \\ &= \mathbf{\Sigma}_{x\xi'}\mathbf{\Gamma}' \end{aligned} \tag{5.24}$$

Employing this result and the formula for $\mathbf{\Gamma}^{*\prime}$, the relation of $\mathbf{\Gamma}^{*\prime}$ to $\mathbf{\Gamma}'$ is

$$\begin{aligned}\mathbf{\Gamma}^{*\prime} &= \mathbf{\Sigma}_{xx'}^{-1}\mathbf{\Sigma}_{xy'} \\ &= \mathbf{\Sigma}_{xx'}^{-1}\mathbf{\Sigma}_{x\xi'}\mathbf{\Gamma}' \end{aligned} \tag{5.25}$$

Equation (5.25) shows that $\mathbf{\Gamma}^{*\prime}$ based on the observed variables differs from $\mathbf{\Gamma}'$ by the factor $\mathbf{\Sigma}_{xx'}^{-1}\mathbf{\Sigma}_{x\xi'}$. The latter factor is the population analog of the formula for the regression coefficients of an equation in which the latent exogenous variables $\boldsymbol{\xi}$ are regressed on the observed variables $\mathbf{x}$. If there is no error, then $\mathbf{x}$ equals $\boldsymbol{\xi}$, $\mathbf{\Sigma}_{xx'}^{-1}\mathbf{\Sigma}_{x\xi}$ equals $\mathbf{I}$, and $\mathbf{\Gamma}^{*}$ equals $\mathbf{\Gamma}$. More generally, the relation of $\mathbf{\Gamma}'$ to $\mathbf{\Gamma}^{*\prime}$ depends on the relation of $\boldsymbol{\xi}$ to $\mathbf{x}$.

The preceding results refer to population covariance matrices, but analogous results hold when estimators based on sample data are applied. For a multiple regression equation, the ordinary least squares estimator of $\mathbf{\Gamma}$, which is equivalent to the maximum likelihood estimator when ζ is normally distributed, is

$$\hat{\mathbf{\Gamma}}' = \hat{\mathbf{\Phi}}^{-1}\mathbf{S}_{\xi\eta'} \tag{5.26}$$

where $\hat{\mathbf{\Phi}}^{-1}$ and $\mathbf{S}_{\xi\eta'}$ are the unbiased sample estimators of the covariance matrices. Taking probability limits (plim) as N goes to infinity of $\hat{\mathbf{\Phi}}^{-1}\mathbf{S}_{\xi\eta'}$

leads to $\boldsymbol{\Phi}^{-1}\boldsymbol{\Sigma}_{\xi\eta'}$. We already saw that this equals $\boldsymbol{\Gamma}'$ in (5.23), so $\hat{\boldsymbol{\Gamma}}'$ is a consistent estimator of $\boldsymbol{\Gamma}'$.

For an analysis that ignores the measurement error in y and $\mathbf{x}$, the OLS estimator of the structural coefficients is

$$\hat{\boldsymbol{\Gamma}}^{*\prime} = \mathbf{S}_{xx'}^{-1}\mathbf{S}_{xy'} \tag{5.27}$$

Taking probability limits of $\mathbf{S}_{xx'}^{-1}\mathbf{S}_{xy'}$ leads to $\boldsymbol{\Sigma}_{xx'}^{-1}\boldsymbol{\Sigma}_{xy'}$. Equation (5.25) shows that generally $\hat{\boldsymbol{\Gamma}}^{*\prime}$ is not a consistent estimator of $\boldsymbol{\Gamma}'$ but rather that $\hat{\boldsymbol{\Gamma}}^{*\prime}$ is a consistent estimator of $\boldsymbol{\Sigma}_{xx'}^{-1}\boldsymbol{\Sigma}_{x\xi'}\boldsymbol{\Gamma}'$. Thus, when we move from population to sample estimators, analogous results hold.

The form of $\boldsymbol{\Sigma}_{xx'}^{-1}\boldsymbol{\Sigma}_{x\xi'}$ can lead to complex relations between $\boldsymbol{\Gamma}^*$ and $\boldsymbol{\Gamma}$. As an illustration, I treat the simplest situation where all explanatory variables are perfectly measured except one. I begin with the analytic results and follow this with an empirical example. The measurement model in this case is

$$\begin{aligned} x_i &= \xi_i + \delta_i, \qquad \text{for } i = 1 \\ x_i &= \xi_i, \qquad \text{for } i = 2, 3, \ldots, q \end{aligned} \tag{5.28}$$

The equation linking the latent variables is

$$\eta = \gamma_1\xi_1 + \gamma_2\xi_2 + \cdots + \gamma_q\xi_q + \zeta \tag{5.29}$$

To simplify the notation, I do not use double subscripts for the γ's, nor do I subscript η or ζ. The estimated equation is

$$\eta = \gamma_1^* x_1 + \gamma_2^*\xi_2 + \cdots + \gamma_q^*\xi_q + \zeta^* \tag{5.30}$$

The plim($\hat{\gamma}_1^*$) equals

$$\operatorname{plim}(\hat{\gamma}_1^*) = b_{\xi_1 x_1 \cdot \xi_2 \ldots \xi_q}\gamma_1 \tag{5.31}$$

In (5.31) $b_{\xi_1 x_1 \cdot \xi_2 \xi_3 \ldots \xi_q}$ is the partial regression coefficient associated with x_1 when ξ_1 is regressed on all the explanatory variables in (5.30). Levi (1973) shows that the absolute value of this coefficient ranges from zero to one. This means that in a multiple regression with all but one of the explanatory variables perfectly measured, the coefficient associated with the observed variable containing measurement error is asymptotically biased toward zero, and thus underestimates ξ_1's effect on η.

Interestingly, the $\hat{\gamma}_1^*$ coefficient is not the only one affected by the random measurement error in ξ_1. Unless the omitted latent variable (ξ_1) is uncorrelated with the other latent explanatory variables, or the effect of ξ_1 is zero, then the $\hat{\gamma}^*$'s associated with the other ξ's are likely to be inconsistent estimators. Consider $\hat{\gamma}_2^*$ from (5.30). Its probability limit is

$$\text{plim}(\hat{\gamma}_2^*) = \gamma_2 + b_{\xi_1\xi_2 \cdot x_1\xi_3 \ldots \xi_q}\gamma_1 \tag{5.32}$$

As you can see, $\hat{\gamma}_2^*$ consists of both the true effect of ξ_2 and the relation of ξ_1 to ξ_2, holding constant $x_1, \xi_3, \ldots, \xi_q$, multiplied by γ_1. Thus $\hat{\gamma}_2^*$ is not a consistent estimator of γ_2. The direction of asymptotic bias is difficult to predict, since it depends on the magnitude and sign of the second term on the right-hand side of (5.32). Analogous formulas can be derived for the other coefficients in (5.30).

I illustrate these results with a model that relates air pollution to mortality. The data and model specification come from Lave and Seskin (1977, 1970). Bollen and Schwing (1987) examine the consequences of random error for this model. The sample is 108 of the major Standard Metropolitan Statistical Areas (SMSA) in the United States. Mortality is estimated as a linear function of seven variables:[5]

$$\eta = \sum_{i=1}^{7} \gamma_i \xi_i + \zeta \tag{5.33}$$

where

η = crude death rate

ξ_1 = sulfate pollutant

ξ_2 = particulate pollutant

ξ_3 = population density

ξ_4 = percentage of the population over 65 years old

ξ_5 = percentage of nonwhite population

ξ_6 = percentage of families with income below poverty line

ξ_7 = natural log of the population

Lave and Seskin (1977, 1970) propose two types of air pollutants that may increase mortality. These are sulfate and particulate pollutants as measured at monitoring sites in each urban area. I represent these as ξ_1 and

[5]This seven-variable model is based on the specification of Lave and Seskin (1977). I use the sample of cases suggested by Gibbons and McDonald (1980) which omits outliers present in Lave and Seskin's data. The correlation matrix is analyzed.

Table 5.1 Regression Coefficients for Lave and Seskin's Seven-Variable Model, Assuming Varying Amounts of Measurement Error in x_1, the Measure of Sulfate Pollutant (N = 108)

Reliability of	Regression Coefficients							
x_1	ξ_1	ξ_2	ξ_3	ξ_4	ξ_5	ξ_6	ξ_7	R^2
1.0	0.107	0.090	0.064	1.008	0.370	−0.063	−0.076	0.839
0.9	0.123	0.086	0.060	1.003	0.369	−0.062	−0.072	0.840
0.7	0.173	0.072	0.047	0.986	0.363	−0.056	−0.056	0.845
0.5	0.291	0.039	0.017	0.947	0.350	−0.044	−0.020	0.855

ξ_2 in the above model. They also include the control variables ξ_3 to ξ_7 because of their influence on mortality and their potential association with pollutant levels. More detailed descriptions of all measures are in Lave and Seskin (1977, 1970). For the purposes of this example, I assume that all the variables except the sulfate pollutant measure are perfect indicators of their corresponding latent variable (e.g., $x_2 = \xi_2$, $x_3 = \xi_3, \ldots$). Though it is clear that the sulfate pollutant measure (x_1) contains random measurement error, the magnitude of error is unknown. I assume a range of values for the reliability of x_1 to explore the sensitivity of the coefficients under different assumptions of the extent of measurement error in x_1.

Table 5.1, adapted from Bollen and Schwing (1987), shows the regression estimates obtained for each explanatory variable varying the reliability of $x_1(\rho_{x_1x_1})$, the measure of the sulfate pollutant. The first row of the body of the table lists the coefficients that result when all variables, including x_1, are perfectly measured. These are what would be obtained if no corrections for measurement error were made. The rows beneath it provide the coefficients correcting for various degrees of error in x_1; all other variables are measured without error. I begin with the case where half of the variance in x_1 is attributable to random measurement error. Taking this into account yields an estimate of γ_1 equal to 0.291 (see the last row of Table 5.1). If the measurement error in x_1 were ignored or implicitly assumed to be absent, the estimated effect of $\xi_1(\hat{\gamma}_1^*)$ is 0.107. Thus we see a specific example of the general result treated earlier: $\hat{\gamma}_1^*$ is considerably less than $\hat{\gamma}_1$. A comparison of the other $\hat{\gamma}_i$ to $\hat{\gamma}_i^*$, for i = 2 to 7 confirms that $\hat{\gamma}_i$ does not equal $\hat{\gamma}_i^*$. This is despite the fact that in this illustration I have assumed $\xi_2, \ldots, \xi_7$ to be perfectly measured by $x_2, \ldots, x_7$, respectively. For instance, consider $\hat{\gamma}_2^*$ the influence of particulate pollutants on mortality. When the error in x_1 is ignored, the $\hat{\gamma}_2^*$ coefficient is 0.090. If measurement error in x_1 is taken into account, however, the coefficient drops to 0.039 for $\rho_{x_1 x_1}$ equal to 0.50. Indeed, Table 5.1 shows several coefficient values to be quite sensitive to changes in the reliability of x_1. This indicates that random measurement error can undermine attempts to estimate the effects of one variable on another even if all but one of the variables are free of error.

It is worth noting that all of the derivations in this section have used simplifying assumptions: (1) only single indicators for the latent variable; (2) no differences in scaling between the latent and observed variables. If these assumptions are relaxed, the relation between $\hat{\Gamma}^*$ and Γ becomes quite complicated. The preceding results should suffice to show that random measurement error in the explanatory variables generally leads to $\hat{\Gamma}^*$ being an inconsistent estimator of Γ.

A possible response to indicators that contain random measurement error is to exclude them and to only use the explanatory variables with no or negligible measurement error. McCallum (1972) and Wickens (1972) point to the limits of this strategy, since the asymptotic bias of the OLS estimator for the coefficients of the remaining variables in the equation that omits an indicator with error is greater than that for the estimator of the equation that includes the indicator. Aigner (1974) qualifies their work in that it is possible for the *variance* of the estimator omitting the proxy variable to be less than that of including the proxy. Judge et al. (1980, 518) note that Aigner's results apply under rather special conditions. In general, it is better to include an indicator with error than to use no indicator of a latent variable that belongs in an equation. Dropping or keeping a faulty indicator are not the only choices. As I soon demonstrate, we can construct models that explicitly allow measurement error in indicators and that lead to consistent estimators.

Standardized regression coefficients are common in some social science disciplines. In the population the standardized coefficient for an element of Γ is defined as

$$\text{Standardized } \gamma_i \equiv \gamma_i \sqrt{\frac{\phi_{ii}}{\text{VAR}(\eta)}} \tag{5.34}$$

The corresponding element of Γ^* is:

$$\begin{aligned}\text{Standardized } \gamma_i^* &\equiv \gamma_i^* \sqrt{\frac{\text{VAR}(x_i)}{\text{VAR}(y)}} \\ &= \gamma_i^* \sqrt{\frac{\phi_{ii} + \text{VAR}(\delta_i)}{\text{VAR}(\eta) + \text{VAR}(\epsilon)}}\end{aligned} \tag{5.35}$$

Comparing (5.34) to (5.35) points to two components of bias in γ_i^*. Not only do γ_i and γ_i^* diverge, but the "standardizing" factors also are not equivalent. Thus the standardized regression coefficients in a model that ignores error in variables differ from the corresponding standardized coefficients in the model with the errorless variables.

The final comparison I draw is between the squared multiple correlation coefficient for regression equations with and without explanatory variables measured with error. In the last section I showed that the squared correlation coefficient ($\hat{\rho}_{xy}^2$) of the observed variables is an inconsistent estimator of the squared correlation between the latent variables ($\rho_{\xi\eta}^2$). Specifically, the plim($\hat{\rho}_{xy}^2$) is less than $\rho_{\xi\eta}^2$. A similar result obtains for the squared multiple correlation coefficient commonly estimated in multiple regressions, Dhrymes' (1978, 261–266) results imply that

$$\operatorname{plim}(R^2) \geq \operatorname{plim}(R^{*2}) \tag{5.36}$$

where R^2 and R^{*2} are the squared multiple correlation coefficients for the equations containing variables without and with measurement error, respectively. According to (5.36), as N goes to infinity, the squared multiple correlation coefficient for the equation containing variables with error (R^{*2}) converges to a value less than that for the same equation with error-free variables (R^2). Random measurement error attenuates the multiple correlation. Table 5.1 provides corroborative evidence in the column labeled R^2. For the first row, where x_1 is assumed to have zero random error, the multiple correlation is estimated as 0.839. If in fact x_1 only has a reliability of 0.5, the R^2 is estimated at 0.855, a slightly higher value. In this case the change in reliability has a minor effect on the R^2. As the other examples will show, this is not always the case.

In this section I examined the consequences of random measurement error for regression coefficients and multiple correlation coefficient in multiple regression. The relation between the coefficients and correlations for models with and without measurement error is more complex than that for simple regression. For instance, in simple regression with the observed and latent variables set to the same scale, the regression coefficient is attenuated to the extent of the explanatory variable's reliability. In multiple regression this generalization does not hold. If only one explanatory variable is subject to error, its regression coefficient is attenuated, but the extent of attenuation depends on the relation of the latent variable to the measured variable as well as its relation to the other explanatory variables. Furthermore the coefficients of the variables free of error are affected as well. Thus error even in a single variable can make itself felt through all of the coefficients. The air pollution-mortality model illustrated these findings.

The situation is further complicated if more than one variable is measured with error. Here any generalizations are difficult. In particular, measurement error does not always attenuate regression coefficients. Coefficients from equations that ignore error in variables can be higher, lower, or even the same as the true coefficients. In contrast, the population multiple correlation coefficient for an equation is less than that containing the same variables without error, a result that parallels that for the bivariate correlation coefficient.

CORRELATED ERRORS OF MEASUREMENT

A simplifying assumption maintained up to this point is that the measurement errors ($\boldsymbol{\delta}$, ϵ) are uncorrelated. Suppose that they are correlated. How does this change the previous results? I consider first the case when the covariances of ϵ with at least one of the unique factors in $\boldsymbol{\delta}$ is nonzero. Now the covariance vector of $\mathbf{x}$ and y' is

$$\begin{aligned}\Sigma_{xy'} &= \text{COV}(\mathbf{x}, y') \\ &= \Sigma_{x\xi}\Gamma' + \Sigma_{\delta\epsilon}\end{aligned} \tag{5.37}$$

This equals equation (5.24), except for the addition of the covariance vector between $\boldsymbol{\delta}$ and ϵ. The $\Gamma^{*\prime}$ equals:

$$\begin{aligned}\Gamma^{*\prime} &= \Sigma_{xx}^{-1}\Sigma_{xy} \\ &= \Sigma_{xx}^{-1}\Sigma_{x\xi}\Gamma' + \Sigma_{xx}^{-1}\Sigma_{\delta\epsilon}\end{aligned} \tag{5.38}$$

Compare this to (5.25), where $\Gamma^{*\prime} = \Sigma_{xx}^{-1}\Sigma_{x\xi}\Gamma'$, and we see that the $\Gamma^{*\prime}$ to Γ' relation is further complicated by the covariance of the unique factors.

In the case of simple regression, (5.38) is

$$\gamma^* = \gamma\rho_{xx} + \frac{\text{COV}(\epsilon, \delta)}{\text{VAR}(x)} \tag{5.39}$$

When the measurement errors were uncorrelated, $|\gamma^*|$ was less than $|\gamma|$ for reliability coefficients less than one. Equation (5.39) reveals that with $\text{COV}(\epsilon, \delta)$ the same sign as γ, this no longer need be true. Alternatively, if γ and $\text{COV}(\epsilon, \delta)$ have opposite signs, the attenuation of γ^* relative to γ is worse than when ϵ and δ are uncorrelated. Indeed, γ^* and γ could conceivably have opposite signs. More generally, the relation between $\Gamma^{*\prime}$ and Γ' is further complicated when ϵ is correlated with $\boldsymbol{\delta}$. In addition, on average the squared multiple correlation coefficient need no longer be greater in the model with the perfectly measured variables.

Return to the assumption that ϵ and δ are uncorrelated but assume that elements of $\boldsymbol{\delta}$ are intercorrelated (i.e., $E(\boldsymbol{\delta}\boldsymbol{\delta}') = \boldsymbol{\Theta}_\delta$, $\boldsymbol{\Theta}_\delta$ not diagonal). Under this condition $\boldsymbol{\Gamma}^{*\prime}$ still equals $\boldsymbol{\Sigma}_{xx}^{-1}\boldsymbol{\Sigma}_{x\xi}\boldsymbol{\Gamma}'$ so that the previous results hold.

CONSEQUENCES IN MULTIEQUATION SYSTEMS

Having examined the consequences of random measurement error in bivariate and multiple regression, I now turn to its consequences in multiequation systems. As before my focus is on coefficients of variables and multiple correlations in the same model with and without measurement error. The main difference is that I examine a system of equations rather than a single equation. I present the comparisons by way of two examples rather than with formal derivations. I also include a subsection that treats methods of estimating the measurement error if the reliability is unknown.

Union Sentiment Example

I introduced McDonald's and Clelland's (1984) data in Chapter 4. Their study of southern textile workers concerned the determinants of employee attitudes toward supervisors, activism, and unions. Figure 5.2 shows the recursive model of union sentiment. The two exogenous variables are years in the mill (x_1) and age (x_2). The three endogenous variables are deference (η_1), activism (η_2), and union sentiment (η_3). When I presented the model in Chapter 4, I implicitly assumed that measurement error was absent from all variables. Such an assumption is difficult to justify, particularly for the endogenous variables that are attitudinal. Fortunately, McDonald and Clelland (1984) report estimates of reliability for three of these variables: 0.57 for deference (y_1), 0.77 for worker activism (y_2), and 0.84 for union sentiment (y_3). I treat the reliability estimates as if they were population values to illustrate the consequences of random measurement error for the union sentiment model. I set VAR(ϵ_i) to $(1 - \hat{\rho}_{y_iy_i})\text{var}(y_i)$. The justification for this is that $(1 - \hat{\rho}_{y_iy_i})$ is the proportion of the variance of y_i due to measurement error.

Table 5.2 lists the ML and ULS estimates both with and without the reliability information.[6] Columns (1) and (3) reproduce the estimates of

[6]The results in Table 5.2 are approximate since in incorporating the reliability estimates, I have not taken into account the sampling fluctuations of the estimates of reliability. See Fuller and Hidiroglous (1978).

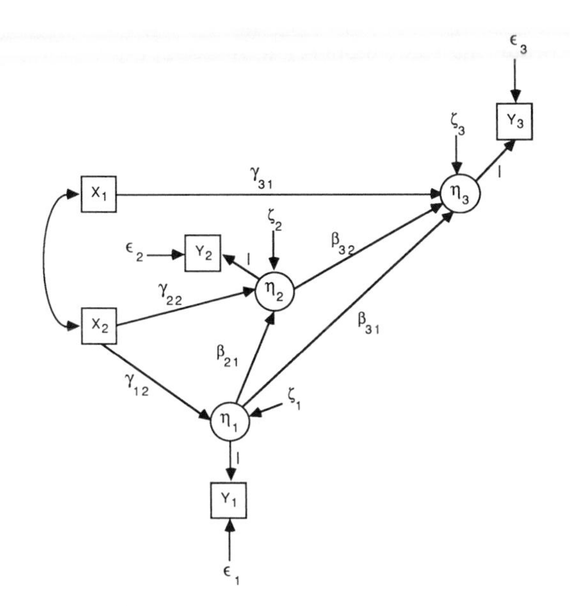

Figure 5.2 Union Sentiment Model with Measurement Error in Y_1, Y_2, and Y_3

Table 4.3 where all measures were treated as error-free. Columns (2) and (4) list the estimates with the error variances for y_1 to y_3 set in accordance with the reliability of the measures. A comparison of the first and the third columns and the second with the fourth reveals few differences for ML versus ULS estimates. The sharpest differences are for the estimates that allow for nonzero error variances [columns (2) and (4)] and those that do not [columns (1) and (3)]. For instance, when measurement error is incorporated into the model, the ML estimate of the influence of deference on activism ($\hat{\beta}_{21}$) changes from -0.29 to -0.55, the effect of age on activism ($\hat{\gamma}_{22}$) declines from 0.06 to 0.03, and the effect of activism on union sentiment ($\hat{\beta}_{32}$) shifts form 0.85 to 1.11. In addition the $R^2_{\eta_i}$'s increase markedly for two out of the three endogenous variables (η_1, η_2). For instance, the squared multiple correlation coefficient for activism is 0.42 when measurement error is allowed, compared to 0.23 when it is not. The coefficient of determination increases slightly from 0.21, without allowing measurement error to 0.26 when it is included. Similar contrasts exist for the ULS estimates.

This example illustrates several points. First, it shows that reliability estimates provide a means of estimating error variances that you can incorporate into a model. Second, it points out that the squared multiple correlation coefficient can increase dramatically when this is done. This will not always occur as the modest increases in the R^2 for the air pollution–mortality model showed, but the potential for increases is present. Table 5.2 also demonstrates that some coefficient estimates shift significantly, while others are more stable.

Table 5.2 The ML and ULS Estimates for Union Sentiment Model: Columns (1) and (3) without and Columns (2) and (4) with Random Measurement Error in y_1, y_2, and y_3

	ML Estimate (standard error)		ULS Estimate	
Parameter	(1)	(2)	(3)	(4)
γ_{12}	−0.087	−0.087	−0.087	−0.087
	(0.019)	(0.019)		
γ_{22}	0.058	0.034	0.058	0.034
	(0.016)	(0.019)		
γ_{31}	0.860	0.707	1.332	0.595
	(0.340)	(0.346)		
β_{21}	−0.285	−0.554	−0.284	−0.556
	(0.062)	(0.132)		
β_{31}	−0.218	−0.219	−0.192	−0.202
	(0.097)	(0.237)		
β_{32}	0.850	1.113	0.846	1.133
	(0.112)	(0.214)		
ψ_{11}	12.961	6.695	12.959	6.689
	(1.398)	(1.397)		
ψ_{22}	8.489	4.971	8.475	4.977
	(0.915)	(1.017)		
ψ_{33}	19.454	11.687	18.838	11.820
	(2.098)	(2.169)		
ϕ_{11}	1.021	1.021	0.774	1.130
	(0.110)	(0.110)		
ϕ_{21}	7.139	7.139	7.139	7.135
	(1.256)	(1.256)		
ϕ_{22}	215.662	215.662	215.663	215.662
	(23.255)	(23.255)		
$\text{VAR}(\epsilon_1)$	0.0^c	6.282^c	0.0^c	6.282^c
	(—)	(—)		
$\text{VAR}(\epsilon_2)$	0.0^c	2.534^c	0.0^c	2.534^c
	(—)	(—)		
$\text{VAR}(\epsilon_3)$	0.0^c	5.115^c	0.0^c	5.115^c
	(—)	(—)		
$R^2_{\eta_1}$	0.113	0.197	0.113	0.197
$R^2_{\eta_2}$	0.229	0.415	0.229	0.416
$R^2_{\eta_3}$	0.390	0.564	0.410	0.560
Coefficient of Determination	0.205	0.258	0.229	0.253

Note: c = constrained parameter.

For recursive models with ζ multinormally distributed, the ML estimator is the ordinary least squares estimator for each equation. Because of this, the analytic results from the section on the consequences of measurement error in multiple regression apply to the ML results for the union sentiment or any recursive model. Thus some of the changes in Table 5.2 found for the ML estimates could have been anticipated from the earlier analytical results.

Unknown Reliabilities

In most situations the reliabilities of measures are unknown. Nevertheless, it may be possible to estimate the reliability or measurement error variance of one or more variables even if only one indicator per latent variable is available. I illustrate this possibility with a simple hypothetical example. In the next subsection a more complicated empirical example builds on this approach. The key idea is that for some overidentified models, the error variance can be identified if treated as a free parameter. For instance, the simple causal chain $x_1 \to y_1 \to y_2$ is overidentified. For this model I can allow and identify measurement error in y_1. I represent this possibility in Figure 5.3 for which the equations are

$$\begin{aligned} y_2 &= \beta_{21}\eta_1 + \zeta_2 \\ \eta_1 &= \gamma_{11}x_1 + \zeta_1 \\ y_1 &= \eta_1 + \epsilon_1 \end{aligned} \tag{5.40}$$

The variables ζ_1, ζ_2, and ϵ_1 are uncorrelated with each other and with x_1. The y_1 is an indicator of η_1 that contains random measurement error (ϵ_1). The variance of ϵ_1 is what we wish to estimate. To determine if $\text{VAR}(\epsilon_1)$ and the other unknown parameters are identified, I must determine if they can be written as functions of the variances and covariances of the observed variables. The only observed variables are x_1, y_1, and y_2 which leads to six ($= \frac{1}{2}(3)(4)$) nonredundant variances and covariances. The six unknown parameters are: γ_{11}, β_{21}, ϕ_{11}, ψ_{11}, ψ_{22}, and $\text{VAR}(\epsilon_1)$. The variances and covariances of x_1, y_1, and y_2 in terms of the parameters are

$$\begin{aligned} &\text{VAR}(x_1) = \phi_{11}, && \text{VAR}(y_1) = \gamma_{11}^2\phi_{11} + \psi_{11} + \text{VAR}(\epsilon_1) \\ &\text{COV}(x_1, y_1) = \gamma_{11}\phi_{11}, && \text{COV}(y_1, y_2) = \beta_{21}\gamma_{11}^2\phi_{11} + \beta_{21}\psi_{11} \\ &\text{COV}(x_1, y_2) = \beta_{21}\gamma_{11}\phi_{11}, && \text{VAR}(y_2) = \beta_{21}^2\gamma_{11}^2\phi_{11} + \beta_{21}^2\psi_{11} + \psi_{22} \end{aligned} \tag{5.41}$$

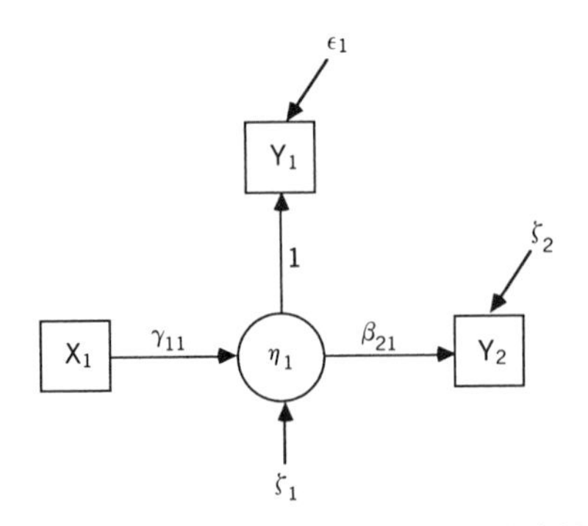

Figure 5.3 Causal Chain Model with Error in Y_1 and All Model Parameters Identified

From (5.41) each of the unknown parameters can be written as functions of the variances and covariances of the observed variables, so, for instance,

$$\phi_{11} = \text{VAR}(x_1), \qquad \gamma_{11} = \frac{\text{COV}(x_1, y_1)}{\text{VAR}(x_1)}, \qquad \beta_{21} = \frac{\text{COV}(x_1, y_2)}{\text{COV}(x_1, y_1)},$$

$$\text{VAR}(\epsilon_1) = \text{VAR}(y_1) - \frac{\text{COV}(y_1, y_2)\text{COV}(x_1, y_1)}{\text{COV}(x_1, y_2)},$$

and so on. Thus the error variance in y_1, $\text{VAR}(\epsilon_1)$, is identified as are the other model parameters. This illustrates that sometimes measurement error can be included in an analysis even if the reliability of an indicator is unknown. The next example shows a more elaborate model where this also is true, though in general without knowing the reliability of measures or having multiple indicators per latent variable, the measurement error variances are not identified.

Objective and Subjective Socioeconomic Status (SES)

In Chapter 4 I introduced a model that explored the relation between objective and subjective dimensions of socioeconomic status. I return to this example to provide a more complex demonstration of the consequences of measurement error for multiequation systems and to estimate the measurement error in some of the variables previously assumed to be error-free. The path diagram of the model is shown in Figure 5.4. The data come from a study by Kluegel et al. (1977). The model is nonrecursive, with actual income (x_1) directly influencing subjective income (η_1) and actual occupational prestige (x_2) affecting subjective prestige (η_2). Subjective income and subjective occupational prestige are reciprocally related, and these both in

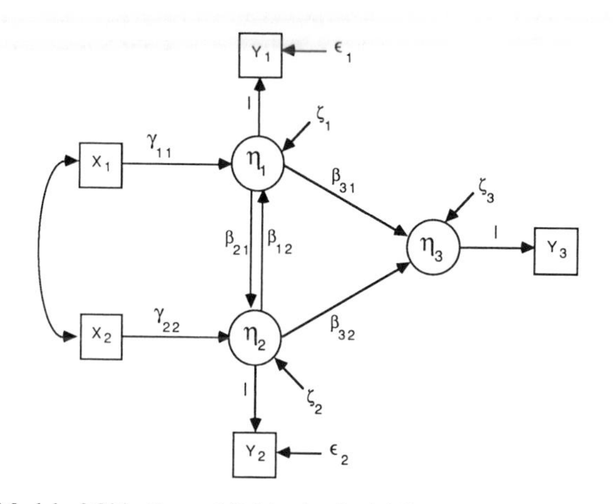

Figure 5.4 A Model of Objective and Subjective Social Status with Measurement Error in Y_1 and Y_2

turn determine subjective overall status (y_3). This model differs from that in Chapter 4 in two ways. One is that the nonsignificant $\hat{\psi}_{ij}(i \neq j)$ elements from Chapter 4 are set to zero. The second is that I allow for random measurement error in the indicators for subjective income (y_1) and subjective occupational prestige (y_2). This can be done because the VAR(ϵ_1) and the VAR(ϵ_2) are identified for the current model structure.[7] Unfortunately, a model that allows random measurement error in y_3 would not be identified. Only the sum of the error variance and disturbance variance is identified for y_3.

Table 5.3 provides the ML and ULS estimates for the objective and subjective status model. (See Appendix 5A for the LISREL and EQS programs.) Columns (1) and (3) refer to a model assuming perfect measurement for y_1 and y_2 [VAR(ϵ_1) = VAR(ϵ_i) = 0]. The ML and ULS estimates in columns (2) and (4) relax this assumption: VAR(ϵ_1) and VAR(ϵ_2) are free parameters. Looking first at the ML estimates [columns (1) and (2)], the largest coefficient changes associated with the allowance for measurement error are for $\hat{\beta}_{32}$, $\hat{\beta}_{12}$, and $\hat{\beta}_{21}$. The estimates more than double for the first two coefficients. This indicates that substantial portions of the effects of the subjective variables on one another were underestimated due to the random measurement error in y_1 and y_2. The only estimate in $\hat{\mathbf{B}}$ that did not

[7]The latent variables corresponding to y_1 and y_2 are set to the same scale as y_1 and y_2. This is necessary to provide meaningful units for the latent variables. Scaling conventions were discussed earlier in this chapter.

Table 5.3 The ML and ULS Estimates for Objective and Subjective Class Model: Columns (1) and (3) without Measurement Errors and Columns (2) and (4) with Measurement Error in Subjective Income (y_1) and Subjective Occupational Prestige (y_2).

	ML Estimate (standard error)		ULS Estimate	
Parameter	(1)	(2)	(3)	(4)
β_{21}	0.24	0.33	0.23	0.30
	(0.10)	(0.09)		
β_{12}	0.12	0.29	0.22	0.25
	(0.11)	(0.17)		
β_{31}	0.41	0.40	0.40	0.42
	(0.04)	(0.14)		
β_{32}	0.26	0.54	0.55	0.55
	(0.04)	(0.20)		
γ_{11}	0.11	0.10	0.11	0.10
	(0.01)	(0.02)		
γ_{22}	0.007	0.007	0.007	0.007
	(0.001)	(0.001)		
ϕ_{11}	4.83	4.83	4.83	4.83
	(0.33)	(0.33)		
ϕ_{21}	13.65	13.66	13.66	13.66
	(2.34)	(2.35)		
ϕ_{22}	452.71	452.71	452.71	452.71
	(30.84)	(30.84)		
ψ_{11}	0.36	0.22	0.29	0.23
	(0.03)	(0.08)		
ψ_{22}	0.33	0.11	0.24	0.12
	(0.02)	(0.05)		
ψ_{33}	0.26	0.21	0.16	0.20
	(0.02)	(0.02)		
VAR(ϵ_1)	0.0^c	0.10	0.0^c	0.10
	(—)	(0.06)		
VAR(ϵ_2)	0.0^c	0.21	0.0^c	0.21
	(—)	(0.05)		
$R^2_{\eta_1}$	0.21	0.52	0.29	0.48
$R^2_{\eta_2}$	0.18	0.73	0.25	0.71
$R^2_{\eta_3}$	0.34	0.47	0.60	0.49
Coefficient of Determination	0.24	0.88	0.32	0.86

Note: c = constrained parameter.

change is $\hat{\beta}_{31}$, the influence of subjective income on subjective overall status. Important decreases in the equation error variances also appear. For instance, $\hat{\psi}_{22}$ drops two-thirds, moving from 0.33 to 0.11. The $R^2_{\eta_i}$'s increase from 0.18 to 0.34, ignoring measurement error to 0.47 to 0.73 when measurement error is included. Similarly, the coefficient of determination changes from 0.24 to 0.88. The ULS estimates differ in a similar fashion as the ML estimates, except that the changes in $\hat{\mathbf{B}}$ are smaller than that for the ML estimates. The most unusual difference is that the R^2 for y_3 decreases when measurement error is allowed in y_1 and y_2. The standard errors (s.e.) for columns (1) and (2) are similar, except for increases in the standard errors for $\hat{\beta}_{12}$, $\hat{\beta}_{31}$, $\hat{\beta}_{32}$, $\hat{\gamma}_{11}$, $\hat{\psi}_{11}$, and $\hat{\psi}_{22}$.

Overall the allowance for measurement error, leads to coefficients that are very similar to those obtained without measurement error for ULS with a few exceptions. The differences in ML estimates with and without measurement error are more pronounced. These results—the air pollution–mortality models and the union sentiment example—illustrate the difficulty of making generalization about the consequences of errors of measurement for even moderately complex models.

SUMMARY

In this chapter I have examined the consequences of random measurement error for univariate and bivariate statistics, multiple regression, and multiequation systems. As we move from the simple statistics to the more complicated ones, generalizations about the influences of measurement error are more difficult to make. Parameter estimates may stay nearly the same, increase, or decrease, as illustrated by the empirical examples and analytic results. Since it is difficult to predict which of these consequences might follow, the need to take into account measurement error becomes increasingly important. In some situations estimates of reliability of the measures can be incorporated into the model, as was true for the union sentiment example. In other cases it is possible to estimate the measurement error variance within the model, as was done for the objective and perceived socioeconomic status model. If neither of these options is available, a range of reasonable reliability values might be tried to assess the sensitivity of the model to random measurement error, as shown with the air pollution–mortality regression equation.

Throughout this chapter I made several simplifying assumptions. For instance, I assumed that each latent variable was measured by a single indicator, often giving the latent and observed variables the same scale. In addition I restricted each indicator to depend on one and only one latent variable. In practice, we often have more than one measure for a latent concept. Some of these indicators will have scales very different from those chosen for the latent variable. We also may have indicators that are affected by more than one latent variable. One or more of these situations are commonly encountered in empirical research. They call for a more sophisticated modeling of the relation of the observed variables to the latent variables than I employed here. I consider these issues and measurement, in general, in the next two chapters.

APPENDIX 5A ILLUSTRATIONS OF LISREL AND EQS PROGRAMS

In Appendix 4D of Chapter 4 I gave the LISREL and EQS programs for generating the estimates for the objective-subjective socioeconomic status example. In this appendix I list the programs to run the same problem, except I set VAR(ϵ_1) and VAR(ϵ_2) to be free parameters, Ψ to be a diagonal matrix, and I use the ULS fitting function. The programming strategy is the same as that detailed in Appendix 4D. I list the new programs and limit my comments to the new elements of it. As before, the numbers in parentheses at the far right refer to the explanations that follow the program.

LISREL PROGRAM

```
TITLE OBJEC.-SUBJ. STATUS FR TE 1 1 TE 2 2 ULS
INPUT PROGRAM
NUMERIC DUMMY
END FILE
END INPUT PROGRAM
USER PROC NAME = LISREL
OBJEC.-SUBJ. STATUS FR TE 1 1 TE 2 2 ULS
DA NI = 5 NO = 432 MA = CM
KM
*
1.000
 .292    1.000
 .282     .184    1.000
 .166     .383     .386    1.000
 .231     .277     .431     .537    1.000
```

```
SD
*
21.277    2.198    0.640    0.670    0.627
LA
*
'OCC' 'INC' 'SUBOCC' 'SUBINC' 'SUBGEN'
SE
4 3 5 2 1/
MO NK = 2 NX = 2 NE = 3 NY = 3 GA = FI LY = ID LX = ID BE = FU TE = FI TD = FI PS = DI   (1)
FR GA 1 1 GA 2 2
FR BE 1 2 BE 2 1 BE 3 1 BE 3 2
FR TE 1 1 TE 2 2                                                                        (2)
MA BE                                                                                   (3)
*
        0     .22    0
      .23       0    0
      .40     .55    0
MA GA
*
  .11    0
  0      .007
  0      0
MA PH
*
   4.8    13.6
  13.6    453
MA PS
*
  .340    .329    .256
OU SE SS TO NS UL                                                                       (4)
END USER
```

(1) The model (MO) card differs from that in Chapter 4 by declaring Ψ a diagonal matrix. All its off-diagonal elements are fixed to zero.

(2) The free statement converts $\text{VAR}(\epsilon_1)$ and $\text{VAR}(\epsilon_2)$ to free parameters.

(3) The MA card allows the user to specify the starting values for a whole matrix. Following MA is the two-letter abbreviation for the matrix whose values are to be initialized. The next card is the format, with an * indicating free format. This is followed by rowwise lists of values at which to begin the estimation. I take the final solution from the model without measurement errors as the starting values. I do this because the automatic starting values provided in LISREL VI do not work well in this example, so it is necessary to supply my own.

(4) The only new output options relative to that in Appendix 4D are NS and UL. NS indicates that the user-supplied starting values should be used, and UL requests the ULS fitting function.

EQS PROGRAM

```
/TITLE
 OBJEC.-SUBJ. STATUS WITH MEAS. ERROR IN Y1 AND Y2 ULS
/SPEC
 CAS=432; VAR=5; ME=ULS; MA=COV;                              (1)
/LAB
 V1=OCC; V2=INC; V3=SUBOCC; V4=SUBINC;
 V5=SUBGEN;
/EQU
 V3=F3 + E3; V4=F4 + E4;                                      (2)
 F3=.007*V1 + .2*F4 + D3;
 F4=.1*V2 + .2*F3 + D4;
 V5=.4*F4 + .5*F3 + D5;
/VAR
 D3 TO D5=.3*; V1=453*; V2=4.8*;
 E3 TO E4=0*;                                                 (3)
/COV
 V1, V2=13.6*;
/MAT
 1.000
  .292    1.000
  .282     .184    1.000
  .166     .383     .386    1.000
  .231     .277     .431     .537    1.000
/STA
 21.277    2.198    .640    .670    .627
/END
```

(1) The SPECIFICATION card is the same as in Chapter 4, except that ULS estimates are requested.

(2) The EQUATION card adds two measurement equations for V3 and V4. The E3 and E4 are their respective error terms. The F3 and F4 variables represent the latent variables for V3 and V4, respectively. The implicit coefficients for F3 in the V3 equation and for F4 in the V4 equation are fixed at one.

(3) The initial estimate for the error variances for E3 and E4 is zero.

CHAPTER SIX

Measurement Models: The Relation between Latent and Observed Variables

In the preceding chapter I demonstrated the consequences of estimating the relations between observed variables when they are imperfect measures of their corresponding latent variables. Generally, ignoring measurement error leads to inconsistent estimators and to inaccurate assessments of the relation between the underlying latent variables. To correct these problems, we need to understand the process of measurement. More specifically, we must incorporate the relation between the observed variables and latent variables into structural equation models.

This chapter explores measurement models from a structural equation perspective. The focus is specification (not estimation) of measurement models. Estimation, identification, and other statistical aspects are postponed until Chapter 7. In this chapter I begin with an introduction to the nature of measurement. This is followed by two sections, one on validity and the other on reliability. I contrast traditional definitions and measures of validity and reliability with new ones that are more closely aligned with structural equations models. The final section before the summary, is on cause indicators.

MEASUREMENT MODELS

In this section I discuss the meanings of concepts, theoretical and operational definitions, and measurement models. I also identify and illustrate the four major steps in measurement.

Measurement is the process by which a concept is linked to one or more latent variables, and these are linked to observed variables. The concept can vary from one that is highly abstract, such as intelligence, economic development, or expectations, to one that is more concrete, such as age, sex, or race. One or several latent variables may be needed to represent the concept. The observed variables can be responses to questionnaire items, census figures, meter readings, or any other observable characteristics.

The measurement process begins with the concept. A concept is an idea that unites phenomena (e.g., attitudes, behaviors, traits) under a single term. Anger, for instance, provides the common element tying together diverse characteristics such as an individual screaming, throwing objects, having a blood-flushed face, or behaving in an agitated way. The concept of anger acts as a summarizing device to replace a list of specific traits that an individual may exhibit. Other concepts play a similar role.

Do concepts really exist? Concepts have the same reality or lack of reality as other ideas. They are created by people who believe that some phenomena have something in common. The concept identifies that thing or things held in common. Latent variables are the representations of concepts in measurement models. Once a concept is selected or devised, the four steps in the measurement process are to (1) give the meaning of the concept, (2) identify the dimensions and latent variables to represent it, (3) form measures, and (4) specify the relation between the measures and the latent variables.

The first step is accomplished by developing a theoretical definition. *A theoretical definition explains in as simple and precise terms as possible the meaning of a concept.* It performs several useful functions. One is that a theoretical definition couples a term and a concept by detailing the specific denotation assigned to a term. Second, it clarifies the dimensions of a concept. Dimensions are the distinct aspects of a concept. They are components that cannot easily be subdivided into additional components. Since many concepts have numerous possible dimensions, a definition is critical to set the limit on the dimensions a researcher selects. We need one latent variable per dimension. Third, a theoretical definition provides guidance in the selection of measures. Whether devising survey questions or collecting census information, we need to know the phenomena that are encompassed or excluded by a concept. This can help us evaluate whether a measure is valid.

The term "terrorism" illustrates these points. In the absence of a theoretical definition, the term has many connotations. To some the term terrorism suggests images of violent nongovernmental groups hijacking planes and taking hostages. To others car bombs that kill civilians in London or Beirut or gangster-style killings in the United States may come to mind. Some would add to the list of terrorism, acts of state violence such as the Nazi's slaughter of Jews during World War II, the Turks' attacks on Armenians early in this century, or the "disappearances" of civilians in several Latin America countries. We cannot say whether these are instances of terrorism until we have a theoretical definition to link the term terrorism to a specific concept.

An example of a theoretical definition is provided by the U.S. Central Intelligence Agency's 1981 report "Patterns of International Terrorism." They defined terrorism as: "the threat or use of violence for political purposes by individuals or groups, whether acting for, or in opposition to, established governmental authority, when such actions are intended to shock or intimidate a target group wider than the immediate victims." This definition brings us closer to understanding the concept linked to the term terrorism. It does not mean that this is the correct definition, that all parts of the definition are clear, or, that we would agree with their classification of acts. But it is a clear improvement over providing no definition and relying on an intuitive and an unspecified meaning of terrorism.

With the above definition I can assess whether the earlier examples represent terrorist acts. Plane hijackings are an interesting case since not all would satisfy the above definition. If the motive for the hijacking is predominantly monetary, then this would not qualify as terrorism since the definition requires political purposes. Car bombings that are politically motivated and that are intended to threaten a group beyond the immediate victims would be terrorist acts whereas gangster-style killings with financial motives would not be. Similarly, other events could be compared to the criteria laid out in the definition, and they could be judged as terrorist or nonterrorist acts. Thus the theoretical definition can guide us in forming measures or counts of terrorism.

The preceding definition also identifies dimensions of terrorism. Cross-cutting whether an act is anti- or pro-government with whether it is perpetrated by an individual or a group leads to four dimensions. We should have one latent variable to represent each dimension which means a total of four latent variables to stand for the concept. Thus, as this example shows, a theoretical definition provides the meaning of the concept, links a term to a specific concept, identifies its dimensions and the number of latent variables, and sets a standard by which to select measures.

The next step in measurement, to form measures, depends on the theoretical definition. This is sometimes referred to as the operational definition. *The operational definition describes the procedures to follow to form measures of the latent variable(s) that represent a concept.* In some situations the latent variable(s) are operationalized as the responses to questionnaire items. Other measures are based on statistics collected by the government such as census population figures or vital statistics on births or deaths. An operational definition or measure is appropriate to the extent that it leads to an observed variable that corresponds to the meaning assigned to a concept.

In the case of terrorism, newspapers, magazines, journals, and other sources are the basis of collecting information. Researchers search these sources for acts that have the characteristics stipulated in the theoretical definition. Often the information is not complete, and a considerable element of judgment must enter classification decisions. The diversity of news sources examined, the accuracy of reporters, and the accessibility of geographic areas also affect the accuracy of the measures. For instance, Western news sources are more likely to label acts as terrorist by countries or groups to which they are hostile than they are to label as terrorist similar acts by friendly groups. If analysts rely exclusively on U.S. sources, an undercount of terrorist acts is likely. These and other factors lead to random and nonrandom errors in measures.

Virtually all measures that we employ contain such errors. It is the task of the fourth step in measurement to formalize these types of errors. The fourth step is to construct the measurement model. *A measurement model specifies a structural model connecting latent variables to one or more measures or observed variables*. The latent variable is the formal representation of a concept. The measurement model describes the relation between the measure and latent variables. This relation can be in an equation or in a path diagram. For example, as stated earlier, terrorism has four dimensions and therefore four latent variables. To simplify matters, I focus on the latent variable for anti-government terrorism by groups. Suppose that two different researchers independently formed measures of this dimension of terrorism. A simple measurement model for the latent variable's influence on the two measures is

$$\begin{aligned} x_1 &= \lambda_{11}\xi + \delta_1 \\ x_2 &= \lambda_{21}\xi + \delta_2 \end{aligned} \tag{6.1}$$

where ξ represents the latent variable of anti-government terrorism by groups, λ_{11} and λ_{21} are constants showing the expected number of unit changes in the observed variables for a one unit change in the true level of ξ, and δ_1 and δ_2 are errors of measurement with expected values of zero and uncorrelated with ξ and with each other. All variables are in deviation form so that intercepts terms do not enter the equations.[1]

[1]Throughout this chapter I simplify the examples by using deviation scores. I refer the reader to Chapters 5, 7, and 8 for sections on using raw scores in analyses.

Suppose that we were willing to assume that on average a one unit increase in the true level of anti-government group terrorism led to a one unit increase in x_1, even though x_1 contained random errors of measurement. This would lead λ_{11} to a value of one. Similarly, if for every ten terrorist acts the sources for x_2 revealed on average only eight such acts, then λ_{21} would be 0.8. The extent of random measurement error in each measure is shown by the variance of δ_i. Equation (6.1) would then describe the relation between this dimension of terrorism and two measures of it. To measure the three other dimensions of terrorism, we would need three more latent variables, with measures of each and equations like (6.1).

In many social science examples the scale for the latent variable is ambiguous. For example, we might measure the anxiety level of individuals with two measures that are scaled quite differently. The problem is to select a scale for the latent variable. Anxiety or any other concept has no inherent scale. The scales employed emerge from consensus, and for many concepts a consensus is absent. In these situations we can assign a scale by setting one of the λ_{ij}'s equal to one. This yields a latent variable that has the same scale as the ith observed variable.

In some situations an incorrect choice of the scale for latent variables can produce misleading results. For instance, suppose that in two populations we use the same indicator to scale the same latent variable in both groups. To do this, the appropriate λ coefficients are set to one. If in reality the relation between the indicator and latent variable is different in each group, this incorrect scaling can lead to inaccurate estimates of the other model parameters. Thus the scaling assumptions can have important consequences, and social science research could benefit from more attention to calibrating measures (see Bielby 1986).

An example from the physical sciences helps to illustrate the preceding points. A theoretical definition of temperature is that it is the average kinetic energy contained in the random motion of the molecules in a substance. Thermometer readings of a substance can be used to measure temperature. Suppose that we utilize two thermometers; one is in Fahrenheit degrees, while the other is in Celsius degrees. The measurement model is represented the same as in equation (6.1), but now x_1 is the Celsius thermometer reading, x_2 is the Fahrenheit reading, and ξ is the latent variable of temperature. By scientific convention the scale for temperature is measured in Kelvin degrees. A one unit change in Kelvin degrees is equal to a one unit change in Celsius so that we could incorporate this knowledge into the model by setting λ_{11} equal to one. This scaling of ξ results from the agreement of scientists to use a certain unit. There is no inherent scale for temperature. We could specify this model further by incorporating the known conversion factor of 9/5 in moving from the Celsius units to

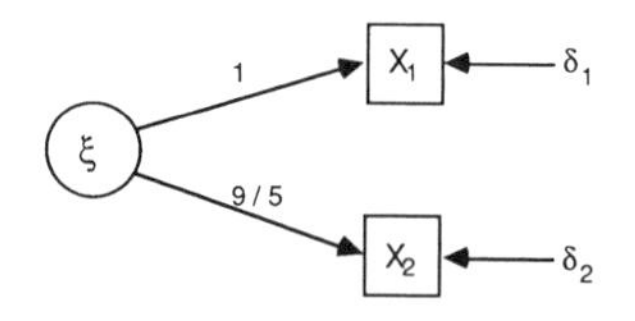

Figure 6.1 A Path Model of Temperature (ξ) with Two Indicators: A Celsius Thermometer Reading (X_1) and a Fahrenheit Thermometer Reading (X_2)

Fahrenheit units. Thus λ_{21} could be set to $9/5\ \lambda_{11}$ or 9/5.[2] A path diagram of this model is shown in Figure 6.1.

Although there are many formal similarities between the measurement model for temperature and those for social science concepts, there are a couple of important differences. First, the theoretical definition of temperature is more widely agreed upon than are those for most social science concepts. These differences in definitions may lead to greater dispersion in the selection of measures for social science concepts than in measures of temperature. Second, the scale on which to measure temperature is agreed upon by most scientists, while no universally agreed-upon scale for most social science latent variables exists. Plus the calibration of the thermometer scales to the latent variable of temperature is well studied. Consequently, the degree of measurement error is greater in the social science measures than in most thermometer readings. Concepts such as inflation, socioeconomic development, intelligence, social class, and anxiety, cannot be perfectly measured with current techniques. Explicit measurement models are needed to better understand the relation between the latent variables, which stand for concepts, and the observed variables that measure them.

In sum, the four steps in measurement are to give meaning, identify dimensions and latent variables, to form measures, and to specify a model. The theoretical definition assigns meaning to a term and the concept associated with it. Based on this definition, we know a concept's dimensions. Each dimension is represented by one latent variable. Guided by theoretical definitions, we form measures, hopefully two or more measures per latent variable. Finally, we formulate the structural relation between indicators and latent variables in the measurement model. Two important properties of measures are their validity and reliability, the topics of the next two sections.

VALIDITY

Validity is concerned with whether a variable measures what it is suppose to measure. Do measures based on information taken from mass media

[2] The complete conversion of Celsius to Fahrenheit degrees is $°F = \frac{9}{5}°C + 32°$. The constant of 32° is ignored since I am using deviation scores in this discussion.

sources really measure terrorism? Does an IQ test measure intelligence? Does the reported gross national product measure the actual value of goods and services produced in a country? These are questions of validity. They can never be answered with absolute certainty. Although we can never prove validity, we can develop strong support for it. Traditionally, psychologists have distinguished four types of validity: *content validity*, *criterion validity*, *construct validity*, and *convergent and discriminant validity*. Each attempts to show whether a measure corresponds to a concept, though their means of doing so differ. Content validity is largely a "conceptual test," whereas the last three are empirically based. If a measure truly corresponds to a concept, we would expect that all four types of validity would be satisfied. Unfortunately, it is possible that a valid measure will fail one or more of these tests or that an invalid measure will pass some of these. In the four subsections that describe each type of validity, I explain how this can occur. The final subsection on validity provides alternative tests of validity that overcome some of the problems with the traditional ones.

Content Validity

Content validity is a qualitative type of validity where the domain of a concept is made clear and the analyst judges whether the measures fully represent the domain. To the extent that they do, content validity is met. A key question is, How do we know a concept's domain? For the answer we must return to the first step in the measurement process discussed in the last section. That is, to know the domain of a concept, we need a theoretical definition that explains the meaning of a concept. Ideally, the theoretical definition should reflect the meanings associated with a term in prior research so that a general rather than an idiosyncratic domain results. In addition the theoretical definition should make clear the dimensions of a concept. For content validity each dimension of a concept should have one or more measures. The concept of terrorism illustrates these points. The theoretical definition states the domain and the dimensions of the concept. For an act to be terrorist requires that the threat or use of violence be present, that the predominant purpose be political, and that the target group be wider than the immediate victims. The definition also highlights that its dimensions include pro- and anti-government acts by individuals and pro- or anti-government behavior by groups. To adequately represent the domain, we need four latent variables and measures for each dimension. If not, the content validity of the measure is questionable. Suppose, for instance, that all the terrorist acts we list are anti-government acts committed by groups. Since our domain includes three other dimensions, the content of the concept is not fully tapped.

Does it matter if our measures lack content validity? In general, the answer is yes. Just as a nonrepresentative sample of people can lead to mistaken inferences to the population, a nonrepresentative sample of measures can distort our understanding of a concept. For instance, the dimension of anti-government terrorism by groups may react differently to government crackdowns than do anti-government terrorist acts by individuals. Thus, to study a concept, we need measures that fully represent its dimensions. Alternatively, we must make clear the dimensions to which we are restricting our analysis.

The major limitation of content validity stems from its dependence on the theoretical definition. For most concepts in the social sciences, no consensus exists on theoretical definitions. The domain of content is ambiguous. In this situation the burden falls on researchers not only to provide a theoretical definition accepted by their peers but also to select indicators that fully cover its domain and dimensions. In sum, content validity is a qualitative means of ensuring that indicators tap the meaning of a concept as defined by the analyst.

Criterion Validity

Criterion validity is the degree of correspondence between a measure and a criterion variable, usually measured by their correlation. To assess criterion validity, we need a variable that is a standard to which to compare our measure. Suppose that in a survey we ask each employee in a corporation to report his or her salary. If we had access to the actual salary records, we could assess the validity of the survey measure by correlating the two. In this case employee records represent an ideal, or nearly ideal, standard of comparison.

When the criterion exists at the same time as the measure, this is called *concurrent validity*. If the criterion occurs in the future, this is *predictive validity*. The previous example of salary represents concurrent validity. An example of predictive validity is if test scores such as SATs are correlated with Grade Point Average in college some years later. Another example is if the index of leading economic indicators is correlated with a measure of future economic activity such as GNP.

The absolute value of the correlation between a measure and a criterion sometimes is referred to as the validity coefficient (see Lord and Novick 1968, 261). Does this correlation of a measure and a criterion reveal the validity of a measure? To gain insight into this question, I employ a structural equation approach. If we represent the measure as x_1 and the criterion as C_1, the validity coefficient may be represented as $\rho_{x_1C_1}$. A simple

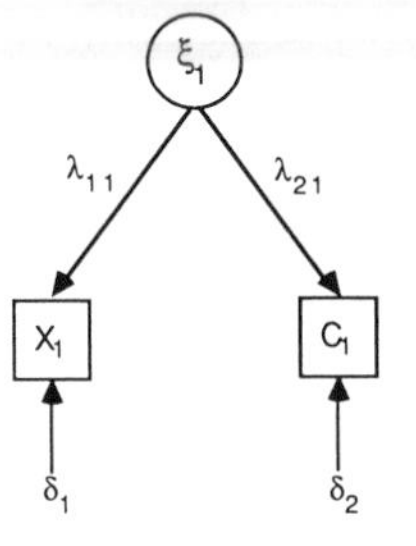

Figure 6.2 Measurement Model with Measure (X_1) and Criterion (C_1) a Function of a Latent Variable

model of the relation between x_1 and C_1, and the latent variable ξ_1 that they measure appears in Figure 6.2 and in the following equations:

$$
\begin{aligned}
x_1 &= \lambda_{11}\xi_1 + \delta_1 \\
C_1 &= \lambda_{21}\xi_1 + \delta_2
\end{aligned}
\tag{6.2}
$$

where δ_1 and δ_2 are uncorrelated with each other and with ξ_1, $E(\delta_1) = E(\delta_2) = 0$, and all variables are in deviation form. If C_1 is a perfect criterion, then $C_1 = \xi_1$.[3] That is, C_1 and ξ_1 are the same, with no scaling differences and no random measurement error in C_1. This would be represented in Figure 6.2 and equation (6.2) by setting λ_{21} to one and δ_2 to zero. For now I use the less restricted form, without constraining these values.

Using covariance algebra and the definition of a correlation, $\rho_{x_1C_1}$ equals

$$
\rho_{x_1C_1} = \frac{\lambda_{11}\lambda_{21}\phi_{11}}{[\mathrm{VAR}(x_1)\mathrm{VAR}(C_1)]^{1/2}}
\tag{6.3}
$$

As (6.3) reveals, the magnitude of $\rho_{x_1C_1}$ depends on factors other than the "closeness" of x_1 and ξ_1. This is made clearer if we standardize x_1, C_1, and ξ_1 to variances of one. In this case $\rho_{x_1C_1}$ equals $\lambda_{11}\lambda_{21}$, the correlation of x_1 and ξ_1 equals λ_{11}, and the correlation of C_1 and ξ_1 equals λ_{21}. The validity coefficient, $\rho_{x_1C_1}$, is affected by not only $\rho_{x_1\xi_1}$ $(= \lambda_{11})$ but also by $\rho_{C_1\xi_1}$ $(= \lambda_{21})$. Even if the correlation of x_1 with ξ_1 stays at 0.5 the validity coefficients would be 0.45, 0.35, or 0.25 if the correlation of C_1 and ξ_1, is 0.9, 0.7, or 0.5. Thus, even with no change in x_1's association with ξ_1, we are

[3] I am using deviation scores here as earlier. Strictly speaking, perfect validity would also mean that the raw criterion variable and raw ξ_1 have the same mean.

led to different values of validity, depending on the criterion's relation to ξ_1. To illustrate, suppose that we develop a new scale to measure self-esteem. As a criterion we use an older self-esteem scale. The correlation of the new scale (x_1) and old scale (C_1) would provide the validity coefficient. A low correlation between x_1 and C_1 could result from a low correlation between the old scale (C_1) and ξ_1 rather than a great deal of error in the new one. The situation is further complicated since frequently analysts may use several criteria. These criteria generally will vary in their correspondence to ξ_1 so that the $\rho_{x_1C_1}$ will not equal $\rho_{x_1C_2}$ nor $\rho_{x_1C_3}$. Thus the validity coefficients can differ depending on the criterion and the degree of error in it, even though the measurement characteristics of x_1 have stayed the same. Of course, if a criterion is a perfect measure of ξ_1 (i.e., $C_1 = \xi_1$) and the assumptions of Figure 6.2 hold, then C_1 can simply be used in place of ξ_1. In this rare case x_1 could be regressed on C_1, and its measurement properties directly studied. This is related to the Platonic true score model (Lord and Novick 1968) where it is theoretically possible to obtain a criterion C_1 that directly and exactly matches ξ_1. For many social science concepts, such criteria are not feasible.

In sum, criterion validity as measured by $\rho_{x_1C_1}$, the validity coefficient, has several undesirable characteristics as a means to assess validity. It is not only influenced by the degree of random measurement error variance in x_1 but also by the error in the criterion. Furthermore different criteria lead to different "validity coefficients" for the same measure, leaving uncertainty as to which is an accurate reading of a measure's validity. Finally, for many measures no criterion is available.

Construct Validity

Construct validity is a third type of validity. Many of the concepts in the social science are incompletely formulated and defined so that content validity is difficult to apply. As mentioned earlier, appropriate criteria for some measures often do not exist. This prevents the computation of criterion validity coefficients. In these common situations construct validity is used instead. In this section I describe construct validity and critique its traditional applications which rely on correlations of measures.

Construct validity assesses whether a measure relates to other observed variables in a way that is consistent with theoretically derived predictions. Hypotheses may suggest positive, negative, or no association between constructs. If we examine the relation between a measure of one construct to other observed variables indicating other constructs, we expect their empirical association to parallel the theoretically specified associations. To the extent that they do, construct validity exists.

To illustrate, suppose that we have developed a measure of the construct of anti-government terrorism by groups for a number of countries. We hypothesize that the extent of this form of terrorism is greatest in democratic countries. To test the construct validity of the terrorism measure, we correlate it with measures of political democracy and expect to find positive correlation coefficients. If the correlations are positive and significant, this provides evidence for the construct validity. If negative or near zero correlations result, this undermines construct validity.

No one empirical test determines construct validity. Establishing construct validity is a long process, with each test providing information and suggesting revisions that can aid the next empirical test. The major steps in the process begin with postulating theoretical relations between constructs. Then the associations between measures of the constructs or concepts are estimated. Based on these associations, the measures, the constructs, and the postulated associations are reexamined.

Some of the difficulties with construct validity can be illustrated with a structural equation approach. As a simple example, consider Figure 6.3. In Figure 6.3, I am assuming two constructs, ξ_1 and ξ_2. Each has one measure represented as x_1 and x_2. As usual δ_1 and δ_2 are random errors of measurement with expected values of 0, uncorrelated with each other and with ξ_1 and ξ_2. Suppose that the construct validity of x_1 is of interest. We hypothesize that the two constructs (ξ_1 and ξ_2) are positively correlated ($\phi_{12} > 0$). To test construct validity, we would compute the correlation between x_1 and x_2.

In Chapter 5 I showed the correlation of two observed variables compared to the correlation of two latent variables in equation (5.13). Making the appropriate substitutions of x_1 and x_2 leads to

$$\rho_{x_1x_2} = \left(\rho_{x_1x_1}\rho_{x_2x_2}\right)^{1/2}\rho_{\xi_1\xi_2} \tag{6.4}$$

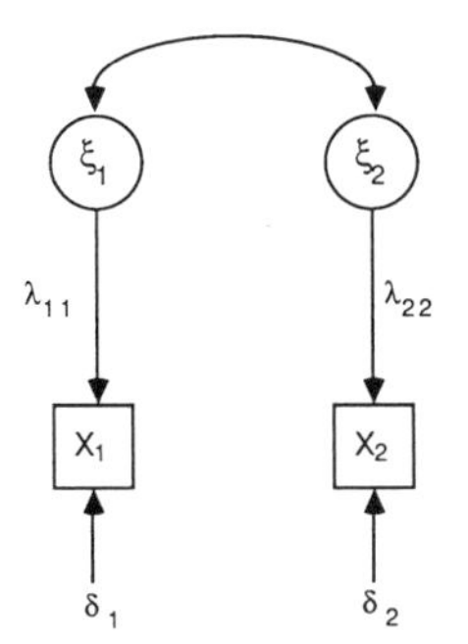

Figure 6.3 Two Constructs (or Latent Variables) with One Measure Each

where $\rho_{x_i x_i}$ is the reliability of x_i. I will have more to say about reliability later in this chapter, but for now we can view it as the squared correlation between x_i and ξ_i. According to (6.4) the correlation of the two observed variables depends not only on the correlation of x_1 and ξ_1 but also on the correlation between the constructs ξ_1 and ξ_2, and the correlation of x_2 and ξ_2. Because of this, the interpretation of construct validity based on $\rho_{x_1 x_2}$ is seriously complicated. For instance, if the correlation between ξ_1 and ξ_2 is near zero, counter to what is hypothesized, then $\rho_{x_1 x_2}$ is near zero regardless of how valid a measure x_1 is of ξ_1. Or, suppose that $\rho_{\xi_1 \xi_2}$ is relatively large and that x_1 has very high reliability ($\rho_{x_1 x_1}$) but x_2 has low reliability ($\rho_{x_2 x_2}$). This would reduce $\rho_{x_1 x_2}$, raising doubts about the construct validity of x_1. In the terrorism and democracy example a low correlation of our measures could result not from low validity for the terrorism measure but from terrorism and democracy not being highly correlated or because our measure of democracy is poor.

The same result could occur with measures from other constructs that are hypothesized to be associated with ξ_1. The difficulties are even greater if x_1 depends on ξ_1 and ξ_2. In this case $\rho_{x_1 x_2}$ may be large because x_1 is a positive function of both ξ_1 and ξ_2 while x_2 depends on ξ_2. Part of $\rho_{x_1 x_2}$'s magnitude results from x_1 and x_2's joint dependence on ξ_2. If correlated errors of measurement exist, then this too affects $\rho_{x_1 x_2}$.

In sum, the construct validity of a measure depends on whether that measure correlates with other measures of other constructs. If the constructs are associated, we expect the measures to correlate. A zero association of the constructs should lead to no relation between the measures. As the preceding results demonstrate, the correlation between observed variables $\rho_{x_1 x_i}$ $(i \neq 1)$ may not be a very good determinant of whether x_1 validly measures ξ_1. This correlation depends on several factors that have little to do with x_1's validity. These include the correlation of ξ_1 with the other construct, the reliability of the measure for the other construct, the presence of correlated measurement errors, and whether latent variables besides ξ_1 affect x_1.

Convergent and Discriminant Validity

Closely related to the idea of construct validity is Campbell and Fiske's (1959) proposal to examine convergent validity. They suggest a multitrait-multimethod design where indicators of two or more traits (or concepts) are constructed by two or more methods. And, the methods are uncorrelated with the traits. For example, we may be interested in social liberalism and pro-business attitudes. Each attitude might be measured with Guttman scaling and Likert "agree–disagree" type of indicators. The two traits are

Table 6.1 Correlation Matrix of Two-Traits Measured with Two Methods

	Method 1		Method 2	
	x_1	x_2	x_3	x_4
Method 1				
x_1	1			
x_2	$\rho_{x_1x_2}$	1		
Method 2				
x_3	$\boxed{\rho_{x_1x_3}}$	$\rho_{x_2x_3}$	1	
x_4	$\rho_{x_1x_4}$	$\boxed{\rho_{x_2x_4}}$	$\rho_{x_3x_4}$	1

Note: x_1, x_3 measure trait one; x_2, x_4 measure trait two; and convergent validity correlations are boxed.

the two attitudes, and the two scale types are the two methods. The researcher forms a correlation matrix of variables and the correlations help in the assessment of validity. The simplest example is two traits measured with two methods. In practice, we would need a more elaborate model, but this one helps to illustrate the basic features of the method. Table 6.1 contains the correlation matrix of the four variables where x_1 and x_3 measure one trait and x_2 and x_4 measure another. The x_1 and x_2 variables are measured with the first method, and x_3 and x_4 are measured with the second. For convergent validity, Campbell and Fiske require that the correlations of the different measures of the same trait should be statistically significant and sufficiently large. In Table 6.1 these convergent validity correlations are $\rho_{x_1x_3}$ and $\rho_{x_2x_4}$.

Assessment of discriminant validity proceeds with at least two comparisons.[4] First, the convergent validity correlations should be greater than the correlations between one variable with any other variable with which it shares neither trait nor method. So $\rho_{x_1x_3}$ and $\rho_{x_2x_4}$ should exceed $\rho_{x_1x_4}$ and $\rho_{x_2x_3}$. Second, the convergent validity correlations should be larger than the correlations of different traits measured with the same method. For Table 6.1, this means that $\rho_{x_1x_3}$ and $\rho_{x_2x_4}$ should be greater than $\rho_{x_1x_2}$ and $\rho_{x_3x_4}$.

[4]An additional comparison for discriminant validity is possible if three or more traits are available. That is, the pattern of intertrait correlations should be the same whether the variables are from the same or different methods.

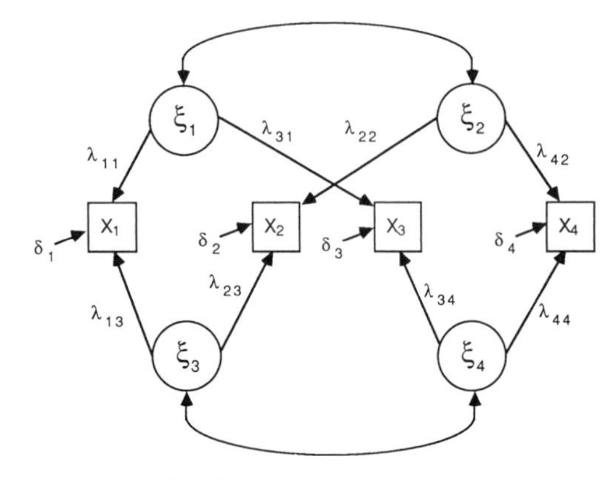

Figure 6.4 Multitrait-Multimethod Hypothetical Example with Two Traits (ξ_1 and ξ_2) and Two Methods (ξ_3 and ξ_4).

Some of the ambiguities that arose in using correlations for other types of validity also hold for the correlational approach to convergent and discriminant validity. These correlations can be influenced by factors other than those which we typically associate with validity. A structural equation approach (Werts and Linn 1970; Althauser and Heberlein 1970) helps to highlight these problems. Staying with the hypothetical example in Figure 6.4 and Table 6.1, one test of convergent validity is that $\rho_{x_1 x_3}$ should be relatively large and significant. In terms of the structural equations implied by the path diagram, this correlation is

$$\rho_{x_1 x_3} = \frac{\text{COV}(x_1, x_3)}{[\text{VAR}(x_1)\text{VAR}(x_3)]^{1/2}}$$
$$= \lambda_{11}\lambda_{31} + \lambda_{13}\lambda_{34}\rho_{\xi_3\xi_4} \tag{6.5}$$

Equation (6.5) follows by assuming standardized variables (all variances of x_i and ξ_j equal to one) and using covariance algebra. It shows that $\rho_{x_1 x_3}$ is a function not only of the ξ_1 trait's influence on x_1 and x_3 but also the correlation between the two methods' factors ξ_3 and ξ_4 and their respective influences on x_1 and x_3. The λ_{11} and λ_{31} coefficients indicate the relation of the ξ_1 trait to the x_1 and x_3 indicators. Yet a low convergent validity correlation might occur with moderate λ_{11} and λ_{31} coefficients if the method factors have a strong inverse correlation ($\rho_{\xi_3\xi_4} < 0$) with λ_{13} and λ_{34} positive and moderate. On the other hand, a low $\rho_{x_1 x_3}$ also might result because λ_{11} and λ_{31} are weak as are $\rho_{\xi_3\xi_4}$, λ_{13}, and λ_{34}. A third possibility is that $\rho_{x_1 x_3}$ is strong because λ_{13}, λ_{34}, and $\rho_{\xi_3 \xi_4}$ are positive and strong. Other combinations are possible. The point is that the correlation that gauges convergent validity can be high or low for many reasons, some of which are not strongly related to the correspondence between the trait and the measure. Decompositions for the other correlations show similar complications that cloud the interpretation of multitrait-multimethod correlation comparisons.

An alternative is to estimate the structural equations such as those represented in Figure 6.4 and to examine the parameter estimates to determine measurement validity. An empirical example from Schneider (1970) helps illustrate this approach. He examined various criteria of leadership in small groups for a sample of 240 group members. Three methods—self-ratings, peer ratings, and observer ratings—were employed to measure four traits—prominence, achievement, affiliation, and leadership. He reports a 12 × 12 correlation matrix, which I reproduce in Table 6.2.

Table 6.2 Multitrait-Multimethod Correlation Matrix for 12 Leadership Variables ($N = 240$)

		Self-rating				Peer Ratings				Observer Ratings			
		x_1	x_2	x_3	x_4	x_5	x_6	x_7	x_8	x_9	x_{10}	x_{11}	x_{12}
						Self-ratings							
Prominence	(x_1)	1.00											
Achievement	(x_2)	0.50	1.00										
Affiliation	(x_3)	0.41	0.48	1.00									
Leader	(x_4)	0.67	0.59	0.40	1.00								
						Peer Ratings							
Prominence	(x_5)	0.45	0.33	0.26	0.55	1.00							
Achievement	(x_6)	0.36	0.32	0.31	0.43	0.72	1.00						
Affiliation	(x_7)	0.25	0.21	0.25	0.30	0.59	0.72	1.00					
Leader	(x_8)	0.46	0.36	0.28	0.51	0.85	0.80	0.69	1.00				
						Observer Ratings							
Prominence	(x_9)	0.53	0.41	0.34	0.56	0.71	0.58	0.43	0.72	1.00			
Achievement	(x_{10})	0.50	0.45	0.29	0.52	0.59	0.55	0.42	0.63	0.84	1.00		
Affiliation	(x_{11})	0.36	0.30	0.28	0.37	0.53	0.51	0.43	0.57	0.62	0.57	1.00	
Leader	(x_{12})	0.52	0.43	0.31	0.59	0.68	0.60	0.46	0.73	0.92	0.89	0.63	1.00

Source: From Schneider (1970). Reprinted with permission of the Helen Dwight Reid Educational Foundation. Published by Heldref Publications, 4000 Albemarle St., N.W., Washington, D.C. 20016. Copyright © 1970.

Estimates[5] for this model are in Table 6.3 where I allow the traits (ξ_1 to ξ_4) to correlate and the methods (ξ_5 to ξ_7) to correlate, but the traits and methods are uncorrelated. The body of the table shows the effects of the traits and methods on each measure. For instance, ξ_1 is the prominence trait, and it has an expected 0.52 change in the self-rating prominence, whereas the self-rating method factor ξ_5 has a 0.58 effect on the same measure. A great deal more information can be gleaned from this table as I show in the next section.

Alternatives to Classical Validity Measures

Thus far I have reviewed four common types of validity: content, criterion, construct, and convergent and discriminant validities. Content validity is largely a theoretical approach to validation. There are no empirical substitutes for substantive and logical arguments that help define a concept, its dimensions, and the indicators needed to capture it, so this remains an important component of validity assessment.

Criterion validity is largely an empirical means to validate. Construct validity and convergent-discriminant validity are both theoretical and empirical. They are theoretical in the sense that theory suggests which constructs should correlate and which should not. The empirical aspect relies on the correlation of observed measures. The limits of these concern the empirical applications. One problem is that they rely on correlations rather than structural coefficients to test validity. Criterion validity examines the correlation of the criterion and the observed measure. Construct validity and convergent-discriminant validity are based on the correlation of measures of the same and different constructs. As I have demonstrated, these correlations may have little to do with the validity of a measure. That is, x_1 may truly measure ξ_1, yet the correlation measures of validity may be small for other reasons.

A second problem with these empirical tests is that they use only observed measures, rather than incorporating the latent variables into the analysis. The implicit assumption is that the correlation of two observed variables accurately mirrors an association involving latent variables. In criterion validity the criterion is a proxy for the latent variable, so it is implicitly assumed that the correlation of the criterion and the measure adequately approximates the correlation of the latent variable and the

[5]Estimation of measurement models is treated in the next chapter. These estimates result from the maximum likelihood fitting function. A listing of the LISREL program is in the appendix to this chapter.

Table 6.3 Estimates of Influence of Four Traits and Three Methods on 12 Observed Variables ($N = 240$)

	Traits				Methods			Error Variance
	ξ_1	ξ_2	ξ_3	ξ_4	ξ_5	ξ_6	ξ_7	
				Self-ratings				
Prominence	0.52	0	0	0	0.58	0	0	0.40
Achievement	0	0.41	0	0	0.60	0	0	0.47
Affiliation	0	0	0.35	0	0.47	0	0	0.67
Leader	0	0	0	0.58	0.63	0	0	0.28
				Peer Ratings				
Prominence	0.85	0	0	0	0	0.32	0	0.17
Achievement	0	0.69	0	0	0	0.53	0	0.24
Affiliation	0	0	0.47	0	0	0.81	0	0.12
Leader	0	0	0	0.86	0	0.43	0	0.08
				Observer Ratings				
Prominence	0.78	0	0	0	0	0	0.55	0.09
Achievement	0	0.66	0	0	0	0	0.65	0.15
Affiliation	0	0	0.72	0	0	0	0.27	0.40
Leader	0	0	0	0.75	0	0	0.64	0.03

measure. In construct and convergent-discriminant validities the correlation of observed measures is a proxy for the correlation of the latent constructs. As I have shown, it can be a poor proxy under a number of conditions.

What if we could estimate the correlation between a latent variable and its measure? Would this be a good validity coefficient? Despite the counter-intuitive notion of obtaining such a correlation, I will show in the next chapter that this can be done. The correlation between a latent variable and its measure (e.g., $\rho_{\xi_1 x_1}$) gets around the criticism that the classical tests rely exclusively on observed variables. Plus this correlation has some intuitive appeal. If x_1 is a valid measure of ξ_1, that is, if x_1 really measures ξ_1, then x_1 should correlate with ξ_1. In the extreme case of $x_1 = \xi_1$, $\rho_{\xi_1 x_1}$ must be one.

Despite the intuitive justification of $\rho_{\xi_1 x_1}$ as a validity coefficient, $\rho_{\xi_1 x_1}$ still suffers from my first criticism of the classical validity measures. That is, it is a bivariate correlation measure that neglects the structural linkages that may underlie it. I use a simple example to illustrate some of the difficulties. Suppose that Figure 6.5 is the true measurement model for ξ_1, ξ_2, and x_1.

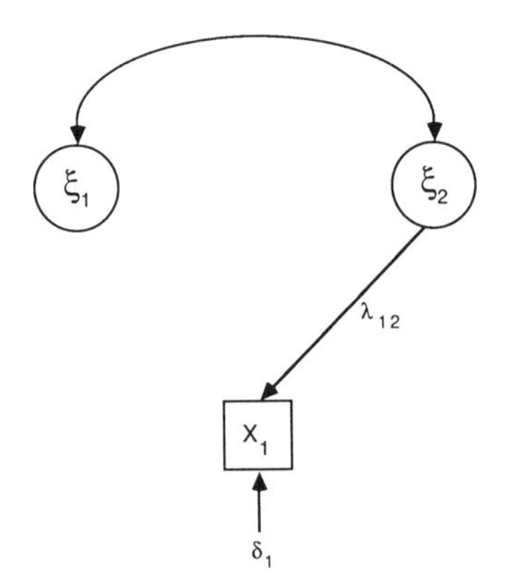

Figure 6.5 Measurement Model for X_1, ξ_1, and ξ_2.

The equation for this model is

$$x_1 = \lambda_{12}\xi_2 + \delta_1 \tag{6.6}$$

where $E(\delta_1) = 0$, $\text{COV}(\xi_1, \delta_1) = 0$, and $\text{COV}(\xi_2, \delta_1) = 0$. Suppose that we mistakenly assume that x_1 measures ξ_1, and to show its validity, we use $\rho_{\xi_1 x_1}$. It is easy to show that $\rho_{\xi_1 x_1}$ is

$$\rho_{\xi_1 x_1} = \frac{\lambda_{12}\phi_{12}}{\left[\text{VAR}(x_1)\phi_{11}\right]^{1/2}} \tag{6.7}$$

If x_1, ξ_1, and ξ_2 are standardized to a variance of one, then

$$\rho_{\xi_1 x_1} = \rho_{\xi_2 x_1}\rho_{\xi_1 \xi_2} \tag{6.8}$$

If ξ_1 and ξ_2 are moderately to highly correlated and the correlation of ξ_2 and x_1 is moderate, the correlation $\rho_{\xi_1 x_1}$ can be moderate to large, even though x_1 does not measure ξ_1 at all. For instance, if $\rho_{\xi_1 \xi_2}$ is 0.7 and $\rho_{\xi_2 x_1}$ is 0.7, $\rho_{\xi_1 x_1}$ is 0.49. With this correlation we could be misled into believing that x_1 has a moderate degree of validity when it has none. As discussed in Chapter 3, the $\rho_{\xi_1 x_1}$ would represent a spurious association where the association of ξ_1 and x_1 is due to the correlation of ξ_1 and ξ_2 and the relation of x_1 to ξ_2. The only situation where $\rho_{\xi_1 x_1}$ might be a useful measure of validity is in the special case where x_1 depends only on ξ_1 and random errors of measurement (δ_1), with ξ_1 and δ_1 uncorrelated. Although in some cases this special model may be correct, in general, $\rho_{\xi_1 x_1}$ is not an appropriate measure of validity.

At this point the reader may well wonder what measures of validity should be used since I demonstrated problems with the most common ones. Part of the difficulty of deriving a sound measure of validity is defining what is meant by validity. As mentioned earlier, validity is commonly defined as the extent to which an indicator measures what it is suppose to measure. A difficulty with this definition is that it is insufficiently clear to provide a way to operationalize it. For instance, what does "measure" mean in the definition? What if the indicator measures more than one latent variable?

To overcome these and other problems, I propose an alternative definition that is close to the standard one but is based on a structural equations approach.[6] *The validity of a measure x_i of ξ_j is the magnitude of the direct structural relation between ξ_j and x_i.* In this definition, for a measure to be valid, the latent and observed variable must have a direct link. I use "direct" in the path analysis sense discussed in Chapters 2 and 3. There must be no intervening variables between x_1 and ξ_1 if x_1 is to be a valid measure. As discussed in Chapter 3, the definition of a direct effect depends on the model, so this definition implies that the researcher has an explicit measurement model from which to evaluate validity.

Another key term in the definition is "structural." I am using structural in the same sense as defined in Chapter 2. Structural refers to an invariant and stable parameter linking ξ_j and x_i. The question of validity boils down to the question of whether a causal relation exists between the latent variable and an observed variable. As such it is subject to the same problems that accompany any attempt to support causal inferences. Many of these problems were discussed in Chapter 3.

With this definition in hand, a natural question is how to measure validity based on it? There is probably no one ideal measure of validity, but I propose several that correspond to this theoretical definition.

1. *Unstandardized Validity Coefficient (λ).* One important gauge of validity—that is, the direct structural relation between an x_i and ξ_j—is λ_{ij} the unstandardized coefficient linking them. For instance, if $x_1 = \lambda_{11}\xi_1 + \delta_1$ [with the usual assumptions $E(\delta_1) = 0$, $\mathrm{COV}(\xi_1, \delta_1) = 0$], then λ_{11} is a structural parameter of the direct link between x_1 and ξ_1. If x_1 measures and depends on ξ_1, then we expect that a change in ξ_1 leads to a change in x_1. It is λ_{11} that provides the expected change in x_1 for a one-unit change in ξ_1. As such λ_{11} provides one measure of the extent of the direct structural relation between x_1 and ξ_1. The λ_{ij} coefficients are in the Λ_x and Λ_y matrices.

[6]My thanks to Allen Wilcox for his suggestions on revising the definition.

As mentioned in the earlier chapters, λ_{ij} may be interpreted as a regression coefficient from the simple regression of x_i on ξ_j. As in *multiple* regression x_i may have a number of explanatory variables. Consider the following measurement model:

$$x_1 = \lambda_{11}\xi_1 + \lambda_{12}\xi_2 + \lambda_{13}\xi_3 + \delta_1 \tag{6.9}$$

where $E(\delta_1) = 0$, $\mathrm{COV}(\delta_1, \xi_j) = 0$ for $j = 1, 2, 3$. In (6.9) x_1 depends on three latent variables rather than the usual assumption of a single latent variable. This could occur if, for example, a survey question measured not only conservatism (ξ_1) but also authoritarianism (ξ_2) and response set effects (ξ_3). Criterion and construct validities measured with correlation coefficients implicitly assume that each measure depends only on one latent variable. They would not be appropriate for a situation such as (6.9). However, the unstandardized validity coefficient λ_{ij} would still be useful.

The validity of x_1 with respect to ξ_1 in (6.9) is indicated by λ_{11}. The λ_{11} coefficient is interpreted as the expected change in x_1 for a one-unit change in ξ_1, holding constant ξ_2 and ξ_3. In addition the validity of x_1 with respect to ξ_2 and ξ_3 can be gauged by λ_{12} and λ_{13}, respectively. Thus the unstandardized validity coefficient λ_{ij} is appropriate for measures that depend on one or more latent variables.

The unstandardized λ_{ij} coefficient also is useful for comparing samples from different populations. For example, the same observed variable may be in samples of males and females, samples from two different countries, or samples of some other groups. A comparison of validity could be made by comparing the corresponding $\hat{\lambda}_{ij}$ coefficients in the separate samples. The advantages of the unstandardized $\hat{\lambda}_{ij}$'s in comparing populations are the same as the advantages of using unstandardized regression coefficients (see Blalock 1967). They better measure the structural relation of the variables and are less influenced by differences in population variances.

There are, however, some disadvantages of λ_{ij} as a validity coefficient. One problem lies with the latent variables. For latent concepts such as social class, self-esteem, and social mobilization, there are few agreements about the units on which these concepts should be measured. To proceed with estimating a model, the latent variable must be assigned a scale. A frequent means of doing this is to set one of the λ_{ij} coefficients leading from the latent variable to one. This sets the latent variable's scale to that of the observed variable, with its λ equal to one. The other λ_{ij}'s leading from the same latent variable are interpretable relative to the unit of the observed variable with a λ of one. For instance, if we had two measures of a latent variable of length with one measure in centimeters and one in inches, we would need to choose the units in which to measure length. The measurement model is

$$x_1 = \lambda_{11}\xi_1 + \delta_1$$
$$x_2 = \lambda_{21}\xi_1 + \delta_2 \tag{6.10}$$

where x_1 is a measure in inches, x_2 is in centimeters, ξ_1 is the actual length, the errors of measurement are δ_1 and δ_2 with $E(\delta_i) = 0$ and $COV(\delta_i, \xi_1) = 0$ for $i = 1, 2$. To provide a unit for length, we could set λ_{11} or λ_{21} to one, setting the scale of ξ_1 to inches or centimeters. If λ_{11} is set to 1, λ_{21} should be close to 2.54 (2.54 cm = 1 in.). If λ_{21} is set to 1, λ_{11} should be close to 1/2.54. As can be seen the magnitudes of the λ_{ij}'s depends on the units assigned to the latent variable.

Since virtually all social science concepts have more ambiguous units than that for length, the meaning of the unstandardized λ_{ij} is even more uncertain for these concepts. The only real solution to this difficulty will occur when a consensus is reached on the most useful units for latent variables.

One related problem in comparing the unstandardized validity coefficients of measures that depend on the same latent variable is that the observed variables may be measured on very different scales. Direct comparisons of the magnitude of λ's to determine the relative validity of measures generally is not appropriate. For instance, in (6.10) λ_{11} and λ_{21} will differ because one measure is in inches and the other in centimeters. This has little to do with the relative validity of the measures. The standardized validity coefficients can help determine the relative validity of measures.

2. *The Standardized Validity Coefficient (λ^s).* The standardized validity coefficient λ^s is defined as[7]

$$\lambda^s_{ij} = \lambda_{ij}\left[\frac{\phi_{jj}}{\mathrm{VAR}(x_i)}\right]^{1/2} \tag{6.11}$$

The λ^s_{ij} coefficient is λ_{ij} times the ratio of the standard deviations for the latent variable, ξ_j, and the observed variable, x_i that depends on it. The λ^s_{ij} is analogous to the standardized regression coefficient. In the case of ξ_j influencing x_i, it gives the "expected" number of standard deviation units

[7]Note that in LISREL VI and earlier versions the standardized solution is only standardized in terms of the latent variables. If the observed variables are not standardized to a variance of one, the "standardized" lambda coefficients from LISREL will differ from those described here.

x_i changes for a one standard deviation change in ξ_j. Like the unstandardized λ_{ij}, it can be used when an observed variable depends on multiple latent variables. Unlike λ_{ij}, λ^s_{ij} is one means to compare the relative influence of ξ_j on several x_i variables. For example, if x_1 and x_2 depend on ξ_j and λ^s_{1j} is 0.8 and λ^s_{2j} is 0.10, this would indicate that x_1 is more responsive to ξ_j than is x_2 in standard deviation units. In addition, if x_i depends on two or more latent variables (e.g., ξ_1 and ξ_2) the relative influence of the latent variables can be compared. The standardized λ^s_{ij} is less useful than λ_{ij} in comparing different populations. This is because it is more subject to the influence of the varying standard deviations of the variables in different populations.

Consider the estimates from Table 6.3. The standardized validity coefficients appear in the columns beneath ξ_1 to ξ_7. For the observed variable of self-rated achievement, the coefficient of the latent trait of achievement (ξ_2) is 0.41, while that for the method factor for the self-ratings (ξ_5) is 0.60. These values show that the self-rated achievement is more responsive to the method factor (ξ_5) than to the trait factor (ξ_2) in terms of shifts in standard deviation units. Thus self-rated achievement and the other observed variables have two components of validity, one due to the trait and the other due to the method. The relative influences of each can be gauged by comparing the standardized coefficients.

3. *Unique Validity Variance ($U_{x_i\xi_j}$).* The unique validity variance, $U_{x_i\xi_j}$, measures that part of explained variance in x_i that is *uniquely* attributable to ξ_j. The formula for $U_{x_i\xi_j}$ is

$$U_{x_i\xi_j} = R^2_{x_i} - R^2_{x_i(\xi_j)} \tag{6.12}$$

where $R^2_{x_i}$ is the squared multiple correlation coefficient or proportion of variance in x_i explained by all variables in a model that have a direct effect on x_i (excluding error terms) and $R^2_{x_i(\xi_j)}$ is the proportion of explained variance in x_i by all variables with a direct effect on x_i *excluding* ξ_j.

To illustrate this measure, consider the case where $x_1 = \lambda_1\xi_1 + \delta_1$. The only variable with a direct effect on x_1 is ξ_1, so the $R^2_{x_1}$ is the squared correlation of x_1 and ξ_1. The $R^2_{x_1(\xi_1)}$, which is the squared correlation removing the influence of ξ_1, is zero since no other ξ affects x_1. Therefore $U_{x_1\xi_1}$ would be $R^2_{x_1}$. In this case and whenever a measure depends solely on one latent variable, $U_{x_1\xi_1}$ is equal to the squared correlation of the latent and observed variables.

The $U_{x_1\xi_1}$ becomes more complicated if x_1 depends on two or more variables. Suppose that

$$x_1 = \lambda_{11}\xi_1 + \lambda_{12}\xi_2 + \delta_1 \tag{6.13}$$

The $R^2_{x_1}$ equals

$$R^2_{x_1} = \frac{\lambda^2_{11}\phi_{11} + 2\lambda_{11}\lambda_{12}\phi_{12} + \lambda^2_{12}\phi_{22}}{\mathrm{VAR}(x_1)} \tag{6.14}$$

To compute $U_{x_1\xi_1}$ requires $R^2_{x_1(\xi_1)}$, which is the squared multiple correlation coefficient for x_1 with all variables having a direct effect on it included except ξ_1. The only other variable besides ξ_1 having an effect is ξ_2, so $R^2_{x_1(\xi_1)}$ simplifies to the squared correlation coefficient of x_1 and ξ_2:

$$\begin{aligned} R^2_{x_1(\xi_1)} &= \frac{[\mathrm{COV}(x_1, \xi_2)]^2}{\mathrm{VAR}(x_1)\phi_{22}} \\ &= \frac{\lambda^2_{11}\phi^2_{12} + 2\lambda_{11}\lambda_{12}\phi_{12}\phi_{22} + \lambda^2_{12}\phi^2_{22}}{\mathrm{VAR}(x_1)\phi_{22}} \\ &= \frac{\lambda^2_{11}\phi^2_{12}/\phi_{22} + 2\lambda_{11}\lambda_{12}\phi_{12} + \lambda^2_{12}\phi_{22}}{\mathrm{VAR}(x_1)} \end{aligned} \tag{6.15}$$

Subtracting (6.15) from (6.14) leads to $U_{x_1\xi_1}$:

$$U_{x_1\xi_1} = \frac{\lambda^2_{11}(\phi_{11} - \phi^2_{12}/\phi_{22})}{\mathrm{VAR}(x_1)} \tag{6.16}$$

This provides the explained variance in x_1 uniquely attributable to ξ_1. Similarly, $U_{x_1\xi_2}$ is

$$U_{x_1\xi_2} = \frac{\lambda^2_{12}(\phi_{22} - \phi^2_{12}/\phi_{11})}{\mathrm{VAR}(x_1)} \tag{6.17}$$

The general formula for $R^2_{x_i}$ is

$$R^2_{x_i} = \frac{\boldsymbol{\sigma}_{x_i\xi}\boldsymbol{\Phi}^{*-1}\boldsymbol{\sigma}'_{x_i\xi}}{\mathrm{VAR}(x_i)} \tag{6.18}$$

where $\boldsymbol{\sigma}_{x_i\xi}$ is a $1 \times d$ vector, with d equal to the number of ξ_i with a direct effect on x_i and its elements equal to the covariances of ξ_j with x_i for $j = 1, 2, \ldots, d$. The $\boldsymbol{\Phi}^*$ matrix is the variance-covariance matrix for those ξ

that directly affect x_i. Similarly, $R^2_{x_i(\xi_j)}$ is

$$R^2_{x_i(\xi_j)} = \frac{\boldsymbol{\sigma}_{x_i\xi_{(j)}} \boldsymbol{\Phi}^{*-1}_{(j)} \boldsymbol{\sigma}'_{x_i\xi_{(j)}}}{\text{VAR}(x_i)} \tag{6.19}$$

where $\boldsymbol{\sigma}_{x_i\xi_{(j)}}$ is $1 \times (d-1)$ vector of covariances of x_i with all ξ's directly influencing it excluding ξ_j. The $\boldsymbol{\Phi}^*_{(j)}$ matrix is $\boldsymbol{\Phi}^*$, removing the row and column corresponding to ξ_j. Finally, $U_{x_i\xi_j}$ is formed by $R^2_{x_i} - R^2_{x_i(\xi_j)}$.

Before illustrating this measure for the leadership multitrait-multimethod data, I review some of its properties. First, unlike λ_i and λ^s_i, $U_{x_i\xi_j}$ always varies between zero and one. If all the variance in x_i is explained only by ξ_j, $U_{x_i\xi_j}$ is one. If no variance is explained, it is zero. Its upper bound is further constrained in that it cannot exceed $R^2_{x_i}$. Second, if only ξ_j has a direct effect on x_i (as shown earlier) $U_{x_i\xi_j}$ equals the squared correlation between ξ_j and x_i [i.e., $U_{x_i\xi_j} = \rho^2_{x_i\xi_j}$ if $x_i = \lambda_{ij}\xi_j + \delta_i$ and $\text{COV}(\delta_i, \xi_j) = 0$]. This means that in the traditional case where a measure is assumed to depend on only one latent variable, $U_{x_i\xi_j}$ is the same as the intuitively pleasing validity measure of the squared correlation between the observed variable and the latent one. But, $U_{x_i\xi_j}$ is more general than $\rho^2_{x_i\xi_j}$ since it allows the observed variable to depend on more than one latent variable and it is zero if ξ_j has no direct effect on x_i. If multiple correlated latent variables underlie x_i, $U_{x_i\xi_i}$ generally will not equal $\rho^2_{x_i\xi_j}$ unless the latent variables are uncorrelated.

Table 6.4 shows the $\hat{U}_{x_i\xi_j}$ for the multitrait-multimethod leadership data from Table 6.3. To illustrate the calculation of the entries in Table 6.4, consider the self-rated prominence variable. The measurement equation for it is

$$x_1 = 0.52\xi_1 + 0.58\xi_5 + \hat{\delta}_1 \tag{6.20}$$

where x_1 is self-rated prominence, ξ_1 is the latent prominence variable, ξ_5 is the self-rating method factor, $\hat{\delta}_1$ is the sample error, and 0.52 and 0.58 are the $\hat{\lambda}_{ij}$ coefficients. From (6.16) the $\hat{U}_{x_1\xi_1}$ is

$$\hat{U}_{x_1\xi_1} = \frac{\hat{\lambda}^2_1\left(\hat{\phi}_{11} - \hat{\phi}^2_{12}/\hat{\phi}_{22}\right)}{\text{var}(x_1)} \tag{6.21}$$

For this example (6.21) simplifies since $\hat{\phi}_{11}$ and $\hat{\phi}_{22}$ are set to one, and the method factor ξ_5 is uncorrelated with the prominence trait ξ_1 so that $\hat{\phi}_{12}$ is zero. Substituting the sample estimates for these quantities into (6.21), we find that $\hat{U}_{x_1\xi_1}$ is 0.27 (see Table 6.4). In an analogous fashion we see that

Table 6.4 Estimates of Unique Validity Variances $\hat{U}_{x_i\xi_j}$ for Traits and Method Factors for Multitrait-Multimethod Leadership Data

	$\hat{U}_{x_i\xi_j}$ for:	
	Traits (ξ_1, ξ_2, ξ_3, or ξ_4)	Method (ξ_5, ξ_6, or ξ_7)
	Self-ratings	
Prominence	0.27	0.34
Achievement	0.17	0.36
Affiliation	0.12	0.22
Leader	0.34	0.40
	Peer Ratings	
Prominence	0.72	0.10
Achievement	0.48	0.28
Affiliation	0.22	0.66
Leader	0.74	0.18
	Observer Ratings	
Prominence	0.61	0.30
Achievement	0.44	0.42
Affiliation	0.52	0.07
Leader	0.56	0.41

$\hat{U}_{x_1\xi_5}$ is 0.34. The other $\hat{U}_{x_i\xi_j}$'s in Table 6.4 are calculated in the same manner.

The $\hat{U}_{x_i\xi_j}$'s show some interesting results. Consider first the self-rating measures. The unique validity variances for the ξ_5 method factor are greater than the validity variances for the four traits. The implication is that the self-ratings of the leadership variables are strongly influenced by the self-rating method. In fact in terms of validity for measuring the traits, the self-ratings fare much worse than the peer and observer ratings for each trait.

The greatest unique validity variance for the prominence trait (ξ_1) is for the peer-rated prominence (= 0.72); for achievement (ξ_2), it is peer-rated achievement (= 0.48) [though the observer rating is close (0.44)]; for affiliation (ξ_3), it is the observer-rated affiliation (= 0.52); and finally, for leadership, it is the peer rating (= 0.74). Overall, the peer ratings have the greatest unique validity variances for three of the four traits. Note also that most of these variables have a nonneglible proportion of their variance that can be attributed to the method factors.

In the case where x_i depends on two or more highly correlated latent variables, the unique validity variance may be quite low. This is one manifestation of the multicollinearity problem in measurement models that I discussed in Chapter 3. In the case where x_1 depends on two highly correlated latent variables ξ_1 and ξ_2, it may be that jointly they explain a large proportion of the variance in x_1. Despite the large $R^2_{x_1}$ we may have small $U_{x_1\xi_1}$ and $U_{x_1\xi_2}$. Small unique validity variances in this case make sense since the collinearity of the latent variables means that we are uncertain which variable is responsible for the joint portion of the explained variance of x_1.

An empirical illustration comes from data on morale and sense of belonging among a random sample of 49 female college students from a liberal arts college. I presented this example in Chapter 3. As part of the survey four measures of morale (x_1 to x_4) and three measures of feelings of belonging (x_5 to x_7) to the college were collected. I estimated a model where the latent variable morale (ξ_1) affects x_1 to x_4, and the sense of belonging (ξ_2) determines x_4 to x_7. The x_4 variable is the only one affected by both latent variables for this illustration. The estimates for the model are in Table 6.5. The $\hat{\phi}_{11}$, $\hat{\phi}_{21}$, and $\hat{\phi}_{22}$ estimates are 2.27, 2.67, and 3.53, respectively. From these last three estimates the correlation of morale (ξ_1) and belonging (ξ_2) is 0.94, indicating a high degree of collinearity between the latent variables.

Since x_4 depends on these highly correlated variables its unique validity variance should be affected even if its $R^2_{x_4}$ is relatively high. The $R^2_{x_4}$ is

$$R^2_{x_4} = \frac{(1.61)^2 2.27 + 2(1.61)(-0.62)(2.67) + (-0.62)^2(3.53)}{2.47}$$

$$= 0.77 \tag{6.22}$$

Thus 77% of x_4's variance is explained by ξ_1 and ξ_2. Based on (6.16), $\hat{U}_{x_4\xi_1}$ is

$$\hat{U}_{x_4\xi_1} = \frac{\hat{\lambda}_1^2\left(\hat{\phi}_{11} - \hat{\phi}_{12}^2/\hat{\phi}_{22}\right)}{\operatorname{var}(x_4)}$$

$$= \frac{(1.61)^2(2.27) - (1.61)^2(2.67)^2/(3.53)}{2.47} = 0.26 \tag{6.23}$$

Table 6.5 Estimates of the Influences of Morale (ξ_1) and Belonging for Females (ξ_2) on x_1 to x_7 ($N = 49$)

Variable	Morale (ξ_1)	Belonging (ξ_2)	Error Variance
x_1	1.00[a]	0.00[a]	1.75
x_2	−1.20[b]	0.00[a]	3.95
x_3	0.95	0.00[a]	0.61
x_4	1.61	−0.62	0.53
x_5	0.00[a]	1.00[a]	0.44
x_6	0.00[a]	0.75	2.64
x_7	0.00[a]	0.70	2.54

Note: a = constrained parameter; *b* = negatively worded item.

The unique validity variance of x_4 for ξ_2 is

$$\hat{U}_{x_4\xi_1} = \frac{\hat{\lambda}_2^2\left(\hat{\phi}_{22} - \hat{\phi}_{12}^2/\hat{\phi}_{11}\right)}{\operatorname{var}(x_4)} = 0.06 \tag{6.24}$$

As $\hat{U}_{x_4\xi_1}$ and $\hat{U}_{x_4\xi_2}$ make clear despite the substantial $R^2_{x_4}$, the unique validity variance that can be attributed to morale (ξ_1) or belonging (ξ_2) is small, with most of the variance in x_4 jointly explained by ξ_1 and ξ_2.

4. *Degree of Collinearity ($R^2_{\xi_j}$).* If x_i depends on only one latent variable or if the latent variables influencing x_i are uncorrelated, $U_{x_i\xi_j}$ equals the squared correlation of x_i and ξ_j and the variance explained by the ξ_j's can be partitioned without ambiguity. In the more general case of x_i depending on correlated latent variables, an exact partitioning is not possible. This problem grows worse the greater the collinearity of the latent variables that influence x_i. One indicator of the degree of collinearity researchers use in regression analysis is appropriate here. It is the squared multiple correlation coefficient resulting when ξ_j is predicted from the other ξ's that have a direct effect on x_i. It is calculated as

$$R^2_{\xi_j} = \frac{\boldsymbol{\sigma}_{\xi_j\xi_{(j)}}\boldsymbol{\Phi}^{*-1}_{(j)}\boldsymbol{\sigma}'_{\xi_j\xi_{(j)}}}{\phi_{jj}} \tag{6.25}$$

In (6.25) $\boldsymbol{\sigma}_{\xi_j\xi_{(j)}}$ is a $1 \times (d-1)$ vector of covariances of ξ_j with all the other ξ's that directly influence x_i except itself, $\boldsymbol{\Phi}^*_{(j)}$ is as defined earlier (the covariance matrix of all ξ's that directly influence x_i except ξ_j), and ϕ_{jj} is the variance of ξ_j.

If there are only two ξ variables, say, ξ_1 and ξ_2, (6.25) simplifies to

$$R^2_{\xi_j} = \frac{\phi^2_{12}}{\phi_{11}\phi_{22}} \tag{6.26}$$

which is the squared correlation of ξ_1 and ξ_2. The morale-sense of belonging example provides an illustration. Considering x_4 which depends on ξ_1 and ξ_2, the $R^2_{\xi_1}$ is formed from (6.26), and substituting in $\hat{\phi}_{11}$ (= 2.27), $\hat{\phi}_{12}$ (= 2.67), and $\hat{\phi}_{22}$ (= 3.53) results in $R^2_{\xi_1}$ equal to 0.89. This high degree of collinearity helps explain the low unique validity variances $\hat{U}_{x_4\xi_1}$ and $\hat{U}_{x_4\xi_2}$.

Summary

In this section I explored the concept of validity. Content validity, criterion validity, construct validity, and convergent-discriminant validity are four traditional validity types that I considered. The first of these emphasizes matching a suitable range of indicators to the meaning of and dimensions of a concept. As such it has the desirable consequence of requiring analysts to explain a concept clearly. Content validity relies more on conceptual arguments than on observing associations. In contrast, the last three validity techniques have a substantial empirical component, most often relying on correlations of measures to evaluate validity. I emphasized the weaknesses of such a correlation strategy. A structural equations approach appears more fruitful. I proposed several measures of validity that correspond to structural equations while also being related to the traditional measures.

RELIABILITY

We have encountered the concept of reliability in the last section and in Chapter 5. There I described it as the squared correlation of a measure and its latent variable. In this section I provide a more detailed description. I present first an overview of the concept. Then I discuss the traditional empirical means to estimate reliability. The last subsection on reliability compares classical test theory, from which the concept of reliability emerged, and the general measurement model we have been using.

Reliability is the consistency of measurement. It is not the same as validity since we can have consistent but invalid measures. To illustrate reliability, suppose that I wish to measure your level of education.[8] I

[8]This illustration is based on a similar one given by Lazarsfeld (1959).

narrowly define education as completed years of formal schooling. I operationalize it by asking: "How many completed years of formal schooling have you had?" Next I record your answer. If I had the ability to erase your memory of the question and the response you gave, I could repeat the same question and again, record your answer. Repeating this process an infinite number of times, I could determine the consistency of your response to the same question. The reliability of this education measure is the consistency in your response over the infinite trials. The greater the fluctuations in your answers, the lower the reliability. The more you tend to give me the same response, the greater is the reliability.

The consistency or reliability of the education measure also could be determined across observations rather than only over time. For example, I could ask an extremely large number of people the same question on completed years of formal schooling. If I could somehow remove their memory of the question and their responses, I could ask them the same question again and record their responses. The reliability is the consistency of the responses across individuals for the two time periods. To the extent that all individuals are consistent, the measure is reliable.[9]

Reliability is that part of a measure that is free of purely random error. Note that nothing in the description of reliability requires that the measure be valid. In fact, it is possible to have a very reliable measure that is not valid. For example, repeatedly weighing yourself on a bathroom scale may provide a reliable measure of your weight but the scale is not valid if it always gives a weight that is 11 lbs (or 5 kg) too light. A more extreme example would be obtaining a measure of intelligence by asking individuals their shoe size. This may provide a very reliable measure, but it lacks validity as an intelligence measure. Thus the distinction between reliability and validity is important.

Much of the social science literature on reliability originates in classical measurement theory from psychology. Since classical measurement theory is basic to some of the empirical tests of reliability, I review some of its relevant features.[10] A fundamental equation of the theory is

$$x_i = \tau_i + e_i \tag{6.27}$$

In (6.27) x_i is the ith observed variable (or "test" score), e_i is the error

[9]To the extent that individuals differ in reliability of responses, the across-individual reliability may differ from the overtime reliability for an individual.

[10]I will examine only a small part of psychological testing and measurement theory. For a more extended treatment, see Lord and Novick (1968), and for a review of developments up to 1980, see Weiss and Davison (1981).

term defined as $x_i - \tau_i$, and τ_i is the *true score* that underlies x_i. It is assumed that the COV(τ_i, e_i) is zero and that $E(e_i) = 0$. The true score τ_i is the variable that results from all the systematic factors in the observed variable. It is what remains when the errors of measurement are removed. Later in this section I explain the relation between τ_i and ξ_i, and δ_i and e_i, but for now note that they need not be the same.

According to classical test theory, the errors of measurement for different items are uncorrelated. The correlation of two measures results from the association of their true scores. Thus the true scores are the systematic components that lead to the association of observed variables.

Parallel, *tau-equivalent*, and *congeneric* measures are the three major types of observed variables in test theory. They can be defined using two measures x_i and x_j as examples:

$$\begin{aligned} x_i &= \alpha_i \tau_i + e_i \\ x_j &= \alpha_j \tau_j + e_j \end{aligned} \tag{6.28}$$

The e_i and e_j terms are uncorrelated. Assume that the true scores are the same ($\tau_i = \tau_j$). If α_i and α_j equal one and the VAR(e_i) equals the VAR(e_j), then x_i and x_j are parallel measures. If α_i and α_j equal one but VAR(e_i) does not equal VAR(e_j), the measures are tau-equivalent. Finally, if α_i does not equal α_j and VAR(e_i) does not equal VAR(e_j), congeneric measures result. Congeneric measures (Jöreskog 1971) are the most general of the three types.

The *reliability* of a measure $\rho_{x_i x_i}$ is defined as

$$\rho_{x_i x_i} = \frac{\alpha_i^2 \,\mathrm{VAR}(\tau_i)}{\mathrm{VAR}(x_i)} \tag{6.29}$$

Since the VAR(x_i) equals α_i^2 VAR(τ_i) + VAR(e_i) and the VAR(e_i) is nonnegative, the reliability can never exceed one. For tau-equivalent or parallel measures, this simplifies to

$$\rho_{x_i x_i} = \frac{\mathrm{VAR}(\tau_i)}{\mathrm{VAR}(x_i)} \tag{6.30}$$

Reliability is the ratio of true score's variance to the observed variable's variance. It equals the squared correlation of the observed variable and the true score:

$$\rho^2_{x_i\tau_i} = \frac{[\mathrm{COV}(x_i, \tau_i)]^2}{\mathrm{VAR}(x_i)\mathrm{VAR}(\tau_i)}$$

$$= \frac{\alpha_i^2[\mathrm{VAR}(\tau_i)]^2}{\mathrm{VAR}(x_i)\mathrm{VAR}(\tau_i)}$$

$$= \frac{\alpha_i^2\,\mathrm{VAR}(\tau_i)}{\mathrm{VAR}(x_i)}$$

$$= \rho_{x_ix_i} \tag{6.31}$$

Thus $\rho_{x_ix_i}$ can be interpreted as the variance of x_i that is explained by τ_i with the remaining variance due to error. The analogy to regression analysis should be evident. In simple regression an observed dependent variable is a function of an observed independent variable and the squared correlation between the variables provides the explained variance. In test theory τ_i is an unobserved independent variable, the measure x_i is the dependent variable, and the reliability coefficient $\rho_{x_ix_i}$ is the squared correlation of τ_i and x_i.

Empirical Means of Assessing Reliability

There are a number of ways that have been proposed to estimate the reliability of measures. I will review the four most common ones: test-retest, alternative forms, split-halves, and Cronbach's alpha. In this subsection I describe each and highlight their merits and limits. This is followed by a general reliability measure.

The *test-retest method* is based on having the same measure ("test") for the same observations at two points in time. The equations for the two measures are

$$\begin{aligned} x_t &= \alpha_t\tau_t + e_t \\ x_{t+1} &= \alpha_{t+1}\tau_{t+1} + e_{t+1} \end{aligned} \tag{6.32}$$

where t and $t+1$ are subscripts referencing the first and second time periods for the x, α, τ, and e. I assume that $E(e_t) = E(e_{t+1}) = 0$, that the true scores (τ_t, τ_{t+1}) are uncorrelated with the errors (e_t, e_{t+1}), and that the errors are uncorrelated $[\mathrm{COV}(e_t, e_{t+1}) = 0]$. In addition this method assumes that x_t and x_{t+1} are parallel measures [$\alpha_t = \alpha_{t+1} = 1$ and $\text{VAR}(e_t) = \text{VAR}(e_{t+1})$] and that the true scores are equal ($\tau_t = \tau_{t+1}$).

The reliability estimate is the correlation of x_t and x_{t+1}. Using the definition of the correlation of two variables and covariance algebra leads to

$$\begin{aligned} \rho_{x_t x_{t+1}} &= \frac{\text{COV}(x_t, x_{t+1})}{[\text{VAR}(x_t)\text{VAR}(x_{t+1})]^{1/2}} \\ &= \frac{\text{VAR}(\tau_t)}{\text{VAR}(x_t)} \\ &= \rho_{x_t x_t} \end{aligned} \tag{6.33}$$

The steps in (6.33) take advantage of the assumptions of the method and show that the correlation of the two parallel measures x_t and x_{t+1} equals $\rho_{x_t x_t}$. By substituting τ_{t+1} for τ_t and the $\text{VAR}(e_{t+1})$ for $\text{VAR}(e_t)$, I could demonstrate that this same correlation also equals $\rho_{x_{t+1} x_{t+1}}$. *In fact, the correlation of any two parallel measures equals their reliability since all parallel measures have identical reliabilities.*

To illustrate the test-retest method, I take an example from Hill (1982). Hill analyzes "government responsiveness" indicators taken from the SRC-CPS 1972 national election study. A random sample of respondents were administered the same questions before and after the U.S. national elections. The time between surveys was only a few weeks for most respondents. Hill formed an index of government responsiveness based on indicators gauging people's attitudes as to whether the government listens to people, whether parties affect policies, whether members of Congress pay attention to constituents, and whether elections affect policy. The correlation of the government responsiveness index for the pre- and postelection period is 0.52. Under the assumptions of the test-retest method, the estimate of reliability of the government responsiveness index is 0.52. This is a moderate degree of reliability.

Despite the intuitive appeal of the test-retest reliability technique, it has several limitations. First, it assumes perfect stability of the true score, τ. In many cases the true score may change over time so that this assumption is not reasonable. Consider Hill's government responsiveness example. The short time interval separating the surveys makes stability more likely than if a longer period were used. However, the election that fell within this interval could have influenced people's perceptions of government responsiveness. Respondents satisfied with the election results might elevate their assessment of government responsiveness, whereas those who voted for losers might see the government as less responsive. If this is true, then τ_t would not equal τ_{t+1}, and the correlation of x_t and x_{t+1} would no longer equal the reliability. If this is the only assumption in error, $\rho_{x_t x_{t+1}}$ underestimates reliability.

Reactivity is another way in which the perfect stability of τ can be disrupted. Reactivity is when the process of measurement induces change in the phenomenon. Such occurrences are most likely with attitude and value measures. The government responsiveness indicators administered before the election could sensitize respondents to campaign information on this issue. Competing candidates may criticize each other as deviating from the will of the people, or they may point to broken campaign promises from the past. An incumbent may highlight his/her responsiveness to the voters. In either case the earlier measures may lead the respondent to pay more attention to information about government responsiveness, and this in turn can affect the very trait we are trying to measure. In short, the x_t measure might affect τ_{t+1} so that τ_t does not equal τ_{t+1}.

If lack of equivalence of τ_t and τ_{t+1} is the only violated assumption, then $\rho_{x_t x_{t+1}}$ is less than the reliability. Other violated assumptions can, however, occur. For instance, memory effects sometimes are present. People's memories of responses during the first interview can influence their responses in a second interview. They may have the tendency to give the same responses to questions on government responsiveness, education, attitudes toward abortions, and so on, as they gave in the initial interview to the same questions. This could be represented as x_t influencing x_{t+1} counter to what is assumed with the test-retest procedure. This factor tends to increase the correlation between x_t and x_{t+1} from what it would be in the absence of memory effects and thereby leads $\rho_{x_t x_{t+1}}$ to be higher than the reliability of the measure.

Another potentially violated assumption is that the covariance of the errors e_t and e_{t+1} is zero. Some of the numerous omitted factors that influence x at time t are likely to be correlated with the omitted factors affecting x at time $t + 1$. This is particularly true if the time period between measures is short, or if the measurement situations are very similar. If, however, the questionnaire is fairly lengthy with many different items, then this may make the memory factor less influential. In Hill's study (1982, 34) the period separating the measures of government responsiveness was short, which would favor memory effects, but he suggests that the complexity of the survey makes this less likely. If positively correlated errors are present, $\rho_{x_t x_{t+1}}$ overestimates reliability.

The assumption of parallel measures in the test-retest method also can be questioned. Employing the same measure at t and $t + 1$ makes the α_t equal to α_{t+1} assumption reasonable in many cases. But the assumption that the error variances are equal $[\text{VAR}(e_t) = \text{VAR}(e_{t+1})]$ raises more serious questions. Though we may be willing to assume that the direct effect of τ on x stays the same, the variability of the error influencing x_t and x_{t+1} may not be so well behaved. For the government responsiveness index, it may well be that the error variance is changed by the election so that greater error variance occurs before rather than after. This would depend on the nature of the election such as campaign tactics, the closeness of the election, and the major issues raised. If we are unwilling to assume a constant error variance, then tau-equivalent rather than parallel measures is a safer assumption. With tau-equivalent measures, $\rho_{x_t x_{t+1}}$ equals

$$\rho_{x_t x_{t+1}} = \frac{\text{VAR}(\tau_t)}{[\text{VAR}(x_t)\text{VAR}(x_{t+1})]^{1/2}} \tag{6.34}$$

which no longer equals the reliability of x_t or x_{t+1}. In short, the test-retest method to estimate reliability has the advantage of simplicity, but it is dependent on one or more assumptions that are unreasonable in practice.

Another technique to estimate reliability is *alternative forms*. It is similar to the test-retest method, except that different measures instead of the same measure are collected at t and $t + 1$. Equation (6.35) represents this situation:

$$\begin{aligned} x_1 &= \tau_t + e_t \\ x_2 &= \tau_{t+1} + e_{t+1} \end{aligned} \tag{6.35}$$

The x_1 variable is a measure of τ at time t, x_2 is a different measure at $t + 1$, and x_1 and x_2 are parallel measures. Like the test-retest method it is assumed that τ_t equals τ_{t+1}, that the expected value of e_t and e_{t+1} are zero, and that the errors are uncorrelated with each other and with τ_t and τ_{t+1}. With these assumptions the correlation of x_1 and $x_2(\rho_{x_1, x_2})$ equals the reliability of both measures. This can be derived by the same steps followed in (6.33), substituting x_1 for x_t and x_2 for x_{t+1}. An example would be if different indexes of government responsiveness were collected in the pre- and postelection surveys. The correlation of the two indexes would be the estimate of reliability.

Several of the limits of the test-retest method are true for the alternative forms technique. The often unrealistic assumption that τ_t equals τ_{t+1} is still present. The process of measurement still may induce change in the phenomenon itself. The assumption that the error variances are equal

[VAR(e_1) equals VAR(e_2)] is even less likely since x_1 and x_2 are different measures as well as being at different time points. So, as discussed for the test-retest method, these factors can undermine $\rho_{x_1x_2}$ as an estimate of reliability.

The alternative form does have two advantages. One is that compared to the test-retest, the alternative form measures are less susceptible to memory effects since time t and $t+1$ have different scales. Second, the errors of measurement for one indicator are less likely to correlate with a new measure at the second time period. Compared to test-retest, correlated errors of measurement are less likely. In short, though the alternative forms' estimate of reliability overcomes some of the limits of the test-retest approach, several unrealistic assumptions remain.

A third means to estimate reliability is with *split-halves*. The split-halves method assumes that a number of items are available to measure τ. Half of these items are combined to form a new measure, say, x_1, and the other half to form x_2. Note that in contrast to the test-retest and alternative form, x_1 and x_2 are measures of τ in the same time period. It is still assumed that $E(e_1) = E(e_2) = 0$, $\text{COV}(\tau_1, e_1) = \text{COV}(\tau_1, e_2) = 0$, the $\text{COV}(e_1, e_2) = 0$, and that x_1 and x_2 are parallel measures. The equations for x_1 and x_2 are

$$\begin{aligned} x_1 &= \tau_1 + e_1 \\ x_2 &= \tau_1 + e_2 \end{aligned} \tag{6.36}$$

The correlation of x_1 and x_2 equals

$$\begin{aligned} \rho_{x_1x_2} &= \frac{\text{COV}(x_1, x_2)}{[\text{VAR}(x_1)\text{VAR}(x_2)]^{1/2}} \\ &= \frac{\text{VAR}(\tau_1)}{\text{VAR}(x_1)} \\ &= \rho_{x_1x_1} = \rho_{x_2x_2} \end{aligned} \tag{6.37}$$

The denominator in the second step of (6.37) follows because the variances of all parallel measures are equal. Thus, as with all parallel measures, their correlation equals the reliability. The government responsiveness trait has four indicators as described earlier. If I combine the first two and the last two, the correlation of these two halves is 0.57. This is the split-halves estimate of reliability for each *half* of the items.

In many cases the unweighted sum of two halves forms a composite to measure τ_1 so that the reliability of $x_1 + x_2$ should be determined. I showed earlier that, in general, the squared correlation of τ_1 with the observed score is the reliability of a measure. Employing this relation, the squared correlation of τ_1 with $(x_1 + x_2)$ is

$$\begin{aligned}
\rho^2_{\tau_1(x_1+x_2)} &= \frac{[\mathrm{COV}(\tau_1, x_1 + x_2)]^2}{\mathrm{VAR}(\tau_1)\mathrm{VAR}(x_1 + x_2)} \\
&= \frac{4[\mathrm{VAR}(\tau_1)]^2}{\mathrm{VAR}(\tau_1)[\mathrm{VAR}(x_1) + \mathrm{VAR}(x_2) + 2\,\mathrm{COV}(x_1, x_2)]} \\
&= \frac{2\,\mathrm{VAR}(\tau_1)/\mathrm{VAR}(x_1)}{\mathrm{VAR}(\tau_1)/\mathrm{VAR}(x_1) + \mathrm{VAR}(x_1)/\mathrm{VAR}(x_1)} \\
&= \frac{2\rho_{x_1x_1}}{1 + \rho_{x_1x_1}} \qquad (6.38)
\end{aligned}$$

This formula is well known as the *Spearman-Brown Prophecy* formula for gauging the reliability of a full test based on split-halves. The split-halves for the government responsiveness items correlated 0.57. If the assumptions are met, each half has a 0.57 reliability. Substituting this value in for $\rho_{x_1x_1}$ in the last line of (6.38) results in a reliability estimate of 0.73.

The split-halves test has several aspects more desirable than the test-retest and alternative forms methods. For one, the split-halves method does not assume perfect stability of τ since τ is only gauged in one time period. Second, the memory effects that can occur if the same item is asked at two points in time are not operating with this approach. Third, the correlated errors of measurement that are likely in test-retest approaches are less likely for split-halves (though correlated errors across halves can arise from other factors). A practical advantage is that split-halves are often cheaper and more easily obtained than overtime data.

A disadvantage is that the split-halves must be parallel measures. Often we cannot know whether the variance of the measurement errors are equal or whether α_1 and α_2 are equal to one. If they are not parallel, then the derivations in (6.37) and (6.38) do not hold, and the reliability values are in error. Another drawback is that there is a certain arbitrariness in the way that the halves are allocated. There are many ways to divide a set of items in half. Each split could lead to a different reliability estimate.

These limits are evident for the government responsiveness data. Since two different indicators are combined in each half we cannot say with confidence that each half is parallel. The slope that characterizes the first half's relation to τ_1 could differ from the slope for the second half. Or, the error variance can vary between halves. Furthermore my selection of the first two indicators for the first half and the last two for the second was arbitrary. It could have been split with the first and third or the second and third as the first half. Thus the accuracy of the reliability estimate with the split-halves method may be deficient.

Cronbach's (1951) *alpha coefficient* overcomes some of the disadvantages of the split-halves method. Coefficient alpha is the most popular reliability coefficient in social science research. It measures the reliability of a simple sum of tau-equivalent or parallel measures. Given its importance, I derive this coefficient, starting with the previous definition of reliability as the squared correlation of the true score τ_1 with the observed variables. For alpha, the observed variables $x_1, x_2, \ldots, x_q$ are summed. The x_i's should be scored so that they are all positively or all negatively related to τ_1. Call this index H so that $\Sigma_{i=1}^{q} x_i = H$. The squared correlation of τ_1 and H or the reliability of H is

$$
\begin{aligned}
\rho_{\tau_1 H}^2 &= \frac{[\mathrm{COV}(\tau_1, H)]^2}{\mathrm{VAR}(\tau_1)\mathrm{VAR}(H)} \\
&= \frac{[\mathrm{COV}(\tau_1, x_1 + x_2 + \cdots + x_q)]^2}{\mathrm{VAR}(\tau_1)\mathrm{VAR}(H)} \\
&= \frac{[\mathrm{COV}(\tau_1, q\tau_1 + \Sigma_{i=1}^{q} e_i)]^2}{\mathrm{VAR}(\tau_1)\mathrm{VAR}(H)} \\
&= \frac{[q\,\mathrm{VAR}(\tau_1)]^2}{\mathrm{VAR}(\tau_1)\mathrm{VAR}(H)} \\
&= \frac{q^2\,\mathrm{VAR}(\tau_1)}{\mathrm{VAR}(H)} \\
&= \rho_{HH}
\end{aligned}
\tag{6.39}
$$

Equation (6.39) provides a general formula for the reliability of the unweighted sum of q tau-equivalent or parallel measures. As (6.40) shows, this can be manipulated so that it appears as the typical formula for Cronbach's alpha:

$$\rho_{HH}^2 = \frac{q^2\,\mathrm{VAR}(\tau_1)}{\mathrm{VAR}(H)}$$

$$= \frac{q(q-1)q\,\mathrm{VAR}(\tau_1)}{(q-1)\mathrm{VAR}(H)}$$

$$= \left(\frac{q}{q-1}\right)\left(\frac{q^2\,\mathrm{VAR}(\tau_1) - q\,\mathrm{VAR}(\tau_1)}{\mathrm{VAR}(H)}\right)$$

$$= \left(\frac{q}{q-1}\right)\left(\frac{q^2\,\mathrm{VAR}(\tau_1) + \Sigma_{i=1}^{q}\,\mathrm{VAR}(e_i) - q\,\mathrm{VAR}(\tau_1) - \Sigma_{i=1}^{q}\,\mathrm{VAR}(e_i)}{\mathrm{VAR}(H)}\right)$$

$$= \left(\frac{q}{q-1}\right)\left\{\frac{\mathrm{VAR}(H) - \left[q\,\mathrm{VAR}(\tau_1) + \Sigma_{i=1}^{q}\,\mathrm{VAR}(e_i)\right]}{\mathrm{VAR}(H)}\right\}$$

$$= \left(\frac{q}{q-1}\right)\left(1 - \frac{\Sigma_{i=1}^{q}\,\mathrm{VAR}(x_i)}{\mathrm{VAR}(H)}\right) \tag{6.40}$$

Since this derivation holds for either parallel or tau-equivalent measures, (6.39) and (6.40) reveal that alpha equals the reliability of H for either type of measure. Furthermore it can be demonstrated that Cronbach's alpha provides a lower bound for the reliability of the simple sum of congeneric measures.

With these features the advantages of alpha over the other reliability measures should be evident. There are no assumptions needed for the stability of τ_1. The measures need not be parallel. The possibility of memory effects are remote since measures for only one time period are applied. There is no problem in selecting splits of the items for testing since all measures can be treated individually. In addition computation of alpha is relatively easy.

For an example of calculating alpha consider Table 6.6. It reproduces the correlation matrix that Hill (1982, 37) provides for the pre-election survey. The alpha for the simple sum of all four measures can be calculated with equation (6.40). The q is equal to 4, the variance of each x_i is one for a correlation matrix, and the variance of the index equals the sum of each element of the correlation matrix. Substituting these values into (6.40), the sample alpha is 0.72 (= (4/3)[1 − (4/8.68)]).

Table 6.6 Correlation Matrix of Four Pre-Election Measures of Government Responsiveness

		x_1	x_2	x_3	x_4
Government Listen to People?	(x_1)	1.00	0.42	0.38	0.43
Parties Affect Policy?	(x_2)	0.42	1.00	0.43	0.34
Elections Affect Policy?	(x_3)	0.38	0.43	1.00	0.34
Members of Congress Attention to Constituents	(x_4)	0.43	0.34	0.34	1.00

Source: From Hill (1982, 37).

The estimated reliability of the government responsiveness index differs for the three methods of reliability calculation: 0.52 for test-retest, 0.73 for split-halves, and 0.72 for coefficient alpha. The split-halves and alpha estimates are close, reflecting their common dependence on indicators at one point in time and the close relation between these procedures. The test-retest figure is considerably lower than these. This finding only can occur if one or more assumptions of at least one of the methods is violated. A likely candidate is the test-retest assumption that government responsiveness is constant over the pre- and postelection period. Since coefficient alpha does not require this assumption, and because of the other advantages of alpha stated above, its estimate is preferable to those from the other methods.

Two drawbacks to alpha are that it underestimates reliability for congeneric measures and that it is not for single indicators. A more general procedure exists. I begin with the reliability of an indicator. It is derived by applying the definition of the reliability as the squared correlation of the true score and observed score. For $x_i = \alpha_i \tau_i + e_i$, the squared correlation is

$$\begin{aligned}\rho^2_{x_i \tau_i} &= \frac{[\mathrm{COV}(x_i, \tau_i)]^2}{\mathrm{VAR}(\tau_i)\mathrm{VAR}(x_i)} \\ &= \frac{\alpha_i^2\,\mathrm{VAR}(\tau_i)}{\mathrm{VAR}(x_i)} \\ &= \rho_{x_i x_i} \end{aligned} \tag{6.41}$$

Equation (6.41) provides the general formula for the reliability of an x_i that depends on τ_i, where x_i can be congeneric, tau-equivalent, or parallel.

In the case of an unweighted sum of measures ($H = \Sigma_{i=1}^{q} x_i$) the reliability is

$$\rho_{H\tau_i}^2 = \frac{\left[\mathrm{COV}(\tau_i, H)\right]^2}{\mathrm{VAR}(\tau_i)\mathrm{VAR}(H)}$$

$$= \frac{\left[\mathrm{COV}(\tau_i, \alpha_1\tau_i + e_1 + \alpha_2\tau_i + e_2 + \cdots + \alpha_2\tau_i + e_q)\right]^2}{\mathrm{VAR}(\tau_i)\mathrm{VAR}(H)}$$

$$= \frac{(\Sigma_{i=1}^{q}\alpha_i)^2\left[\mathrm{VAR}(\tau_i)\right]^2}{\mathrm{VAR}(\tau_i)\mathrm{VAR}(H)}$$

$$= \frac{(\Sigma_{i=1}^{q}\alpha_i)^2\mathrm{VAR}(\tau_i)}{\mathrm{VAR}(H)}$$

$$= \rho_{HH} \tag{6.42}$$

I assume that items are scored so that all α_i have the same sign. Equation (6.42) can be applied to parallel, tau-equivalent, or congeneric measures. Thus it, along with (6.41), is more general than the other reliability formulas. Even with its generality there are situations where its appropriateness is questionable. One such case is when the errors of measurement are correlated contrary to the initial assumptions. To the extent that the correlation of e_i and e_j represent stable and consistent relations, an argument could be made that such factors should be a part of the reliability. Another ambiguous case is if x_i depends on more than one true score. The derivations in this section have assumed that only τ_i underlies the measures. This is not a flexible enough condition to fit the circumstances encountered in practice.

Alternatives to Classical Reliability Measures

In the discussion of validity I treated more general measurement models that allowed multiple latent variables to influence a measure and that allowed correlated errors. This suggests that assessments of reliability might be better treated within such a framework. A natural question is what is the relation between classical test theory introduced in this section and the general measurement models discussed earlier?

Fundamental to understanding the links between classical test theory and the general measurement model is the relation of the true score, τ, to the latent variable ξ. If we represent an $n \times 1$ vector of true scores as $\boldsymbol{\tau}$, an $n \times 1$ vector of observed scores as $\mathbf{x}$, and the $n \times 1$ vector of errors of measurement as $\mathbf{e}$, the observed scores in classical test theory can be written as

$$\mathbf{x} = \boldsymbol{\tau} + \mathbf{e} \tag{6.43}$$

where $E(\mathbf{e})$ is zero and $\mathbf{e}$ is uncorrelated with $\boldsymbol{\tau}$. Note that the implicit assumption of (6.43) is that $\mathbf{x}$ consists of parallel or tau-equivalent measures. The true scores $\boldsymbol{\tau}$ in turn depends on the latent variables in $\boldsymbol{\xi}$ such that

$$\boldsymbol{\tau} = \boldsymbol{\Lambda}_x \boldsymbol{\xi} + \mathbf{s} \tag{6.44}$$

where $\boldsymbol{\Lambda}_x$ is $n \times q$ matrix of coefficients and $\mathbf{s}$ is an $n \times 1$ vector of specific variance components uncorrelated with $\boldsymbol{\xi}$ and with $\mathbf{e}$.[11] Substituting equation (6.44) into (6.43) leads to

$$\mathbf{x} = \boldsymbol{\Lambda}_x \boldsymbol{\xi} + \mathbf{s} + \mathbf{e} \tag{6.45}$$

If we define $\boldsymbol{\delta} = \mathbf{s} + \mathbf{e}$, (6.45) becomes

$$\mathbf{x} = \boldsymbol{\Lambda}_x \boldsymbol{\xi} + \boldsymbol{\delta} \tag{6.46}$$

which is the measurement model for $\mathbf{x}$. Equation (6.45) shows that the error term $\boldsymbol{\delta}$ is composed of two parts, $\mathbf{s}$ and $\mathbf{e}$.

One potential source of confusion is that both $\boldsymbol{\delta}$ and $\mathbf{e}$ are errors of measurement in $\mathbf{x}$. However, they are different types of errors. The $\boldsymbol{\delta}$ is the

[11] Some researchers differ on the relation between true scores ($\boldsymbol{\tau}$) and latent variables ($\boldsymbol{\xi}$). For example, Alwin and Jackson (1979, 80) suggest that classical test theory assumes that specific variance is absent, but Lord and Novick (1968, 243) include specific variance as part of the true score. Jöreskog (1974, 24–25) has the specific component part of the true score with the same specific variance present in two or more observed scores that depend on τ_i. I mention these differences to alert the reader to the several ways that the $\boldsymbol{\tau}$ to $\boldsymbol{\xi}$ relation can be conceptualized.

error in **x** when measuring $\boldsymbol{\xi}$. The **e** is the error in **x** when measuring $\boldsymbol{\tau}$. But $\boldsymbol{\delta}$ is *not* the error in measuring $\boldsymbol{\tau}$ since $\boldsymbol{\delta}$ contains **s** and **s** is part of $\boldsymbol{\tau}$. To emphasize this difference between $\boldsymbol{\delta}$ and **e**, some refer to $\boldsymbol{\delta}$ as the unique component of **x**, while reserving the term of error for **e**.

The specific variance, **s** plays a role in reliability. For example, earlier I defined the reliability of a parallel or congeneric measure as $\text{VAR}(\tau_1)/\text{VAR}(x_1)$. If $\tau_1 = \lambda_{11}\xi_1 + s_1$ the reliability is

$$\rho_{x_1x_1} = \frac{\text{VAR}(\tau_1)}{\text{VAR}(x_1)}$$

$$= \frac{\lambda_{11}^2\phi_{11} + \text{VAR}(s_1)}{\text{VAR}(x_1)} \tag{6.47}$$

Equation (6.47) shows that the reliability is influenced by the variance of the specific component. If we ignore the $\text{VAR}(s_1)$ and use only $\lambda_{11}^2\phi_{11}$ in the numerator, we underestimate a measure's reliability. Unfortunately, we do not typically know the specific variance of a measure. The specific component of x_i, called s_i, is uncorrelated with ξ_1 and with the specific components of the other **x** variables in an analysis. But s_i might contribute to the correlation between x_i and other variables not included in a model. As such, s_i is considered a consistent and *reliable* component of x_i that is not captured by any of the variables in $\boldsymbol{\xi}$.

Since we rarely know $\text{VAR}(s_i)$, it follows that most of the time we cannot know the reliability of a measure as defined in (6.47). There are at least three ways to react to this difficulty. One is to assume that the specific variance is zero (or nearly so) and to write the reliability as

$$\rho_{x_1x_1} = \frac{\lambda_{11}^2\phi_{11}}{\text{VAR}(x_1)} \tag{6.48}$$

A second approach is to consider (6.48) as a conservative estimate of reliability that underestimates the true reliability since the $\text{VAR}(s_i)$ is ignored.

A third reaction is to find a different approach to reliability. Traditionally, reliability is defined to be the consistency or repeatability of a measure. Consistency or repeatability are difficult to operationalize, particularly given the obstacles to estimating the specific variance. Further problems, which I mentioned earlier, are that classical test theory reliability makes no allowances for correlated errors of measurement, nor does it treat indicators influenced by more than one latent variable.

As an alternative definition, I define the *reliability of* x_i *as the magnitude of the direct relations that all variables (except* δ*'s) have on* x_i. Like the previous structural equation definition of validity, this definition presupposes an explicit measurement model. The ξ variables with direct effects on x_i are the systematic components of x_i. All else is error. The stronger is the systematic component, the greater is x_i's reliability.

A straightforward measure of this reliability is the squared multiple correlation coefficient for x_i, $R^2_{x_i}$. This ranges from zero to one, with values closer to one indicating higher reliability. Unlike the traditional definition, this definition and $R^2_{x_i}$ apply when x_i has multiple latent (or observed) causes or when the error term for x_i correlates with other error terms. The specific variance only makes a contribution to reliability if s_i is explicitly incorporated and identified in the measurement model. That is, we must be able to estimate the specific variance, to count it as a systematic influence on x_i. When the specific variance is assumed to be zero, and x_1 depends on a single factor, ξ_1, the $R^2_{x_1}$ is the squared correlation of x_1 and ξ_1. Indeed, if we assume that $\xi_1 = \tau_1$, then all of the previous results with τ_1 hold for ξ_1. In the more general case $R^2_{x_1}$ provides a gauge of the systematic variance in x_1 that can be explained by the variables in the measurement model.

How does reliability differ from validity? Comparing this structural equation definition of reliability to the earlier one for validity, we can see that reliability measures all influences of variables—valid and invalid—on x_i. Validity measures the strength of the direct effect of a particular ξ_j on x_i. As with the traditional definitions, an indicator can have high reliability but low or zero validity. Also the reliability measure $R^2_{x_i}$ sets an upper limit on the unique validity variance, $U_{x_i\xi_j}$. In this sense a measure's validity cannot exceed its reliability. For instance, consider the x_4 indicator from the morale and belonging example in the validity section [see eqs. (6.22) to (6.24)]. The $R^2_{x_4}$ (reliability) is 0.77, while $U_{x_4\xi_1}$ $(= 0.26)$ and $U_{x_4\xi_2}$ $(= 0.06)$ are smaller.

Summary

Reliability is the consistency of measurement. It is a concept that grew out of classical test theory which assumes that a single true score underlies a measure. The main techniques of estimating reliability are the test-retest, alternative forms, split-halves, and Cronbach's alpha. Of these approaches, Cronbach's alpha coefficient makes the least restrictive assumptions. However, alpha does underestimate the reliability of congeneric measures. To circumscribe this problem, even more general formulas for the reliability of a measure or of an index composed of several measures were derived. Finally, I called attention to the ambiguous meaning of reliability when errors of measurement are correlated or when a measure is affected by more than one latent variable. Coupled with this is the concept of specific variance that may be a reliable component but elusive to separate estimation. A viable alternative is a structural equations approach to reliability that has $R^2_{x_i}$ as a reliability estimate. This measures the proportion of variance in a measure that is explained by the variables that directly affect x_i. It is appropriate under very general conditions, and in simple cases it is equal to some of the traditional measures.

CAUSE INDICATORS

Throughout the preceding sections I assumed that the measures *depend on* one or more latent variables or on a true score. As such the measures are *effect* indicators. In Chapter 3 I introduced a second class of variables that influence latent variables. These are cause indicators. For instance, time spent with family and time spent with friends are cause indicators of the latent variable of time in social interaction. Race and sex are cause indicators of exposure to discrimination. Unfortunately, traditional validity assessments and classical test theory do not cover cause indicators. However, many of the structural equations ideas from the earlier sections carry through to these measures with some changes in their interpretations.

Consider the idea of validity first. I defined validity as the strength of the direct structural relation between a measure and a latent variable. This definition applies to cause indicators, with the understanding that the indicator has a direct effect on the latent variable rather than vice versa. The validity measures also apply. For instance, the unstandardized and standardized validity coefficients show the direct effect of x_i on the latent variable(s) it influences. The unique validity variance reveals that proportion of the variance in the latent variable that is uniquely attributable to the cause indicator. Finally, if the latent variable is determined by more than one variable the degree of collinearity in the cause indicator can be gauged by the squared multiple correlation coefficient of the indicator with any other causes of the latent variable.

The idea of reliability is less easy to apply to cause indicators. We could calculate $R^2_{x_i}$, but this only tells us the systematic component in x_i and says little about the latent variable that x_i affects. Also, if a cause indicator is an exogenous variable, its $R^2_{x_i}$ is zero by the definition of exogenous. Or, we could regard $x_i = \xi_j$ as the implicit measurement model for x_i, and then its $R^2_{x_i}$ is one. None of these options is appealing. Thus the traditional and the new definitions of reliability do not work well for indicators that cause latent variables.[12]

SUMMARY

Measurement is a broad topic in social science research. This chapter emphasized the issues of measurement most relevant to a structural equations approach to measurement models. Most basic is the need to begin with a clear definition of the concept to be measured. Without such a definition, we have little hope of identifying the dimensions and latent variables to represent a concept. The definition also helps in the selection of indicators and in formulating the relation between indicators and latent variables. Validity and reliability are two basic characteristics of measures. Validity addresses the issue of the direct correspondence between a measure and a concept. Reliability assesses the consistency of a measure, regardless of whether it is valid. Many have proposed empirical techniques to estimate validity and reliability. These often are based on correlation coefficients and restrictive assumptions about the properties of the measures. I have shown that several alternative means of estimating validity and reliability are more general than the traditional procedures and that they fit well into a structural equations approach to measurement. With these basic ideas on measurement in hand, the next task is to estimate measurement models with confirmatory factor analysis. I take this topic up in the next chapter.

APPENDIX 6A LISREL PROGRAM FOR THE MULTITRAIT-MULTIMETHOD EXAMPLE

In this appendix I list the LISREL commands used to estimate the leadership multitrait-multimethod example discussed in the chapter. See Jöreskog and Sörbom's (1986) LISREL VI manual for a detailed discussion of the commands.

[12]Cause and effect indicators also differ in whether we should expect indicators of the same latent variable to correlate. If two or more effect indicators are positively related to a single latent variable, then these indicators should have a positive association. The higher this association, the greater is their correlation with the latent variable. On the other hand, two or more cause indicators of a latent variable need not correlate. They can have a zero or even an inverse association (see Bollen 1984).

The program follows:

```
TITLE MULTITRAIT MULTIMETHOD LEADERSHIP ML
DA NI = 12 NO = 240 MA = KM
LA
'SELFPROM' 'SELFACH' 'SELFAFFL' 'SELFLDR'
'PEERPROM' 'PEERACH' 'PEERAFFL' 'PEERLDR'
'OBSPROM' 'OBSACH' 'OBSAFFL' 'OBSLDR'
SE
1 5 9 2 6 10 3 7 11 4 8 12/
KM
1 .50 1 .41 .48 1 .67 .59 .40 1 .45 .33 .26 .55 1
 .36 .32 .31 .43 .72 1 .25 .21 .25 .30 .59 .72 1
 .46 .36 .28 .51 .85 .80 .69 1
 .53 .41 .34 .56 .71 .58 .43 .72 1
 .50 .45 .29 .52 .59 .55 .42 .63 .84 1
 .36 .30 .28 .37 .53 .51 .43 .57 .62 .57 1
 .52 .43 .31 .59 .68 .60 .46 .73 .92 .89 .63 1
MO NX = 12 NK = 7 PH = FI TD = FU,FI
LK
'PROMINENCE' 'ACHIEVEMENT' 'AFFILIATION' 'LEADERSHIP'
'SELFRAT' 'PEERRAT' 'OBSRATE'
FI PH 1 1 PH 2 2 PH 3 3 PH 4 4
FR LX 1 1 LX 2 1 LX 3 1 LX 4 2 LX 5 2 LX 6 2 LX 7 3 LX 8 3
FR LX 9 3 LX 10 4 LX 11 4 LX 12 4
FR LX 1 5 LX 2 6 LX 3 7 LX 4 5 LX 5 6 LX 6 7
FR LX 7 5 LX 8 6 LX 9 7 LX 10 5 LX 11 6 LX 12 7
FR PH 2 1 PH 3 1 PH 3 2 PH 4 1 PH 4 2 PH 4 3
FR PH 6 5 PH 7 5 PH 7 6
FR TD 1 1 TD 2 2 TD 3 3 TD 4 4 TD 5 5 TD 6 6 TD 7 7 TD 8 8
FR TD 9 9 TD 10 10 TD 11 11 TD 12 12
ST 1 PH 1 1 PH 2 2 PH 3 3 PH 4 4
ST 1 PH 5 5 PH 6 6 PH 7 7
MA LX
.5 0 0 0 .2 0 0
.5 0 0 0 0 .3 0
.5 0 0 0 0 0 .2
0 .5 0 0 .3 0 0
0 .5 0 0 0 .2 0
0 .5 0 0 0 0 .3
0 0 .5 0 .3 0 0
0 0 .5 0 0 .2 0
0 0 .5 0 0 0 .3
0 0 0 .5 .3 0 0
0 0 0 .5 0 .2 0
0 0 0 .5 0 0 .3
```

```
MA PH
1
.5 1
.6 .5 1
.4 .5 .5 1
0 0 0 0 1
0 0 0 0 0 1
0 0 0 0 0 0 1
ST .1 TD 1 1 - TD 12 12
OU SE TV MR TO NS
```

CHAPTER SEVEN

Confirmatory Factor Analysis

In the preceding chapter I presented some of the basic aspects of measurement theory and measurement models, setting the stage for the discussion of procedures to estimate measurement models and to evaluate their correspondence to data. Confirmatory factor analysis, which is the topic of the present chapter, is one such technique. This chapter begins with the relation between exploratory and confirmatory factor analysis. It then moves to model specification for confirmatory factor analysis, followed by sections on the implied covariance matrix, identification, estimation, the evaluation of model fit, comparisons of models, diagnostics for misspecified models, and extensions of the model.

EXPLORATORY AND CONFIRMATORY FACTOR ANALYSIS

Spearman is commonly credited with the initial development of factor analysis.[1] In his 1904 article he used this technique to determine whether a general intelligence factor underlies individual performance on tests. His goal was to explain the relationship between a number of observed variables in terms of a single latent variable. Although Spearman's belief in a single latent variable (unifactor) solution soon gave way to solutions with several latent variables (multifactor), in general terms the purpose of factor analysis has remained the same. The primary goal is to explain the covariances or correlations between many observed variables by means of relatively few underlying latent variables. In this sense it is a data reduction technique. Figure 7.1 provides an example. In part (*a*) I show six intercor-

[1]Pearson's (1901) work on principal axes was an important predecessor to Spearman's (1904) factor analysis work.

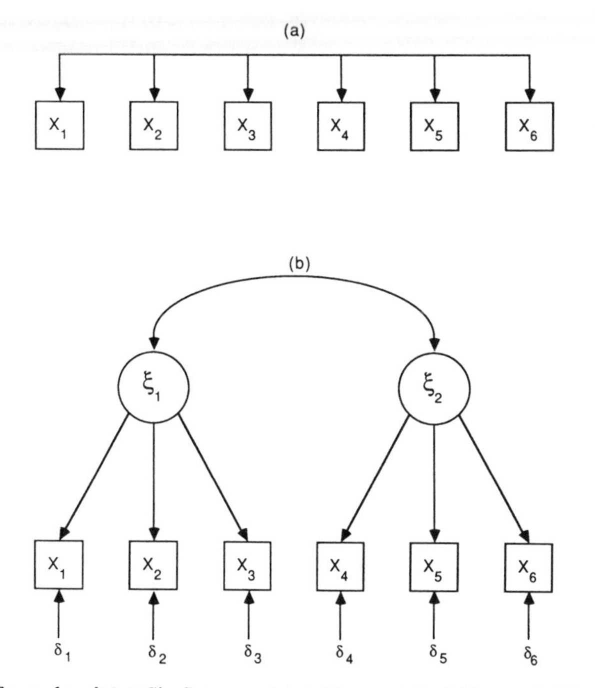

Figure 7.1 Example of (*a*) Six Intercorrelated Observed Variables and (*b*) a Two-Factor Model for the Six Observed Variables

related observed variables. Usually the unanalyzed association is represented by curved arrows connecting each pair of variables. To avoid cluttering the diagram, I show the correlations between all pairs of variables by a single solid line with arrows from the line to all variables. In (*a*) the reason for the associations among the observed variables is unspecified. Figure 7.1(*b*) is a factor analysis model that has two latent variables (ξ_1 and ξ_2) with direct effects on the indicators. According to this model the six observed variables are correlated because they depend on two correlated latent variables. An implication of this model is that if ξ_1 and ξ_2 are held constant, there is no association between any of the observed variables. Thus the correlation of many observed variables is explained by a few latent factors.

Exploratory and *confirmatory* factor analyses are two major approaches to factor analysis.[2] Exploratory factor analyses (EFA) are the more tradi-

[2]An additional distinction sometimes drawn is that of *restricted* and *unrestricted* solutions. See Jöreskog (1979a) and Mulaik (1972) for further discussion.

tional procedures, and they are widely available in popular statistical packages (e.g., SAS and SPSS). Though there are many varieties of EFA, perhaps their most distinctive feature is that a detailed model relating the latent to the observed variables is not specified in advance. In addition in EFA the number of latent variables is not determined before the analysis, all latent variables typically influence all observed variables, the measurement errors ($\boldsymbol{\delta}$) are not allowed to correlate, and underidentification of parameters is common. In contrast, in confirmatory factor analysis (CFA) a model is constructed in advance, the number of latent variables is set by the analyst, whether a latent variable influences an observed variable is specified, some direct effects of latent on observed variables are fixed to zero or some other constant (e.g., one), measurement errors may correlate, the covariance of latent variables can be estimated or set to any value, and parameter identification is required. Figure 7.1(*b*) gives an example of a CFA model. In short, CFA requires a detailed and identified initial model.

In practice the distinction between EFA and CFA is more blurred than the foregoing discussion suggests. For instance, researchers using traditional EFA procedures may restrict their analysis to a group of indicators that they believe are influenced by one factor. By doing so, they test an implicit if not explicit model. Or, researchers with poorly fitting models in a CFA often modify their model in an exploratory way with the goal of improving fit. Thus the labels EFA and CFA refer to ideal types with most applications falling between these extremes.

In substantive areas where little is known, exploratory factor analysis can prove valuable and can suggest underlying patterns in the data. If, however, hypotheses about plausible model structures exist, then exploratory factory analysis can frustrate attempts to test these ideas. As an illustration of exploratory factor analysis (EFA) consider the eight variables in the correlation matrix given in Table 7.1. The first four variables are indicators of political democracy in developing countries as of 1960: freedom of the press (x_1), freedom of group opposition (x_2), fairness of elections (x_3), and the elective nature of the legislative body (x_4). The second four variables (x_5 to x_8) are the same variables from the same sources for 1965.[3] The reader may recognize these variables as part of the example in the notation section of Chapter 2. There I also included three indicators of industrialization, which I ignore here. In Chapter 2 I postulated that a separate political democracy factor exists for each time point and that the 1960 latent variable only influences the 1960 indicators, while the same is true for the 1965 democracy factor and indicators. Furthermore

[3]For a detailed description of these measures, see Bollen (1979, 1980).

Table 7.1 Correlation Coefficients and Standard Deviations for Four 1960 Indicators (x_1 to x_4) and Four 1965 Indicators (x_5 to x_8) of Political Democracy for 75 Developing Countries

	x_1	x_2	x_3	x_4	x_5	x_6	x_7	x_8
x_1	1.000							
x_2	0.604	1.000						
x_3	0.679	0.451	1.000					
x_4	0.693	0.719	0.609	1.000				
x_5	0.739	0.543	0.576	0.652	1.000			
x_6	0.650	0.705	0.427	0.659	0.565	1.000		
x_7	0.674	0.581	0.650	0.680	0.678	0.609	1.000	
x_8	0.666	0.606	0.530	0.737	0.630	0.753	0.712	1.000
Standard Deviation	2.623	3.947	3.281	3.349	2.613	3.373	3.286	3.246

I specified some correlated measurement errors. I ignore these hypotheses for the moment to demonstrate what a standard EFA approach would show.

A first step in traditional EFA is to prepare the correlation matrix for analysis by replacing its main diagonal with initial estimates of "communalities."[4] Each variable's communality is that part of its variance explained by the factors. The most popular initial communality estimate is the squared multiple correlation coefficient obtained when each variable is regressed on the remaining observed variables. The next step in EFA is to select the number of factors. A widely used principal axis method does this by examining the eigenvalues (see Appendix A) of this modified correlation matrix of the observed variables. Each eigenvalue is associated with a factor, and the bigger the eigenvalue is, the greater is the capability of the underlying factor to "account for" the correlations of the observed variables. Typically, researchers rank the eigenvalues and use a cutoff value of one or a sharp drop in the size of the eigenvalues to determine the number of factors. For the political democracy data, the three largest eigenvalues are 5.90, 0.26, and 0.06. Thus a one-factor solution seems appropriate.

How does this result correspond with a priori knowledge about the data? The eight variables are four measures at two distinct points in time. A one-factor solution raises some interpretive difficulties. First, if one factor exists, then it must be a latent variable in the 1960 (or possibly earlier) period. If it occurred later than 1960, this would violate temporal priority

[4]Readers unfamiliar with EFA can refer to Harman (1976) or Mulaik (1972) for good introductions.

(see Chapter 3). This follows since four 1960 observed variables cannot be caused by a latent variable in the post-1960 period.

The "factor loadings," or in our terms the λ_{ij} coefficients, provide the direct effects of the factor on the observed variables. If the latent variable comes in the 1960 or earlier period, we might expect that its influences are greatest on the variables that are closest to it in time. So for each pair of 1960 and 1965 indicators the coefficient (or "loading") for the 1960 measure should exceed that of the 1965 measure. I find a mixed pattern with the coefficients for x_1 and x_4 larger than their 1965 counterparts of x_5 and x_8 but the reverse pattern for x_2 and x_6 and x_3 and x_7. It is possible that we could construct an explanation for this result, but as it stands, it runs counter to what we expect for a circa 1960 latent variable.

Alternatively, it might be that political democracy is perfectly correlated from 1960 to 1965 so that the two factors cannot be distinguished from one factor. Though this is possible, we can tell that the rankings of the developing countries in the sample according to democracy was not identical in 1960 and 1965 by examining the history of some of these countries over this period. For instance, Brazil's rank in 1965, following a military coup in 1964, was lower than its rank in 1960. I would expect a high correlation between the 1960 and 1965 democracy factors, but it would be preferable to estimate it rather than assuming a correlation of one as is done by representing the two factors as one.

One could object that the problem with EFA arises from reliance on an arbitrary statistical criterion (i.e., eigenvalues > 1) to determine the number of factors. If the researcher were to specify the number of factors, then the exploratory factor analytic technique might be adequate. Although this argument has some intuitive appeal, it neglects the fact that most EFAs are applied to help determine the number of factors. Furthermore, by specifying the number of factors, the analysis moves from exploratory toward more confirmatory research.

Ambiguities remain even with the number of factors set in advance. To illustrate this, I reanalyze Table 7.1 with the same procedure except that I force a two-factor solution. Figure 7.2 shows a path diagram of the resulting model,[5] and Table 7.2 presents the factor pattern. The coefficients of factor loadings in Table 7.2 indicate the degree of relation between each factor, ξ_1 and ξ_2, and the x_i variables; the larger the coefficient, the stronger is the relation. Most x variables load on both ξ_1 and ξ_2, and

[5]I use the same procedure as the last one, except I specify two factors and oblique rotation of the factors using PROMAX (Mulaik 1972). In a one-factor solution the rotation problem is not relevant.

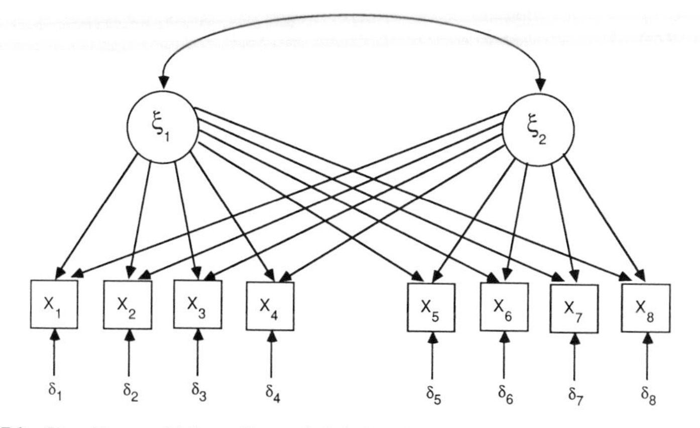

Figure 7.2 Two-Factor Oblique Rotated Solution for Indicators of Political Democracy in 1960 (X_1 to X_4) and 1965 (X_5 to X_8)

Table 7.2 Factor Patterns for Oblique Rotated Solution for Indicators of Political Democracy in 1960 (x_1 to x_4) and 1965 (x_5 to x_8)

Variable	Factor Pattern ξ_1	ξ_2
x_1	0.240	0.672
x_2	0.745	0.067
x_3	−0.054	0.810
x_4	0.542	0.376
x_5	0.200	0.645
x_6	0.859	−0.005
x_7	0.279	0.605
x_8	0.647	0.255

the variables that are most closely related to ξ_1 or ξ_2 are not always from the same time period. For instance, x_2, x_4, x_6, and x_8 have the highest loadings on ξ_1. Half of these are 1960 measures, and the other half are their 1965 counterparts. Furthermore, if we assume that one factor should represent democratic political structure in 1960 and the second the same factor in 1965, ξ_1 and ξ_2 do not seem to meet this criterion. A post hoc explanation of the latent factors and the pattern exhibited by the x_i's, ξ_1 and ξ_2 might be possible, but this would be spinning theory to meet the constraints of a statistical procedure.

This example points to several limits of a standard exploratory factor analysis (EFA). First, the technique does not allow the analyst to constrain some of the factor loadings to zero. For instance, if we conceptualize ξ_1 in Figure 7.2 as a latent variable in 1960 that only affects 1960 measures, we should be able to constrain its effects on the 1965 measures (x_5 to x_8) to zero. Similarly, the effects of ξ_2, the 1965 factor on the 1960 indicators should be zero. In the standard EFA each factor influences *all* of the variables.[6]

Second, EFA does not allow correlated errors of measurement. This is problematic, especially in the current example: the errors in measuring an indicator might correlate over time. In other situations measurement errors may correlate because indicators come from the same source, because of response set bias in answering survey questions, or for some other reasons. EFA confounds the correlated measurement errors with the latent factors, potentially leading to ambiguous and misleading solutions. This may be one of the problems in the EFA for the political democracy example.

A third problem with EFA is that of determining the number of factors. If statistical criteria are relied on, such as only retaining factors associated with an eigenvalue greater than one, then problematic solutions may emerge as illustrated earlier. Most EFA procedures allow the user to specify the number of factors, though many researchers rely on statistical criteria for these decisions. If more than one factor emerges or is specified, EFA has the restriction that *all* factors are uncorrelated or *all* factors are correlated. Typically, analysts cannot specify that some factors are associated but others are not, and they cannot set the degree of association to a particular value. If only two factors are present, as in the current example, this is not as much of a problem. But in the more typical case of three or more factors, it is a real limitation.

All of these limits reflect the inability of EFA to accommodate our theoretical and substantive knowledge. Confirmatory factor analysis overcomes these shortcomings. But to do so, researchers must employ their expertise and formalize their ideas into a model. Once a model is constructed, it can be estimated, and its fit to the data can be assessed. The process begins with the specification of the measurement model. The next section discusses the general forms of measurement models.

[6]A researcher might choose to avoid these problems by analyzing the 1960 and 1965 variables in separate groups. Though this would eliminate the possibility of latent variables in one year influencing indicators in another year, it would not be without costs. For instance, this type of analysis would not allow correlated measurement errors across time periods, the correlation of the 1960 and 1965 measure could not be estimated, and equality constraints for factor loadings would not be feasible.

MODEL SPECIFICATION

The general model for confirmatory factor analysis is little more than the measurement models that were introduced in Chapter 2:

$$\mathbf{x} = \mathbf{\Lambda}_x \boldsymbol{\xi} + \boldsymbol{\delta} \tag{7.1}$$

$$\mathbf{y} = \mathbf{\Lambda}_y \boldsymbol{\eta} + \boldsymbol{\epsilon} \tag{7.2}$$

In (7.1) and (7.2) **x** and **y** are observed variables, $\boldsymbol{\xi}$ and $\boldsymbol{\eta}$ are latent factors, and $\boldsymbol{\delta}$ and $\boldsymbol{\epsilon}$ are errors of measurement. The factor analysis model can be represented with **x**, $\boldsymbol{\xi}$, and $\boldsymbol{\delta}$ as in (7.1) or with **y**, $\boldsymbol{\eta}$, and $\boldsymbol{\epsilon}$ as in (7.2). The basic models in (7.1) and (7.2) are the same. The observed variables depend on one or more latent variables and an errors of measurement vector. The errors of measurement are uncorrelated with the latent variables. The coefficients describing the effects of the latent variables on the observed variables are in $\mathbf{\Lambda}_x$ or $\mathbf{\Lambda}_y$. To simplify matters, I use (7.1) throughout this chapter, although I also could have used (7.2). The assumptions accompanying equation (7.1) are that $E(\boldsymbol{\delta}) = 0$, and $E(\boldsymbol{\xi}\boldsymbol{\delta}') = 0$. By convention, all variables in **x** and $\boldsymbol{\xi}$ are written as deviations from their respective means. In the usual factor analysis model the $\boldsymbol{\delta}$ term is composed of two components:

$$\boldsymbol{\delta} = \mathbf{s} + \mathbf{e} \tag{7.3}$$

In (7.3) **s** represents the specific variance associated with each variable and **e** is the remaining random component in **x**. Together they form the "unique factor" of **x**. Since both components are errors in **x** with respect to measuring $\boldsymbol{\xi}$ and both are uncorrelated with $\boldsymbol{\xi}$ and with each other, I continue to refer to $\boldsymbol{\delta}$ as random errors of measurement.

To illustrate the movement from theoretical beliefs and ideas to model specifications, consider the political democracy panel data example from the last section. Four indicators of political democracy are available for 1960 and 1965 (i.e., x_1 to x_4 and x_5 to x_8). I hypothesize that the four indicators for 1960 are linearly dependent on a single political democracy factor (ξ_1) in 1960 and the four indicators for 1965 are linearly affected by the same latent variable in 1965 (ξ_2.) The 1960 factor has no effects on the 1965 indicators, and the 1965 factor does not influence the 1960 indicators. In addition each indicator contains an error of measurement (δ_i) term that is assumed uncorrelated with the latent variables. Equation (7.4) represents these relations:

$$\mathbf{x} = \mathbf{\Lambda}_x \boldsymbol{\xi} + \boldsymbol{\delta}$$

$$\begin{bmatrix} x_1 \\ x_2 \\ x_3 \\ x_4 \\ x_5 \\ x_6 \\ x_7 \\ x_8 \end{bmatrix} = \begin{bmatrix} \lambda_{11} & 0 \\ \lambda_{21} & 0 \\ \lambda_{31} & 0 \\ \lambda_{41} & 0 \\ 0 & \lambda_{52} \\ 0 & \lambda_{62} \\ 0 & \lambda_{72} \\ 0 & \lambda_{82} \end{bmatrix} \begin{bmatrix} \xi_1 \\ \xi_2 \end{bmatrix} + \begin{bmatrix} \delta_1 \\ \delta_2 \\ \delta_3 \\ \delta_4 \\ \delta_5 \\ \delta_6 \\ \delta_7 \\ \delta_8 \end{bmatrix} \tag{7.4}$$

$$\text{COV}(\xi_i, \delta_j) = 0, \quad \text{for all } i \text{ and } j$$

$$E(\delta_j) = 0, \quad \text{for all } j$$

Each column of Λ_x corresponds to one latent variable; the first column to ξ_1, and the second to ξ_2. The double subscript of λ_{ij} indicates the row and column position in Λ_x. A zero in Λ_x indicates that the corresponding observed variable is not influenced by the latent variable in that column. The number of nonzero coefficients in each row shows the number of latent variables that affect the observed variable. This number is sometimes referred to as the *factor complexity* of a variable. The Λ_x in (7.4) shows that all eight variables have a factor complexity of one.

The λ_{ij}'s in Λ_x are sometimes referred to as the "factor loadings" because they indicate which variable "loads" on which factor. In the last chapter the λ_{ij}'s were sometimes interpreted as unstandardized validity coefficients since they describe the direct structural relation between a latent and observed variable. The λ_{ij} also may be viewed as regression coefficients. In either case, for $x_i = \lambda_{ij}\xi_i + \delta_i$, λ_{ij} is the number of units x_i is expected to change for a one-unit change in ξ_j. If more than one ξ affects x_i, λ_{ij} is the expected change holding constant the other latent variables.

I hypothesize that the relation between the latent variable and each indicator is the same in 1960 as in 1965. For example, x_1 and x_5 represent a measure of freedom of the press derived from the same source for 1960 and 1965. I believe that λ_{11} relating the 1960 "free press" measure (x_1) to the 1960 political democracy factor (ξ_1) is equivalent to the λ_{52} of the 1965 relation of free press (x_5) to political democracy (ξ_2). The same is true for the remaining three pairs of 1965 and 1960 indicators. This substantive idea

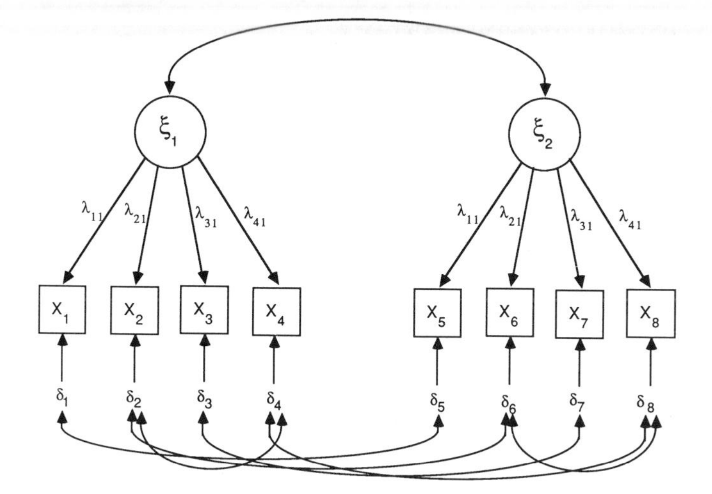

Figure 7.3 Specification of Two-Factor Model for Indicators of Political Democracy in 1960 (X_1 to X_4) and 1965 (X_5 to X_8)

is translated into the model by the following constraints: $\lambda_{11} = \lambda_{52}$, $\lambda_{21} = \lambda_{62}$, $\lambda_{31} = \lambda_{72}$, and $\lambda_{41} = \lambda_{82}$.

For these panel data I also hypothesize that the 1965 and 1960 political democracy latent variables are strongly associated so that the covariance of ξ_1 and ξ_2 should be estimated. Since these are panel data, I expect that the errors of measurement for the same indicator at two points in time are correlated. For instance, the error of measurement in the free press 1960 indicator is likely to correlate with the error of measurement for the 1965 free press indicator since both measures come from the same source and some of the error in one year probably appears in the other years. Also I expect a correlation between δ_2 and δ_4 and between δ_6 and δ_8 because each pair of the corresponding observed variables is from the same data source for the same year. These relations are incorporated into Figure 7.3. The curved arrow between ξ_1 and ξ_2 stands for the association between political democracy in 1965 and 1960. The curved arrows between δ_i and δ_{i+4} ($i = 1, 2, 3, 4$), δ_2 and δ_4, and δ_6 and δ_8 indicate correlated errors of measurement. Note also that I have replaced λ_{52} to λ_{82} with λ_{11} to λ_{41} respectively, to show the hypothesized constraints on these parameters.

The model can be estimated with CFA, but it cannot be estimated with EFA procedures. Before estimation it is necessary to understand the relation of the covariance matrix of the observed variables to the structural parameters of the model.

IMPLIED COVARIANCE MATRIX

In Chapter 4 I showed the covariance matrix of the observed variables for structural equations with only observed variables and the relation between the structural parameters of the model and each element of the covariance matrix. In this section I undertake a similar task for the covariance matrix of $\mathbf{x}$ and the measurement model parameters. As was true in Chapter 4, the estimation is based on choosing the values of the structural parameters to reproduce the covariance matrix. Thus it is important to understand the relation of the covariance elements to the parameters.

Since the $\mathbf{x}$ variables are deviations from their respective means, the covariance matrix for $\mathbf{x}$, is equal to the expected value of $\mathbf{xx}'$. I write the covariance matrix of $\mathbf{x}$ as a function of $\boldsymbol{\theta}$ and represent it as $\boldsymbol{\Sigma}(\boldsymbol{\theta})$:

$$\begin{aligned}\boldsymbol{\Sigma}(\boldsymbol{\theta}) &= E(\mathbf{xx}')\\ &= E\left[(\boldsymbol{\Lambda}_x\boldsymbol{\xi} + \boldsymbol{\delta})(\boldsymbol{\xi}'\boldsymbol{\Lambda}_x' + \boldsymbol{\delta}')\right]\\ &= \boldsymbol{\Lambda}_x E(\boldsymbol{\xi}\boldsymbol{\xi}')\boldsymbol{\Lambda}_x' + \boldsymbol{\Theta}_\delta\\ &= \boldsymbol{\Lambda}_x\boldsymbol{\Phi}\boldsymbol{\Lambda}_x' + \boldsymbol{\Theta}_\delta \end{aligned} \tag{7.5}$$

The last line of equation (7.5), shows that the covariance matrix of $\mathbf{x}$ for the general factor analysis model (7.1) may be decomposed in terms of the parameters of $\boldsymbol{\Lambda}_x$, $\boldsymbol{\Phi}$, and $\boldsymbol{\Theta}_\delta$, where $\boldsymbol{\Phi}$ is the covariance matrix of the latent factors $\boldsymbol{\xi}$, and $\boldsymbol{\Theta}_\delta$ is the covariance matrix for the errors of measurement $\boldsymbol{\delta}$.

The meaning of (7.5) may be clarified with an example. Consider a simple one-factor, three-indicator model:

$$\begin{aligned} x_1 &= \lambda_{11}\xi_1 + \delta_1\\ x_2 &= \lambda_{21}\xi_1 + \delta_2\\ x_3 &= \lambda_{31}\xi_1 + \delta_3 \end{aligned} \tag{7.6}$$

$$E(\delta_i) = 0$$

$$\text{COV}(\xi_1, \delta_i) = 0, \qquad \text{for } i = 1, 2, 3$$

$$\text{COV}(\delta_i, \delta_j) = 0, \qquad \text{for } i \neq j$$

The relevant matrices are

$$\mathbf{x} = \begin{bmatrix} x_1 \\ x_2 \\ x_3 \end{bmatrix}, \qquad \boldsymbol{\Lambda}_x = \begin{bmatrix} \lambda_{11} \\ \lambda_{21} \\ \lambda_{31} \end{bmatrix}, \qquad \boldsymbol{\xi} = [\xi_1], \qquad \boldsymbol{\Phi} = [\phi_{11}]$$
$$\boldsymbol{\delta} = \begin{bmatrix} \delta_1 \\ \delta_2 \\ \delta_3 \end{bmatrix}, \qquad \boldsymbol{\Theta}_\delta = \begin{bmatrix} \mathrm{VAR}(\delta_1) & & \\ 0 & \mathrm{VAR}(\delta_2) & \\ 0 & 0 & \mathrm{VAR}(\delta_3) \end{bmatrix} \tag{7.7}$$

To determine $\boldsymbol{\Sigma}(\boldsymbol{\theta})$ for (7.6), the relevant matrices from (7.7) are substituted into (7.5):

$$\boldsymbol{\Sigma}(\boldsymbol{\theta}) = \begin{bmatrix} \lambda_{11}^2\phi_{11} + \mathrm{VAR}(\delta_1) & & \\ \lambda_{21}\lambda_{11}\phi_{11} & \lambda_{21}^2\phi_{11} + \mathrm{VAR}(\delta_2) & \\ \lambda_{31}\lambda_{11}\phi_{11} & \lambda_{31}\lambda_{21}\phi_{11} & \lambda_{31}^2\phi_{11} + \mathrm{VAR}(\delta_3) \end{bmatrix} \tag{7.8}$$

In terms of variances and covariances of the observed variables $\boldsymbol{\Sigma}$ may be written as:

$$\boldsymbol{\Sigma} = \begin{bmatrix} \mathrm{VAR}(x_1) & & \\ \mathrm{COV}(x_2, x_1) & \mathrm{VAR}(x_2) & \\ \mathrm{COV}(x_3, x_1) & \mathrm{COV}(x_3, x_2) & \mathrm{VAR}(x_3) \end{bmatrix} \tag{7.9}$$

The elements above the main diagonal of $\boldsymbol{\Sigma}(\boldsymbol{\theta})$ and $\boldsymbol{\Sigma}$ are redundant with the elements below the diagonal, so they are not shown. If each element in (7.9) were matched to the corresponding element in (7.8), the variances and covariances are decomposed into their structural parameters. For instance, the $\mathrm{VAR}(x_1)$ is

$$\mathrm{VAR}(x_1) = \lambda_{11}^2\phi_{11} + \mathrm{VAR}(\delta_1) \tag{7.10}$$

Equation (7.10) shows that the variance of x_1 will vary if λ_{11}, ϕ_{11}, or the $\mathrm{VAR}(\delta_1)$ take different values. Similarly, I could write an equation for the other variance and covariance elements in $\boldsymbol{\Sigma}$. The important point is that once researchers specify a measurement model, they can write the variances and covariances of the observed variables as a function of the structural parameters of the measurement model.

As another illustration consider the political democracy example of the last section. For this model the Λ_x, Φ, and Θ_δ matrices are

$$\Lambda_x = \begin{bmatrix} \lambda_{11} & 0 \\ \lambda_{21} & 0 \\ \lambda_{31} & 0 \\ \lambda_{41} & 0 \\ 0 & \lambda_{11} \\ 0 & \lambda_{21} \\ 0 & \lambda_{31} \\ 0 & \lambda_{41} \end{bmatrix}, \qquad \Phi = \begin{bmatrix} \phi_{11} & \\ \phi_{21} & \phi_{22} \end{bmatrix} \tag{7.11}$$

$$\Theta_\delta = \begin{bmatrix} \mathrm{VAR}(\delta_1) & & & & & & & \\ 0 & \mathrm{VAR}(\delta_2) & & & & & & \\ 0 & 0 & \mathrm{VAR}(\delta_3) & & & & & \\ 0 & \mathrm{COV}(\delta_4, \delta_2) & 0 & \mathrm{VAR}(\delta_4) & & & & \\ \mathrm{COV}(\delta_5, \delta_1) & 0 & 0 & 0 & \mathrm{VAR}(\delta_5) & & & \\ 0 & \mathrm{COV}(\delta_6, \delta_2) & 0 & 0 & 0 & \mathrm{VAR}(\delta_6) & & \\ 0 & 0 & \mathrm{COV}(\delta_7, \delta_3) & 0 & 0 & 0 & \mathrm{VAR}(\delta_7) & \\ 0 & 0 & 0 & \mathrm{COV}(\delta_8, \delta_4) & 0 & \mathrm{COV}(\delta_8, \delta_6) & 0 & \mathrm{VAR}(\delta_8) \end{bmatrix} \tag{7.12}$$

Substituting equations (7.11) and (7.12) into (7.5) [i.e., $\Sigma(\theta) = \Lambda_x \Phi \Lambda_x' + \Theta_\delta$] provides an equation showing that each covariance and variance element in the covariance matrix of x_1 to x_8 is a function of the structural parameters of the model. For instance, the $\mathrm{VAR}(x_1)$ is $\lambda_{11}^2 \phi_{11} + \mathrm{VAR}(\delta_1)$. The variance of the 1960 free press indicator (x_1) depends on λ_{11}, the coefficient linking it to the 1960 political democracy factor (ξ_1), the variance (ϕ_{11}) of this latent democracy variable, and the measurement error variance [$\mathrm{VAR}(\delta_1)$] of the free press variable. The $\mathrm{COV}(x_1, x_5)$ is $\lambda_{11}^2 \phi_{12} + \mathrm{COV}(\delta_1, \delta_5)$. The remaining variances and covariances can be decomposed by forming $\Lambda_x \Phi \Lambda_x' + \Theta_\delta$ and matching this to the corresponding elements in Σ.

If the measurement model is correct then knowing the structural parameters tells us the variances and covariances of the **x** variables. Of course, in practice, we do not know all the values of the structural parameters. The unknown parameters must be estimated. First, however, we must determine if the model parameters are identified.

IDENTIFICATION

I presented the identification problem for structural equations with observed variables in Chapter 4. The identification problem for these models is whether a unique solution exists for each parameter in $\mathbf{\Phi}$, $\mathbf{B}$, $\mathbf{\Gamma}$, and $\mathbf{\Psi}$. For confirmatory factor analysis though the matrices differ, the problem is the same: Does a unique solution exist for the structural parameters in $\mathbf{\Lambda}_x$, $\mathbf{\Phi}$, and $\mathbf{\Theta}_\delta$?

The $t \times 1$ $\boldsymbol{\theta}$ vector contains all of the unknown and unconstrained parameters of the model. The parameters in $\boldsymbol{\theta}$ are identified if no vectors $\boldsymbol{\theta}_1$ and $\boldsymbol{\theta}_2$ exist such that $\boldsymbol{\Sigma}(\boldsymbol{\theta}_1) = \boldsymbol{\Sigma}(\boldsymbol{\theta}_2)$, unless $\boldsymbol{\theta}_1 = \boldsymbol{\theta}_2$. The parameter vectors $\boldsymbol{\theta}_1$ and $\boldsymbol{\theta}_2$ are any specific values of $\boldsymbol{\theta}$. I assume that $\boldsymbol{\theta}$ excludes imaginary numbers and improper values such as negative variances or undefined solutions. The $\boldsymbol{\Sigma}(\boldsymbol{\theta}_1)$ and $\boldsymbol{\Sigma}(\boldsymbol{\theta}_2)$ are the implied covariance matrices [see (7.5)] evaluated at $\boldsymbol{\theta}_1$ and $\boldsymbol{\theta}_2$. If two sets of different values for the parameters lead to the same value for the implied covariance matrices, then the model is not identified.

The $\boldsymbol{\theta}$ from a confirmatory factor analysis model is identified if it is possible to show that the elements of $\boldsymbol{\theta}$ are uniquely determined functions of the elements of $\boldsymbol{\Sigma}$. As in Chapter 4, solving for $\boldsymbol{\theta}$ in terms of the known-to-be-identified elements of $\boldsymbol{\Sigma}$ establishes identification. Clearly, without any restriction on the elements of $\mathbf{\Lambda}_x$, $\mathbf{\Phi}$, and $\mathbf{\Theta}_\delta$, the model is underidentified. To eliminate the indeterminacy requires constraints on some of the parameters. One might set these parameters to fixed, known constraints, most commonly zero. For example, in $\mathbf{\Theta}_\delta$ the hypothesis of uncorrelated errors of measurement constrains the off-diagonal elements of $\mathbf{\Theta}_\delta$ to zero. Or, if a latent variable does not influence an observed variable, one sets the corresponding λ_{ij} coefficient to zero. Another type of constraint is to set parameters equal. For instance, if we analyze three parallel measures (see Chapter 6), the variances of the errors of measurement are equal, and the λ_{ij}'s also would be equal to one.

Some constraints must be introduced to provide a scale for the latent variables of a model. I introduced this issue in chapter four. There the only "latent variables" are in $\boldsymbol{\zeta}$, the errors in the equation, and by convention each ζ_i is given a scale by fixing its coefficient to one in the structural equation in which it appears. For CFA models, $\boldsymbol{\xi}$ and $\boldsymbol{\delta}$ are both vectors of latent variables. The model $\mathbf{x} = \mathbf{\Lambda}_x\boldsymbol{\xi} + \boldsymbol{\delta}$ implicitly specifies that the $\boldsymbol{\delta}$ coefficient matrix is an identity matrix that provides $\boldsymbol{\delta}$ with a scale in the same way as $\boldsymbol{\zeta}$. The CFA model leaves open the scaling options for the substantive latent variables in $\boldsymbol{\xi}$. Latent variables such as social class, ambition, political instability, and many others have no universal scales. Typically, researchers set the variances of these latent variables to one or give them a scale in the same units as one of their indicators by constraining a λ_{ij} to one. The meaning of giving ξ_j the same scale as the observed variable may not be clear. Consider first the case of $x_i = \xi_j$ ($\lambda_{ij} = 1$, $\delta_i = 0$). Clearly, x_i and ξ_j have identical scales and ranges. A one-unit change in ξ_j exactly corresponds to a one-unit change in x_i. Next consider the more typical situation of $x_i = \xi_j + \delta_i$. Here ξ_j equals $x_i - \delta_i$. An exact one-unit to one-unit relation between ξ_j and x_i is absent. The ξ_j and x_i can take different values and have different ranges. However, they do have the same scale in the more limited sense that *on average* a one-unit shift of ξ_j leads to a one unit shift in x_i. It is in this sense that you should interpret the statement that a latent variable is assigned the scale of an observed variable.

Once the scaling and other constraints are in place, $\boldsymbol{\theta}$ is identified if each element of $\boldsymbol{\theta}$ has a unique solution in terms of the elements of $\boldsymbol{\Sigma}$. To demonstrate this can be very difficult, however. To further explore this means of identification, I examine a one-factor model with two and three indicators. The equations for a two indicator model are

$$\begin{aligned} x_1 &= \lambda_{11}\xi_1 + \delta_1 \\ x_2 &= \lambda_{21}\xi_1 + \delta_2 \end{aligned} \tag{7.13}$$

$$E(\delta_i)=0, \qquad \mathrm{COV}(\xi_1, \delta_i) = 0, \qquad \text{for } i = 1, 2$$

$$\mathrm{COV}(\delta_1, \delta_2) = 0$$

The covariance matrix of x_1 and x_2 has three $(= (1/2)(q)(q+1))$ unique elements. The $\boldsymbol{\Sigma}$ and $\boldsymbol{\Sigma}(\boldsymbol{\theta})$ matrices follow:

$$\boldsymbol{\Sigma} = \begin{bmatrix} \mathrm{VAR}(x_1) & \\ \mathrm{COV}(x_2, x_1) & \mathrm{VAR}(x_2) \end{bmatrix} \tag{7.14}$$

$$\boldsymbol{\Sigma}(\boldsymbol{\theta}) = \begin{bmatrix} \lambda_{11}^2\phi_{11} + \mathrm{VAR}(\delta_1) & \\ \lambda_{21}\lambda_{11}\phi_{11} & \lambda_{21}^2\phi_{11} + \mathrm{VAR}(\delta_2) \end{bmatrix} \tag{7.15}$$

With only three known values and five unknown ones, the model is not identified. Even assigning a scale by setting λ_{11} or λ_{21} to one does not identify the model. To illustrate the underidentification, suppose that

$$\boldsymbol{\Sigma} = \begin{bmatrix} 10 & \\ 5 & 10 \end{bmatrix} \tag{7.16}$$

Even setting λ_{11} to one, an infinite number of $\boldsymbol{\theta}$'s satisfy the equation $\boldsymbol{\Sigma} = \boldsymbol{\Sigma}(\boldsymbol{\theta})$, where $\boldsymbol{\theta}' = [\phi_{11} \quad \lambda_{21} \quad \mathrm{VAR}(\delta_1) \quad \mathrm{VAR}(\delta_2)]$. Elements of a pair of vectors $\boldsymbol{\theta}_1$ and $\boldsymbol{\theta}_2$ that leads to $\boldsymbol{\Sigma}(\boldsymbol{\theta}_1) = \boldsymbol{\Sigma}(\boldsymbol{\theta}_2)$ and which both equal the above Σ are

$$\begin{aligned} \phi_{11} &= 5, & \phi_{11} &= 2.5 \\ \lambda_{21} &= 1, & \lambda_{21} &= 2 \\ \mathrm{VAR}(\delta_1) &= 5, & \mathrm{VAR}(\delta_1) &= 7.5 \\ \mathrm{VAR}(\delta_2) &= 5, & \mathrm{VAR}(\delta_2) &= 0 \end{aligned} \tag{7.17}$$

We cannot tell which of the preceding parameter values, or which of the many other possible parameter values are the true ones.

Suppose that I know the reliability of $x_1(\rho_{x_1x_1})$. With the scaling convention that $\lambda_{11} = 1$, this provides enough information to identify the model. This follows since $\rho_{x_1x_1} = \phi_{11}/\mathrm{VAR}(x_1)$, $\phi_{11} = \rho_{x_1x_1}\mathrm{VAR}(x_1)$, and $\lambda_{21} = \mathrm{COV}(x_2, x_1)/[\rho_{x_1x_1}\mathrm{VAR}(x_1)]$. With these values we can find solutions for $\mathrm{VAR}(\delta_1)$ and $\mathrm{VAR}(\delta_2)$. Thus knowing an indicator's reliability can aid identification.

Suppose that x_1 and x_2 are tau-equivalent measures. This means that λ_{11} and λ_{21} are equal to one. Now there are three knowns and three unknowns, and $\Sigma(\theta)$ is

$$\Sigma(\theta) = \begin{bmatrix} \phi_{11} + \mathrm{VAR}(\delta_1) & \\ \phi_{11} & \phi_{11} + \mathrm{VAR}(\delta_2) \end{bmatrix} \tag{7.18}$$

Here, ϕ_{11} equals the $\mathrm{COV}(x_1, x_2)$, the $\mathrm{VAR}(\delta_1)$ equals the $[\mathrm{VAR}(x_1) - \mathrm{COV}(x_1, x_2)]$, and the $\mathrm{VAR}(\delta_2)$ equals the $[\mathrm{VAR}(x_2) - \mathrm{COV}(x_1, x_2)]$. Thus the two-indicator model (7.13) with one latent variable and uncorrelated measurement errors is identified for tau-equivalent measures.

Next I take up the case of three indicators and one latent variable. This model is represented in equations (7.7) to (7.9) in the last section. Without further restrictions the model is not identified since there are seven unknowns in θ and six nonredundant elements in the covariance matrix of x_1, x_2, and x_3. If λ_{11} is set equal to 1 to scale ξ_1, the $\Sigma(\theta)$ is

$$\Sigma(\theta) = \begin{bmatrix} \phi_{11} + \mathrm{VAR}(\delta_1) & & \\ \lambda_{21}\phi_{11} & \lambda_{21}^2\phi_{11} + \mathrm{VAR}(\delta_2) & \\ \lambda_{31}\phi_{11} & \lambda_{21}\lambda_{31}\phi_{11} & \lambda_{31}^2\phi_{11} + \mathrm{VAR}(\delta_3) \end{bmatrix} \tag{7.19}$$

Writing each element in the covariance matrix of x_1, x_2, and x_3 in terms of the corresponding structural parameters leads to six equations in six unknowns. Some algebraic manipulation leads to unique solutions for the unknowns:

$$\lambda_{21} = \frac{\mathrm{COV}(x_2, x_3)}{\mathrm{COV}(x_1, x_3)}$$

$$\lambda_{31} = \frac{\mathrm{COV}(x_2, x_3)}{\mathrm{COV}(x_1, x_2)} \tag{7.20}$$

$$\phi_{11} = \frac{\mathrm{COV}(x_1, x_2)\,\mathrm{COV}(x_1, x_3)}{\mathrm{COV}(x_2, x_3)}$$

$$\mathrm{VAR}(\delta_1) = \mathrm{VAR}(x_1) - \phi_{11}$$

$$\mathrm{VAR}(\delta_2) = \mathrm{VAR}(x_2) - \lambda_{21}^2\phi_{11}$$

$$\mathrm{VAR}(\delta_3) = \mathrm{VAR}(x_3) - \lambda_{31}^2\phi_{11}$$

These results show that the three-indicator, single latent variable model is identified when one of the λ_{ij}'s is set to 1 (or $\phi_{11} = 1$).

Staying with this example, but assuming that all three indicators are tau-equivalent ($\lambda_{11} = \lambda_{21} = \lambda_{31} = 1$), the $\Sigma(\theta)$ for (7.19) now would show three solutions for ϕ_{11}: $\phi_{11} = \mathrm{COV}(x_1, x_2)$, $\phi_{11} = \mathrm{COV}(x_1, x_3)$, and $\phi_{11} = \mathrm{COV}(x_2, x_3)$. The parameter is overidentified. For a correct model, the multiple solutions for an overidentified parameter are equal so the parameter has a unique solution. If all unknown parameters in a CFA are identified or overidentified, the model is identified.

With even moderately complex measurement models, attempts to establish the identification of all parameters through algebraic means as I have done above are extremely difficult. This is analogous to problems faced in Chapter 4 for structural equations with observed variables. Again, I turn to identification rules to help determine whether a model is identified. I present several rules appropriate to CFA models: the *t*-rule, the three-indicator rule, and the two-indicator rules. For all cases I assume that each latent variable is assigned a scale. Otherwise, identification is not possible.

t-Rule

To understand the basis of the *t*-rule, consider $\Sigma(\theta)$ as defined in (7.5):

$$\Sigma(\theta) = \Lambda_x \Phi \Lambda_x' + \Theta_\delta$$

The Λ_x matrix is $q \times n$ which means it has qn elements, $\mathbf{\Phi}$ has $\frac{1}{2}(n)(n+1)$ nonredundant parameters, and $\mathbf{\Theta}_\delta$ has $\frac{1}{2}(q)(q+1)$ unique parameters. Thus $\mathbf{\Sigma}(\boldsymbol{\theta})$ is decomposed into $qn + \frac{1}{2}(n)(n+1) + \frac{1}{2}(q)(q+1)$ parameters. With q variables in $\mathbf{x}$, $\mathbf{\Sigma}$ has $\frac{1}{2}(q)(q+1)$ known-to-be-identified elements. If we know none of the parameters in Λ_x, $\mathbf{\Phi}$, or $\mathbf{\Theta}_\delta$, then it is impossible to solve for $qn + \frac{1}{2}(n)(n+1) + \frac{1}{2}(q)(q+1)$ elements in $\boldsymbol{\theta}$ using only the $\frac{1}{2}(q)(q+1)$ known elements in the covariance matrix of $\mathbf{x}$.

The t-rule requires that

$$t \leq \tfrac{1}{2}(q)(q+1) \tag{7.21}$$

where t is the number of free parameters in $\boldsymbol{\theta}$. In other words, the number of free parameters (t) must be less than or equal to the number of unique elements in the covariance matrix of $\mathbf{x}$ [i.e., $\frac{1}{2}(q)(q+1)$]. The t-rule is a *necessary but not sufficient condition* for identification.[7]

To illustrate the t-rule, consider the three-factor measurement model with the following Λ_x, $\mathbf{\Phi}$, and $\mathbf{\Theta}_\delta$:

$$\Lambda_x = \begin{bmatrix} 1 & 0 & 0 \\ \lambda_{21} & 0 & 0 \\ 0 & 1 & \lambda_{33} \\ 0 & 0 & \lambda_{43} \\ 0 & 0 & 1 \end{bmatrix}, \qquad \mathbf{\Phi} = \begin{bmatrix} \phi_{11} & & \\ \phi_{12} & \phi_{22} & \\ \phi_{13} & \phi_{23} & \phi_{33} \end{bmatrix}$$

$$\mathbf{\Theta}_\delta = \begin{bmatrix} \mathrm{VAR}(\delta_1) & & & & \\ 0 & \mathrm{VAR}(\delta_2) & & & \\ 0 & \mathrm{COV}(\delta_2, \delta_3) & \mathrm{VAR}(\delta_3) & & \\ 0 & 0 & 0 & \mathrm{VAR}(\delta_4) & \\ 0 & 0 & 0 & \mathrm{COV}(\delta_4, \delta_5) & \mathrm{VAR}(\delta_5) \end{bmatrix} \tag{7.22}$$

A count of the unknown parameters in Λ_x, $\mathbf{\Phi}$, and $\mathbf{\Theta}_\delta$ leads to a t of 16. The number of known elements in the covariance matrix for x_1 to x_5 equals $\frac{1}{2}(q)(q+1)$, or 15. Since t is greater than 15, the model does not meet the necessary condition for identification, and it is not identified. That a model

[7]As in the earlier chapters I ignore situations where inequality constraints are placed on parameters to aid in identification.

does meet the t-rule criterion does not guarantee identification, however. It only means that a necessary condition has been met.

Three-Indicator Rules

I demonstrated the three-indicator rule for one factor earlier in this section. A sufficient condition to identify a one-factor model is to have at least three indicators with nonzero loadings and $\boldsymbol{\Theta}_\delta$ diagonal. With more than three indicators, the unifactor model is overidentified. A multifactor model is identified when it has (1) three or more indicators per latent variable, (2) each row of $\boldsymbol{\Lambda}_x$ with one and only one nonzero element, and (3) a diagonal $\boldsymbol{\Theta}_\delta$. There are no restrictions on $\boldsymbol{\Phi}$. I can establish the identification of λ_{ij}, ϕ_{jj}, and $\text{VAR}(\delta_i)$ for any factor ξ_j and its three indicators in the model by using the covariance matrix of these indicators and the equations (7.20) given earlier for a one-factor model. Repeating this for every ξ demonstrates that $\boldsymbol{\Lambda}_x$, $\boldsymbol{\Theta}_\delta$, and diag $\boldsymbol{\Phi}$ are identified. The only remaining parameters are the off-diagonal elements of $\boldsymbol{\Phi}$. They are identified since ϕ_{jk} $(j \neq k)$, the covariance of ξ_j and ξ_k, equals the covariance between the two x variables that scale (have λ's set to one) ξ_j and ξ_k.

The three-indicator rule is sufficient but not necessary for identification. An even less restrictive means of identifying models with fewer indicators is next.

Two-Indicator Rules

The two-indicator rule is an alternative sufficient condition for measurement models with more than one ξ (see, e.g., Wiley 1973; Kenny 1979, 142–143). Like the three-indicator rules, $\boldsymbol{\Theta}_\delta$ is assumed to be diagonal. Each latent variable is scaled (e.g., one λ_{ij} set to 1 for each ξ_j). Under these conditions, having two indicators per latent variable is sufficient to identify the measurement model provided that the factor complexity of each x_i is one and that there are no zero elements in $\boldsymbol{\Phi}$. Consider the simplest case with two latent variables. The $\boldsymbol{\Theta}_\delta$ matrix for this model is diagonal. The $\boldsymbol{\Lambda}_x$ and $\boldsymbol{\Phi}$ matrices are

$$\boldsymbol{\Lambda}_x = \begin{bmatrix} 1 & 0 \\ \lambda_{21} & 0 \\ 0 & 1 \\ 0 & \lambda_{42} \end{bmatrix}, \qquad \boldsymbol{\Phi} = \begin{bmatrix} \phi_{11} & \\ \phi_{21} & \phi_{22} \end{bmatrix} \tag{7.23}$$

Since the factor complexity for each observed variable is one, the first condition is met. As long as none of the elements in $\mathbf{\Phi}$ are zero, the second condition is also met. The model is identified. I can prove that the two-indicator rule is a sufficient condition for identification by writing each unknown parameter as a function of the identified parameters of $\mathbf{\Sigma}$. The starting point is the covariance structure hypothesis $\mathbf{\Sigma} = \mathbf{\Sigma}(\boldsymbol{\theta})$ and equation (7.5), which defines $\mathbf{\Sigma}(\boldsymbol{\theta})$ for CFA models. Substituting (7.23) and the elements for $\mathbf{\Theta}_\delta$ into (7.5), I find that

$$\mathbf{\Sigma}(\boldsymbol{\theta}) = \begin{bmatrix} \phi_{11} + \mathrm{VAR}(\delta_1) & & & \\ \lambda_{21}\phi_{11} & \lambda_{21}^2\phi_{11} + \mathrm{VAR}(\delta_2) & & \\ \phi_{12} & \lambda_{21}\phi_{12} & \phi_{22} + \mathrm{VAR}(\delta_3) & \\ \lambda_{42}\phi_{12} & \lambda_{21}\lambda_{42}\phi_{12} & \lambda_{42}\phi_{22} & \lambda_{42}^2\phi_{22} + \mathrm{VAR}(\delta_4) \end{bmatrix} \tag{7.24}$$

The easiest proof of identification is for ϕ_{12} which equals $\mathrm{COV}(x_3, x_1)$. From the main diagonals of $\mathbf{\Sigma}$ and $\mathbf{\Sigma}(\boldsymbol{\theta})$ it follows that

$$\mathrm{VAR}(\delta_1) = \mathrm{VAR}(x_1) - \phi_{11}, \qquad \mathrm{VAR}(\delta_3) = \mathrm{VAR}(x_3) - \phi_{22} \tag{7.25}$$

$$\mathrm{VAR}(\delta_2) = \mathrm{VAR}(x_2) - \lambda_{21}^2\phi_{11}, \qquad \mathrm{VAR}(\delta_4) = \mathrm{VAR}(x_4) - \lambda_{42}^2\phi_{22} \tag{7.26}$$

If I can show that ϕ_{11}, ϕ_{22}, λ_{21}, and λ_{42} are identified, then the above elements of $\mathbf{\Theta}_\delta$ are identified. The equations below do just that:

$$\lambda_{21} = \frac{\mathrm{COV}(x_3, x_2)}{\mathrm{COV}(x_3, x_1)}, \qquad \phi_{11} = \frac{\mathrm{COV}(x_2, x_1)\,\mathrm{COV}(x_3, x_1)}{\mathrm{COV}(x_3, x_2)} \tag{7.27}$$

$$\lambda_{42} = \frac{\mathrm{COV}(x_4, x_1)}{\mathrm{COV}(x_3, x_1)}, \qquad \phi_{22} = \frac{\mathrm{COV}(x_4, x_3)\,\mathrm{COV}(x_3, x_1)}{\mathrm{COV}(x_4, x_1)} \tag{7.28}$$

So all nine elements of $\boldsymbol{\theta}$ for this model are identified as the two-indicator rule claims. The same procedure applies to CFA models with three or more latent variables and no zero elements in $\mathbf{\Phi}$.

I generalize this rule even further by loosening the requirements for $\mathbf{\Phi}$. The following four conditions are sufficient for model identification: (1) each row of $\mathbf{\Lambda}_x$ has one and only one nonzero value, (2) there are at least two indicators per latent variable, (3) each row of $\mathbf{\Phi}$ has at least one nonzero off-diagonal element, and (4) $\mathbf{\Theta}_\delta$ is diagonal. The major difference from the first two-indicator rule is that some off-diagonal elements of $\mathbf{\Phi}$ can be zero, but the rule still applies.

The proof of this rule has its basis as follows. Choose a pair of indicators that measures a latent variable, say, ξ_j. Choose another pair of indicators that measures another latent variable, say, ξ_k, that is correlated with ξ_j (i.e., $\phi_{jk} \neq 0$). The covariance matrix and the implied covariance matrix for these four observed variables are the same as in (7.24) except for a change in subscripts for the parameters. No additional information from the other observed variables is necessary to solve for the subset of $\boldsymbol{\theta}$ that corresponds to the unknown parameters in this part of the CFA model. Since the same can be done for any subset of four variables that meet the above conditions, all parameters in $\boldsymbol{\theta}$ are identified. This rule should prove useful for a range of models with at least two-indicators per latent variable. Its major limitations are the requirements of a diagonal $\boldsymbol{\Theta}_{\delta}$ and a factor complexity of one for all x_i, and that it is a sufficient but not necessary condition for identification.

Summary of Rules

Table 7.3 summarizes the rules of identification. All three determine if the model as a whole is identified, as indicated in the second column. Even if the model is not identified, individual parameters or equations that are part of the model may be identified. The third column summarizes the rule, and the fourth and fifth columns indicate whether the rule provides a necessary or sufficient condition for identification.

An Insufficient Condition of Identification

An issue in factor analysis that differs from model identification, but is sometimes confused with it, is whether factor loadings are *unique under rotation*. "A solution is unique if all linear transformations of the factors that leave the fixed parameters unchanged also leave the free parameters unchanged" (Jöreskog 1979b, 23). Jöreskog (1979b) and Howe (1955) stated conditions that are sufficient for uniqueness under factor rotation for $\boldsymbol{\Lambda}_x$. Some authors have stated that these same conditions are sufficient for CFA model identification. This is not correct. I refer the reader to Bollen and Jöreskog (1985) for further details.

Empirical Tests of Identification

Chapter 4 and this chapter present identification rules to aid model identification. Unfortunately, the rules do not cover all models. In Chapter 4, for instance, nonrecursive models with restrictions on $\boldsymbol{\Psi}$ or models with equality constraints across equations do not fall under the rules. For the CFA models of this chapter, none of the available rules are necessary and

Table 7.3 Summary of Identification Rules for Confirmatory Factor Analysis ($\mathbf{x} = \Lambda_x \xi + \delta$)

Identification Rule	Identifies:	Summary of Rules[a]	Necessary Condition	Sufficient Condition
t-Rule	model	$t \leq \frac{1}{2}(q)(q+1)$	yes	no
Three-Indicator Rule	model	$n \geq 1$ one nonzero element per row of Λ_x three or more indicators per factor Θ_δ diagonal	no	yes
Two-Indicator Rules				
Rule 1	model	$n > 1$ $\phi_{ij} \neq 0$ for all i and j one nonzero element per row of Λ_x two or more indicators per factor Θ_δ diagonal	no	yes
Rule 2	model	$n > 1$ $\phi_{ij} \neq 0$ for at least one pair of i and j, where $i \neq j$ one nonzero element per row of Λ_x two or more indicators per factor Θ_δ diagonal	no	yes

[a] it is assumed that all latent variables are given a scale.

sufficient conditions for identification. Furthermore several of the rules are not derived for the frequent situation of correlated disturbances (nondiagonal Θ_δ). Algebraic solutions for the unknown parameters in terms of the elements of Σ can be tried. However, with even moderately complex models, solving the set of nonlinear equations for the unknown parameters can prove virtually impossible.

Given this situation, researchers often turn to empirical tests of identification. These are applicable to all the structural equation models of the book. Yet the nature of empirical identification tests needs to be understood before an analyst applies them. To discuss them, I need to distinguish two types of identification: *global* and *local identification*. The empirical procedures test local identification. The prior definition of identification is for global identification: a parameter vector θ is globally identified if there are no vectors θ_1 and θ_2 such that $\Sigma(\theta_1) = \Sigma(\theta_2)$ unless $\theta_1 = \theta_2$. Local identification is a weaker concept of uniqueness: a *parameter vector* θ *is locally identified at a point* θ_1, *if in the neighborhood of* θ_1 *there is no vector* θ_2 *for which* $\Sigma(\theta_1) = \Sigma(\theta_2)$ *unless* $\theta_1 = \theta_2$. Although the definitions may sound similar, they have differences. Local identification assesses identification at a specific point of θ, θ_1, and determines whether the implied covariance matrix changes with small changes in θ_1. Global identification is not so limited. Any pair of vectors that are possible values of θ are legitimate to consider. If any of these have the same implied covariance matrix, then θ is not globally identified. It still could be locally identified.

Global identification implies local identification. Local identification is necessary but not sufficient for global identification. Also for many nonlinear models such as covariance structure models, local identification at one point does not mean that θ is locally identified at other points. Given these properties the reader may well wonder why examine local identification? Probably three reasons account for the widespread use of local identification. First, the failure to achieve local identification tells us that global identification also fails. It provides a means to detect some models that are not globally nor locally identified with a method that is generally applicable. Second, empirical tests of local identification are readily available in programs to estimate covariance structure models. The major covariance structure programs provide a test as a part of the analysis. Third, viable alternative means to establish global identification are not available.

Two closely related empirical tests of local identification are the best known. One is based on the work of Wald (1950). Wald's rank rule holds that the $t \times 1$ parameter vector θ is locally identified at a point $\theta = \theta_1$, if and only if the rank of $\partial\sigma(\theta)/\partial\theta$ evaluated at θ_1 is t, where $\sigma(\theta)$ is a vector that contains the nonredundant elements of $\Sigma(\theta)$ stacked into a vector. Closely related to this (see Wiley 1973, 82–83) is a second test recommended by Keesling (1972), Wiley (1973), Jöreskog and Sörbom (1986), and others. They suggest that the information matrix for θ can help determine local identification.[8] The information matrix is minus the expected value of the second-order partial derivatives of the fitting functions with respect to each parameter in θ (see the discussion of ML in Chapter 4). The θ parameter is locally identified at some point θ_1, if and only if the inverse of the information matrix exists (Rothenberg 1971).

[8]Some discussions of this test do not clearly state that it usually is a test of local not global identification.

The inverse of the estimated information matrix is routinely computed in programs such as LISREL or EQS since its inverse is the basis for computing standard errors of $\hat{\boldsymbol{\theta}}$. Thus the failure of the information matrix to have an inverse means $\boldsymbol{\theta}$ at $\hat{\boldsymbol{\theta}}$ is not locally identified. Based on the previous discussion, this in turn means that $\boldsymbol{\theta}$ is not globally identified. If the information matrix has an inverse, $\boldsymbol{\theta}$ is locally identified at the evaluated point.

Two sources of uncertainty underlie this test (McDonald and Krane 1979). One is that numerical means must be relied on to evaluate the singularity or existence of the inverse of the information matrix. Numerical methods have limits to their accuracy. As a result of truncation a "numerical" inverse may be calculated, although an inverse would not exist if the numbers entering the calculation were recorded at a higher degree of accuracy. A second problem [except when elements of $\boldsymbol{\Sigma}(\boldsymbol{\theta})$ are linear functions of $\boldsymbol{\theta}$] is that the goal is to evaluate local identification at the true population parameter value, say, $\boldsymbol{\theta}_0$, whereas the empirical test is for the estimate, $\hat{\boldsymbol{\theta}}$.

A model that is locally identified at the true population parameter $\boldsymbol{\theta}_0$ may fail an empirical test at $\hat{\boldsymbol{\theta}}$ or a locally unidentified $\boldsymbol{\theta}$ at $\boldsymbol{\theta}_0$ may pass it. Experiences to date, have not shown this to be a practical problem, but it is worthwhile to be aware of this possibility. To understand how such a situation might arise, remember that local identification is sought for population parameters at $\boldsymbol{\theta}_0$ and the empirical tests are based on *sample* estimates of parameters at $\hat{\boldsymbol{\theta}}$. It is possible that the sample estimates take values such that the parameter estimates lead to a singular *estimated* information matrix and the model appears underidentified. On the other hand, the population parameters may take values that lead to a locally unidentified model, but the sample estimates may not, and the model would "pass" an empirical identification test.

I take a simple, one-factor, three-indicator model for illustration. Suppose the model is

$$\begin{bmatrix} x_1 \\ x_2 \\ x_3 \end{bmatrix} = \begin{bmatrix} \lambda_{11} \\ \lambda_{21} \\ \lambda_{31} \end{bmatrix} [\xi_1] + \begin{bmatrix} \delta_1 \\ \delta_2 \\ \delta_3 \end{bmatrix} \tag{7.29}$$

where $\boldsymbol{\Theta}_\delta$ diagonal and ξ_1 is uncorrelated with δ_i $(i = 1, 2, 3)$. If for scaling purposes I set λ_{11} to one, then $\boldsymbol{\Sigma}(\boldsymbol{\theta})$ is (7.19). As shown in (7.20), this model is exactly identified in the sense that each unknown in $\boldsymbol{\theta}$ [i.e., $\phi_{11}, \lambda_{21}, \lambda_{31}, \text{VAR}(\delta_1), \text{VAR}(\delta_2), \text{VAR}(\delta_3)$] may be written as functions of the variances and covariances of x_1, x_2, and x_3 (i.e., $\boldsymbol{\Sigma}$). The equations for

λ_{21}, λ_{31} and ϕ_{11} are

$$\lambda_{21} = \frac{\text{COV}(x_2, x_3)}{\text{COV}(x_1, x_3)}$$
$$\lambda_{31} = \frac{\text{COV}(x_2, x_3)}{\text{COV}(x_1, x_2)} \tag{7.30}$$
$$\phi_{11} = \frac{\text{COV}(x_1, x_2)\,\text{COV}(x_1, x_3)}{\text{COV}(x_2, x_3)}$$

Suppose that λ_{21} is zero; that is, ξ_1 and x_2 are not related. If $\lambda_{21} = 0$, then (7.19) indicates that $\text{COV}(x_1, x_2)$ and $\text{COV}(x_2, x_3)$ are zero. In addition (7.30) implies undefined solutions for λ_{31} and ϕ_{11} since the denominators are zero. In the earlier discussions I restricted $\boldsymbol{\theta}$ so that points like these are excluded. I lift this restriction for this example. Under these conditions a zero value of λ_{21} creates an unidentified model at $\boldsymbol{\theta}_0$.

Would empirical tests detect this? The answer is not always. Since the covariance matrix is based on sample estimates, it is likely that the sample covariances of x_2 with x_1 and x_3 are not exactly zero [$\text{cov}(x_1, x_2) \neq 0$, $\text{cov}(x_2, x_3) \neq 0$] and that parameter estimates may be computed with the sample variances and covariances. If as a consequence the estimated information matrix can be inverted, an identification problem may not appear to exist, even though in the population the model is not identified. However, if the sample covariances are near zero, then large standard errors for the parameter estimates may result and alert researchers to the problem.

Another possibility is that in the population $\boldsymbol{\theta}$ at $\boldsymbol{\theta}_0$ is locally identified, yet the variances and covariances for a particular sample take values that give the appearance of an unidentified model. For instance, suppose that in the one-factor, three-indicator model the sample values $\text{cov}(x_1, x_2)$ or $\text{cov}(x_2, x_3)$ are approximately zero, although λ_{21}, λ_{31}, ϕ_{11}, and $\text{VAR}(\delta_i)$ may be nonzero. It may not be possible to calculate sample estimates of the population parameters or to form and invert an estimated information matrix. This may lead the researcher to believe that the model is not locally identified at $\boldsymbol{\theta}_0$ even when this is false.

The general points of this subsection are, first, local identification is a more limited definition of identification than is global identification. Second, the empirical test on the information matrix (or Wald's rank rule) is a test of local identification. Third, the procedures test the local identification of $\boldsymbol{\theta}$ at $\hat{\boldsymbol{\theta}}$, the sample estimate, but we want to know the local identification of $\boldsymbol{\theta}$ at $\boldsymbol{\theta}_0$, the true population value. Finally, since the test is numerically based, it is limited by the numerical accuracy of the calculations. Even with these limitations, the empirical tests are valuable tools in exploring identification (McDonald 1982). Typically, models known to be underidentified have singular information matrices, but known-to-be-identified ones do not (Jöreskog and Sörbom 1986). However, we should not take a casual attitude toward these tests since we have more to learn about their properties and fallibilities.

Recommendations for Checking Identification

Until a general necessary and sufficient condition for identification is available, identification remains a challenge. In the meantime how best can we proceed? As a first step, I recommend application of any of the identification rules that are appropriate. If some question remains because the model meets a necessary condition but not a sufficient one, then if at all possible, each unknown should be solved in terms of the elements of the covariance matrix of the observed variables. As an additional check, the empirical test on the information matrix should be performed. This is automatically calculated in programs such as LISREL VI and EQS. Finally, Jöreskog and Sörbom (1986, I.24) suggest that if doubt still remains about identification status, another test is possible. The first step is to analyze $\mathbf{S}$ as usual and to save $\hat{\Sigma}$ where $\hat{\Sigma}$ is the predicted covariance matrix based on the estimates of the model parameters. Next, substitute $\hat{\Sigma}$ for $\mathbf{S}$ and rerun the same program. If the model is identified, the new estimates should be identical to the first ones that generated $\hat{\Sigma}$. Of course, if $\hat{\Sigma} = \mathbf{S}$ as is common when $t = \frac{1}{2}(q)(q+1)$, this test is not informative since the analysis has the same covariance matrix as the previous run and necessarily has the same final estimates. Another empirical test is to estimate the model with different starting values to see if it converges to the same parameter estimates each time. Finally, researchers can run the empirical identification tests separately for random subsamples to see if the test results are consistent.

The empirical tests of identification apply to all structural equation models and not just the CFA models of this chapter.

Political Democracy Example

The CFA of political democracy in 1965 and 1960 provides an example of establishing model identification. The path diagram of this model is given in Figure 7.3 and the corresponding $\mathbf{\Lambda}_x$, $\mathbf{\Phi}$, and $\mathbf{\Theta}_\delta$ are in equations (7.11) and (7.12). Since $\mathbf{\Theta}_\delta$ is not diagonal, only the t-rule can be used. After scaling ξ_1 and ξ_2 by setting $\lambda_{11} = 1$, the $t(= 20)$ unknowns in $\boldsymbol{\theta}$ are λ_{21},

λ_{31}, λ_{41}, ϕ_{11}, ϕ_{12}, ϕ_{22}, VAR(δ_1), VAR(δ_2), VAR(δ_3), VAR(δ_4), VAR(δ_5), VAR(δ_6), VAR(δ_7), VAR(δ_8), COV(δ_1, δ_5), COV(δ_2, δ_6), COV(δ_3, δ_7), COV(δ_4, δ_8), COV(δ_4, δ_2), and COV(δ_8, δ_6). With eight variables, 36 [$= \frac{1}{2}(8)(9)$] nonredundant known elements are in Σ. With $t(= 20)$ less than 36, this necessary condition for identification is met.

At this point I turn to algebraic means to establish identification. To begin with, each element of Σ may be written as a function of the structural parameters since $\Sigma(\theta) = \Lambda_x \Phi \Lambda'_x + \Theta_\delta$. There are 36 equations in all. To save space, I report only those equations needed to demonstrate that each unknown may be solved.

The following equations provide solutions for λ_{21}, λ_{31}, and λ_{41}:

$$\begin{aligned} \text{COV}(x_1, x_6) &= \lambda_{21}\phi_{12}, & \text{COV}(x_2, x_7) &= \lambda_{21}\lambda_{31}\phi_{12} \\ \text{COV}(x_1, x_7) &= \lambda_{31}\phi_{12}, & \text{COV}(x_2, x_8) &= \lambda_{21}\lambda_{41}\phi_{12} \end{aligned} \tag{7.31}$$

The solutions are

$$\lambda_{21} = \frac{\text{COV}(x_2, x_7)}{\text{COV}(x_1, x_7)}$$

$$\lambda_{31} = \frac{\text{COV}(x_2, x_7)}{\text{COV}(x_1, x_6)}$$

$$\lambda_{41} = \frac{\text{COV}(x_2, x_8)}{\text{COV}(x_1, x_6)}.$$

Solutions for ϕ_{11}, ϕ_{12}, and ϕ_{22} can be obtained from

$$\begin{aligned} \text{COV}(x_1, x_4) &= \lambda_{41}\phi_{11} \\ \text{COV}(x_3, x_5) &= \lambda_{31}\phi_{12} \\ \text{COV}(x_5, x_6) &= \lambda_{21}\phi_{22} \end{aligned} \tag{7.32}$$

If the previously derived values of λ_{41}, λ_{31}, and λ_{21} are substituted into (7.32), ϕ_{11}, ϕ_{12}, and ϕ_{22} may be written as functions of the covariances.

Specifically,

$$\phi_{11} = \frac{\text{COV}(x_1, x_4)\,\text{COV}(x_1, x_6)}{\text{COV}(x_2, x_8)}$$

$$\phi_{12} = \frac{\text{COV}(x_3, x_5)\,\text{COV}(x_1, x_6)}{\text{COV}(x_2, x_7)}$$

$$\phi_{22} = \frac{\text{COV}(x_5, x_6)\,\text{COV}(x_1, x_7)}{\text{COV}(x_2, x_7)}$$

Next the error variances, VAR(δ_i), may be derived. The general form of the equation for the variance of x_1 to x_4 is $\text{VAR}(x_i) = \lambda_{i1}^2\phi_{11} + \text{VAR}(\delta_i)$ for $i = 1$ to 4. For x_5 to x_8 the general form is $\text{VAR}(x_{i+4}) = \lambda_{i1}^2\phi_{22} + \text{VAR}(\delta_{i+4})$, $i = 1$ to 4. Since λ_{11} to λ_{41}, ϕ_{11}, and ϕ_{22} are now known to be identified, they can be substituted into the eight variance equations and the VAR(δ_i) (for $i = 1$ to 8) easily determined. Finally, the six equations involving covariances of the error terms are

$$\begin{aligned}
\text{COV}(x_1, x_5) &= \phi_{12} + \text{COV}(\delta_1, \delta_5) \\
\text{COV}(x_2, x_6) &= \lambda_{21}^2\phi_{12} + \text{COV}(\delta_2, \delta_6) \\
\text{COV}(x_3, x_7) &= \lambda_{31}^2\phi_{12} + \text{COV}(\delta_3, \delta_7) \\
\text{COV}(x_4, x_8) &= \lambda_{41}^2\phi_{12} + \text{COV}(\delta_4, \delta_8) \\
\text{COV}(x_4, x_2) &= \lambda_{41}\lambda_{21}\phi_{11} + \text{COV}(\delta_4, \delta_2) \\
\text{COV}(x_8, x_6) &= \lambda_{41}\lambda_{21}\phi_{22} + \text{COV}(\delta_8, \delta_6)
\end{aligned} \tag{7.33}$$

With λ_{21}, λ_{31}, λ_{41}, ϕ_{11}, ϕ_{22}, and ϕ_{12} in hand, derivation of the covariances of the measurement error is straightforward.

Since the CFA of political democracy in 1960 and 1965 is overidentified, other equations in terms of the covariances of the observed variables can be used to solve for the parameters. There is no gain in reviewing these alternative equations. We need only one solution for each parameter to show that the model is identified. With identification established, the next step is estimation.

ESTIMATION

In Chapter 4 I explained the ML, GLS, and ULS methods of estimation for structural equations with observed variables. The same methods are appropriate for CFA. The goal of estimation is identical: choose values for the unknown parameters that lead to an implied covariance matrix, $\hat{\boldsymbol{\Sigma}}$, as close to the sample covariance matrix, $\mathbf{S}$, as possible.[9] Since the material on estimation in Chapter 4 is appropriate here, I will not repeat it, except to list the ML, GLS, and ULS fitting functions that I refer to in this chapter:

$$F_{ML} = \log|\boldsymbol{\Sigma}(\boldsymbol{\theta})| + \operatorname{tr}[\mathbf{S}\boldsymbol{\Sigma}^{-1}(\boldsymbol{\theta})] - \log|\mathbf{S}| - q \tag{7.34}$$

$$F_{GLS} = (\tfrac{1}{2})\operatorname{tr}\{[\mathbf{I} - \boldsymbol{\Sigma}(\boldsymbol{\theta})\mathbf{S}^{-1}]^2\} \tag{7.35}$$

$$F_{ULS} = (\tfrac{1}{2})\operatorname{tr}\{[\mathbf{S} - \boldsymbol{\Sigma}(\boldsymbol{\theta})]^2\} \tag{7.36}$$

In Chapter 9 I present alternative estimators.

Nonconvergence

The fitting functions generally require iterative numerical procedures to obtain solutions. This process was described in Appendix 4C. When the values for the unknown parameters in two consecutive steps differ by less than some preset criterion, the iterative process stops. Nonconvergence occurs if the values are insufficiently close after repeated iterations. Whether estimates converge or not depends on several factors. First, the convergence criterion affects convergence. It defines what is "insufficiently close." In EQS the default is that the average of the absolute differences between parameter estimates between two iterations must be 0.001 or less (Bentler

[9] In the structural equations with observed variables, $\boldsymbol{\Sigma}$ is a function of $\mathbf{B}$, $\boldsymbol{\Gamma}$, $\boldsymbol{\Phi}$, and $\boldsymbol{\Psi}$ whereas in CFA $\boldsymbol{\Sigma}$ equals $\boldsymbol{\Lambda}_x\boldsymbol{\Phi}\boldsymbol{\Lambda}_x' + \boldsymbol{\Theta}_\delta$. The difference in the structural matrices that form $\boldsymbol{\Sigma}$ does not matter from the point of view of estimation.

1985, 82). This value can be altered, but the researcher should be aware that the smaller the number, the more iterations typically required.

Second, convergence is affected by the number of iterations allowed. In LISREL VI, for instance, 250 iterations are tried before stopping. In EQS the default for the ML, GLS, and ULS estimators is 30 iterations, though this default can be reset up to a maximum of 500. If the iteration count reaches these limits, the programs stop and report that the estimates have not converged.

A third factor making nonconvergent solutions more or less likely are the starting values for the unknown parameters. The closer these values are to the final estimates, the fewer steps needed to converge. LISREL VI provides automatic starting values with an instrumental variable technique, which I describe in Chapter 9. EQS also is planning an noniterative estimator (see Bentler 1982) to provide the initial values. Often these estimates are quite good and lead to convergence in a few iterations. Other times they are less helpful, and the analyst needs to supply the values. Analysts can select starting values in several ways. When a prior run does not converge, the values for the parameters from the last iteration can be the starting values for a new series of iterations. Or, if there are other analyses of similar data, these may provide good starting values. For instance, a published article may have many of the same or similar variables. The estimates from the prior work may provide reasonable starting values. Knowledge of the variances and covariances of the observed variables and the direction and approximate magnitude of the relations hypothesized between variables also can help. For instance, scaling a latent variable to one of the observed variables by setting its λ to one leads to information about other initial values. The variance of the latent variable should not exceed that of the observed variable to which it is scaled. If other indicators load on the same factor and have a similar range of values as the scaling indicator, then reasonable starting points for their λ's are near one. Starting values for covariances of measurement errors should not lead to correlations of the errors that exceed one. Similarly, values for $\mathbf{\Phi}$ should keep the correlations for the latent variables within the minus-to-plus-one limits. (The discussion on selecting starting values in Appendix 4C may be helpful.)

Other causes of nonconvergence include poorly specified models and sampling fluctuations in the variances and covariances of the observed variables. In Monte Carlo experiments, Boomsma (1982) and Anderson and Gerbing (1984) have found that nonconvergent solutions are greatest for CFA on samples with N's less than 150 and for CFA with only two indicators per factor. In their simulations the models are correctly specified, and good starting values are given. These findings imply that if analysts cannot reach convergence and they are convinced that the model specification, the convergence criterion, and the starting values are adequate, then they should increase the sample size or the number of indicators, if possible. Estimates from nonconvergent runs should not be used or given substantive interpretations. However, when the iterative procedure successfully converges, we can begin to evaluate the model. It is this topic to which I now turn.

MODEL EVALUATION

The estimated coefficients and the strength of associations require close examination. Are they of the right sign? Do the magnitudes of effects conform with previous research? If they do not, are there good reasons to explain why they differ? Are the estimated effects substantively as well as statistically significant? To help in the evaluation of a model, a number of statistical measures of fit have been proposed. These fit measures are general and can be applied to all the models treated in the book, including the observed variable models of Chapter 4. I group these statistics into overall and individual measures.

Overall Model Fit Measures

The covariance structure hypothesis is that $\Sigma = \Sigma(\boldsymbol{\theta})$. The overall fit measures help to assess whether this is valid, and if not, they help to measure the departure of Σ from $\Sigma(\boldsymbol{\theta})$. Being population parameters, Σ and $\Sigma(\boldsymbol{\theta})$ are unavailable, so researchers examine their sample counterparts $\mathbf{S}$ and $\Sigma(\hat{\boldsymbol{\theta}})$. The $\mathbf{S}$ is the usual sample covariance matrix, and $\Sigma(\hat{\boldsymbol{\theta}})$ is the implied covariance matrix evaluated at the estimate of $\boldsymbol{\theta}$ which minimizes either F_{ML}, F_{GLS}, or F_{ULS}. I will abbreviate $\Sigma(\hat{\boldsymbol{\theta}})$ to $\hat{\Sigma}$.

Virtually all measures of overall model fit involve functions of $\mathbf{S}$ and $\hat{\Sigma}$. These fit indices gauge the "closeness" of $\mathbf{S}$ to $\hat{\Sigma}$, though this closeness is measured in many different ways. The principal advantages of the overall model fit measures are that they evaluate the whole model and that they can indicate inadequacies not revealed by the fit of the model components (e.g., equations and parameter estimates). One limitation is that they are inapplicable to exactly identified models. In this situation $\mathbf{S}$ always equals $\hat{\Sigma}$ so that overall fit is not an issue. With overidentification it is possible for $\mathbf{S}$ to differ from $\hat{\Sigma}$, so we can test whether the restrictions leading to overidentification seem valid. A second limitation is that overall fit measures can differ from the fit of components of the model. For example, the overall fit may be good, but parameter estimates may not be statistically significant or may have signs opposite to that predicted. Furthermore the overall match of $\hat{\Sigma}$ to $\mathbf{S}$ does not tell us how well the explanatory variables predict the observed dependent variables. That is, the overall fit measures do not summarize the R^2's for all model equations. For these reasons the overall fit measures should not be used in isolation from the component fit measures that I describe later.

Residuals

The maintained hypothesis call H_0, is that $\Sigma = \Sigma(\boldsymbol{\theta})$. When the covariance structure hypothesis is true, the population residual covariance matrix, $\Sigma - \Sigma(\boldsymbol{\theta})$ is a zero matrix. Any nonzero population residual means that the model specification is in error. Although we do not have the population matrices to directly examine $\Sigma - \Sigma(\boldsymbol{\theta})$, we do have its sample counterpart $\mathbf{S} - \hat{\Sigma}$. The $\hat{\Sigma}$ may be from either the F_{ML}, F_{GLS}, or F_{ULS} solutions.

The residual matrix is perhaps the simplest function of $\mathbf{S}$ and $\hat{\Sigma}$ for assessing overall model fit. The individual sample residual covariances are $(s_{ij} - \hat{\sigma}_{ij})$ where s_{ij} is the ijth element in $\mathbf{S}$ and $\hat{\sigma}_{ij}$ is the corresponding element in $\hat{\Sigma}$. A positive residual means that the model underpredicts the covariance between two variables, whereas a negative one means the predicted covariance is too high. The individual residuals and the mean (or median) of their *absolute* magnitudes can help in assessing model fit.[10] The average residuals include only the lower half and main diagonal elements because of the redundancy of the upper half of the residual matrix.

Ideally, all residuals should be near zero for a "good" model. But the sample residuals are affected by several factors: (1) the departure of Σ from $\Sigma(\boldsymbol{\theta})$, (2) the scales of the observed variables, and (3) sampling fluctuations. We are most interested in using the residuals to detect (1), that is, whether $\Sigma = \Sigma(\boldsymbol{\theta})$. When $\Sigma \neq \Sigma(\boldsymbol{\theta})$, one or more of the covariances or variances of

[10] A closely related summary statistic for the residuals is the root mean-square residual (RMR) proposed by Jöreskog and Sörbom (1986):

$$\mathrm{RMR} = \left[2 \sum_{i=1}^{q} \sum_{j=1}^{i} \frac{(s_{ij} - \hat{\sigma}_{ij})^2}{q(q+1)} \right]^{1/2}$$

Assessments of fit using RMR should be similar to those using the mean absolute value of the unstandardized residuals.

the observed variables are not exactly predicted by the model. The lack of congruity can manifest itself in the sample residuals, with the largest population residuals more likely to be detectable than the smaller ones.

The departure of $\boldsymbol{\Sigma}$ from $\boldsymbol{\Sigma}(\boldsymbol{\theta})$ is not alone in determining $\mathbf{S} - \hat{\boldsymbol{\Sigma}}$. Consider (2), the scales of the observed variables. The magnitudes of the individual and mean of the residuals are altered if the observed variables are measured in different units. A big residual can be due to an observed variable with scale units that have a much larger range than that of the other variables. If sufficiently large, it can distort the mean of the absolute values of the residuals and give a misleading impression. In addition, unless the analyst is very familiar with the variables' scales, it may be difficult to evaluate the magnitude of individual residuals.

A simple solution to this is to calculate *correlation residuals*. Each correlation residual is[11]

$$r_{ij} - \hat{r}_{ij} \tag{7.37}$$

where r_{ij} is the sample correlation of the ith and jth variables, and $\hat{r}_{ij}$ is the model predicted correlation. Using the definition of a correlation, $\hat{r}_{ij}$ is $\hat{\sigma}_{ij}/(\hat{\sigma}_{ii}\hat{\sigma}_{jj})^{1/2}$, where $\hat{\sigma}_{ij}$ is the ijth element of $\hat{\boldsymbol{\Sigma}}$ and $\hat{\sigma}_{ii}$ and $\hat{\sigma}_{jj}$ are the iith and jjth variances in this matrix. Individually $(r_{ij} - \hat{r}_{ij})$ gauges how well a correlation (or a standardized variance for $i = j$) is reproduced. Since r_{ij} and $\hat{r}_{ij}$ range from -1 to $+1$, the correlation residual has a theoretical range from -2 to $+2$. However, a value anywhere near these limits indicates a serious problem in fit. A correlation residual should be fairly close to zero for most well-fitting models. In addition to the individual residuals, the mean or median of the absolute value of the nonredundant correlation residuals gives insight into the overall magnitude of these.

The third factor affecting the sample residuals and correlation residuals, is sampling error. Even when $\boldsymbol{\Sigma} = \boldsymbol{\Sigma}(\boldsymbol{\theta})$ is true, the expected magnitude of the sample residuals depends on N. Under fairly general conditions $\mathbf{S}$ converges to $\boldsymbol{\Sigma}$ and $\hat{\boldsymbol{\Sigma}}$ converges to $\boldsymbol{\Sigma}(\boldsymbol{\theta})$ as N grows large. Similarly, $\mathbf{S} - \hat{\boldsymbol{\Sigma}}$ converges to $\boldsymbol{\Sigma} - \boldsymbol{\Sigma}(\boldsymbol{\theta})$ as $N \to \infty$. For a given model $(s_{ij} - \hat{\sigma}_{ij})$ tends to be smaller, the bigger is the sample (Browne 1982, 79–80). The implication is that in judging the residuals for small samples, we should expect (on average) bigger residuals than when examining residuals in large samples, when the model is true in both samples.

[11]In the common case when $\hat{\sigma}_{ii} = s_{ii}$ and $\hat{\sigma}_{jj} = s_{jj}$, these residual correlations are the same as Bentler's (1985, 92–93) standardized residuals. When these equalities do not hold, the standardized residuals can take a different range than (7.37).

Jöreskog and Sörbom (1986, I.42) propose a *normalized residual* that provides an approximate correction for such sample size effects and for scaling differences. The formula is:

$$\text{Normalized residual} = \frac{(s_{ij} - \hat{\sigma}_{ij})}{\left[(\hat{\sigma}_{ii}\hat{\sigma}_{jj} + \hat{\sigma}_{ij}^2)/N\right]^{1/2}} \tag{7.38}$$

The numerator is the residual. The denominator is the square root of its estimated asymptotic variance. The largest absolute values of the normalized residuals indicate the s_{ij} elements that are most poorly fit by the model. Recently, Jöreskog and Sörbom (1988) have stated that the estimated asymptotic variance in the denominator of (7.38) tends to be too high. This implies that the normalized residuals are smaller than a standardized normal variable.

I illustrate some of the preceding ideas with an example taken from a social psychological experiment by Reisenzein (1986). His full experiment was on helping behavior where he uses a hypothetical story about a person collapsing and lying on a subway floor. As part of the experimental manipulation half the subjects were told that the person was drunk, while the others were told that the person was ill. Reisenzein tests whether a subject's feelings of sympathy and anger mediate the likelihood of their helping the victim. I focus on the sympathy and anger latent variables which Reisenzein measures with three indicators apiece. The indicator questions for sympathy and anger are in Table 7.4. The path model is in Figure 7.4. The sample covariance matrix for an N of 138 is[12]

$$\mathbf{S} = \begin{bmatrix} 6.982 & & & & & \\ 4.686 & 6.047 & & & & \\ 4.335 & 3.307 & 5.037 & & & \\ -2.294 & -1.453 & -1.979 & 5.569 & & \\ -2.209 & -1.262 & -1.738 & 3.931 & 5.328 & \\ -1.671 & -1.401 & -1.564 & 3.915 & 3.601 & 4.977 \end{bmatrix} \tag{7.39}$$

[12] I checked for outliers using the procedure described in Chapter 2. I identified 10 cases that had unusual patterns of responses for these variables. However, including or excluding these cases had little effect on the CFA results.

Table 7.4 Reisenzein's (1986) Indicator Questions for Sympathy and Anger

	Sympathy
x_1	How much sympathy would you feel for that person? (1 = none at all, 9 = very much)
x_2	I would feel pity for this person. (1 = none at all, 9 = very much)
x_3	How much concern would you feel for this person? (1 = none at all, 9 = very much)
	Anger
x_4	How angry would you feel at that person? (1 = not at all, 9 = very much)
x_5	How irritated would you feel by that person? (1 = not at all, 9 = very much)
x_6	I would feel aggravated by that person. (1 = not at all, 9 = very much so)

Source: Adapted from Reisenzein (1986, 1126).

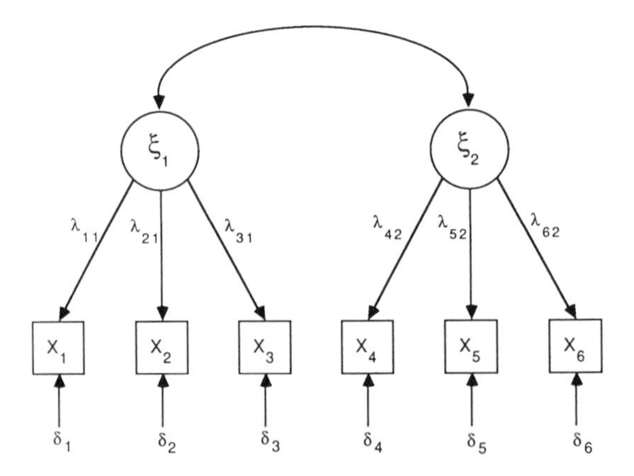

Figure 7.4 Confirmatory Factor Analysis Model for Sympathy (ξ_1) and Anger (ξ_2). Each Measured with Three Indicators (X_1 to X_6)

The residual matrix $(\mathbf{S} - \hat{\boldsymbol{\Sigma}})$ after applying F_{ML} is

$$\mathbf{S} - \hat{\boldsymbol{\Sigma}} = \begin{bmatrix} 0.000 & & & & & \\ 0.029 & 0.000 & & & & \\ -0.016 & -0.027 & 0.000 & & & \\ -0.074 & 0.248 & -0.390 & 0.000 & & \\ -0.172 & 0.299 & -0.280 & -0.030 & 0.000 & \\ 0.337 & 0.137 & -0.127 & 0.013 & 0.020 & 0.000 \end{bmatrix} \tag{7.40}$$

None of the elements seems large. The residuals are a mixture of positive and negative signs. The covariances of x_3 with other variables are overpredicted, but most of these negative elements are not large. The mean absolute value of the residuals is 0.105. Relative to the magnitude of the elements in **S**, this is small. Some elements of a covariance matrix are exactly predicted for a given model and fitting function regardless of the sample covariance matrix. The zero elements deflate the mean residual and can give an overly optimistic assessment of fit. Though it is not always easy to determine which elements are trivially fit for a particular combination of a fitting function and a model, the zero elements of $\mathbf{S} - \hat{\boldsymbol{\Sigma}}$ are likely candidates. Reporting the mean of the absolute value of the *nonzero* residuals removes the effect of the zero values. Doing so for (7.40) leads to a mean of 0.147.

This average still seems relatively small. To control for the scaling factors, I form the correlation residual matrix:

$$\begin{bmatrix} 0.000 & & & & & \\ 0.004 & 0.000 & & & & \\ -0.003 & -0.005 & 0.000 & & & \\ -0.012 & 0.043 & -0.074 & 0.000 & & \\ -0.028 & 0.053 & -0.055 & -0.006 & 0.000 & \\ 0.057 & 0.025 & -0.025 & 0.002 & 0.004 & 0.000 \end{bmatrix} \tag{7.41}$$

The absolute values of all the correlation residuals are less than 0.1. The mean absolute value of residual correlations is 0.019 for all the lower-half elements and 0.026 when the zero elements are removed.

The normalized residual matrix is

$$\begin{bmatrix} 0.000 & & & & & \\ 0.043 & 0.000 & & & & \\ -0.026 & -0.049 & 0.000 & & & \\ -0.131 & 0.482 & 0.829 & 0.000 & & \\ -0.314 & 0.597 & -0.611 & -0.052 & 0.000 & \\ 0.636 & 0.282 & -0.286 & 0.023 & 0.037 & 0.000 \end{bmatrix} \quad (7.42)$$

None of the residuals is even larger than one. If there are large discrepancies between Σ and $\Sigma(\boldsymbol{\theta})$, the unstandardized, correlation, and normalized sample residuals do not reflect them.

What are the sample size effects? By current standards an N of 138 is a small to moderate sample. Given that we expect larger residuals for smaller samples (other things equal), the small residuals and residual correlations reinforce the impression of a good fit. In addition the normalized residuals, which take into account N, are small. To illustrate the effects of sample size on the sample residuals, I randomly selected 60 out of the original 138 cases. For this smaller sample, the mean of the absolute values of the residuals is 0.123 and 0.172 when the diagonal elements are removed. Compared to the means of 0.105 and 0.147 for the N of 138, the residuals for an N of 60 are larger.

In sum, the sample residual matrix $\mathbf{S} - \hat{\Sigma}$ can identify the sample variance or covariance elements poorly fit by a model. The mean (or median) of the absolute value of the lower half of this matrix and the mean (or median) of the same elements, except without the zero elements, are summary measures of model fit. To standardize for the influence of the different scales of the observed variables, I recommend the same measures for correlation residuals. The difficulty with all these fit indicators is that they are affected by sample size. The bigger is N, the smaller tends to be the absolute magnitude of the correlation and unstandardized residuals.

It would be helpful to have a test of whether the residuals depart from the population values of zero. The normalized residuals provide a very rough test of statistical significance, but without prior specification of the individual element that is tested and with the tendency of these residuals to be too small, the probability levels associated with the test are not accurate. It would be extremely helpful to have a simultaneous significance test of the hypothesis that all population residuals in $[\Sigma - \Sigma(\boldsymbol{\theta})]$ are zero. The test should take into account N and df and should be based on the sample residuals. The next section presents such a significance test.

A Chi-Square (χ^2) Test

I introduced the chi-square test for overidentified models in Chapter 4. Under the conditions I describe here $(N-1)F_{\mathrm{ML}}$ or $(N-1)F_{\mathrm{GLS}}$ provide chi-square estimators to test H_0: $\boldsymbol{\Sigma} = \boldsymbol{\Sigma}(\boldsymbol{\theta})$.[13] Since H_0 is equivalent to the hypothesis that $\boldsymbol{\Sigma} - \boldsymbol{\Sigma}(\boldsymbol{\theta}) = 0$, the chi-square test is a simultaneous test that all residuals in $\boldsymbol{\Sigma} - \boldsymbol{\Sigma}(\boldsymbol{\theta})$ are zero.

The chi-square test also is a test of overidentification in the following sense. An overidentified model implies one or more restrictions that must hold in the population. For instance, suppose I have two parallel measures such that $x_1 = \xi_1 + \delta_1$, $x_2 = \xi_1 + \delta_2$, and the $\mathrm{VAR}(\delta_1) = \mathrm{VAR}(\delta_2)$. The model is overidentified, and there are two ways to write $\mathrm{VAR}(\delta_1)$ [or $\mathrm{VAR}(\delta_2)$] as a function of $\boldsymbol{\Sigma}$: $\mathrm{VAR}(\delta_1) = \mathrm{VAR}(x_1) - \mathrm{COV}(x_1, x_2)$ and $\mathrm{VAR}(\delta_1) = \mathrm{VAR}(x_2) - \mathrm{COV}(x_1, x_2)$. These equations imply the restriction that $\mathrm{VAR}(x_1) = \mathrm{VAR}(x_2)$. If the model is valid, this restriction will be true in the population. In any given sample we do not expect it to exactly hold because of sampling fluctuations. But the more the sample $\mathrm{var}(x_1)$ departs from $\mathrm{var}(x_2)$, the less likely it is that the equality restriction between $\mathrm{VAR}(x_1)$ and $\mathrm{VAR}(x_2)$ is valid and the less likely that the model is true. Thus the chi-square test evaluates whether the restrictions are valid.

Browne (1974, 1982, 1984) provides the justification for using $(N-1)F_{\mathrm{GLS}}$ or $(N-1)F_{\mathrm{ML}}$ as chi-square estimators based on generalized least squares principles. Here I explain the likelihood ratio rationale for the asymptotic chi-square distribution of $(N-1)F_{\mathrm{ML}}$. The null hypothesis, H_0 is $\boldsymbol{\Sigma} = \boldsymbol{\Sigma}(\boldsymbol{\theta})$; that is, the specification of the fixed, free, and constrained parameters in $\boldsymbol{\Lambda}_x$, $\boldsymbol{\Phi}$, and $\boldsymbol{\Theta}_\delta$ is valid. Under H_0 I have ML estimators of the free and constrained parameters in these matrices that together with the fixed parameters comprise the estimated matrices $\hat{\boldsymbol{\Lambda}}_x$, $\hat{\boldsymbol{\Phi}}$, and $\hat{\boldsymbol{\Theta}}_\delta$. Next I form $\hat{\boldsymbol{\Sigma}}[=\boldsymbol{\Sigma}(\hat{\boldsymbol{\theta}})]$, which is the sample predicted covariance matrix of the observed variables under H_0. Corresponding to H_0 is the log of the likelihood function. I represent this as $\log L_0$ and drop from this function the irrelevant constants (see Chapter 4) that do not affect the choice of $\hat{\boldsymbol{\theta}}$. When evaluated at $\mathbf{S}$ and $\hat{\boldsymbol{\Sigma}}$, the $\log L_0$ is

$$\log L_0 = -\frac{N-1}{2}\left\{\log|\hat{\boldsymbol{\Sigma}}| + \mathrm{tr}(\hat{\boldsymbol{\Sigma}}^{-1}\mathbf{S})\right\} \tag{7.43}$$

This is the log of the numerator for the likelihood ratio test.

[13]As I mentioned in Chapter 4, Browne (1982, 1984) provides a means to obtain a chi-square test based on F_{ULS}. Since this chi-square estimator is more complicated to form and rarely applied, I restrict my discussion to the chi-square estimators formed from $(N-1)F_{\mathrm{ML}}$ and $(N-1)F_{\mathrm{GLS}}$.

To illustrate $\hat{\Sigma}$ and (7.43), I use the prior CFA example (see Figure 7.4) of sympathy and anger measured with six indicators.

The ML estimates $\hat{\Lambda}_x$, $\hat{\Phi}$, and $\hat{\Theta}_\delta$ are

$$\hat{\Lambda}_x = \begin{bmatrix} 1.00 & 0.00 \\ 0.77 & 0.00 \\ 0.72 & 0.00 \\ 0.00 & 1.00 \\ 0.00 & 0.92 \\ 0.00 & 0.90 \end{bmatrix}, \qquad \hat{\Phi} = \begin{bmatrix} 6.08 & \\ -2.22 & 4.32 \end{bmatrix} \tag{7.44}$$

$$\operatorname{diag}(\hat{\Theta}_\delta) = [0.90 \quad 2.48 \quad 1.92 \quad 1.25 \quad 1.69 \quad 1.45]$$

The $\hat{\Sigma}(= \hat{\Lambda}_x \hat{\Phi} \hat{\Lambda}'_x + \hat{\Theta}_\delta)$ is

$$\hat{\Sigma} = \begin{bmatrix} 6.982 & & & & & \\ 4.657 & 6.047 & & & & \\ 4.351 & 3.334 & 5.037 & & & \\ -2.220 & -1.701 & -1.589 & 5.569 & & \\ -2.037 & -1.561 & -1.458 & 3.961 & 5.328 & \\ -2.007 & -1.538 & -1.437 & 3.903 & 3.581 & 4.977 \end{bmatrix} \tag{7.45}$$

Under H_0, (7.45) is the $\hat{\Sigma}$ we would predict. For this sample, $\log L_0$ is

$$\begin{aligned} \log L_0 &= -\frac{137}{2}\left\{\log|\hat{\Sigma}| + \operatorname{tr}(\hat{\Sigma}^{-1}\mathbf{S})\right\} \\ &= -890.3 \end{aligned} \tag{7.46}$$

To form the likelihood ratio requires an alternative hypothesis, H_1. I choose an H_1 for which the corresponding log of the likelihood function, $\log L_1$, is at a maximum. The least restrictive H_1 possible is that Σ is any positive definite matrix (see Appendix A). If I set $\hat{\Sigma}$ to $\mathbf{S}$ the sample covariance matrix, the $\log L_1$ is at its maximum value. Again, ignoring the irrelevant constants the likelihood function for H_1, $\log L_1$, is

$$\begin{aligned} \log L_1 &= -\frac{(N-1)}{2}\left\{\log|\mathbf{S}| + \operatorname{tr}(\mathbf{S}^{-1}\mathbf{S})\right\} \\ &= -\frac{(N-1)}{2}\left\{\log|\mathbf{S}| + q\right\} \end{aligned} \tag{7.47}$$

This is the log of the denominator for the likelihood ratio.

For the sympathy and anger CFA example, $\mathbf{S}$ is shown in (7.39). The $\log L_1$ is

$$\log L_1 = -\frac{137}{2}\{\log|\mathbf{S}| + \mathrm{tr}[\mathbf{I}_4]\}$$
$$= -885.5 \qquad (7.48)$$

The H_1 hypothesis is not informative in and of itself. There is no single model that it represents in the sense that H_0 does. Selecting H_1 so that $\hat{\boldsymbol{\Sigma}} = \mathbf{S}$ provides a standard of perfect fit against which to compare H_0. The H_1 could be any exactly identified model since $\hat{\boldsymbol{\Sigma}}$ equals $\mathbf{S}$ in this case. The best we can hope is that $\hat{\boldsymbol{\Sigma}} = \hat{\boldsymbol{\Lambda}}_y\hat{\boldsymbol{\Phi}}\hat{\boldsymbol{\Lambda}}'_y + \hat{\boldsymbol{\Theta}}_\delta$ (based on H_0) exactly reproduces $\mathbf{S}$. Since H_1 sets $\hat{\boldsymbol{\Sigma}}$ to $\mathbf{S}$, comparing $\log L_1$ to $\log L_0$ evaluates H_0 vis-à-vis a perfect fit, H_1. The natural logarithm of the likelihood ratio, $\log(L_0/L_1)$, when multiplied by -2 is distributed as chi-square variate when H_0 is true and $(N-1)$ is large. In this case

$$-2\log\left(\frac{L_0}{L_1}\right) = -2\log L_0 + 2\log L_1$$
$$= (N-1)\left[\log|\hat{\boldsymbol{\Sigma}}| + \mathrm{tr}(\hat{\boldsymbol{\Sigma}}^{-1}\mathbf{S})\right] - (N-1)(\log|\mathbf{S}| + q)$$
$$= (N-1)\left(\log|\hat{\boldsymbol{\Sigma}}| + \mathrm{tr}(\hat{\boldsymbol{\Sigma}}^{-1}\mathbf{S}) - \log|\mathbf{S}| - q\right) \qquad (7.49)$$

The last line of the right-hand side of (7.49) should look familiar. The quantity within parentheses is the fitting function F_{ML} of (7.34) evaluated at $\mathbf{S}$ and $\hat{\boldsymbol{\Sigma}}$. Expression (7.49) thus shows that $(N-1)$ times the fitting function F_{ML} evaluated at $\hat{\boldsymbol{\theta}}$ is approximately distributed as a chi-square variate. Its degrees of freedom are $\frac{1}{2}(q)(q+1) - t$, where the first term is the number of nonredundant elements in $\mathbf{S}$ given q observed variables, and t is the number of free parameters in $\boldsymbol{\theta}$.

The logic of significance testing here is different than that usual in testing, say, the statistical significance of the explained variance in a regression equation. In the latter situation the null hypothesis is set such that it runs counter to our theoretical hypothesis, and our hope is to reject the null hypothesis. For instance, an F-test for a regression equation may be based on a null hypothesis that all regression coefficients in the population are zero. We reject the null hypothesis if the F-statistic exceeds a critical value determined by an α level and the degrees of freedom. Rejection of the null hypothesis does not prove the alternative hypothesis that the coefficients are nonzero, but it makes it highly unlikely that they all are zero. In contrast, for the chi-square test of the confirmatory factor analysis the null hypothesis H_0 is that the constraints on $\boldsymbol{\Sigma}$ implied by the model are valid [i.e., $\boldsymbol{\Sigma} = \boldsymbol{\Sigma}(\boldsymbol{\theta})$]. The standard of comparison is the perfect fit of $\hat{\boldsymbol{\Sigma}}$ equal to $\mathbf{S}$. The probability level of the calculated chi-square is the probability of obtaining a χ^2 value larger than the value obtained if H_0 is correct. The higher the probability of the χ^2, the closer is the fit of H_0 to the perfect fit.

I return to the sympathy and anger example to illustrate these ideas. The null hypothesis H_0 is rejected if $-2\log(L_0/L_1)$ is greater than the critical value of χ^2, which is determined by the degrees of freedom (df) and a prespecified α value. A good fit is indicated when we cannot reject H_0 at α, with α generally chosen at 0.10 or 0.05. From (7.46) and (7.48) we know the $\log L_0$ is -890.3 and that the $\log L_1$ is -885.5. If we substitute these values into (7.49), $-2\log(L_0/L_1)$ is

$$\begin{aligned} -2\log(L_0/L_1) &= -2\log L_0 + 2\log L_1 \\ &= 9.6 \end{aligned} \tag{7.50}$$

The right-hand side of (7.50) equals $(N-1)F_{\mathrm{ML}}$ evaluated at $\mathbf{S}$ and $\hat{\boldsymbol{\Sigma}}$. For this problem, N is 138 and $(N-1)$ is 137. The $F_{\mathrm{ML}}(\mathbf{S}, \hat{\boldsymbol{\Sigma}})$ is

$$\begin{aligned} F_{\mathrm{ML}}(\mathbf{S}, \hat{\boldsymbol{\Sigma}}) &= \log|\hat{\boldsymbol{\Sigma}}| + \operatorname{tr}(\hat{\boldsymbol{\Sigma}}^{-1}\mathbf{S}) - \log|\mathbf{S}| - q \\ &= 0.0698 \end{aligned} \tag{7.51}$$

This value times 137 equals the quantity in (7.50), and it is the estimated chi-square value for testing the model H_0 against the general alternative model H_1. The degrees of freedom are $21 - 13$, or 8. The probability of obtaining a χ^2 value larger than the value obtained, if H_0 is true, is about 0.3. At conventional significance levels (e.g., 0.05) I cannot reject the null hypothesis H_0, that the proposed model has a fit that is the same as H_1. Based on this significance test, the model has quite a good fit. Another way of viewing this result is that we *cannot* reject the hypothesis that all residuals in $\boldsymbol{\Sigma} - \boldsymbol{\Sigma}(\boldsymbol{\theta})$ are zero.

There are several reasons for exercising caution in the use of the chi-square estimates. The chi-square approximation assumes that (1) $\mathbf{x}$ has no kurtosis, (2) the covariance matrix is analyzed, (3) the sample is sufficiently large, and (4) the H_0: $\boldsymbol{\Sigma} = \boldsymbol{\Sigma}(\boldsymbol{\theta})$ holds exactly. In practice, one or more of these assumptions are violated. Consider (1) first. Browne (1974, 1982) shows (see Chapter 4) that $(N-1)F_{\mathrm{ML}}$ and $(N-1)F_{\mathrm{GLS}}$ are chi-square estimators when $\mathbf{x}$ has no excessive kurtosis. The best-known distribution with no kurtosis is the multinormal In practice, nonnormal observed variables with excessive kurtosis do occur. Browne (1984, 81) suggests that leptokurtic ("more peaked" than normal) distributions result in too many rejections of H_0, and he speculates that platykurtic distributions will lead to too low of estimates of chi-square. Boomsma's (1983) simulation work suggests that high degrees of skewness lead to excessively large chi-square estimates, but it is unclear whether this is due to the kurtosis that often accompanies skewed variables.

Fortunately, research on nonnormal data and the alternatives to the usual chi-square estimators is developing rapidly. One line of research (e.g., Satorra and Bentler 1987) reveals some conditions where the usual chi-square test is robust under nonnormality even with excessive kurtosis. Another trend is the development of elliptical and arbitrary distribution estimators (Browne 1982, 1984) that provide alternatives to the usual F_{ML}, F_{GLS}, and F_{ULS} estimators and lead to accurate test statistics. Chi-square estimators insensitive to violations of the distributional assumptions of an estimator also are under development (Arminger and Schoenberg 1987). I discuss some of these options and tests for the multivariate kurtosis and skewness of the observed variables in Chapter 9. Suffice it to say here that when the observed variables are nonnormal with low or high kurtosis, there are alternatives to pursue for a correct chi-square estimator, but the usual ones may not be accurate.

The second factor that can influence the chi-square estimator is whether the covariance or correlation matrix is analyzed. The impact is subject to some dispute. Jöreskog and Sörbom (1986, I.39) state that the chi-square estimate is not accurate when a correlation matrix is analyzed. Based on his simulation work, Boomsma (1983, 112) states: "Under conditions of invariance standardizing the observed variables to a variance of one does not affect the chi-square estimate for goodness of fit." The "invariance condition" is violated when, for instance, factor loadings or error variances are constrained to be equal. This is true for the political democracy panel data because of the equality constraints for the 1965 and 1960 factor loadings. However, it is not true for the sympathy and anger CFA model. Analysis of either the correlation or covariance matrix of invariant models, such as the sympathy and anger one, leads to the same chi-square estimate in a given sample. The issue is whether this is an accurate estimator if the correlation matrix is analyzed across independent, random samples.

A third condition for $(N-1)F_{ML}$ to approximate a chi-square variate is that the sample be sufficiently large. Here again, much work remains to be done to determine the minimum necessary sample size. Boomsma's (1983, 119) simulation work suggests that the chi-square estimator $(N-1)F_{ML}$ is not accurate for samples smaller than 50 and recommends 100 or more cases. Anderson and Gerbing (1984) in their simulations also find the greatest departures of $(N-1)\ F_{ML}$ from a χ^2 variate in smaller samples (< 100). In both studies this departure from a χ^2 generally led to too frequent of rejections of H_0. This suggests that in small samples the chi-square test statistic tends to be too large. Another guideline for sample size is that the greater the number of free parameters in a model, the greater N should be. Though I know of no hard and fast rule, a useful suggestion is to have at least several cases per free parameter.

Finally, the fourth factor affecting the chi-square approximation is that it assumes that $\Sigma = \Sigma(\boldsymbol{\theta})$ is exactly true. In virtually all cases we do not expect to have a completely accurate description of reality. The goal is more modest. If the model that leads to $\Sigma(\boldsymbol{\theta})$ helps us to understand the relation between variables and does a "reasonable" job of matching the data, we may judge it as partially validated. The assumption that we have identified the exact process generating the data would not be accepted. Yet the chi-square test derives from a comparison of the hypothesized model H_0 to H_1, a model of perfect fit. A perfect fit may be an inappropriate standard, and a high chi-square estimate may indicate what we already know—that H_0 holds approximately, not perfectly.

The power (i.e., the probability of rejecting a false H_0) of the chi-square test to detect discrepancies between Σ and $\Sigma(\boldsymbol{\theta})$ partially depends on sample size. The estimator of chi-square is $(N-1)F_{ML}$ [or $(N-1)F_{GLS}$], so it is obvious that for the same value of F_{ML} (or F_{GLS}), the estimate of chi-square increases in direct proportion to $(N-1)$ and the power of the test increases as N increases. This situation is analogous to rejecting the null hypothesis of a zero correlation coefficient between two variables with a sample correlation of, say, 0.03 but an extremely large sample. The large sample may give us confidence that the population correlation is unlikely to be zero, but the substantive significance of a 0.03 correlation in most situations is nil. So it is with the chi-square test. A large sample may increase our confidence that the residual matrix $[\Sigma - \Sigma(\boldsymbol{\theta})]$ is not zero, but the substantive significance of the difference may be negligible. Alternatively, a small sample can reduce the power of the test so that with a small N we cannot detect even large differences between Σ and $\Sigma(\boldsymbol{\theta})$. In the next chapter I will show how to assess the statistical power of the chi-square test against specific alternative models.

In sum, the chi-square test has both advantages and disadvantages as a summary measure of model fit. When the sample size is fairly large, the covariance matrix is analyzed, the observed variables have no excessive kurtosis (e.g., when $\mathbf{x}$ is multinormal), and H_0 is true, then $(N-1)F_{ML}$ [or $(N-1)F_{GLS}$] is a good approximation to a chi-square variable suitable for tests of statistical significance. However, if one or more of the preceding conditions is violated, then the χ^2 test loses some of its value.

Given these limitations, does the estimated chi-square of 9.6 (8 df) for the sympathy and anger model represent an acceptable fit? As in any example with real data, the answer is complex. Like most survey items, the distributions of the observed variables is not a multinormal one. In addition the model is an approximation, so a perfect fit is unlikely and this would increase the chi-square estimate. The covariance instead of the correlation matrix is analyzed, so this is not an issue. Any adverse effect of the sample size of 138 is not obvious since it is neither conspicuously small nor large. Overall, I judge the low chi-square estimate to indicate a reasonable fit of the model $\boldsymbol{\Sigma} = \boldsymbol{\Sigma}(\boldsymbol{\theta})$. I also base this on the relatively small sample residuals discussed in the last section and additional measures of overall fit, which I describe next.

Earlier I said that all fit indices are functions that include $\mathbf{S}$ and $\hat{\boldsymbol{\Sigma}}$. The role of $\mathbf{S}$ and $\hat{\boldsymbol{\Sigma}}$ in the chi-square estimator follows since F_{ML} and F_{GLS} are functions of $\mathbf{S}$ and $\hat{\boldsymbol{\Sigma}}$ and the chi-square estimators are $(N-1)F_{\mathrm{ML}}$ and $(N-1)F_{\mathrm{GLS}}$. Thus the fitting functions provide scalar measures of the discrepancy between $\mathbf{S}$ and $\hat{\boldsymbol{\Sigma}}$. Some properties of F (where F is F_{ML}, F_{GLS}, or F_{ULS}) include (1) F's minimum is zero, (2) when H_0 is true and the distributional assumptions hold, F is inversely related to N with $F \to 0$ as $N \to \infty$, and (3) for a given sample and model, adding free parameters can never increase the value of F. By itself, a value of F is not easy to interpret. Multiplying F_{ML} or F_{GLS} by $(N-1)$ leads to a chi-square scale as I described earlier. The incremental fit indices that I treat next rescale F to provide *alternative* measures of the closeness of $\mathbf{S}$ to $\hat{\boldsymbol{\Sigma}}$.

Incremental Fit Indices

In this section I present several incremental fit indices. I begin with the normed fit index, Δ_1, proposed by Bentler and Bonett (1980):

$$\Delta_1 = \frac{F_b - F_m}{F_b} = \frac{\chi_b^2 - \chi_m^2}{\chi_b^2} \tag{7.52}$$

In (7.52) F_b is the fitting function value of a "baseline model" and F_m is the value of the fitting function for the maintained or hypothesized model. The alternative formula in χ^2's is sometimes computationally more convenient. It is equivalent to the other formula, with χ_b^2 being the chi-square estimate for the baseline model and χ_m^2 being that for the maintained model.

The baseline model is the simplest, most restrictive model that is a reasonable standard to which to compare the less restrictive maintained model. For example, in factor analysis a common baseline is one that suggests that no factors underlie the observed variables and that the covariances (or correlations) between observed indicators are zero in the population. The variances of the observed variables are not restricted. One way to obtain the χ_b^2 estimate for this uncorrelated variables baseline is to specify a model such that $q = n$, $\mathbf{x} = \boldsymbol{\xi}$, $\boldsymbol{\Theta}_\delta = \mathbf{0}$, $\boldsymbol{\Lambda}_x = \mathbf{I}$, and $\boldsymbol{\Phi}$ is a diagonal, free matrix. The F_b is $\chi_b^2/(N-1)$. Of course, other baseline models may be more appropriate (Cudeck and Browne 1983; Sobel and Bohrnstedt 1985).

A rationale for Δ_1 is as follows: A problem with F_m evaluated at $\mathbf{S}$ and $\hat{\boldsymbol{\Sigma}}$ is that its size cannot be easily judged since we have no maximum value to which to compare it. The normed index provides a maximum, F_b. The best possible fit is when F_m is zero, which leads to a Δ_1 of one. The worst fit is if the maintained model fits no better than the baseline model so that $F_m = F_b$ and Δ_1 is zero. So $0 \leq \Delta_1 \leq 1$ and the closer Δ_1 is to one, the better is the model fit. The Δ_1 measures the proportionate reduction in the fitting function or chi-square values when moving from the baseline to the maintained model. We also can view this as the "incremental" improvement in fit for the maintained model relative to the baseline one. Another property of Δ_1 is that $\text{plim}(\Delta_1) = c$, where $0 \leq c \leq 1$. This holds provided that $\text{plim}(F_b)$ and $\text{plim}(F_m)$ exist and are finite real values. When the model is valid, $c = 1$ so that $\text{plim}(\Delta_1) = 1$.

A limitation of Δ_1 is that it does not control for degrees of freedom. The value of F_m can be reduced by adding parameters. This is analogous to increasing the R^2 for a regression equation by including more explanatory variables. Though the R^2 may improve, the degrees of freedom decrease, and the model becomes more complex. An adjusted R^2 that corrects for degrees of freedom can reveal that a more parsimonious equation has a superior fit. Similarly, the χ_m^2 (or F_m) for a CFA with many estimated parameters and much complexity is never higher and nearly always lower than the χ_m^2's for more parsimonious models that contain subsets of these free parameters. The complex model would have a higher Δ_1 even though it has fewer degrees of freedom and may be "overfitting" the data.

Another influence is sample size. In general, we can distinguish two types of sample size effects. One is whether N enters the *calculation* of a fit measure. Specifically, for given values of F_m, df_m , F_b, and df_b, does changing the value of N alter the value of the fit index? Measures of fit whose calculated values are not changed by N are a useful way to rescale F_m and F_b (or χ_m^2 and χ_b^2) to a more interpretable metric than the original scales.

The other sample size effect is whether the *mean of the sampling distribution* of the fit index is associated with N. To understand this effect, imagine that we have the same model estimated in a large number of samples of varying sizes. If we choose one sample size, say, $N = 50$, we could form the sampling distribution of the fit index for all samples of size 50. Similarly, we could form the sampling distributions of the fit index for sample sizes of $51, 52, 53, \ldots,$ and for every sample size. For each N, we could compute a mean of the fit indices and then determine if these means are related to the size of the sample. If so, we say that the mean of the sampling distribution of the fit index is related to N.

Fit indices whose means are not associated with N are helpful when comparing models from samples with unequal N's. They also lead to fit measures whose average magnitude should be about the same for a valid model regardless of the sample size. None of the fit measures have both calculated values uninfluenced by N *and* means of their sampling distributions unrelated to N.

For $\Delta_1[= (F_b - F_m)/F_b]$, N does not enter the calculation of Δ_1. If F_b and F_m keep the same values, Δ_1 is the same for a sample of 100 as it is for one of 10,000. However, the key phrase is *if* F_b *and* F_m *keep the same values*. As I mentioned in the last section, the fitting function value is related to sample size. When the maintained model $\Sigma = \Sigma(\boldsymbol{\theta})$ is valid, the mean of the sampling distribution of F_m tends to grow smaller as N gets larger. Specifically, if $(N-1)F_m$ follows a chi-square distribution, the mean of F_m should approximate $\mathrm{df}_m/(N-1)$. The behavior of F_b depends on the choice of the baseline, but it seems likely that if the mean of the sampling distribution F_b declines with N, it does so at a slower rate than does the mean of F_m. This implies that the *mean of the sampling distribution* of Δ_1 is larger for bigger samples than it is for smaller ones. Thus comparing Δ_1 in two samples of different sizes can give the impression that the large sample has a better fit than the smaller one, even if the identical model holds for both samples. Bearden et al.'s (1982) simulations support this idea. Later, I report a small simulation that further corroborates the positive association between the mean of the sampling distributions of Δ_1 and N.

I propose a modification of Δ_1 that lessens the dependence of its mean on N and takes account of df_m (Bollen 1988):

$$\Delta_2 = \frac{F_b - F_m}{F_b - [\mathrm{df}_m/(N-1)]} = \frac{\chi_b^2 - \chi_m^2}{\chi_b^2 - \mathrm{df}_m} \tag{7.53}$$

The rationale for Δ_2 is as follows. The expected value of χ_m^2 is the standard of best fit. This equals df_m when the chi-square approximation holds for the maintained model. For a correct maintained model, the numerator, $(\chi_b^2 - \chi_m^2)$ is *on average* $(\bar{\chi}_b^2 - \mathrm{df}_m)$, where $\bar{\chi}_b^2$ is the "average" value of the baseline chi-square. In general, the χ_b^2 estimator does not follow a chi-square distribution since we know that the baseline is overly restrictive. Under some conditions it may follow a noncentral chi-square distribution (Steiger, Shapiro, and Browne 1985), but this is not essential to my argument. If on average $(\bar{\chi}_b^2 - \mathrm{df}_m)$ is what we expect for a correct model, this provides an alternative standard to which to compare $(\chi_b^2 - \chi_m^2)$. A correct model should have both of these equal on average. The χ_b^2 replaces $\bar{\chi}_b^2$ in the denominator of Δ_2 since it is the only estimator of $\bar{\chi}_b^2$ available.

The mean of the sampling distribution of Δ_2 should be less related to sample size than is Δ_1. The *calculation* of Δ_2 is influenced by N so that for given values of F_b, F_m, and df_m, Δ_2 is larger for small samples than for big ones. This kind of adjustment is needed to counteract the tendency for the mean of Δ_1 to be lower in small samples than in large. Also Δ_2 has an adjustment for df_m such that if two maintained models have the same χ_b^2 and χ_m^2, the one with the fewest parameters has the bigger Δ_2. Among the properties of Δ_2 for a correct maintained model are the following: (1) it should tend toward one for different sample sizes, (2) it is not confined to a zero to one range, (3) values of Δ_2 much greater than one may be due to overfitting data, and (4) when the numerator and denominator of Δ_2 are positive, $\Delta_1 < \Delta_2$. As N grows large the $[\mathrm{df}_m/(N-1)]$ term in (7.53) goes to zero, and Δ_2 comes closer to Δ_1 [i.e., $\mathrm{plim}(\Delta_1 - \Delta_2) = 0$]. Thus on average the biggest differences between Δ_1 and Δ_2 should be in small to moderate sample sizes with much smaller differences with big N's. Finally, the $\mathrm{plim}(\Delta_2) = c$, and c is one when the model is valid.

Another fit measure proposed in Bollen (1986) is

$$\rho_1 = \frac{(F_b/\mathrm{df}_b) - (F_m/\mathrm{df}_m)}{(F_b/\mathrm{df}_b)}$$

$$= \frac{(\chi_b^2/\mathrm{df}_b) - (\chi_m^2/\mathrm{df}_m)}{(\chi_b^2/\mathrm{df}_b)} \tag{7.54}$$

The ρ_1 index takes the F (or χ^2) values in Δ_1 and divides them by their respective degrees of freedom. The rationale for ρ_1 is similar to that of Δ_1. For Δ_1, the comparison is the fit of the maintained model relative to the baseline model. For ρ_1, the comparison is one of the fit *per* degrees of freedom for the baseline and maintained model. Since introducing additional parameters lowers df, it is possible for ρ_1 to stay the same or to decrease for more complex specifications. Maintained models that have a lower fitting function value with relatively few parameters have higher ρ_1 values than models with the same fit value for a more complicated specification. The lowest possible value for F_m is still zero, so the maximum of ρ_1 is one. It is possible that (χ_m^2/df_m) exceeds (χ_b^2/df_b) and that ρ_1 is less than zero, but this is unlikely in practice. The ρ_1 is like Δ_1 in that its *calculated* value is unaltered by changing N as long as F_m, df_m, F_b, and df_b are the same. But the mean of the sampling distribution of ρ_1 increases with N. Provided the $\text{plim}(F_m)$ and $\text{plim}(F_b)$ exist and are finite, $\text{plim}(\rho_1) = d$, where $d \leq 1$. If the maintained model is valid, $\text{plim}(\rho_1) = 1$.

An index of fit that predates ρ_1 and lessens the dependency of the mean of ρ_1 on N was proposed by Tucker and Lewis (1973) and Bentler and Bonett (1980):

$$\rho_2 = \frac{F_b/\text{df}_b - F_m/\text{df}_m}{F_b/\text{df}_b - [1/(N-1)]}$$

$$= \frac{\chi_b^2/\text{df}_b - \chi_m^2/\text{df}_m}{\left(\chi_b^2/\text{df}_b\right) - 1} \tag{7.55}$$

Like ρ_1, ρ_2 corrects for the df of the baseline and maintained models. Its only difference is in the denominator. The justification for ρ_2 differs from ρ_1 in the definitions of the "best possible fitting model." For ρ_1, the best fit for the maintained model is zero, the minimum value for F_m or χ_m^2. At this minimum ρ_1 is one, its maximum value. For ρ_2, the best fit is defined as the *expected value of* (χ_m^2/df_m). This equals one when the assumptions underlying the chi-square approximation are satisfied for the maintained model, since the expected value of a chi-square variate is its df. The denominator of ρ_2 contrasts the worst fit (χ_b^2/df_b) to this new definition of best fit (1), leading to $(\chi_b^2/\text{df}_b - 1)$. The numerator of ρ_2 compares the worst fit to the sample fit for the maintained model, and it is $(\chi_b^2/\text{df}_b - \chi_m^2/\text{df}_m)$. When χ_m^2/df_m is one, ρ_2 is one, and this is an ideal fit. Values of ρ_2 much lower than one may indicate misspecification, whereas ρ_2 values much greater than one may signify overfitting.

Tucker and Lewis (1973) suggest that ρ_2 should tend toward one for a valid model regardless of sample size. A simulation study by Anderson and Gerbing (1984) finds that the means of the sampling distributions of ρ_2 are only weakly related to sample size and that they do tend toward one for a valid model. Also in their study ρ_2 had considerable sampling variability and a tendency to produce extreme values. Other properties of ρ_2 are that it can be less than zero or greater than one, and in the typical case where ρ_2's numerator and denominator are positive, $\rho_2 > \rho_1$. The $\text{plim}(\rho_2) = d$, and $\text{plim}(\rho_1 - \rho_2)$ is zero. This implies that the differences between ρ_1 and ρ_2 grow smaller as N grows larger. If the maintained model is valid, both ρ_1 and ρ_2 converge to one.

There is no unambiguous answer to how large Δ_1, Δ_2, ρ_1, or ρ_2 must be to indicate an "adequate" fit. Bentler and Bonett (1980, 600) suggest that for Δ_1 and ρ_2: "In our experience models with overall fit indices of less than 0.9 can usually be improved substantially." Although this provides a rough guideline, several factors can affect the incremental fit measures that may lead to other cutoffs. In small samples Δ_1 and ρ_1 may provide an overly pessimistic assessment of model fit. The Δ_2 and ρ_2 are less subject to this problem. Other factors affecting their magnitude include: (1) the choice of baseline, (2) the standards set by prior work, (3) the relations between the fit indices, and (4) the selection of the fitting function.

The choice of the baseline affects all four measures. The more restrictive the baseline model is, the better is the fit of a given maintained model relative to the baseline. This tends to inflate the incremental fit indices compared to what they would be for a less restrictive baseline. For example, two researchers analyzing identical data with an identical two-factor model will have different Δ's or ρ's if one chooses a no-factor baseline and the other chooses a one-factor baseline model.

One way to lessen this ambiguity is to always report the fit with a no-factor baseline *along with* these indices with any other appropriate baselines. This recommendation is analogous to the routine practice of reporting the R^2 for a regression equation. The "baseline model" in regression is that all explanatory variables have a zero slope so that the mean of the dependent variables is the best predictor. This is an overly restrictive "baseline," but it provides a useful standard. Similarly, reporting Δ_1, Δ_2, ρ_1, and ρ_2 with a no-factor baseline can provide a standard of comparison, even when alternative fit measures with different baselines are also calculated.

A second factor influencing the choice of cutoffs for fit indices is the standards set by prior work. If other analyses of the same or similar variables with the same baseline, typically report Δ's or ρ's of 0.95 or higher, then a 0.85 or 0.90 may be unacceptable. Alternatively, if models in another research area have fit indices typically below 0.80, values of 0.85 or higher can represent an important improvement over existing work. Also the "reasonableness" of the model, its contribution to a substantive field, and its fit as measured by the other empirical means make it difficult to require a specific cutoff for "good" or "bad" fits.

The differences between the incremental fit measures can lead to separate cutoffs for each. For instance, consider the relation between ρ_1 and Δ_1. Algebraically, manipulate the ρ_1 formula so that

$$\rho_1 = \frac{\chi_b^2 - (\mathrm{df}_b/\mathrm{df}_m)\chi_m^2}{\chi_b^2} \tag{7.56}$$

Equation (7.56) shows that ρ_1 is the same as Δ_1 except that χ_m^2 is multiplied by the ratio $\mathrm{df}_b/\mathrm{df}_m$. The degrees of freedom for the baseline model are virtually always greater than that for the maintained model, so $(\mathrm{df}_b/\mathrm{df}_m) > 1$, and the numerator for ρ_1 is smaller than that for Δ_1. It follows that $\rho_1 < \Delta_1$. The extent to which it is lower depends on the chi-square estimate of the baseline model and the relative number of parameters in the baseline compared to the maintained model.

Finally, the selection of the fitting function must be accounted for when judging the magnitude of the incremental fit indices. Generally, F_b and F_m from GLS estimation need not equal the corresponding quantities from ML estimation, even for the identical model and sample. The same can be said for the other fitting functions I introduce in Chapter 9. This means that the Δ's and ρ's magnitudes may differ across estimators. A related point is that the rationale for Δ_2 and ρ_2 is based on $(N-1)F_m$ approximating a chi-square variate. In a case like ULS this is not true so that the properties of Δ_2 and ρ_2 may change. Overall, selecting a rigid cutoff for the incremental fit indices is like selecting a minimum R^2 for a regression equation. Any value will be controversial. Awareness of the factors affecting the Δ's and ρ's and good judgment are the best guides to evaluating their size.

I illustrate these fit indices for the sympathy and anger CFA example. The chi-square estimate for the maintained model is $\chi_m^2 = 9.6$ and $\mathrm{df}_m = 8$, and for the no-factor baseline, it is $\chi_b^2 = 469.8$ and $\mathrm{df}_b = 15$. The Δ_1 and ρ_1 are

$$\hat{\Delta}_1 = \frac{469.8 - 9.6}{469.8}, \qquad \hat{\rho}_1 = \frac{(469.8/15) - (9.6/8)}{(469.8/15)}$$

$$= 0.98 \qquad\qquad = 0.96 \tag{7.57}$$

Since these statistics have a ceiling of one, these estimates indicate an excellent fit. To adjust for sample size effects, I also calculate Δ_2 and ρ_2:

$$\hat{\Delta}_2 = \frac{469.8 - 9.6}{469.8 - 8}, \qquad \hat{\rho}_2 = \frac{(469.8/15) - (9.6/8)}{(469.8/15) - 1}$$

$$= 1.00 \qquad\qquad = 0.99 \tag{7.58}$$

Although $\hat{\Delta}_2$ and $\hat{\rho}_2$ can exceed 1.00, values this high represent an exceptional fit.

All four fit indices have an even more general form. For instance, Bentler and Bonett's (1980) Δ_1 is

$$\Delta_1 = \frac{\chi_r^2 - \chi_m^2}{\chi_b^2} \tag{7.59}$$

where χ_r^2 is a model more restrictive than the maintained model but less restrictive than the baseline one. It measures the improvement in fit for the maintained model (χ_m^2) relative to the fit of the restrictive model (χ_r^2) as a proportion of the fit of the baseline model (χ_b^2). Analogous formulas for Δ_2, ρ_1, and ρ_2 are straightforward to write.

Additional Measures of Overall Model Fit

Researchers also use several other ways to calibrate the match of $\mathbf{S}$ to $\hat{\boldsymbol{\Sigma}}$. Jöreskog and Sörbom (1986) propose a Goodness of Fit Index (GFI) and an Adjusted GFI for models fitted with F_{ML}:

$$\text{GFI}_{\text{ML}} = 1 - \frac{\operatorname{tr}\left[(\hat{\boldsymbol{\Sigma}}^{-1}\mathbf{S} - \mathbf{I})^2\right]}{\operatorname{tr}\left[(\hat{\boldsymbol{\Sigma}}^{-1}\mathbf{S})^2\right]} \tag{7.60}$$

$$\text{AGFI}_{\text{ML}} = 1 - \left[\frac{q(q+1)}{2\,\text{df}}\right]\left[1 - \text{GFI}_{\text{ML}}\right] \tag{7.61}$$

The GFI_{ML} measures the relative amount of the variances and covariances in $\mathbf{S}$ that are predicted by $\hat{\boldsymbol{\Sigma}}$. The AGFI_{ML} adjusts for the degrees of freedom of a model relative to the number of variables. For given values of GFI_{ML} and q, AGFI_{ML} rewards simpler models with fewer parameters. Both indices reach their maximum of one when $\mathbf{S} = \hat{\boldsymbol{\Sigma}}$. They typically are greater than zero, though it is possible for them to be negative.

Jöreskog and Sörbom (1986) propose analogous measures for models fitted with F_{ULS}:

$$\text{GFI}_{\text{ULS}} = 1 - \frac{\text{tr}\left[(\mathbf{S} - \hat{\boldsymbol{\Sigma}})^2\right]}{\text{tr}(\mathbf{S}^2)} \tag{7.62}$$

$$\text{AGFI}_{\text{ULS}} = 1 - \left[\frac{q(q+1)}{2\,\text{df}}\right][1 - \text{GFI}_{\text{ULS}}] \tag{7.63}$$

whereas Tanaka and Huba (1985) propose GLS versions:

$$\text{GFI}_{\text{GLS}} = 1 - \frac{\text{tr}\left[(\mathbf{I} - \hat{\boldsymbol{\Sigma}}\mathbf{S}^{-1})^2\right]}{q} \tag{7.64}$$

$$\text{AGFI}_{\text{GLS}} = 1 - \left[\frac{q(q+1)}{2\,\text{df}}\right][1 - \text{GFI}_{\text{GLS}}] \tag{7.65}$$

With regard to the two types of sample size influences, the *calculations* of the GFIs and AGFIs are not affected by N, but Anderson and Gerbing's (1984) simulation study suggest that the means of the sampling distributions of GFI_{ML} and the AGFI_{ML} tend to increase as sample size increases. They also find that these values decrease as the number of indicators per factor, or the number of factors, increase, especially for smaller sample sizes.

Hoelter (1983) proposes a Critical N (CN) statistic:[14]

$$\text{CN} = \frac{\text{Critical } \chi^2}{F} + 1 \tag{7.66}$$

where the critical χ^2 is the critical value of the chi-square distribution with df equal to the maintained model's df and at a selected alpha level (e.g., 0.05). The F is the value of F_{ML}, or F_{GLS}, at $\mathbf{S}$ and $\hat{\boldsymbol{\Sigma}}$. CN gives the sample size at which the F value would lead to the rejection of H_0 [i.e., $\boldsymbol{\Sigma} = \boldsymbol{\Sigma}(\boldsymbol{\theta})$] at a chosen alpha level. Hoelter (1983) suggests a tentative cutoff of $\text{CN} \geq 200$ and illustrates this measure with large and small samples. Since N does not enter the formula for CN, its *calculated* value is the same for a given critical χ^2 and F value for all sample sizes. However, when H_0 is valid, F goes to zero as N grows larger and the *mean of the sampling distribution* of CN increases with N. Bollen and Liang (1988) illustrate that

[14] Hoelter (1983, 330–331, 341) uses a formula that involves an approximation for a χ^2 variate when the df are very large. For the more common case where df do not exceed the tabulated χ^2's, (7.66) is more accurate (see Matsueda and Bielby 1986).

this positive association also can occur for some misspecified models. Thus CN may lead to an overly pessimistic assessment of fit for small samples.

A rather ad hoc fit measure is the chi-square estimator divided by its degrees of freedom. Its justification appears to be that the expected value of a chi-square variate is its degrees of freedom. So χ^2/df estimates how many times larger the chi-square estimate is than its expected value when $(N-1)F_{\text{ML}}$ or $(N-1)F_{\text{GLS}}$ approximate chi-square variates. There is no consensus on what represents a "good" fit, with recommendations ranging from ratios of 3, 2, or less (Carmines and McIver 1981) to as high as 5. Some researchers have the impression that χ^2/df lessens the problems of the possible excessive power of the usual chi-square test when N is large. However, since the chi-square ratios are $(N-1)F_{\text{ML}}/\text{df}$ or $(N-1)F_{\text{GLS}}/\text{df}$, they have a similar relation to N as do the usual chi-square estimators and do not correct the excessive power problem.

An estimator for a standardized chi-square is

$$\text{Standardized } \chi^2 = \frac{\chi^2 - \text{df}}{\sqrt{2\,\text{df}}} \tag{7.67}$$

This follows since the variance of a χ^2 variate is $2\,\text{df}$, so its standard deviation is $\sqrt{2\,\text{df}}$. The numerator is the difference between the chi-square estimator and its mean. Thus the standardized χ^2 measures the deviation of the chi-square estimator from its expected value in standard deviation units. Its advantage over χ^2/df is that it controls for the standard deviation of the chi-square variate. But it suffers the same limitations in that it has an ambiguous cutoff point for a "good fit" and in that it has the same relation to sample size as χ^2/df.

Cudeck and Browne (1983) advocate a cross-validation method to assess overall model fit. It begins by randomly splitting a sample in half and forming two sample covariance matrices, $\mathbf{S}_1$ and $\mathbf{S}_2$. Then a model is fitted to $\mathbf{S}_1$ in the usual way, resulting in an estimate of the implied covariance matrix, $\hat{\mathbf{\Sigma}}_1$. The cross-validation step does not involve estimation of a new model but, instead, calculates F when $\mathbf{S}_2$ and $\hat{\mathbf{\Sigma}}_1$ are substituted for $\mathbf{S}$ and $\mathbf{\Sigma}$ in the fitting function. This procedure is repeated for several models. The model with the smallest value of F in the validation half of the sample has the best fit. Optionally, the technique can be done again by reversing the roles of the first and second half samples. That is, the second half becomes the fitted sample, and the first half the validation one. In a small simulation study Cudeck and Browne (1983) note that with small N's, cross-validation tends to favor models with few parameters, whereas with large N's models with many parameters tend to cross-validate the best.

Two alternative indices of fit for comparing models do not require split samples. The first, based on Akaike (1974), is

$$A = \chi_m^2 + 2t \tag{7.68}$$

where t is the number of free parameters and χ_m^2 is the value of the chi-square estimator for a maintained model. The second derives from Schwarz's (1978) modification of Akaike's criterion and is

$$B = \chi_m^2 + t\log(N) \tag{7.69}$$

The recommendation is to choose that model with the smallest A or B value. Both indices penalize models with many rather than few free parameters, and B also weights the free parameters by the log of sample size. In Cudeck and Browne's (1983) simulations they find that compared to cross-validation procedures, B tends to reach a minimum for models with fewer free parameters. The A values tend to be minimized with more free parameters than B. It is not known whether these results generalize beyond Cudeck and Browne's simulations.

Empirical and Simulation Examples

Below are the estimates of most of the overall fit measures for the sympathy and anger CFA models:

Fit Index	χ^2 (df = 8)	Δ_1	Δ_2	ρ_1	ρ_2	GFI	AGFI	CN ($\alpha = 0.05$)	χ^2/df	Standardized χ^2
Estimate	9.6	0.98	1.00	0.96	0.99	0.98	0.94	222	1.2	0.4

All measures indicate an excellent match of $\hat{\mathbf{\Sigma}}$ to $\mathbf{S}$. The only measure even close to a recommended cutoff is Hoelter's CN. As I mentioned, any fixed cutoff for CN often favors large samples. Given the small to moderate $N(= 138)$ for this sample, it is impressive that a potentially high cutoff is satisfied.

As a final illustration of the overall fit measures, I use a small Monte Carlo simulation example with two factors and five indicators. Table 7.5 reports the major overall fit measures for a model with the following population structure:

$$\mathbf{\Lambda}_x = \begin{bmatrix} 1 & 0 \\ 1 & 0 \\ 1 & 0 \\ 0 & 1 \\ 0 & 1 \end{bmatrix}, \qquad \mathbf{\Phi} = \begin{bmatrix} 1 & \\ 1 & 2 \end{bmatrix} \tag{7.70}$$

$$\operatorname{diag} \mathbf{\Theta}_\delta = \begin{bmatrix} 1 & 1 & 1 & 1 & 1 \end{bmatrix} \tag{7.71}$$

Table 7.5 Simulation Results for CFA Model for Three Sample Sizes (75, 150, 300) and 14 Replications for Each Sample Size

Overall Fit Measure	Mean Value (Standard Deviation) N		
	75	150	300
$\chi^2(\mathrm{df} = 4)$	3.77	3.48	3.87
	(2.12)	(2.57)	(4.11)
Δ_1	0.967	0.985	0.992
	(0.018)	(0.012)	(0.008)
Δ_2	1.003	1.002	1.000
	(0.019)	(0.012)	(0.008)
ρ_1	0.918	0.963	0.980
	(0.046)	(0.030)	(0.020)
ρ_2	1.008	1.005	1.001
	(0.052)	(0.030)	(0.021)
GFI	0.983	0.990	0.995
	(0.010)	(0.007)	(0.005)
AGFI	0.928	0.963	0.981
	(0.037)	(0.024)	(0.018)
CN	269.8	749.2	1628.5
($\alpha = 0.05$)	(182.5)	(583.8)	(1276.1)

All observed variables are normally distributed. The simulations are for three different sample sizes (75, 150, 300) and have 14 replications for each sample size.[15] The table reports the means and standard deviations for each fit index for the F_{ML} estimator. The results are largely consistent with those predicted. The mean chi-square estimate shows no trend across the different sample sizes, though it is slightly less than its expected value of four (df = 4). Since the chi-square estimator is $(N - 1)F_{\mathrm{ML}}$, these results suggest that as N increases, F_{ML} decreases such that the product $(N - 1)F_{\mathrm{ML}}$ approximates a chi-square variate for each sample size. Had the incorrect model been fitted, then the average chi-square estimate would increase with N.

The upward trend in the mean of Δ_1 and ρ_1 as N increases is consistent with the earlier arguments. The mean of the sampling distributions of Δ_2 and ρ_2 measures are near one for all sample sizes. A slight decrease in the third decimal place occurs as N grows but seems negligible. Table 7.5 also

[15]The simulation used the random number generator function NORMAL from SAS to create the variables with this structure. My thanks to the participants in the seminar "Structural Equations with Latent Variables" who helped to create the simulation data.

illustrates the positive association of the means of the sampling distributions for GFI, AGFI, and CN with N.

A notable characteristic of the standard deviations (s.d.) is that for CN the s.d. increases as N grows (Bollen and Liang 1988), whereas the others (except χ^2) decrease with N. Of the incremental fit measures, Δ_1 and Δ_2 have s.d.'s less than half that of ρ_1 and ρ_2. The smallest s.d. is for GFI. AGFI's s.d. is nearly four times as large as GFI's.

Results from Bollen and Stine (1987) suggest that although GFI has the lowest standard deviation of the fit indices, it can maintain high values (> 0.9) even with seriously misspecified models. In these simulations Δ_1 and Δ_2 were more likely to indicate a decline in fit under misspecification, while maintaining relatively small s.d.'s. The ρ_1 and ρ_2 showed bigger declines in fit, but their s.d.'s were larger than those for Δ_1 and Δ_2.

Summary of Overall Fit Measures

We can measure the closeness of $\mathbf{S}$ to $\hat{\boldsymbol{\Sigma}}$ in many ways. The one most amenable to tests of statistical significance is the chi-square estimator where the null hypothesis is $\boldsymbol{\Sigma} = \boldsymbol{\Sigma}(\boldsymbol{\theta})$. The other overall fit measures provide largely descriptive measures of fit since their distributions are unknown. The previous subsections presented some characteristics of these fit indices, but much remains to be discovered. At this stage the safest recommendation is to always report the chi-square estimate along with several of the other fit indices (e.g., residuals, Δ_1, Δ_2, ρ_1, and ρ_2).

Component Fit Measures

Nonsense results for individual parameters can occur in conjunction with good overall fit measures, and these would be missed if a researcher only examined the overall fit. Moreover summary measures are not always applicable. For exactly identified models, the $\hat{\boldsymbol{\Sigma}}$ matrix equals $\mathbf{S}$. The χ^2 in this situation is zero. The GFI and Δ_1 for such a model equal one. With a df of zero for exactly identified models, the AGFI, ρ_1, ρ_2, CN, χ^2/df, and standardized χ^2 are not defined since a zero appears in the denominators of their formulas. Thus an examination of the components of the model is essential.

Parameter Estimates

Analysts typically have some expectations about the sign and sometimes the magnitude of the unknowns in $\boldsymbol{\Lambda}_x$, $\boldsymbol{\Phi}$, and $\boldsymbol{\Theta}_\delta$. A first step in examining parameter estimates is to see if they "make sense." Standardized coefficients such as $\lambda_{ij}[\phi_{jj}/\mathrm{VAR}(x_i)]^{1/2}$ may be useful in comparing relative effects when the observed or latent variables have very different units of measurement. If the unstandardized or standardized parameter estimates deviate far from expectations, a hypothesis may be faulty, or more broadly, the model may be misspecified. Other signs of misspecification appear when the parameter estimates are improper solutions. Improper solutions refers to sample estimates that take values that are impossible in the population. One common case is when estimates of variances are negative (called "Heywood cases"). Another occurs if the correlation of two variables is greater than one.

Improper solutions can be caused by several factors. The population parameter may be a value that is acceptable but close to the boundary of admissible values. For instance, a population correlation may be near one, or an error variance may be near zero. In these situations a sample estimate may assume an inadmissible value due to sampling fluctuations. A rough check for this possibility is to test whether the sample estimate differs statistically from a value of the population parameter that is acceptable (e.g., a correlation less than one or an error variance greater than or equal to zero).[16] If it is not very different from admissible values, the issue becomes whether such values are "reasonable" for the given parameter. For instance, does it "make sense" for the error variance of a measure to be near zero? This means that the indicator is a "nearly perfect" measure of the latent variable(s) (ignoring any differences in scaling). Or, is it likely that two or more factors have a perfect correlation? If researchers cannot answer questions such as these in the affirmative, or if they reject the null hypothesis that the estimates represent population parameters that take admissible values, then at least three other explanations may account for such results.

First, the analyst may have been extremely unlucky with this sample and in a more "typical" one the estimates would be acceptable. Second, the covariance (correlation) matrix analyzed may have outliers or influential observations that lead to distorted measures of association for the observed variables, which in turn affect the parameter estimates. The cloud cover data introduced in Chapter 2 illustrate this (Bollen 1987). Recall that the three variables, x_1 to x_3, represent three judges' estimates of the percent of cloud cover for the identical view of the sky at different times and days ($N = 60$). Screening the data with procedures described in Chapter 2, I found three outliers. The sample covariance matrices with (**S**) and without

[16]This is a rough check since the calculation of standard errors assumes that the parameter estimates are at admissible values.

($\mathbf{S}_{(i)}$) the outliers are

$$\mathbf{S} = \begin{bmatrix} 1301 & & \\ 1020 & 1463 & \\ 1237 & 1200 & 1404 \end{bmatrix}$$

$$\mathbf{S}_{(i)} = \begin{bmatrix} 1129 & & \\ 1170 & 1494 & \\ 1149 & 1313 & 1347 \end{bmatrix}$$

Assume that a single latent variable, perceived cloud cover, underlies the three estimates. The one-factor rule establishes model identification, provided that I scale the latent variable (e.g., $\lambda_{11} = 1$). Substituting the elements of $\mathbf{S}$ in (7.20), the parameter estimates are

$$\hat{\Lambda}'_x = [1.00 \quad 0.97 \quad 1.18]$$

$$\hat{\phi} = [1052]$$

$$\text{diag}\, \hat{\Theta}_\delta = [249 \quad 474 \quad -51]$$

The var(δ_3) is -51. We could react to this value in several ways. One is to re-estimate the model with an inequality constraint on the diag $\hat{\Theta}_\delta$ so that none of the error variances are negative. Bentler's (1985) EQS program automatically imposes this restriction. Another strategy is to drop the x_3 variable, which has the negative error variance. Finally, we could ignore the negative value and consider it essentially zero. Each of these strategies has its limits. An inequality constraint may mask a more serious problem with the data. Dropping the third indicator leads to an underidentified model, unless additional restrictions are added. Most important, none of these alternatives eliminates the cause of the problem. The outliers have created the Heywood case as is demonstrated when the elements of $\mathbf{S}_{(i)}$ are substituted in (7.20). The new estimates of the parameters are

$$\hat{\Lambda}'_x = [1.00 \quad 1.14 \quad 1.12]$$

$$\hat{\phi} = [1023]$$

$$\text{diag}\, \hat{\Theta}_\delta = [106 \quad 157 \quad 58]$$

Dropping the outliers eliminates the negative error variance, and all estimates now seem reasonable. As this example shows, it is wise to check for outliers before using a covariance matrix, particularly when the number of observations is small. Of course, the cause of the outliers should be sought.

A third cause of improper solutions is a fundamental fault of specification in the model. As such the model requires reconstruction. There is no formula for how to respecify a model. This must depend on the researcher's substantive knowledge and is specific to each problem.

Boomsma's (1982) and Anderson and Gerbing's (1984) simulation work found that for correctly specified models, implausible values were most likely under some of the same conditions that led to nonconvergence. Specifically, small sample sizes and two indicators per factor frequently led to negative error variances. Anderson and Gerbing (1984, 171) recommend sample sizes of 150 or more and three or more indicators per factor to lessen the chances of these improper values.

A related question in evaluating parameter estimates is the small sample behavior of $\hat{\theta}$. We know that $\hat{\theta}$ from F_{ML}, F_{GLS}, or F_{ULS} is a consistent estimator, but this does not mean that $\hat{\theta}$ works well for small samples. I turn again to Boomsma (1983) and Gerbing and Anderson (1985) for simulation evidence on this issue. Boomsma (1983) generally finds low degrees of bias in estimates of parameters from F_{ML}, except with relatively small samples and low indicator reliability. Boomsma suggests that the bias is negligible for N greater than 100. Gerbing and Anderson's (1985) simulations with N's from 50 to 300 find that $\hat{\theta}$ from F_{ML} is unbiased, with the only exception being in a model with only two indicators per factor. Thus the simulation evidence to date suggests a largely unbiased estimator for most sample sizes.

Table 7.6 reports the ML estimates of parameters for the model of sympathy and anger. The λ_{11} to λ_{31} coefficients and the VAR(δ_1) to VAR(δ_3) correspond to the three measures of sympathy and λ_{42} to λ_{62} and VAR(δ_4) to VAR(δ_6) correspond to the indicators of anger. The ϕ_{11} represents the variance of the latent sympathy variable, ϕ_{22} is the variance of the latent anger variable, and ϕ_{21} is their covariance. The λ_{11} and λ_{42} coefficient are fixed at 1.00 to establish scales for ξ_1 and ξ_2, respectively. The other parameters are free and estimated. There is no evidence of Heywood cases. The variances of sympathy and anger are 6.08 and 4.32, and their covariance is negative ($= -2.22$). The error variances range from 0.90 to 2.48. As expected, all λ_{ij}'s are positive, and the covariance of sympathy and anger is negative. The indicators of anger have loadings on the anger latent variable that are closer to one than are the indicators for

Table 7.6 Parameter Estimates of Sympathy and Anger CFA with Two Factors and Six Indicators ($N = 138$)

Parameter	ML Estimate (standard error)
λ_{11}	1.00^c
	(—)
λ_{21}	0.77
	(0.07)
λ_{31}	0.72
	(0.07)
λ_{42}	1.00^c
	(—)
λ_{52}	0.92
	(0.08)
λ_{62}	0.90
	(0.08)
ϕ_{11}	6.08
	(0.92)
ϕ_{21}	−2.22
	(0.53)
ϕ_{22}	4.32
	(0.70)
VAR(δ_1)	0.90
	(0.39)
VAR(δ_2)	2.48
	(0.38)
VAR(δ_3)	1.92
	(0.31)
VAR(δ_4)	1.25
	(0.28)
VAR(δ_5)	1.69
	(0.29)
VAR(δ_6)	1.45
	(0.26)
$\chi^2 = 9.6$	df = 8

Note: c = constrained parameter.

the sympathy indicators. Overall, none of the estimates seem unusual, nor do they suggest any problems.

Asymptotic Standard Errors

As discussed in Chapter 4, the variance-covariance matrix of the ML estimators is derived from the inverse of the information matrix. The main diagonal of the covariance matrix contains the asymptotic variances of the parameter estimators. When the sample estimates are substituted for the unknown parameters, the square root of the main diagonal of the asymptotic covariance matrix provides estimated asymptotic standard errors. If the assumptions of the ML method are valid, the asymptotic standard errors may be used to test whether the population parameters equal certain constants. If $\hat{\theta}_i$ denotes a parameter estimator, with θ_i the hypothesized value of the parameter and $S_{\hat{\theta}_i}$ the asymptotic standard error of the estimator, then a test statistic is formed as

$$\frac{\hat{\theta}_i - \theta_i}{S_{\hat{\theta}_i}} \tag{7.72}$$

The use of this is the same as in Chapter 4.

The simulation studies of Boomsma (1983) and Gerbing and Anderson (1985) also evaluated the estimated standard errors from the ML method compared to the observed standard deviations of the parameter estimates for simulated data. Both find that the estimates of standard errors are close to the actual standard deviations under most conditions. The sampling variability of parameter estimates for small samples is considerable, but the standard errors reflect this.

Table 7.6 presents estimated asymptotic standard errors for the model of sympathy and anger, where the standard errors are in parentheses beneath the parameter estimates. All the parameter estimates are considerably larger than their standard errors. Thus the ratios of the estimates to their standard errors do not suggest problems with the fit of the individual components.

A frequent situation where the standard errors are not accurate is when the correlation rather than the covariance matrix is analyzed. After studying this problem, Boomsma (1983, 113) concludes that "the analysis of correlation matrices leads to unprecise estimates of the variances and covariances of parameter estimates. Therefore, even when the sample size is as large as 400, the results of a LISREL analysis may be seriously distorted. For this reason it is not recommended to analyze correlation matrices." Lawley and Maxwell (1971), Jennrich and Thayer (1973), and Browne (1982) suggest corrections for the standard errors when correlations or standardized coefficients are analyzed.

Asymptotic Correlation Matrix of Parameter Estimates

A transformation of the estimated asymptotic variance-covariance matrix of the parameter estimates provides the asymptotic correlation matrix of the estimated parameters. The transformation draws on the definition of the correlation as the covariance of two variables divided by the product of their standard deviations:

$$\text{asym } \hat{\rho}_{\hat{\theta}_i \hat{\theta}_j} = \frac{\text{acov}(\hat{\theta}_i, \hat{\theta}_j)}{\sqrt{\text{avar}(\hat{\theta}_i)\text{avar}(\hat{\theta}_j)}} \tag{7.73}$$

where

$$\hat{\theta}_i = \text{the } i\text{th estimated parameter}$$

$$\hat{\theta}_j = \text{the } j\text{th estimated parameter}$$

If (7.73) is applied to all parameter estimates, the estimated correlation matrix results. This matrix is analogous to the estimated correlation of regression coefficient estimates available in regression analysis. (These are estimated asymptotic correlations of estimates, not correlations of the explanatory variables in an equation.) Correlations that are extremely large indicate that the estimates of the two-parameter estimators are very closely associated. High correlations sometimes are a symptom of severe collinearity.

The estimated asymptotic correlation matrix of the parameter estimates for the $\hat{\lambda}_{ij}$'s and $\hat{\phi}_{ij}$ of the sympathy and anger model is in Table 7.7. None of the correlations is very large. The biggest ones are those between $\hat{\phi}_{21}$ and $\hat{\phi}_{22}$ $(= -0.516)$ and $\hat{\phi}_{21}$ and $\hat{\phi}_{11}$ $(= -0.490)$. This shows a moderately negative association between the estimates of the variances of the latent variables and their covariance. A few other moderately sized elements also are present, but these results do not indicate any correlation sufficiently high to suggest serious problems.

The warning about analyzing correlation matrices applies here. The asymptotic correlation matrix is accurate for the analysis of covariance matrices but not for the analysis of correlation matrices.

Table 7.7 Asymptotic Correlation Matrix of Parameter Estimates for Sympathy and Anger Model

	$\hat{\lambda}_{21}$	$\hat{\lambda}_{31}$	$\hat{\lambda}_{52}$	$\hat{\lambda}_{62}$	$\hat{\phi}_{11}$	$\hat{\phi}_{21}$	$\hat{\phi}_{22}$
$\hat{\lambda}_{21}$	1.000						
$\hat{\lambda}_{31}$	0.399	1.000					
$\hat{\lambda}_{52}$	−0.000	−0.000	1.000				
$\hat{\lambda}_{62}$	0.000	−0.000	0.452	1.000			
$\hat{\phi}_{11}$	−0.365	−0.380	0.000	0.000	1.000		
$\hat{\phi}_{21}$	0.085	0.089	0.146	0.151	−0.490	1.000	
$\hat{\phi}_{22}$	0.000	0.000	−0.447	−0.464	0.113	−0.516	1.000

$R^2_{x_i}$ *for Observed Variables*

Another measure of component fit is the squared multiple correlation coefficient for each x_i variable. It is calculated as

$$R^2_{x_i} = 1 - \frac{\operatorname{var}(\delta_i)}{\hat{\sigma}_{ii}} \tag{7.74}$$

where $\hat{\sigma}_{ii}$ = variance of x_i predicted by the model. The $R^2_{x_i}$ is analogous to the squared multiple correlation coefficient, with x_i as the dependent variable and the latent variables (ξ) as the explanatory variables. The smaller the error variance [var(δ_i)] is relative to the x_i's variance ($\hat{\sigma}_{ii}$), the higher will be the $R^2_{x_i}$. In general, the goal is to find and use measures with high $R^2_{x_i}$'s.[17]

For the sympathy and anger indicators the $R^2_{x_i}$'s are

	x_1	x_2	x_3	x_4	x_5	x_6
$R^2_{x_i}$	0.87	0.59	0.62	0.78	0.68	0.71

All $R^2_{x_i}$'s are moderate to large. The highest is for x_1, which indicates that a large proportion (0.87) of its variance is "accounted for" by the latent sympathy variable. The magnitude of all the $R^2_{x_i}$ suggest that these indicators are "good" measures of the latent variables. By current standards of survey data, the error variances are a small proportion of their indicators' total variances. The coefficient of determination for measurement models

[17]The $R^2_{x_i}$ in (7.74) differs from that of Jöreskog and Sörbom (1986) in that I use $\hat{\sigma}_{ii}$ in place of s_{ii}. For many CFA fitted with ML, $s_{ii} = \hat{\sigma}_{ii}$ so that this change makes no difference.

(Jöreskog and Sörbom 1986) is

$$\text{Coefficient of determination} = 1 - \frac{|\hat{\Theta}_\delta|}{|\hat{\Sigma}|} \tag{7.75}$$

This summarizes the joint effect of the latent variables on the observed ones in CFA. It has the same properties and limitations as the analogous measure described in Chapter 4. The coefficient of determination is 0.987 for the sympathy and anger model.

Does the Sympathy and Anger Model Fit?

I introduced the sympathy and anger model to illustrate the summary and component fit measures for assessing the adequacy of confirmatory factor analyses. Now that I have completed the introduction of these various fit measures, the fundamental question is whether the model is consistent with the data. Considering the component fit measures, they did not show any obvious abnormalities, and in fact they looked quite good. If I judged the model solely on component fit, I would be tempted to conclude that it is adequate.

The summary fit measures also were quite favorable. Virtually all the overall model fit measures showed a good to excellent fit. The sample residuals and the standardized and normalized residuals were small, and the chi-square estimate was not statistically significant. Thus on virtually all counts the model appears to be an excellent match to the data.

COMPARISON OF MODELS

Researchers sometimes fit more than one model to the same data. The models can range from ones with radically different structures to those with minor differences in one or two parameters. Analysts can compare the overall fit of models by contrasting Δ_1, ρ_1, AGFI, and other fit measures. Those measures that account for degrees of freedom (e.g., ρ_1, ρ_2, AGFI) are preferable to ones that do not since models with more parameters can better fit data (other things equal) by capitalizing on chance fluctuations in the sample. Even with controls for df, the differences in these fit measures are still descriptive statistics; they do not allow tests of *statistical significance*.

The political democracy panel data from the beginning of the chapter illustrates this (see Figure 7.3). It has eight indicators of political democ-

Table 7.8 Summary of Fit Measures for CFA of Uncorrelated Errors and for Political Democracy 1960 to 1965 for Six Pairs of Correlated Errors

	Model	
Summary Fit Measure	Uncorrelated Measurement Errors	Correlated Measurement Errors
χ^2	46.7	15.1
df	22	16
prob.	< 0.002	0.52
	Baseline #1	
ρ_1	0.87	0.94
ρ_2	0.93	1.00
Δ_1	0.90	0.97
Δ_2	0.94	1.00
	Baseline #2	
ρ_1	0.79	0.91
ρ_2	0.88	1.01
Δ_1	0.81	0.94
Δ_2	0.89	1.00

racy, with four circa 1960 and the four circa 1965. Political democracy in 1960 and 1965 are the latent variables (ξ_1 and ξ_2). The measurement error for an indicator in 1960 is expected to correlate with the error for the same measure in 1965. Also the measurement errors for x_2 and x_4 and those for x_6 and x_8 correlate because they are from the same data source in the same year. I also hypothesize equal λ's for the latent variable's influence on an indicator in 1960 and the same indicator in 1965 ($\lambda_{11} = \lambda_{52} = 1$, $\lambda_{21} = \lambda_{62}$, $\lambda_{31} = \lambda_{72}$, $\lambda_{41} = \lambda_{82}$).

I compare this model to a more parsimonious version that is identical, except that none of the errors correlate (i.e., diag Θ_δ). Table 7.8 reports overall goodness of fit measures for the more restrictive, uncorrelated errors of measurement model and the less restrictive model with correlated errors. The covariance matrix is from Table 7.1 ($N = 75$). I have two baselines for the incremental fit indices. Baseline #1 treats all variables as uncorrelated (chi-square estimate = 455, df = 28). Baseline #2 allows the 1960 and 1965 paired measures to correlate, and the other covariances between indicators are zero (chi-square estimate = 247, df = 24). The chi-square estimate for the uncorrelated errors model is 46.7 (df = 22), which is highly significant ($p < 0.002$). With baseline #1, the normed fit Δ_1 is 0.90 and ρ_1 is 0.87. The Δ_2 and ρ_2 are 0.93 and 0.94. Considering ρ_1, ρ_2, Δ_1, and Δ_2 for baseline #2, a less favorable assessment of model fit follows. The other measures of overall fit (mean absolute value of correlation residuals, GFI_{ML}, $AGFI_{ML}$, etc.) also indicate a moderate degree of fit. The component measures of fit are fairly good with no clear signs of problems.

The last column of Table 7.8 gives the overall fit for the original model with correlated errors (see Figure 7.3). The chi-square estimate is low (chi-square estimate = 15.1, df = 16) and would not be statistically significant by conventional standards ($p = 0.52$). With either baseline #1 or #2, the overall fit of this model is excellent. The residuals and the overall fit indices also suggest a very good fit.

To compare this model's fit to the previous one, we could examine the differences in the fit measures. For instance, for baseline #1 the difference in Δ_1's is 0.07, whereas for Δ_2, it is 0.06. This represents some improvement, but it is not possible to assess the statistical significance of these differences since the distributions of the fit measures are unknown. Cross-validation measures or the A and B statistics for these models still would leave open the question of the statistical significance of any differences found.

In what follows I describe three asymptotically equivalent tests of the statistical significance of the differences in chi-square estimators of model fit. All tests are for *nested* models. Before presenting the tests, it is necessary to clarify the meaning of nested. In general, any model which requires that some function of its free parameters equals another free parameter or equals a constant is nested in the identical model that has no such restriction. The simplest case is when the free parameters of one model are a subset of the free parameters in a second. For example, the political democracy model with uncorrelated measurement errors is nested in the model with six pairs of correlated errors. Implicitly, the restrictive model constrains the "missing" parameters to equal zero.

Other types of restrictions also lead to nested models. For instance, a CFA model that forces some λ_{ij}'s to be equal is nested in the same model without this restriction. Or, a more complex restriction involving several parameters set equal to another parameter or a constant [e.g., $\phi_{21}\phi_{52} + \mathrm{VAR}(\delta_1) = 5$] is nested in the identical model that has no such restriction. In brief, the nested model is a special case of the less restrictive model.

The significance tests for nested models are (1) the likelihood ratio (LR) or chi-square difference test, (2) the Lagrangian multiplier (LM) test, and, (3) the Wald (W) test.

Likelihood Ratio (LR) Test

Let $\hat{\theta}_r$ be the ML estimator for the restrictive, nested model, and let $\hat{\theta}_u$ be the ML estimator for the model without restrictions. The LR statistic is

$$\text{LR} = -2[\log L(\hat{\theta}_r) - \log L(\hat{\theta}_u)] \tag{7.76}$$

which has a limiting χ^2 distribution when the restrictive model is valid. The df equal the difference in the df for the two models. In practice, analysts calculate the LR statistic as the difference in the usual chi-square estimators for the restricted and unrestricted models, with df equal to their difference in df. For this reason the LR test is often called a *chi-square difference test*. Equation (7.76) is a generalization of the chi-square test I presented earlier [see (7.49)]. There the unrestricted model was an unspecified exactly identified model such that $\hat{\mathbf{\Sigma}} = \mathbf{S}$ with df = 0, and the restricted model was overidentified with positive df. The difference in log likelihood functions was (7.49).

The political democracy panel data models illustrate the more general form of the LR test. The restrictive model sets all error covariances to zero, whereas the unrestricted version frees six pairs of correlated errors. The chi-square estimates are 46.7 (df = 22) and 15.1 (df = 16), respectively. The LR test statistic is 31.6 (= 46.7 − 15.1) with 6 (= 22 − 16) df. This chi-square estimate is statistically significant ($p < 0.001$), making it highly unlikely that the restrictive, uncorrelated errors model is correct. The simultaneous introduction of the six pairs of correlated errors leads to a significant improvement in fit. Of course, the strict interpretation of the LR test requires the assumptions underlying the ML estimators be valid. So the LR test must be interpreted with the same caution as the usual chi-square test.

The LR test may be written as

$$\begin{aligned} \text{LR} &= (N-1)F_r - (N-1)F_u \\ &= (N-1)(F_r - F_u) \end{aligned} \tag{7.77}$$

where F_r is F_{ML} evaluated at $\hat{\theta}_r$ and F_u is F_{ML} at $\hat{\theta}_u$. This equation shows that the LR test, like the usual chi-square test, has higher power to detect a false restrictive model in large samples than in small ones. The same is true for the two other tests I present.

A computational disadvantage of the LR test is that two models must be estimated for every pair of model comparisons. The Lagrangian multiplier and Wald tests overcome this limitation.

Lagrangian Multiplier (LM) Test

The Lagrangian multiplier (LM) test compares the fit of restrictive to less restrictive models. It only requires estimating the restrictive model. The LM statistic is based on $\partial \log L(\boldsymbol{\theta})/\partial \boldsymbol{\theta}$, where $\log L(\boldsymbol{\theta})$ is the *unrestricted* log likelihood function. This partial derivative vector, called the score $\mathbf{s}(\boldsymbol{\theta})$, gives the change in the log likelihood for changes in $\boldsymbol{\theta}$. The unrestricted elements of $\hat{\boldsymbol{\theta}}_r$ should have a zero partial, since the ML estimator solves for $\hat{\boldsymbol{\theta}}_r$ by setting these partials to zero. However, the elements of $\mathbf{s}(\boldsymbol{\theta})$ corresponding to restricted parameters only are zero when the restrictions hold exactly. Substituting $\hat{\boldsymbol{\theta}}_r$ for $\boldsymbol{\theta}$ in $\mathbf{s}(\boldsymbol{\theta})$ and examining the magnitude of the departures from zero for the restricted parameters indicates the validity of the restrictions. Of course, sample fluctuations lead to some nonzero values even when the restrictive model is valid so we need to take account of this variability. The LM test statistic that does this is

$$\mathrm{LM} = \left[\mathbf{s}(\hat{\boldsymbol{\theta}}_r)\right]' \mathbf{I}^{-1}(\hat{\boldsymbol{\theta}}_r)\mathbf{s}(\hat{\boldsymbol{\theta}}_r) \tag{7.78}$$

where $\mathbf{I}^{-1}(\hat{\boldsymbol{\theta}}_r) = \{-E[\partial^2 \log L(\boldsymbol{\theta})/\partial\boldsymbol{\theta}\, \partial\boldsymbol{\theta}']\}^{-1}$ evaluated at $\hat{\boldsymbol{\theta}}_r$ and $\mathbf{s}(\hat{\boldsymbol{\theta}}_r)$ is $\mathbf{s}(\boldsymbol{\theta})$ evaluated at $\hat{\boldsymbol{\theta}}_r$. The limiting distribution of LM is a chi-square variate with degrees of freedom equal to the difference in df for the restrictive and unrestrictive models. Given the relation between F_{ML} and $\log L(\boldsymbol{\theta})$, the test statistic also can be written as

$$\mathrm{LM} = \frac{(N-1)}{2}\left(\frac{\partial F_{\mathrm{ML}}}{\partial \boldsymbol{\theta}}\right)'\left[E\left(\frac{\partial^2 F_{\mathrm{ML}}}{\partial\boldsymbol{\theta}\,\partial\boldsymbol{\theta}'}\right)\right]^{-1}\left(\frac{\partial F_{\mathrm{ML}}}{\partial \boldsymbol{\theta}}\right) \tag{7.79}$$

where $\boldsymbol{\theta}$ is evaluated at $\hat{\boldsymbol{\theta}}_r$.

I return to the political democracy panel data to illustrate the LM test. The restrictive model has uncorrelated errors while the LM test is for the improvement in fit by freeing the six pairs of error covariances. Estimating only the restrictive model, the LM chi-square estimate is 32.3 with 6 df($p < 0.001$). Therefore loosening the restrictions on the nested uncorrelated errors model leads to a statistically significant improvement in fit. The LM estimate of 32.3 is close to the LR one of 31.6. The biggest difference in obtaining these, is that only the restrictive model was estimated for the LM test.

Wald (W) Test

Suppose I represent the constraints placed on the unrestricted model by an $r \times 1$ vector $\mathbf{r}(\boldsymbol{\theta})$, where r is less than the row dimension of $\boldsymbol{\theta}$. The $\mathbf{r}(\boldsymbol{\theta})$ can be any function of the parameters. The political democracy panel data provides a simple example:

$$\mathbf{r}(\boldsymbol{\theta}) = 0$$

$$\begin{bmatrix} \text{COV}(\delta_1, \delta_5) \\ \text{COV}(\delta_2, \delta_6) \\ \text{COV}(\delta_3, \delta_7) \\ \text{COV}(\delta_4, \delta_8) \\ \text{COV}(\delta_4, \delta_2) \\ \text{COV}(\delta_8, \delta_6) \end{bmatrix} = \begin{bmatrix} 0 \\ 0 \\ 0 \\ 0 \\ 0 \\ 0 \end{bmatrix} \tag{7.80}$$

The restrictive, nested model sets the six error covariances to zero. For the restrictive model $\mathbf{r}(\hat{\boldsymbol{\theta}}_r)$ is zero by construction. If the restrictive model is valid, then $\mathbf{r}(\hat{\boldsymbol{\theta}}_u)$ also should be zero within sampling error. But, if the restrictive model is false and the unrestrictive one is true, the elements of $\mathbf{r}(\hat{\boldsymbol{\theta}}_u)$ can be far from zero.

The Wald (W) test determines the extent to which $\hat{\boldsymbol{\theta}}_u$ departs from the restrictions imposed by the nested model. The test statistic is

$$\text{W} = \left[\mathbf{r}(\hat{\boldsymbol{\theta}}_u)\right]' \left\{ \left[\frac{\partial \mathbf{r}(\hat{\boldsymbol{\theta}}_u)}{\partial \hat{\boldsymbol{\theta}}_u}\right]' \left[\text{acov}(\hat{\boldsymbol{\theta}}_u)\right] \left[\frac{\partial \mathbf{r}(\hat{\boldsymbol{\theta}}_u)}{\partial \hat{\boldsymbol{\theta}}_u}\right] \right\}^{-1} \left[\mathbf{r}(\hat{\boldsymbol{\theta}}_u)\right]$$

where $\mathbf{r}(\hat{\boldsymbol{\theta}}_u)$ is $\mathbf{r}(\boldsymbol{\theta})$ evaluated at $\hat{\boldsymbol{\theta}}_u$ and $\text{acov}(\hat{\boldsymbol{\theta}}_u)$ is an estimate of the asymptotic covariance matrix of $\hat{\boldsymbol{\theta}}_u$. The term in braces $\{\,.\,\}$ is an estimate of the asymptotic covariance matrix of $\mathbf{r}(\hat{\boldsymbol{\theta}}_u)$. Thus W is the inverse of the estimated asymptotic covariance matrix of the constraints imposed by the restrictive model, pre- and postmultiplied by $\mathbf{r}(\boldsymbol{\theta})$ evaluated at $\hat{\boldsymbol{\theta}}_u$. Asymptotically, W follows a chi-square distribution, with degrees of freedom equal to the number of constraints in $\mathbf{r}(\boldsymbol{\theta})$ when the restrictive model is valid.

When $\mathbf{r}(\boldsymbol{\theta})$ consists of a single constraint of $\theta_1 = 0$, W is

$$\text{W} = \frac{\hat{\theta}_1^2}{\text{avar}(\hat{\theta}_1)} \tag{7.81}$$

where $\text{avar}(\hat{\theta}_1)$ is the estimated asymptotic variance of $\hat{\theta}_1$ and W has one df. Equation (7.81) is the square of the usual Z ratio for testing the statistical significance of a parameter estimate. Thus we can view W as a generalization of the usual Z-test.

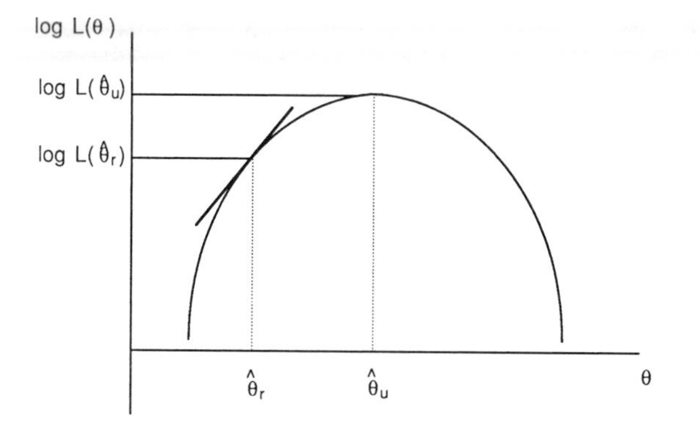

Figure 7.5 Relation between LR, LM, and W Tests for a Single Parameter

For the political democracy example, the W is 27.6 with 6 df for testing the restriction that all six error covariances in (7.76) are zero. Though this is a lower chi-square estimate than those from LR and LM, it still is highly significant and suggests that restricting all six errors to be uncorrelated leads to a significant decline in model fit.

Summary of LR, LM, and W Tests

The LR, LM, and W tests provide three asymptotically equivalent ways of testing nested models. The relations between these tests for a single parameter are illustrated in Figure 7.5.[18] The curve is the likelihood function. The abscissa is the value of θ and the ordinate is the value of $\log L(\theta)$. The $\log L(\theta)$ reaches its maximum at $\hat{\theta}_u$, while it is $\log L(\hat{\theta}_r)$ for the restrictive hypothesis. The LR test compares the $\log L(\hat{\theta}_u)$ with $\log L(\hat{\theta}_r)$ with greater differences reducing the probability that the restrictive model is true. The LM test depends on the slope of the likelihood function at $\hat{\theta}_r$. The further from zero this is, the less likely that the restriction is true. Finally, the W test is determined by the squared distance of $\hat{\theta}_r$ from $\hat{\theta}_u$, with greater discrepancies implying the restrictive model is false. All the comparisons are weighted by the curvatures of the log likelihood.

The W and LM tests do not require separate estimation of restrictive and unrestrictive models. They do have different purposes. The W test evaluates whether restrictions can be imposed on the estimated model, whereas LM tests whether restrictions can be removed. For W, the unrestrictive model is estimated to test the constraints imposed by the restrictive model. For LM,

[18]For further discussion, see Buse (1982) and Engle (1984).

Table 7.9 Applications and Comparisons of the LR, LM, and W Tests

	Restrictions		Estimates Required	
Test	Imposed	Removed	θ_u	θ_r
LR	√	√	√	√
LM	—	√	—	√
W	√	—	√	—

the restrictive model is estimated, and loosening restrictions is tested. Table 7.9 summarizes these characteristics.

RESPECIFICATION OF MODEL

Often an initial model does not fit the data well. Many potential causes for low measures of overall fit exist, but a common cause is a misspecified model. The error can range from the incorrect inclusion or exclusion of a parameter to using a fundamentally flawed model. A common response to a poorly fitting model is to respecify it. Since the respecification often is based on the results of the initial model, the analysis becomes exploratory. A consequence is that the probability levels for the tests of statistical significance for new models must be regarded as approximations. And the need to replicate the final model for an independently drawn sample becomes even more important (see Cliff 1983).

In this section I discuss several means to respecify models and to test the fit of the revised models. I cover first the use of theory or substantive expertise. Next I explain the exploratory applications of the LR, LM, and W statistics. Finally, I review several miscellaneous empirical means of respecification.

Theory and Substantive-Based Revisions

Researchers with inadequate models have many ways—in fact, too many ways—in which to modify their specification. An incredible number of major or minor alterations are possible, and the analyst needs some procedure to narrow the choices. The empirical means can be helpful, but they can also lead to nonsensical respecifications. Furthermore empirical means work best in detecting simple alterations and are less helpful when major changes in structure are needed. The potentially richest source of ideas for respecification is the theoretical or substantive knowledge of the researcher.

Would not a researcher use all of their knowledge in the initial specification? If so, then the recommendation to turn to these sources is not helpful. Though this may be true in some situations, there are many cases where it is not. For instance, analysts may be unsure whether a set of additional parameters (e.g., omitted paths or correlated disturbances) are needed, so they begin with a more parsimonious model. If this fails to fit, then the secondary parameters can be introduced. In other cases the initial models do summarize the researchers' current thinking, but the lack of fit can stimulate researchers to reconsider their ideas. This rethinking process is necessarily specific to the problem at hand, so I can provide only broad guidelines on how to proceed. Relevant questions include: Could additional latent variables underlie the observed variables? Are some of the single factors really two or more latent constructs? Do some of the measures load on more than one factor? If some of the observed variables are taken from the same source or constructed in the same fashion, is a method factor likely? Are there other reasons to expect correlated errors of measurement? Do some of the indicators have direct causal effects on the latent variables or on other indicators?

As an illustration, I use a CFA of perceived air quality. These data, collected by Cermak (see Cermak and Bollen 1982; Bollen 1982), are from an environmental study of the relation between objective and subjective air quality. I treat only the four subjective measures that are from a survey administered in Shenadoah National Park in Virginia. The variables are respondents' assessments of overall air quality (x_1), the clarity (x_2), color (x_3), and odor (x_4) of the air. I hypothesize that a single latent variable, subjective air quality (ξ_1), underlies these four measures. I discussed the plausibility of the causal priority of ξ_1 over x_1 to x_4 in Chapter 3. Initially, I assume that all errors are uncorrelated. The covariance matrix ($N = 57$) of x_1 to x_4 is

$$\mathbf{S} = \begin{bmatrix} 0.331 & & & \\ 0.431 & 1.160 & & \\ 0.406 & 0.847 & 0.898 & \\ 0.216 & 0.272 & 0.312 & 0.268 \end{bmatrix} \tag{7.82}$$

The CFA model is

$$x_i = \lambda_{i1}\xi_1 + \delta_i, \qquad \text{for } i = 1, 2, 3, 4$$

with diag Θ_δ, COV$(\delta_i, \xi_1) = 0$, $E(\delta_i) = 0$, and $\lambda_{11} = 1$ to provide a scale for ξ_1. The overall fit of this model is marginal at best ($\chi^2 = 16.0$, df $= 2$, $p = 0.0003$, $\rho_1 = 0.70$, $\rho_2 = 0.73$, $\Delta_1 = 0.90$, $\Delta_2 = 0.91$, with a baseline model of uncorrelated variables). An exploration of alternative specifications could be helpful.

I considered a number of different causes for the inadequate fit. For example, a separate latent variable might exist for the color, clarity, and odor of the air. The overall measure could be a summary assessment of these three latent components. Such a model seems plausible but cannot be estimated with only four imperfect indicators. A less radical alternative suggested itself after further investigation of the data collection procedure. The person who administered the questionnaire revealed that some respondents had trouble distinguishing the difference between the color and clarity questions. This seemed reasonable since we would expect clear air to be without color and, conversely, if the color of the air is bad, so too would be the clarity rating. This may have led the measurement error in clarity to correlate with the measurement error in color since the error in responding to one indicator is likely to correlate with the error in a partially indistinguishable alternative. Thus a possible respecification is to free a parameter for this covariance, which I expect to be positive. Before doing so, I turn to the empirical means of respecification.

Exploratory LR, LM, and W Tests

In an earlier section I illustrated the LR, LM, and W Tests for testing hypotheses that are specified in advance. These same test statistics also are useful in empirical attempts to respecify models. For instance, one of the simplest modifications that could be made is to introduce a single new parameter (i.e., remove a restriction). An analyst could estimate all possible identified models that eliminate one restriction at a time and compute a chi-square estimate for each. If none of the new models fits, then all possible two-at-time restrictions could be freed and chi-squares estimated. Theoretically, this process could continue until no identified alternative models remain. Alternatively, a researcher could introduce restrictions one at a time, two at a time, etc., to determine if a simpler model with a good fit exists. Comparisons of the nested versions of models is possible with the LR test statistic. Of course, the preceding options are hardly practical, given the tremendous computational burden of estimating chi-square values for all possible restrictions to be freed or imposed.

The LM statistic provides an asymptotically equivalent alternative for removing restrictions that reduces the number of estimated models. The univariate LM test is

$$\mathrm{LM} = \left[\partial \log L(\boldsymbol{\theta})/\partial\theta_i\right]^2\left[\mathbf{I}^{-1}(\hat{\boldsymbol{\theta}}_r)\right]_{ii} \tag{7.83}$$

The ii subscript refers to the ith diagonal element of $\mathbf{I}^{-1}(\hat{\boldsymbol{\theta}}_r)$. The restricted and unrestricted models differ only by the constraint on θ_i. Equation (7.83) gives a chi-square estimate with one df to assess the improvement in fit that results when θ_i is freed. It is equivalent to the *modification index* proposed by Jöreskog and Sörbom (1986).[19] Of course, the LM statistic could also be estimated for all possible two-at-a-time, three-at-a-time, etc., freed restrictions. Though the LM statistic does drastically reduce the number of estimated models, it still involves many LM estimates if a researcher allows all possible combinations of restrictions.

Because of this, alternative search strategies have arisen. The most common is to examine the univariate LM (modification indices), to free the restriction that leads to the largest reduction in the chi-square estimate, and to repeat this process with the revised model until an adequate fit is developed. This is analogous to some variants of stepwise regression where the variable that adds most to the explained variation of the dependent variable is added first and then the variable that adds most to the revised equation is added next, until no other variables lead to a significant increment in R^2 (see Bentler 1986b).

I return to the subjective air quality example to illustrate the univariate LM statistic. The $\boldsymbol{\Phi}$ matrix is a free scalar, and the only restriction in $\boldsymbol{\Lambda}_x$ for this model is $\lambda_{11} = 1$ which is necessary to scale the latent variable. Therefore the only restrictions to consider are the off-diagonal elements of $\boldsymbol{\Theta}_\delta$. The univariate LM (modification indices) statistics for $\hat{\boldsymbol{\Theta}}_\delta$ are

$$\begin{bmatrix} 0.00 & & & \\ 0.19 & 0.00 & & \\ 7.47 & 12.28 & 0.00 & \\ 12.28 & 7.47 & 0.19 & 0.00 \end{bmatrix} \tag{7.84}$$

The main diagonal that corresponds to the free parameters in $\boldsymbol{\Theta}_\delta$ are zero.

[19]Prior to the univariate LM statistic, the $\partial F_{\mathrm{ML}}/\partial\theta_i$, which is related to the numerator of (7.83) was widely used to identify restrictions that should be freed (Sörbom 1975). The justification was that $\partial F_{\mathrm{ML}}/\partial\theta_i$ identifies which single newly freed parameter leads to the greatest decrease in F_{ML}, and thus the greatest improvement in fit. The LM statistic (modification index) is an improvement over the first order partials since LM takes into account the variance of the estimated partials. Other limitations of $\partial F_{\mathrm{ML}}/\partial\theta_i$ are the same as the LM statistic discussed in this section.

The elements for the covariances of δ_1 and δ_4 ("overall" and "odor") and δ_2 and δ_3 ("clear" and "color") have the largest LM's of 12.28. As a chi-square estimate with one df, each is highly significant by conventional standards. Without the other information, there would be no easy way to know which restriction to lift. Here the earlier discussion suggests that $COV(\delta_2, \delta_3)$ should be freed. Thus not only is substantive knowledge helpful by itself, it also can help in evaluating information from the empirical tests. There is little use in examining the LM statistics for the model that frees $COV(\delta_2, \delta_3)$ since any further loosening of restrictions leads to a zero chi-square, if the model is identified. This follows since only one df remains.

The W test also has applications in an exploratory mode. It identifies which new restrictions (e.g., parameters set to zero) lead to the smallest declines in the chi-square estimate. Like the LM test, it is possible to compute W estimates for all one-at-a-time, two-at-a-time, etc., restrictions and to find which combination is best. But this is not practical. One strategy is to compute univariate W statistics for each free parameter and to see which W is lowest. When the restriction is setting a free parameter to zero, the univariate W tests are not needed since they are equivalent to the square of the usual Z-tests for single-parameter estimates. Once the restriction with the least significant W (or Z) estimate is imposed, the lowest W estimate for a second restriction is sought. This process continues until no further restrictions can be added without a significant increase in the incremental chi-square estimates. Again, the analogy to stepwise regression is clear.

I illustrate the W test by returning to the political democracy panel data. Although I found that the CFA model with six pairs of correlated errors had a much better fit than the model without them, three covariances for the overtime measurement errors were considerably less than twice their standard errors. To explore further the significance of the correlated errors, I calculated a multivariate W test for these parameters with Bentler's (1985) EQS. The results are summarized in Table 7.10.

I restricted my search to the six correlated errors parameters. The test begins by selecting the least significant parameter estimate of the six. Next the parameter estimate that makes the smallest increment to the chi-square is chosen. This process continues until all six covariances for the errors are entered. Table 7.10 indicates that three of the overtime error covariances [$COV(\delta_4, \delta_8)$, $COV(\delta_3, \delta_7)$, and $COV(\delta_1, \delta_5)$] can be set to zero without much degradation in the fit. However, constraining the remaining three parameters leads to a significantly worse fit.

The LR and particularly the LM and W tests have merits as exploratory tools, but their limits also must be recognized. One is that the order in

Table 7.10 Cumulative Multivariate W Test for Six Pairs of Correlated Measurement Errors for the Political Democracy Panel Data

Parameter	Chi-square	Probability
COV(δ_4, δ_8)	1.09	0.30
COV(δ_3, δ_7)	2.68	0.26
COV(δ_1, δ_5)	5.94	0.11
COV(δ_2, δ_4)	11.33	0.02
COV(δ_6, δ_8)	18.37	0.00
COV(δ_2, δ_6)	27.62	0.00

which parameters are freed or restricted can affect the significance tests for the remaining parameters. The problem is that the parameter estimates are correlated so that the freed or fixed status of other parameters affects the significance of the LM or W test (Saris, de Pijper, and Zegwaart 1979). It can happen that some combination of restrictions introduced or removed would lead to a better model than that developed through stepwise search procedures.

A second difficulty mentioned earlier is that the probability levels associated with the W and LM statistics in the stepwise procedures are not likely to be accurate. The difficulty is that estimation and model modifications are based on the same data. The importance of replicating or cross-validating the model on an independent sample is clear. When the sample is sufficiently large, it is possible to split the sample randomly and to use one part for exploratory analysis and the other to validate the resulting model.

Third, these tests assess the changes in chi-square estimates, not the size of changes in parameter estimates. It is possible that large changes in parameter estimates may be associated with small LM or W tests statistics, and vice versa. Saris, Satorra, and Sörbom (1987) show that the parameter estimate change associated with the univariate LM equals the univariate LM estimate divided by the first-order partial derivative for the restricted parameter.[20] A prudent strategy is to examine the estimated parameter change as well as the change in the chi-square.

[20] Both the univariate LM estimate (modification index) and the first-order partial are available from LISREL VI. Because of the fitting function for LISREL, the "first-order partial" should be multiplied by $-(N-1)$ before dividing the LM estimate by it [Saris et al. (1987), 120]. Bentler's (1986b) EQS automatically reports the estimated parameter change associated with loosening a restriction.

A fourth limitation for the LM test was raised by Byron (1972) and others. If a model is invalid, the LM tests are biased versions of the LM tests that would be obtained for the same restrictions under the correct model. The LM estimate for a single restriction in an incorrect model can be a function of several parameters rather than only the single parameter associated with the restriction (Dijkstra 1981, 32). Thus restricted parameters with the largest LM estimates will not always be the one that should be freed.

A fifth characteristic of the W and LM tests is that they are most useful when the misspecifications are minor. Fundamental changes in structure may be needed but are not detectable with these procedures. For instance, if a model requires additional latent variables or the direction of causation between an indicator and construct is wrong, these are unlikely to show with these exploratory techniques.

A final limitation of the empirical search procedures is that the freed parameters introduced into a model may have no clear interpretation. For instance, introducing correlated errors or an additional path may improve statistical fit, but without an explanation for them, the substantive gain of introducing them is ambiguous. So part of model evaluation should include the interpretability of the estimated parameters.

Much remains to be learned about the performance of the empirical search procedures in practice. To date only limited simulation work is available. Herbing and Costner (1985) combine the univariate LM statistic with normalized residuals to respecify a poorly fitting measurement model with simulated data. They find that these diagnostics work well for detecting correlated measurement error for indicators from different constructs, but they are unsatisfactory for revealing pairs of correlated errors within a construct. The ability of these diagnostics to detect omitted factor loadings is affected by the strength of the omitted path, being less helpful when the omitted path is strong. Also difficult to identify are cause indicators that are treated as effects.

MacCallum (1986) undertakes a more elaborate simulation that focuses on using the univariate LM (modification index) test for specification searches in latent variable models. I will discuss latent variable models more fully in the next chapter, but his major findings are worth describing here. In his misspecified models he omitted one or two paths and sometimes included one irrelevant path. His *unrestricted* search procedure followed four steps: (1) fit the initial misspecified model, (2) free the parameter with the biggest LM estimate, (3) repeat (1) and (2) until there are no more significant ($\alpha = 0.01$) LM estimates, and (4) identify the nonsignificant parameter estimates. This search matches the conditions of a researcher who uses no substantive knowledge to guide the modifications. He also devised a restricted search where no zero true parameters are freed and no nonzero true parameters are forced to zero. Overall, MacCallum (1986) found a surprising number of the restricted searches did not lead to the true model. Factors that seemed to enhance the chances of a successful search were (1) the closeness of the initial model to the true one, (2) a big N, (3) use of a restrictive rather than unrestrictive search, and (4) continuing to search beyond a nonsignificant ($\alpha = 0.05$) chi-square for a modified model.

Thus the simulation work to date provides a sobering assessment of the chances of successfully finding a correct model by purely empirical means and highlights the importance of guidance from substantive knowledge in specification searches.

Other Empirical Procedures

Other empirical results to help in respecification are the residual matrix, component fit measures, and piecewise fitting strategies. Consider first the residual matrix. The residuals reveal the elements of $\mathbf{S}$ that are poorly reproduced by $\hat{\Sigma}$. Intuitively, large positive residuals seemingly suggest that some parameter omitted from the initial model could, if included, better match s_{ij} to $\hat{\sigma}_{ij}$. Small residuals create the impression that the structure of the model that concerns the corresponding covariances is sound. Unfortunately, though residuals can provide hints of where a specification is in error, they can be misleading. Costner and Schoenberg (1973) illustrate that correlated errors for indicators of a single construct or two constructs fitted as one can create a very misleading pattern of residuals (see also Herting and Costner 1985). The subjective air quality CFA provides an example. In the initial, uncorrelated error model, the largest normalized residual (= 1.1) was between "overall" (x_1) and "odor" (x_4), whereas that for "clear" (x_2) and "color" (x_3) was small (= 0.114). This occurred despite the substantive arguments that errors for clear and color are correlated and the significant improvement in fit that results when $\text{COV}(\delta_2, \delta_3)$ is introduced.

The problem with residuals is compounded since incorrect respecifications can improve the overall statistical fit of a model. Part of the difficulty is that elements of the covariance matrix usually are functions of several parameters, as is clear from the implied covariance matrix [see (7.5)]. The omission or inclusion of one parameter can make itself felt in more than one covariance element, sometimes in a counterintuitive fashion. Since F_{ML}, F_{GLS}, and F_{ULS} are full-information procedures, specification error in one part of the system can be spread to other parts in ways that are difficult to trace. Though residuals can suggest areas of misspecification, they should not be employed in isolation of other substantive and empirical information.

Elements of the component fit may be useful in locating problems. Improper solutions or parameter estimates that are implausible (e.g., $R^2_{x_i}$ that are near zero or far higher than is likely, signs of coefficients counter to that predicted) can call attention to troublesome sectors of a model. Like residuals though, the problem can be located in a part different from where the suspicious component of fit appears.

Another broad strategy to respecification is what I term "piecewise model fitting." As the name suggests, the procedures involve estimating components of the entire model in the attempt to isolate the sources of misspecification. Costner and Schoenberg (1973) proposed a procedure where all two-indicator and three-indicator submodels are estimated. A difficulty with this approach is the extremely large number of submodels that must be estimated for moderately complex specifications.

A related piecewise approach is to break a poorly fitting model into components and reestimate each part separately. For instance, suppose a CFA model has three factors, each measured with four indicators. If the CFA for the whole model is not adequate, then the analyst could estimate a separate factor analysis for one factor and its four indicators at a time. Suppose that two of the three CFAs have a good fit, but one does not. The poorly fitting CFA is a possible source of error for the full model, and this model may need respecification. If all three CFAs fit well, then the researcher could take combinations of two factors at a time. The process continues until the analysts has narrowed down the sectors that require the most attention. The other empirical diagnostics may prove helpful to further pinpoint problems. The main benefit of this procedure is that it can aid finding that part of a complex model with a poor fit. The main limitations are that it cannot unambiguously identify the error within the problem sector, and it may involve the estimation of many submodels when the initial model is elaborate. It also is possible that the problem with the model only is evident when the complete model is fitted.

Finally, Glymour et al. (1987) have suggested an automated model search procedure based on the tetrad differences of the covariances (correlations) of the observed variables implied by the model. Since the search procedure is a programmed algorithm (TETRAD), it is faster and it covers more models than is usually possible in an unautomated search. The properties of this method compared to the other search strategies are not known.

Summary of Respecification Procedures

Respecification like initial specification seems to work best when closely guided by a researcher's substantive expertise. This does not deny the potential usefulness of the empirical procedures reviewed earlier. Ideally, subject matter expertise should be combined with several empirical guides such as the LM and W tests, the residuals, and piecewise fitting strategies. Reliance solely on empirical means is hazardous. Researchers should use the empirical tests as "advisors" not the "supervisors" of their model modification.

Once a final model with adequate fit is derived, the probability levels and estimates must be viewed with suspicion. Cross-validation or replication for an independent sample is an important step in building confidence in the new specification. Lastly, it should be clear that the procedures discussed in this section have applications for the observed variable models of Chapter 4 as well as the general models I present in Chapter 8.

EXTENSIONS

Factor Score Estimation

Sometimes researchers wish to know the values of the latent variable for individual observations. The best we can do is to *estimate* these by forming some weighted function of the observed variables. Perhaps the most popular function derives from the regression method of factor score estimation:

$$\hat{\boldsymbol{\xi}} = \hat{\boldsymbol{\Phi}}\hat{\boldsymbol{\Lambda}}_x'\hat{\boldsymbol{\Sigma}}^{-1}\mathbf{x}$$

where $\hat{\boldsymbol{\xi}}$ is the estimate of $\boldsymbol{\xi}$. The weight that premultiplies $\mathbf{x}$ is the OLS estimator of regression coefficients from the "hypothetical" regression of $\boldsymbol{\xi}$ on $\mathbf{x}$. Other methods of factor score estimation also are available (McDonald and Burr 1967; Saris, de Pijper, and Mulder 1978). In practice, the different methods lead to estimates of $\boldsymbol{\xi}$ that often are highly correlated.

Regardless of the method, the factor score estimate does not equal the factor (i.e., $\hat{\boldsymbol{\xi}} \neq \boldsymbol{\xi}$ for $n < q$). This factor indeterminancy is due to having more latent variables and errors of measurement than observed variables ($n + q$ vs. q). That is, we have only q equations for $\mathbf{x}$ in $n + q$ unknown latent and error variables. An analogous situation would be in a multivariate regression model, $\mathbf{y} = \boldsymbol{\Gamma}\mathbf{x} + \boldsymbol{\zeta}$, where we are given only $\mathbf{y}$ and $\hat{\boldsymbol{\Gamma}}$ and told to estimate $\mathbf{x}$. Clearly, as long as $\boldsymbol{\zeta} \neq 0$, we could not find an $\hat{\mathbf{x}}$ that perfectly matches $\mathbf{x}$ with only this information.

The implication is that we can form many possible estimates of factor scores and an individual observation need not have identical rankings on two factor score estimates for the same latent variable. Thus researchers should refrain from too fine comparisons of standings on factor scores, for they may be asking more from the factor score estimates than they can provide.

Some researchers use factor score estimates in observed variable models (e.g., regression models). The implicit assumption is that the factor score estimates remove the problems with measurement error present in the unadjusted variables. Though using $\hat{\boldsymbol{\xi}}$ in place of single indicators for $\boldsymbol{\xi}$ can reduce measurement error, it does not remove it. Since $\hat{\boldsymbol{\xi}}$ is a weighed combination of $\mathbf{x}$ and since $\hat{\boldsymbol{\xi}} \neq \boldsymbol{\xi}$, we can regard $\hat{\boldsymbol{\xi}}$ as an *indicator* of $\boldsymbol{\xi}$ that contains measurement error. So in most cases using factor score estimates to replace latent variables and then employing classical econometric (regression) procedures on these estimates still leads to inconsistent coefficient estimators (see Chapter 5).

Latent Variable Means and Equation Intercepts

Throughout the chapter I have ignored the mean values of latent variables and the intercepts of measurement equations. All variables were assumed to be deviated from their means. There are situations where researchers wish to know these means and intercepts. Most often interest lies in comparing means for the same latent variable in different groups. Chapter 8 treats these cases.

Even when a single sample is analyzed, the mean values or equation intercepts may be sought. In panel data a contrast between the means of the same latent variable at two time points reveals if a group's average has shifted. Or, if the same group of indicators are employed at two or more time points, we can test whether the equation intercepts are stable over time. Two latent variables that are distinct but are assigned the same units also can be compared. For instance, researchers can test whether the mean level of permanent income and permanent consumption are the same if both are scaled to the same dollar unit.

To estimate these models, I need to supplement the existing notation. The equations for the $\mathbf{y}$ and $\mathbf{x}$ measurement equations with intercept terms $\boldsymbol{\upsilon}_y$ and $\boldsymbol{\upsilon}_x$ are

$$\mathbf{y} = \boldsymbol{\upsilon}_y + \boldsymbol{\Lambda}_y\boldsymbol{\eta} + \boldsymbol{\epsilon} \tag{7.85}$$

$$\mathbf{x} = \boldsymbol{\upsilon}_x + \boldsymbol{\Lambda}_x\boldsymbol{\xi} + \boldsymbol{\delta} \tag{7.86}$$

The $\boldsymbol{\upsilon}_y$ vector is $p \times 1$, and $\boldsymbol{\upsilon}_x$ is $q \times 1$. The former contains the expected value of $\mathbf{y}$ when $\boldsymbol{\eta}$ is zero, and the latter is the same for $\mathbf{x}$ when $\boldsymbol{\xi}$ is zero. The mean value of $\boldsymbol{\xi}$ ($E(\boldsymbol{\xi})$) is $\boldsymbol{\kappa}$, an $n \times 1$ vector. The expected value of $\boldsymbol{\eta}$ depends on its relation to $\boldsymbol{\xi}$. I postpone discussion of its mean until Chapter

8, and I focus on the equation for **x**. The expected value of **x** is

$$E(\mathbf{x}) = \boldsymbol{\upsilon}_x + \boldsymbol{\Lambda}_x\boldsymbol{\kappa} \tag{7.87}$$

In the simple case of one factor and three indicators, this reduces to

$$\begin{aligned} E(X_1) &= \upsilon_{X_1} + \kappa_1 \\ E(X_2) &= \upsilon_{X_2} + \lambda_{21}\kappa_1 \\ E(X_3) &= \upsilon_{X_3} + \lambda_{31}\kappa_1 \end{aligned} \tag{7.88}$$

In the first equation of (7.88) λ_{11} is set to one since we know that the model is not identified unless we assign a scale to the latent variables. Variances and covariances of the three observed variables stay the same as when deviation scores are assumed, so $\boldsymbol{\Sigma}(\boldsymbol{\theta})$ is

$$\begin{bmatrix} \phi_{11} + \text{VAR}(\delta_1) & & \\ \lambda_{21}\phi_{11} & \lambda_{21}^2\phi_{11} + \text{VAR}(\delta_2) & \\ \lambda_{31}\phi_{11} & \lambda_{21}\lambda_{31}\phi_{11} & \lambda_{31}^2\phi_{11} + \text{VAR}(\delta_3) \end{bmatrix} \tag{7.89}$$

To understand this equivalency consider the first element of (7.89) the $\text{VAR}(X_1)$:

$$\begin{aligned} \text{VAR}(X_1) &= E\left[(X_1 - E(X_1))^2\right] \\ &= E\left[\left(\upsilon_{X_1} + \xi_1 + \delta_1 - (\upsilon_{X_1} + \kappa_1)\right)^2\right] \\ &= E\left[(\xi_1 - \kappa_1 + \delta_1)^2\right] \\ &= \phi_{11} + \text{VAR}(\delta_1) \end{aligned}$$

A similar series of steps would show the equivalency for the other elements in (7.89). I showed earlier that the λ_{ij}, $\text{VAR}(\delta_i)$, and ϕ_{11} elements are identified in this model. However, (7.88) introduces four new parameters (υ_{X_i} for $i = 1, 2, 3$, and κ_1) in three equations. I have three new sample means of the observed variables, but this is not sufficient to identify the additional parameters. The problem is analogous to the scaling issue. Even when using deviation scores the model could not be identified unless we gave the latent variable a scale. Here the origin or mean of the latent variable is arbitrary, and we must assign it. An example from the physical sciences helps to clarify this point. Suppose that the latent variable were temperature and that X_1, X_2, and X_3 were thermometer readings in Kelvin, Celsius, and Fahrenheit degrees. Without agreement about the scale and origin for temperature, this model is not identified. However, the scientific consensus is that Kelvin degrees has the appropriate scale and origin for the latent variable of temperature. The agreement on scale is represented by setting λ_{11} to one, and the agreement on origin is incorporated by setting υ_{X_1} to zero. With these restrictions, $E(X_1)$ is κ_1, and the remaining parameters are identified.

A major dissimilarity for the social sciences is that little consensus exists on the scales or zero points for latent factors. This means that a "natural choice" for a scale or origin is lacking and thus we must select one. We can do this in many ways. The temperature example suggests one possibility. Select one variable to fix λ_{i1} to one and its intercept, υ_{X_i}, to zero. This provides the latent variable the same scale as X_i and the same mean since $E(X_i)$ equals κ_1. The other intercepts and λ_{i1}'s can be interpreted relative to these values. Another possibility is to choose arbitrarily a value for κ_1 such as zero or some other constant. As long as one of the λ_{i1}'s is set to one (or some other nonzero constant), the remaining parameters in (7.88) are identified.

I illustrate the above points with an empirical example of individuals' perceptions of line lengths. I asked five judges to estimate, to the nearest tenth of an inch, the length of lines drawn on 60 index cards with one line per card. The first subject was more familiar with the metric system, so he estimated to the nearest tenth of a centimeter (i.e., nearest millimeter). I assume that a latent variable, perceived line length, underlies the five judged length indicators. I also assume that the measurement equations are of the following form:[21]

$$X_i = \upsilon_{X_i} + \lambda_{i1}\xi_1 + \delta_i, \qquad \text{for } i = 1, \ldots, 5 \tag{7.90}$$

To identify the model, a scale for the latent variable must be chosen. Even if the judges obtained perfect accuracy in estimating length, their λ_{i1}'s would not be the same since the first judge used the metric system and the others used the English system. Since most of the judges were estimating in inches, I set the scale to one of the last four judges. I arbitrarily chose judge five by setting λ_{51} equal to one. In addition I set υ_{X_5} to zero to provide an origin for perceived length. Thus perceived length (ξ_1) has the same

[21]Steven's power law suggests that a multiplicative form $X_i = \upsilon_{X_i}\xi_1^{\lambda_i}\delta_i$ would be more appropriate. This model was transformed to a linear one by taking the logarithms of both sides. When I examined the log X_i, I found that outliers and nonlinearities had been introduced that were not present in the original data. Therefore I did not employ the logged transformation.

expected value as X_5 and the same scale in the sense that a one-unit change in ξ_1 leads to an *expected* change of one unit in X_5.

The estimation of this model in LISREL VI (and earlier versions) requires a modification of the standard programming. In specific, a constant must be added to the model, parameters introduced for the intercepts and means of the latent variables, and then the moments-around-zero matrix (see Chapter 4) must be analyzed. Also there is a shift in the notation. Equations (7.91) and (7.92) show these changes for a general CFA:

$$\mathbf{y}^{\cdot} = \mathbf{\Lambda}^{\cdot}_y \boldsymbol{\eta}^{\cdot} + \boldsymbol{\epsilon}^{\cdot} \tag{7.91}$$

$$\boldsymbol{\eta}^{\cdot} = \boldsymbol{\Gamma}^{\cdot} \boldsymbol{\xi}^{\cdot} + \boldsymbol{\zeta}^{\cdot} \tag{7.92}$$

I have superscripted the matrices with dots to indicate that their meaning departs from the usual one. The $\mathbf{y}^{\cdot}$ contains the $\mathbf{x}$ variables and has a dimension of $q \times 1$. The $\boldsymbol{\eta}^{\cdot}$ vector is $(n+1) \times 1$ with the n ξ variables in it and the last element being the constant 1. The $\mathbf{\Lambda}^{\cdot}_y$ matrix is $q \times (n+1)$ where the first q rows and n columns correspond to the original $\mathbf{\Lambda}_x$ and the last column contains the q intercept terms, υ_{x_i}. The $\boldsymbol{\epsilon}^{\cdot}$ vector is $q \times 1$ and equal $\boldsymbol{\delta}$ in the original notation. In equation (7.92) $\boldsymbol{\xi}^{\cdot}$ is a scalar of one; $\boldsymbol{\Gamma}^{\cdot}$ is $(n+1) \times 1$ where the first n elements are κ_1 to κ_n and the last is one; $\boldsymbol{\zeta}^{\cdot}$ is $(n+1) \times 1$ with the first n elements equal to $\xi_i - \kappa_i$ for $i = 1, \ldots, n$, and the last element is zero.

For the line length example, the model is

$$\mathbf{y}^{\cdot} = \mathbf{\Lambda}^{\cdot}_y \boldsymbol{\eta}^{\cdot} + \boldsymbol{\epsilon}^{\cdot}$$

$$\begin{pmatrix} X_1 \\ X_2 \\ X_3 \\ X_4 \\ X_5 \end{pmatrix} = \begin{pmatrix} \lambda_{11} & \upsilon_{X_1} \\ \lambda_{21} & \upsilon_{X_2} \\ \lambda_{31} & \upsilon_{X_3} \\ \lambda_{41} & \upsilon_{X_4} \\ 1 & 0 \end{pmatrix} \begin{pmatrix} \xi_1 \\ 1 \end{pmatrix} + \begin{pmatrix} \delta_1 \\ \delta_2 \\ \delta_3 \\ \delta_4 \\ \delta_5 \end{pmatrix} \tag{7.93}$$

$$\boldsymbol{\eta}^{\cdot} = \boldsymbol{\Gamma}^{\cdot} \boldsymbol{\xi}^{\cdot} + \boldsymbol{\zeta}^{\cdot}$$

$$\begin{pmatrix} \xi_1 \\ 1 \end{pmatrix} = \begin{pmatrix} \kappa_1 \\ 1 \end{pmatrix} (1) + \begin{pmatrix} \xi_1 - \kappa_1 \\ 0 \end{pmatrix} \tag{7.94}$$

Initial examination of the five length estimate variables revealed an outlier for one observation. This case was not influential, however, in that the same results were obtained with or without the observation. Also bivariate plots of the **x** variables suggested greater variances at the higher values of **x** than at the lower values. This was largely corrected by a square root transformation of the data. The results of analyzing the square root and the original data are essentially unchanged. I report the analysis with the data in its untransformed state since it is easier to interpret. The covariance matrix and the means of the estimates by the five judges is

$$\mathbf{S} = \begin{bmatrix} 14.19 & & & & \\ 4.38 & 1.46 & & & \\ 7.50 & 2.34 & 4.21 & & \\ 4.55 & 1.42 & 2.45 & 1.50 & \\ 4.01 & 1.24 & 2.17 & 1.31 & 1.20 \end{bmatrix} \tag{7.95}$$

$$\bar{\mathbf{x}} = \begin{bmatrix} 6.76 & 2.23 & 3.43 & 2.10 & 1.92 \end{bmatrix} \tag{7.96}$$

Using **S** and $\bar{\mathbf{x}}$ to generate the moment matrix and applying the model in (7.93) and (7.94) leads to a χ^2 of 3.92 with 5 df and an alpha level of 0.56. The other overall fit measures were quite good. The parameter estimates with their standard errors below them are

$$\hat{\Lambda}_y^{\cdot} = \begin{bmatrix} 3.48 & 0.08 \\ (0.11) & (0.24) \\ 1.09 & 0.14 \\ (0.05) & (0.10) \\ 1.87 & -0.17 \\ (0.07) & (0.15) \\ 1.13 & -0.08 \\ (0.03) & (0.08) \\ 1.00 & 0.00 \\ (-) & (-) \end{bmatrix} \qquad \hat{\Gamma}^{\cdot} = \begin{bmatrix} 1.92 \\ (0.14) \\ 1.00 \\ (-) \end{bmatrix} \qquad \operatorname{diag} \hat{\Psi}^{\cdot} = \begin{bmatrix} 1.13 & 0.00 \\ (0.22) & (-) \end{bmatrix} \tag{7.97}$$

$$\operatorname{diag} \hat{\theta}_{\epsilon}^{\cdot} = \begin{bmatrix} 0.23 & 0.09 & 0.16 & 0.02 & 0.05 \\ (0.06) & (0.02) & (0.03) & (0.01) & (0.01) \end{bmatrix} \tag{7.98}$$

The first column of $\hat{\Lambda}_y^{\cdot}$ lists the scaling differences for the judges relative to judge 5. Since there are 2.54 centimeters to an inch, I would expect $\hat{\lambda}_{11}$ to be larger than the other $\hat{\lambda}_{i1}$'s, as is found. However, his length estimate increases about 3.5 for every unit increase in the perceived length (ξ_1). The estimates of $\hat{\lambda}_{21}$ and $\hat{\lambda}_{41}$ for judges 2 and 4 are similar in scale to judge 5 since they are both near one. In contrast, $\hat{\lambda}_{31}$ for judge 3 shows that a one-unit change in ξ_1 is accompanied by 1.9 change in his length estimate. All $\hat{\lambda}$'s are considerably greater than twice their standard errors. None of the intercept estimates in the second column of $\hat{\Lambda}^{\cdot}_{y}$ is at least twice its standard error. To test whether all five judges have zero intercepts, the model could be reestimated, constraining all five intercepts to zero and comparing this model's fit to the current one. Alternatively, we could form a Wald test of this restriction.

The mean, κ_1, of the latent variable perceived length (ξ_1) is the (1,1) element of $\hat{\Gamma}^{\cdot}$. In the scale of judge 5's perceptions of inches, the average perceived length is 1.92, which is the mean of X_5. The variance of perceived length is the (1,1) element of $\hat{\Psi}^{\cdot}$ and is 1.13. Finally, the error variances associated with the length estimates of each judge are in diag $\hat{\Theta}^{\cdot}_{\epsilon}$. The $R^2_{X_i}$'s are all very high (e.g., $R^2_{X_5} = 0.96$), indicating a close relation between the perceived length ξ_1, and the estimated lengths from the five judges.

The example illustrates the general procedure with which to estimate means and intercepts in CFA. The LISREL program commands for this problem are in Appendix 7A.

Cause Indicators

The measurement model $\mathbf{x} = \mathbf{\Lambda}_x\boldsymbol{\xi} + \boldsymbol{\delta}$ assumes that the indicators depend on the latent variables. In Chapter 3 I gave several examples of indicators as *causes* of latent variables. Analysts can incorporate "cause" or "formative" indicators into the current model structure. To do so, define each cause indicator, x_i, as equal to a "latent" variable ξ_i (i.e., $\mathbf{x} = \mathbf{I}\boldsymbol{\xi}$, where $\mathbf{x}$ is a vector of cause indicators). Next write all latent variables that are influenced by the cause indicators as part of $\boldsymbol{\eta}$, the latent endogenous variables. I represent the dependence of the latent variables on the indicators as $\boldsymbol{\eta} = \mathbf{\Gamma}\boldsymbol{\xi} + \boldsymbol{\zeta}$. Without some effect indicators, $\mathbf{y}$, of $\boldsymbol{\eta}$, such a model generally would not be identified. Assuming that effect indicators are available, their dependence on $\boldsymbol{\eta}$ is $\mathbf{y} = \mathbf{\Lambda}_y\boldsymbol{\eta} + \boldsymbol{\epsilon}$. Collecting these results together, I can represent a model with cause indicators as

$$\begin{aligned} \mathbf{x} &= \mathbf{I}\boldsymbol{\xi} \\ \mathbf{y} &= \mathbf{\Lambda}_y\boldsymbol{\eta} + \boldsymbol{\epsilon} \qquad (7.99) \\ \boldsymbol{\eta} &= \mathbf{\Gamma}\boldsymbol{\xi} + \boldsymbol{\zeta} \end{aligned}$$

If $\mathbf{x}$ contains a mixture of cause and effect indicators, then the first equation is the usual $\mathbf{x} = \mathbf{\Lambda}_x \boldsymbol{\xi} + \boldsymbol{\delta}$, and the λ_{ij}'s and δ_i's that correspond to the cause indicators should be set to one and zero, respectively. I make the usual assumptions that the errors of measurement ($\boldsymbol{\epsilon}$) are uncorrelated with $\boldsymbol{\eta}$, $\boldsymbol{\xi}$, and $\boldsymbol{\zeta}$, that the errors in equations, $\boldsymbol{\zeta}$, are uncorrelated with $\boldsymbol{\xi}$, that the expected values of $\boldsymbol{\epsilon}$ and $\boldsymbol{\zeta}$ are zero, and that the $\boldsymbol{\xi}$ and $\boldsymbol{\eta}$ variables are written as deviations from their means.

Although (7.99) may seem like an awkward way to incorporate cause indicators, there is an advantage to it. The first equation, $\mathbf{x} = \mathbf{I}\boldsymbol{\xi}$, explicitly shows the assumption that the observed $\mathbf{x}$ variables contain no measurement error. It is the observed variable $\mathbf{x}$, not a latent variable per se, that influences $\boldsymbol{\eta}$. This assumption requires careful consideration. If the observed variable does contain error so that it is a proxy for some other underlying variable, then the equation, $\mathbf{x} = \mathbf{I}\boldsymbol{\xi}$, is not adequate. Three alternative strategies to treat the measurement error in $\mathbf{x}$ are (1) find multiple indicators for ξ and use the usual $\mathbf{x} = \mathbf{\Lambda}_x \boldsymbol{\xi} + \boldsymbol{\delta}$ model, (2) use estimates of reliabilities for the x variables, or (3) try a range of plausible reliability values and determine the sensitivity of the estimates to these values. The first strategy employs the CFA procedures of this chapter, and the last two were shown in Chapter 5.

I return to the objective and subjective socioeconomic status (SES) data from Chapter 4 for an example where cause indicators are plausible. Recall that the five observed variables are income (x_1), occupational prestige (x_2), subjective income (y_1), subjective occupational prestige (y_2), and subjective overall status (y_3). Suppose that we conceptualize the first two variables as indicators of objective socioeconomic status (ξ_1) and the last three as measures of subjective socioeconomic status (η_2). One model for these variables is shown in Figure 7.6(a), where all indicators are effects of the latent variables. Though we may be willing to treat the subjective measures y_1 to y_3 as "reflecting" the overall perception of SES, assuming that x_1 and x_2 are determined by objective SES is more difficult to defend (see Chapter 3). An alternative specification with x_1 and x_2 as cause indicators of objective SES ($= \eta_1$) is shown in Figure 7.6(b). The equations for this model follow from (7.99). But without further restrictions, the model is not identified since η_1 has no effect indicators. Constraining β_{21} to one and ψ_{11} to zero would identify it. The first constraint implies that η_1 is "scaled" to η_2 in the sense that a change of one unit in η_1 leads to an *expected* change of one unit in η_2. The second constraint ($\psi_{11} = 0$) defines objective SES to be a linear combination of its cause indicators (i.e., $\eta_1 = \gamma_{11}x_1 + \gamma_{12}x_2$). This makes η_1 a "weighted index" of its indicators. The chi-square estimates for both models in Figure 7.6 [with $\psi_{11} = 0$, $\beta_{21} = 1$ in (b)] are identical (chi-square $= 26.6$, df $= 4$, $p < 0.001$). This illustrates that overidentified models with different structures can have identical fits. It makes

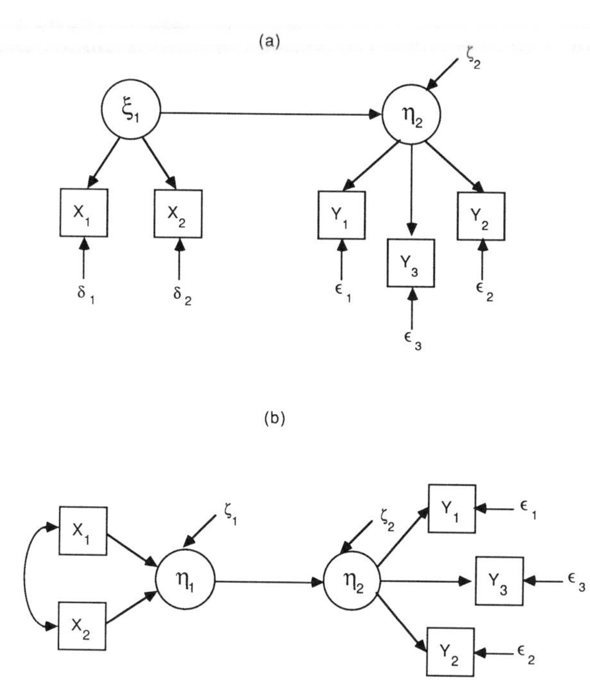

Figure 7.6 Alternative Representations of the Relationships between Objective and Subjective SES Indicators and Latent Variables

clear that examining fit statistics alone cannot determine the validity of a model. In this case neither model adequately matches the data.

Figure 7.6(b) with $\psi_{11} = 0$ and $\beta_{21} = 1$ is equivalent to a model that eliminates η_1 and allows x_1 and x_2 to affect η_2 directly. I discuss this type of "MIMIC model" in Chapter 8. Also we could examine the sensitivity of the model estimates to the constraints that ψ_{11}, VAR(δ_1), and VAR(δ_2) are zero by setting these parameters to a range of fixed values and comparing the new estimates to the original ones.

Higher-Order Factor Analyses

Part of the motivation for the development of factor analysis was the recognition that relatively few underlying latent variables may underlie a large number of indicators. And the latent variables more closely correspond to the concepts of social science theory than do the indicators. Less widely appreciated is that more general and abstract latent variables may determine the "first-order" latent variables. That is, the latent variables directly influencing the observed variables may be influenced by other latent variables that need not have direct effects on the observed variables. Such a model is a higher-order factor analysis.

The potential for structural relations between higher- and lower-order latent variables has long been recognized (Thurstone 1947), but there are still relatively few applications. Gerbing and Anderson (1984) argue that the failure to consider higher-order factors may explain the correlated errors that are common in CFA and that the higher-order factors are more informative than the correlated error representation. For instance, we may find correlated errors between a series of tests that tap different dimensions of intelligence. It may be that a second-order general intelligence factor can explain the association between the first-order intelligence dimensions and thus eliminate the correlated measurement errors. The equations for higher-order factor models are

$$\boldsymbol{\eta} = \mathbf{B}\boldsymbol{\eta} + \boldsymbol{\Gamma}\boldsymbol{\xi} + \boldsymbol{\zeta} \tag{7.100}$$

$$\mathbf{y} = \boldsymbol{\Lambda}_y\boldsymbol{\eta} + \boldsymbol{\epsilon} \tag{7.101}$$

The relations between the first-, second-, and higher-order factors are given by the first equation. The $\boldsymbol{\Gamma}\boldsymbol{\xi}$ term is not needed since the higher-order factors it represents could be defined as part of $\boldsymbol{\eta}$ with their respective coefficients in $\mathbf{B}$. Alternatively, the $\mathbf{B}\boldsymbol{\eta}$ term could be dropped when there are only second-order factors and none of the first-order factors have direct effects on one another. The first-order factor loadings of $\boldsymbol{\eta}$ on $\mathbf{y}$ are in $\boldsymbol{\Lambda}_y$.

An example comes from a study of self-concept by Marsh and Hocevar (1985). Figure 7.7 represents their model. The single second-order factor is nonacademic self-concept (ξ_1) that directly influences four first-order factors: physical ability (η_1), physical appearance (η_2), relations with peers (η_3), and relations with parents (η_4). These first-order factors have direct effects on four indicators apiece. The $\boldsymbol{\Lambda}_y$ is 16×4, with one indicator per construct chosen to scale the latent first-order factors. The $\boldsymbol{\Theta}_\epsilon$ matrix is assumed to be diagonal. For this example I do not need $\mathbf{B}\boldsymbol{\eta}$ in (7.100). The $\boldsymbol{\Gamma}$, $\boldsymbol{\Phi}$, and $\boldsymbol{\Psi}$ are

$$\boldsymbol{\Gamma} = \begin{bmatrix} 1 \\ \gamma_{21} \\ \gamma_{31} \\ \gamma_{41} \end{bmatrix}, \qquad \boldsymbol{\Phi} = \phi_{11}, \qquad \operatorname{diag} \boldsymbol{\Psi} = [\psi_{11} \quad \psi_{22} \quad \psi_{33} \quad \psi_{44}] \tag{7.102}$$

The first element in Γ scales nonacademic self-concept (ξ_1) to η_1 (self-concept of physical ability). The remaining coefficients are the second-order

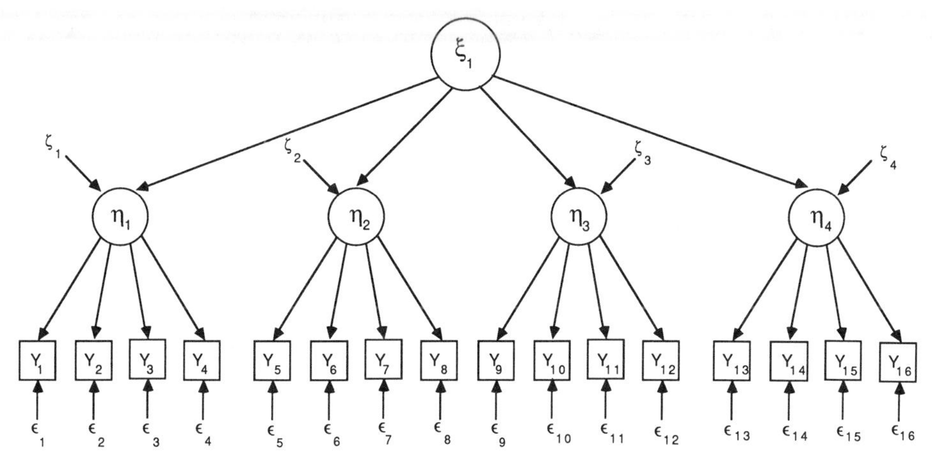

Figure 7.7 A Path Diagram for a Higher-Order Factor Analysis Model

factor loadings that are free. The variance of the second-order factor is ϕ_{11}. The Ψ matrix provides the variance in the first-order factors not explained by the second-order factor.

I analyze Marsh and Hocevar's (1985) data for 251 fifth graders in Sydney, Australia. The EQS program in Appendix 7A contains the data as well as the commands. All first- and second-order factor loadings ($\hat{\Lambda}_y$ and $\hat{\Gamma}$) are statistically significant and positive. The $R^2_{x_i}$'s for the indicators range from 0.27 to 0.77, and $R^2_{\eta_i}$'s are 0.30, 0.54, 0.77, and 0.25, respectively. Though these are somewhat good, the chi-square estimate is 218.6 with 100 df which is highly significant ($p < 0.001$). Also the other measures of overall fit suggest some improvements could be made (e.g., $\Delta_1 = 0.87$, $\Delta_2 = 0.93$). One possibility is that the factor complexity for all indicators is not one as assumed. Another is that nonacademic self-concept should be two second-order factors—physical self-concept and social self-concept. Both of these might depend on a third-order factor of overall nonacademic self-concept. Finally, it could be that the first-order factors determine the second-order factor rather than the reverse. These possibilities illustrate that similar issues of specification are involved in higher-order factor analysis as there are in the more standard CFA.

SUMMARY

The chapter demonstrated the estimation of confirmatory factor analysis models. As should be evident, these techniques allow relations between latent and observed variables that are not possible with traditional, exploratory factor analysis techniques. Furthermore a series of statistics are available to help assess the fit between a model and data. Also obvious is that even the "confirmatory" techniques have an exploratory component when trying to improve a model with a poor fit. Replication of results is needed as much for CFA as for other structural equation models.

The last section on extensions of CFA showed that the line separating measurement models from latent variable models becomes blurred. For instance, in the higher-order factor analysis the equation linking η to ξ is part of the measurement model, yet it also is an equation linking latent variables. The equations to treat cause indicators appears more as a hybrid model than just as a measurement model. In the next chapter I present the complete synthesis of measurement and latent variable models that these examples foreshadow.

APPENDIX 7A EXAMPLES OF PROGRAM LISTINGS

Example 1: Estimating Means and Intercepts

In this appendix I have listed the LISREL program for estimating the line length example with intercepts and means. Note that the sixth input "variable" is the constant "1." This has zero covariances with other five variables and a mean of one. Also note that I provide most of the starting values for this example, since the usual initial estimates do not work well. See Jöreskog and Sörbom (1986) for a more complete explanation of the LISREL commands. See the text for an explanation and interpretation of the results.

```
LINE LENGTH W/ MEANS & INTERCEPTS 5 JUDGES
DA NI=6 NOBS=60 MA=MM
CM
*
14.1852
 4.38314    1.45875
 7.50486    2.34359    4.20911
 4.54647    1.42332    2.4467     1.50202
 4.00586    1.24436    2.17272    1.30718    1.20017
 0          0          0          0          0          0
ME
*
6.75833333   2.23   3.42666667   2.096666667   1.91833333 1
```

```
MO NY=5 FI NX=1 NE=2 TE=SY PS=DI GA=FI
FR LY 1 1 LY 1 2 LY 2 1 LY 2 2 LY 3 1 LY 3 2 LY 4 1 LY 4 2
FI PS 2 2
FR GA 1 1
ST 1 GA 2 1
MA LY
*
2.5 0 1 .1 1 .1 1 .1 1 0
MA TE
*
.1 0 .1 0 0 .1 0 0 0 .1 0 0 0 0 .1
OU TV MI TO
```

Example 2: Estimating a Higher-Order Factor Analysis

The text lists the matrices required to program the higher-order factor analysis example with the LISREL program. The following is a listing of the commands for this problem in EQS.

```
/TITLE
   HIGHER-ORDER FACTOR ANALYSIS OF SELF-CONCEPT WITH ONE
   2ND ORDER, FOUR 1ST ORDER FACTORS 5TH GRADE STUDENTS
   1ST 16 VARS FROM MARSH & HOCEVAR (1985) PSYCHOLOGICAL
   BULLETIN
/SPEC
   CAS=251; VAR=16; ME=ML; MA=COV;
/LAB
   F1=NONACAD; F2=PHYSABIL; F3=APPEAR; F4=PEERS;
   F5=PARENTS;
   V1=PHYAB1; V2=PHYAB2; V3=PHYAB3; V4=PHYAB4;
   V5=APPEAR1; V6=APPEAR2; V7=APPEAR3; V8=APPEAR4;
   V9=PEERREL1; V10=PEERREL2; V11=PEERREL3;
   V12=PEERREL4; V13=PARREL1; V14=PARREL2;
   V15=PARREL3; V16=PARREL4;
/EQU
  F2=1F1 + D2; F3=1*F1 + D3; F4=1*F1 + D4; F5=1*F1 + D5;
  V1=1F2 + E1; V5=1F3 + E5; V9=1F4 + E9; V13=1F5 + E13;
  V2=1*F2 + E2; V3=1*F2 + E3; V4=1*F2 + E4;
  V6=1*F3 + E6; V7=1*F3 + E7; V8=1*F3 + E8;
  V10=1*F4 + E10; V11=1*F4 + E11; V12=1*F4 + E12;
  V14=1*F5 + E14; V15=1*F5 + E15; V16=1*F5 + E16;
/VAR
  F1=1*; D2 TO D5=1*; E1 TO E16=2*;
```

```
/MAT
 1.00
 .31 1.00
 .52 .45 1.00
 .54 .46 .70 1.00
 .15 .33 .22 .21 1.00
 .14 .28 .21 .13 .72 1.00
 .16 .32 .35 .31 .59 .56 1.00
 .23 .29 .43 .36 .55 .51 .65 1.00
 .24 .13 .24 .23 .25 .24 .24 .30 1.00
 .19 .26 .22 .18 .34 .37 .36 .32 .38 1.00
 .16 .24 .36 .30 .33 .29 .44 .51 .47 .50 1.0
 .16 .21 .35 .24 .31 .33 .41 .39 .47 .47 .55 1.00
 .08 .18 .09 .12 .19 .24 .08 .21 .21 .19 .19 .20 1.00
 .01 -.01 .03 .02 .10 .13 .03 .05 .26 .17 .23 .26 .33 1.00
 .06 .19 .22 .22 .23 .24 .20 .26 .16 .23 .38 .24 .42 .40 1.00
 .04 .17 .10 .07 .26 .24 .12 .26 .16 .22 .32 .17 .42 .42 .65 1.0
/STA
 1.84 1.94 2.07 1.82 2.34 2.61 2.48 2.34 1.71 1.93 2.18 1.94
 1.31 1.57 1.77 1.47
/END
```

CHAPTER EIGHT

The General Model, Part I: Latent Variable and Measurement Models Combined

Until now I have treated two basic types of structural equation models. The one most closely linked to regression and econometric techniques concentrates on the relation between observed variables with the only "unobserved" variables being the errors in equations (or disturbance terms). The other emerged from the factor analysis tradition with its emphasis on the link between latent variables and observed ones but with little discussion of the influence of one latent variable on another. The general structural equation model, the topic of this chapter, represents a synthesis of these two model types. It consists of a *measurement model* that specifies the relation of observed to latent variables and a *latent variable model* that shows the influence of latent variables on each other. The models presented in the earlier chapters (especially Chapters 4, 5, and 7) are special cases of this general one, as I will show in the next section. This chapter builds on the earlier ones, and much of the prior materials on estimation, testing, and model fit are appropriate here. The chapter topics are as follows: model specification, implied covariance matrix, identification, estimation and model evaluation, standardized and unstandardized coefficients, means and equation constants, comparing groups, missing values, and the decomposition of effects in path analysis.

MODEL SPECIFICATION

The first component of the structural equations is the latent variable model:

$$\boldsymbol{\eta} = \mathbf{B}\boldsymbol{\eta} + \boldsymbol{\Gamma}\boldsymbol{\xi} + \boldsymbol{\zeta} \tag{8.1}$$

In (8.1), $\boldsymbol{\eta}$, the vector of latent endogenous random variables, is $m \times 1$; $\boldsymbol{\xi}$, the latent exogenous random variables, is $n \times 1$; $\mathbf{B}$ is the $m \times m$ coefficient matrix showing the influence of the latent endogenous variables on each other; $\boldsymbol{\Gamma}$ is the $m \times n$ coefficient matrix for the effects of $\boldsymbol{\xi}$ on $\boldsymbol{\eta}$. The matrix $(\mathbf{I} - \mathbf{B})$ is nonsingular. $\boldsymbol{\zeta}$ is the disturbance vector that is assumed to have an expected value of zero $[E(\boldsymbol{\zeta}) = \mathbf{0}]$ and which is uncorrelated with $\boldsymbol{\xi}$.

The second component of the general system is the measurement model:

$$\mathbf{y} = \boldsymbol{\Lambda}_y \boldsymbol{\eta} + \boldsymbol{\epsilon} \tag{8.2}$$

$$\mathbf{x} = \boldsymbol{\Lambda}_x \boldsymbol{\xi} + \boldsymbol{\delta} \tag{8.3}$$

The $\mathbf{y}$ $(p \times 1)$ and the $\mathbf{x}$ $(q \times 1)$ vectors are observed variables, $\boldsymbol{\Lambda}_y (p \times m)$ and $\boldsymbol{\Lambda}_x (q \times n)$ are the coefficient matrices that show the relation of $\mathbf{y}$ to $\boldsymbol{\eta}$ and $\mathbf{x}$ to $\boldsymbol{\xi}$, respectively, and $\boldsymbol{\epsilon}$ $(p \times 1)$ and $\boldsymbol{\delta}$ $(q \times 1)$ are the errors of measurement for $\mathbf{y}$ and $\mathbf{x}$, respectively. The errors of measurement are assumed to be uncorrelated with $\boldsymbol{\xi}$ and $\boldsymbol{\zeta}$ and with each other. This latter assumption that $\boldsymbol{\epsilon}$ and $\boldsymbol{\delta}$ are uncorrelated is not as restrictive as it seems, and I will show how it can be relaxed in Chapter 9. The expected values of $\boldsymbol{\epsilon}$ and $\boldsymbol{\delta}$ are zero. To simplify matters $\boldsymbol{\eta}$, $\boldsymbol{\xi}$, $\mathbf{y}$, and $\mathbf{x}$ are written as deviations from their means.

It should be clear that (8.2) and (8.3) are the same as (7.1) and (7.2) of the confirmatory factor analysis model in Chapter 7. If we assume that $\boldsymbol{\Lambda}_y = \mathbf{I}_m$, $\boldsymbol{\Lambda}_x = \mathbf{I}_n$, $\boldsymbol{\Theta}_\delta = \mathbf{0}$, and $\boldsymbol{\Theta}_\epsilon = \mathbf{0}$, then (8.1) becomes

$$\mathbf{y} = \mathbf{B}\mathbf{y} + \boldsymbol{\Gamma}\mathbf{x} + \boldsymbol{\zeta} \tag{8.4}$$

which is identical to the structural equations with observed variables of Chapter 4.

The models of Chapter 5 on the consequences of measurement error are another instance of the general model where $\boldsymbol{\Lambda}_x = \mathbf{I}_q$ and $\boldsymbol{\Lambda}_y = \mathbf{I}_p$ and with some nonzero elements of $\boldsymbol{\Theta}_\delta$ and $\boldsymbol{\Theta}_\epsilon$:

$$\boldsymbol{\eta} = \mathbf{B}\boldsymbol{\eta} + \boldsymbol{\Gamma}\boldsymbol{\xi} + \boldsymbol{\zeta} \tag{8.5}$$

$$\mathbf{y} = \boldsymbol{\eta} + \boldsymbol{\epsilon} \tag{8.6}$$

$$\mathbf{x} = \boldsymbol{\xi} + \boldsymbol{\delta} \tag{8.7}$$

Models with causal indicators (see end of Chapter 7) or MIMIC models can be incorporated into (8.1) to (8.3) by setting $\Lambda_x = \mathbf{I}_p$ and $\Theta_\delta = \mathbf{0}$, leading to

$$\boldsymbol{\eta} = \mathbf{B}\boldsymbol{\eta} + \boldsymbol{\Gamma}\mathbf{x} + \boldsymbol{\zeta} \tag{8.8}$$

$$\mathbf{y} = \boldsymbol{\Lambda}_y\boldsymbol{\eta} + \boldsymbol{\epsilon} \tag{8.9}$$

$$\mathbf{x} = \boldsymbol{\xi} \tag{8.10}$$

Finally, the second-order factor analyses of Chapter 7 also are a specialization of (8.1) to (8.3) where there are no $\mathbf{x}$ variables:

$$\boldsymbol{\eta} = \mathbf{B}\boldsymbol{\eta} + \boldsymbol{\Gamma}\boldsymbol{\xi} + \boldsymbol{\zeta} \tag{8.11}$$

$$\mathbf{y} = \boldsymbol{\Lambda}_y\boldsymbol{\eta} + \boldsymbol{\epsilon} \tag{8.12}$$

The many seemingly unique procedures of econometrics and factor analyses fall under one umbrella. There are, however, some restrictions. For instance, $\boldsymbol{\xi}$ cannot directly affect any y's. If the x's and y's contain measurement error, they cannot directly influence one another. I will show in Chapter 9 that even these restrictions can be overcome.

To use the general model, we need to specify the pattern of elements in each of eight matrices—$\mathbf{B}$, $\boldsymbol{\Gamma}$, $\boldsymbol{\Lambda}_y$, $\boldsymbol{\Lambda}_x$, $\boldsymbol{\Phi}$, $\boldsymbol{\Psi}$, $\boldsymbol{\Theta}_\epsilon$, and $\boldsymbol{\Theta}_\delta$. To do this, we must draw upon our substantive knowledge of the research area. The type of questions that require attention are, for instance, what endogenous variables have direct effects on other endogenous variables? Are the errors of one equation (ζ) likely to correlate with errors of another? Similarly, are the errors of measurement for any indicators likely to correlate with those of another measure? Which latent variables are related to which indicators? Realistically, the existing knowledge in most substantive areas is not detailed enough to answer all of these. It may be tempting to "let the data provide the answers" and to not restrict in advance any of the eight matrices. Clearly, such a strategy would produce an underidentified model from which we would learn little. Instead, we should incorporate into the model as much information about the problem as possible.

To illustrate the specification of a general model, I return to the example first presented in Chapter 2. This panel data represents the relation of political democracy circa 1965 and 1960 and industrialization in 1960 for developing countries. In the cross-national study of political democracy, industrialization is often seen as enhancing the chances for a democratic political system. In addition the current extent of democracy is likely to be influenced by the prior level of political democracy. These ideas suggest that the latent variable model should have democracy in 1960 (η_1) and industrialization in 1960 (ξ_1) influencing democracy in 1965 (η_2). Furthermore I allow a contemporaneous effect of industrialization on political democracy in 1960.[1] From this description we can form the elements of **B** and Γ and the latent variable model:

$$\begin{bmatrix} \eta_1 \\ \eta_2 \end{bmatrix} = \begin{bmatrix} 0 & 0 \\ \beta_{21} & 0 \end{bmatrix} \begin{bmatrix} \eta_1 \\ \eta_2 \end{bmatrix} + \begin{bmatrix} \gamma_{11} \\ \gamma_{21} \end{bmatrix} [\xi_1] + \begin{bmatrix} \zeta_1 \\ \zeta_2 \end{bmatrix} \tag{8.13}$$

I assume that the errors in the equation for η_1 (i.e., ζ_1) are uncorrelated with those for η_2 (i.e., ζ_2), making $\boldsymbol{\Psi}$ a diagonal matrix. Also $\boldsymbol{\Phi}$ in this model is a scalar, ϕ_{11}, that is the variance of industrialization (ξ_1).

For the measurement model, I use four indicators, each at two points in time to measure political democracy. They are the same as those employed in Chapter 7: freedom of the press (y_1, y_5), freedom of group opposition (y_2, y_6), fairness of elections (y_3, y_7), and the elective nature and effectiveness of the legislative body (y_4, y_8). As in Chapter 7 I assume that the 1960 latent democracy variable (η_1) only influences the 1960 measures (y_1 to y_4) and that the same pattern holds for the 1965 political democracy variable (η_2) and measures (y_5 to y_8). The scale of η_1 is set to y_1, and that for η_2 is set to y_5. Furthermore the coefficient linking political democracy to freedom of group opposition in 1960 is the same value in 1965. Similarly, the other coefficients for the same indicator at different time points are set equal. The preceding information reveals the pattern of Λ_y. The y measurement equation is

$$\begin{bmatrix} y_1 \\ y_2 \\ y_3 \\ y_4 \\ y_5 \\ y_6 \\ y_7 \\ y_8 \end{bmatrix} = \begin{bmatrix} 1 & 0 \\ \lambda_2 & 0 \\ \lambda_3 & 0 \\ \lambda_4 & 0 \\ 0 & 1 \\ 0 & \lambda_2 \\ 0 & \lambda_3 \\ 0 & \lambda_4 \end{bmatrix} \begin{bmatrix} \eta_1 \\ \eta_2 \end{bmatrix} + \begin{bmatrix} \epsilon_1 \\ \epsilon_2 \\ \epsilon_3 \\ \epsilon_4 \\ \epsilon_5 \\ \epsilon_6 \\ \epsilon_7 \\ \epsilon_8 \end{bmatrix} \tag{8.14}$$

[1]As with all substantive areas, controversy over a model is likely. For instance, some would argue that political democracy in developing countries could diminish industrialization. This suggests a reciprocal relation between industrialization and democracy in 1960. A more elaborate model than the present one would be needed to test this idea since the present one would not be identified with this feedback relation.

To complete the y measurement model, we require $\boldsymbol{\Theta}_\epsilon$. I assume that the only free and nonzero elements in $\boldsymbol{\Theta}_\epsilon$ are in its main diagonal, the i and $i + 4$ (where $i = 1, 2, 3, 4$) elements, and the $(4, 2)$ and $(8, 6)$ elements. These latter elements correspond to the covariances of the measurement errors for the same indicator at two points in time and to the indicators that come from the same data source.

The x equation is simpler. The three measures of 1960 industrialization (ξ_1) are gross national product (GNP) per capita (x_1), energy consumption per capita (x_2), and the percent of the labor force in industrial occupations (x_3):[2] Industrialization (ξ_1) is set to the scale of x_1, whereas coefficients showing ξ_1's influence on x_2 and x_3 are unconstrained. The **x** measurement equation is[3]

$$\begin{bmatrix} x_1 \\ x_2 \\ x_3 \end{bmatrix} = \begin{bmatrix} 1 \\ \lambda_6 \\ \lambda_7 \end{bmatrix} [\xi_1] + \begin{bmatrix} \delta_1 \\ \delta_2 \\ \delta_3 \end{bmatrix} \tag{8.15}$$

The $\boldsymbol{\Theta}_\delta$ for this model is diagonal since I do not have reasons to suspect otherwise.

All of these relations and assumptions are summarized in the path diagram in Figure 8.1. As in the prior chapters, before estimation it is useful to examine the relation of the covariance matrix of the observed variables to the structural parameters of the model.

IMPLIED COVARIANCE MATRIX

Chapters 4 and 7 derived the implied covariance matrix based on structural equations with observed variables and confirmatory factor analysis models, respectively. In this section I show that each of these are special cases of the implied covariance matrix for the general structural equation model.

To begin, I represent $\boldsymbol{\Sigma}_{yy}$ as the covariance matrix of the observed **y** variables, and $\boldsymbol{\Sigma}_{yy}(\boldsymbol{\theta})$ contains the covariances of **y** written as a function of the unknown model parameters that are stacked in the vector, $\boldsymbol{\theta}$. The

[2] I transformed the three variables. The first two were logarithmically transformed, while I used the arcsin of the square root of the percent variable. These transformations led to better approximations to normal distributions than did the original variables.

[3] The λ's in $\boldsymbol{\Lambda}_x$ and $\boldsymbol{\Lambda}_y$ have single subscripts since the double subscript system does not work well when both **x** and **y** indicators are present (e.g., does λ_{21} refer to ξ_1's effect on x_2 or η_1's effect on y_2?) Alternatively, I could add a y or x superscript in conjunction with λ_{ij}, but to avoid complicating the notation I do not.

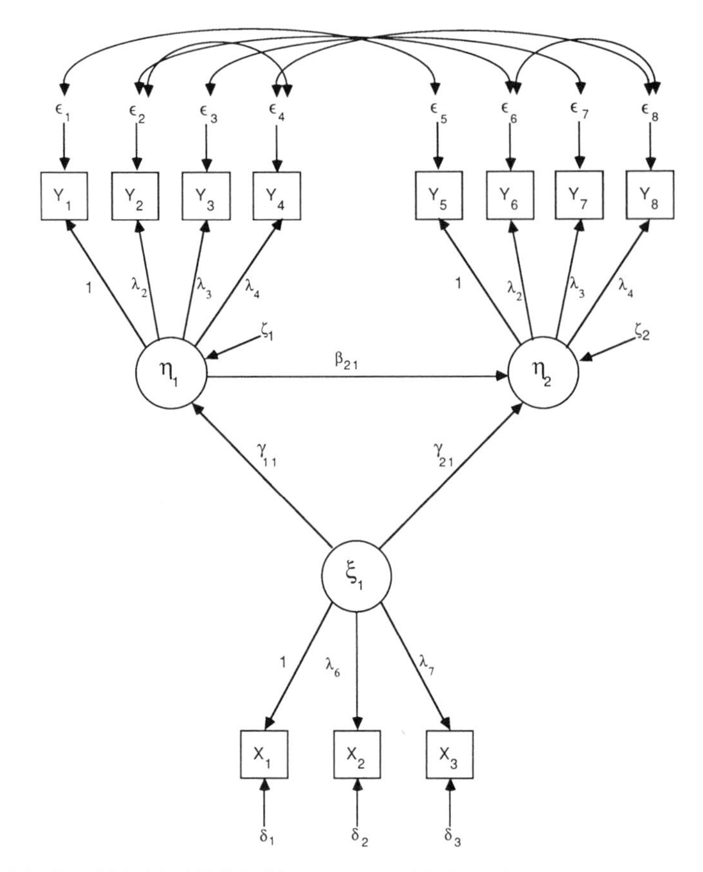

Figure 8.1 Panel Model of Political Democracy and Industrialization for Developing Countries, 1960 to 1965

$\boldsymbol{\Sigma}_{yy}(\boldsymbol{\theta})$ is

$$\begin{aligned}\boldsymbol{\Sigma}_{yy}(\boldsymbol{\theta}) &= E(\mathbf{yy}') \\ &= E\left[(\boldsymbol{\Lambda}_y\boldsymbol{\eta} + \boldsymbol{\epsilon})(\boldsymbol{\eta}'\boldsymbol{\Lambda}_y' + \boldsymbol{\epsilon}')\right] \\ &= \boldsymbol{\Lambda}_y E(\boldsymbol{\eta\eta}')\boldsymbol{\Lambda}_y' + \boldsymbol{\Theta}_\epsilon \qquad (8.16)\end{aligned}$$

The $E(\boldsymbol{\eta\eta}')$ can be further broken down by substituting the reduced form of equation (8.1)—that is, $\boldsymbol{\eta} = (\mathbf{I} - \mathbf{B})^{-1}(\boldsymbol{\Gamma\xi} + \boldsymbol{\zeta})$—for $\boldsymbol{\eta}$ in (8.16) and by simplifying

$$\boldsymbol{\Sigma}_{yy}(\boldsymbol{\theta}) = \boldsymbol{\Lambda}_y(\mathbf{I} - \mathbf{B})^{-1}(\boldsymbol{\Gamma\Phi\Gamma}' + \boldsymbol{\Psi})\left[(\mathbf{I} - \mathbf{B})^{-1}\right]'\boldsymbol{\Lambda}_y' + \boldsymbol{\Theta}_\epsilon \qquad (8.17)$$

Thus the covariance matrix of **y** is a complex function of six of the eight model parameter matrices.

The covariance matrix of **y** with **x** is Σ_{yx}, and when referring to it as a function of the structural parameters, it is $\Sigma_{yx}(\boldsymbol{\theta})$. This equals

$$\begin{aligned}\Sigma_{yx}(\boldsymbol{\theta}) &= E(\mathbf{y}\mathbf{x}') \\ &= E\left[(\Lambda_y\boldsymbol{\eta} + \boldsymbol{\epsilon})(\boldsymbol{\xi}'\Lambda_x' + \boldsymbol{\delta}')\right] \\ &= \Lambda_y E(\boldsymbol{\eta}\boldsymbol{\xi}')\Lambda_x' \end{aligned} \tag{8.18}$$

Again making use of the reduced form of $\boldsymbol{\eta}$ leads to

$$\Sigma_{yx}(\boldsymbol{\theta}) = \Lambda_y(\mathbf{I} - \mathbf{B})^{-1}\Gamma\Phi\Lambda_x' \tag{8.19}$$

Finally the covariance matrix of **x**, Σ_{xx}, written as a function of the structural parameters $[\Sigma_{xx}(\boldsymbol{\theta})]$ is the same as that derived in Chapter 7:

$$\Sigma_{xx}(\boldsymbol{\theta}) = \Lambda_x\Phi\Lambda_x' + \Theta_\delta \tag{8.20}$$

If I assemble (8.17), (8.19), and (8.20) into a single matrix $\Sigma(\boldsymbol{\theta})$, I will have the covariance matrix for the observed **y** and **x** variables as a function of the model parameters:

$$\Sigma(\boldsymbol{\theta}) = \begin{bmatrix} \Sigma_{yy}(\boldsymbol{\theta}) & \Sigma_{yx}(\boldsymbol{\theta}) \\ \Sigma_{xy}(\boldsymbol{\theta}) & \Sigma_{xx}(\boldsymbol{\theta}) \end{bmatrix}$$

$$= \begin{bmatrix} \Lambda_y(\mathbf{I}-\mathbf{B})^{-1}(\Gamma\Phi\Gamma' + \Psi)\left[(\mathbf{I}-\mathbf{B})^{-1}\right]'\Lambda_y' + \Theta_\epsilon & \Lambda_y(\mathbf{I}-\mathbf{B})^{-1}\Gamma\Phi\Lambda_x' \\ \Lambda_x\Phi\Gamma'\left[(\mathbf{I}-\mathbf{B})^{-1}\right]'\Lambda_y' & \Lambda_x\Phi\Lambda_x' + \Theta_\delta \end{bmatrix} \tag{8.21}$$

[The $\Sigma_{xy}(\boldsymbol{\theta})$ matrix in the lower left quadrant of (8.21) is the transpose of $\Sigma_{xy}(\boldsymbol{\theta})$.]

You should note that $\Sigma(\boldsymbol{\theta})$ of equation (8.21) contains the implied covariance matrices treated in Chapters 4 and 7. Consider structural equations with only observed variables where $\Theta_\epsilon = \mathbf{0}$, $\Theta_\delta = \mathbf{0}$, $\Lambda_y = \mathbf{I}_p$, and $\Lambda_x = \mathbf{I}_q$. Substituting these into (8.21) leads to equation (4.10). Similarly, substituting $\mathbf{B} = \mathbf{0}$, $\Gamma = \mathbf{0}$, $\Theta_\epsilon = \mathbf{0}$, $\Lambda_y = \mathbf{0}$, and $\Psi = \mathbf{0}$ leads to the $\Sigma(\boldsymbol{\theta})$ for confirmatory factor analysis [see equation (7.6)] which also is the lower-right quadrant of (8.21). In analogous fashion the $\Sigma(\boldsymbol{\theta})$ for other specializations (e.g., MIMIC models) or for specific examples can be obtained from (8.21). The $\Sigma(\boldsymbol{\theta})$ for the industrialization and political democracy example can be found by substituting the specific form of the eight parameter matrices in (8.21). This would lead to an 11×11 matrix. As in the earlier chapters, estimation proceeds by selecting values for the unknown parameters in $\boldsymbol{\theta}$ so that $\Sigma(\boldsymbol{\theta})$ matches the covariance matrix of the observed variables. But before discussing estimation, we need to consider identification.

IDENTIFICATION

The issue of identification is as important to the general model as it is to the factor analysis and simultaneous equation models treated in the earlier chapters. The parameters in $\boldsymbol{\theta}$ are *globally* identified if no vector $\boldsymbol{\theta}_1$ and $\boldsymbol{\theta}_2$ exist such that $\Sigma(\boldsymbol{\theta}_1) = \Sigma(\boldsymbol{\theta}_2)$ unless $\boldsymbol{\theta}_1 = \boldsymbol{\theta}_2$. One way to establish identification is algebraically. Each element of $\boldsymbol{\theta}$ must be solved for in terms of one or more known-to-be-identified elements of Σ. The last section showed that the covariance structure $\Sigma = \Sigma(\boldsymbol{\theta})$ implies $\frac{1}{2}(p + q)(p + q + 1)$ nonredundant equations of the form $\sigma_{ij} = \sigma_{ij}(\boldsymbol{\theta})(i \leq j)$, where σ_{ij} is the *ij* element of Σ and $\sigma_{ij}(\boldsymbol{\theta})$ is the *ij* element of $\Sigma(\boldsymbol{\theta})$. If an element of $\boldsymbol{\theta}$ can be expressed as a function of one or more σ_{ij}, then this establishes its identification. If all elements of $\boldsymbol{\theta}$ meet this condition, the model is identified.

Consider the model in Figure 8.2. The equations for this path diagram are

$$\begin{gathered}
[\eta_1] = [\gamma_{11}][\xi_1] + [\zeta_1] \\
\begin{bmatrix} y_1 \\ y_2 \end{bmatrix} = \begin{bmatrix} 1 \\ \lambda_2 \end{bmatrix}[\eta_1] + \begin{bmatrix} \epsilon_1 \\ \epsilon_2 \end{bmatrix}, \qquad \text{diag } \boldsymbol{\Theta}_\epsilon \\
\begin{bmatrix} x_1 \\ x_2 \end{bmatrix} = \begin{bmatrix} 1 \\ \lambda_1 \end{bmatrix}[\xi_1] + \begin{bmatrix} \delta_1 \\ \delta_2 \end{bmatrix}, \qquad \text{diag } \boldsymbol{\Theta}_\delta
\end{gathered} \tag{8.22}$$

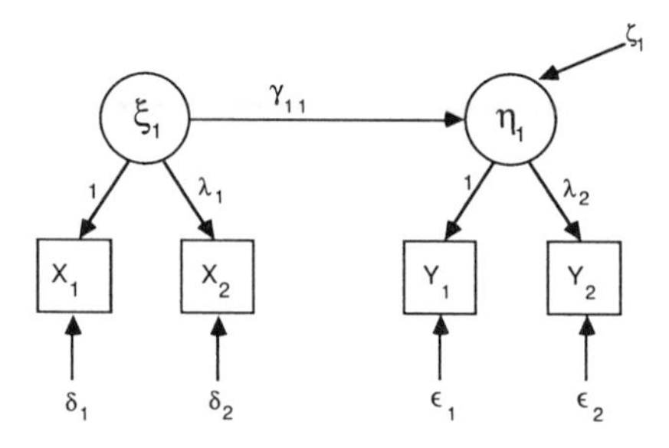

Figure 8.2 A Model of Two Latent Variables (ξ_1, η_1), Each Measured with Two Indicators (X_1, X_2, Y_1, Y_2)

The covariance matrix of the observed variables is

$$\Sigma = \begin{bmatrix} \mathrm{VAR}(y_1) & & & \\ \mathrm{COV}(y_2, y_1) & \mathrm{VAR}(y_2) & & \\ \mathrm{COV}(x_1, y_1) & \mathrm{COV}(x_1, y_2) & \mathrm{VAR}(x_1) & \\ \mathrm{COV}(x_2, y_1) & \mathrm{COV}(x_2, y_2) & \mathrm{COV}(x_2, x_1) & \mathrm{VAR}(x_2) \end{bmatrix} \quad (8.23)$$

Substituting the parameter matrices for the above model (8.22) into the implied covariance matrix derived in the last section [see (8.21)] leads to

$$\Sigma(\theta) = \begin{bmatrix} \gamma_{11}^2\phi_{11} + \psi_{11} + \mathrm{VAR}(\epsilon_1) & & & \\ \lambda_2(\gamma_{11}^2\phi_{11} + \psi_{11}) & \lambda_2^2(\gamma_{11}^2\phi_{11} + \psi_{11}) + \mathrm{VAR}(\epsilon_2) & & \\ \gamma_{11}\phi_{11} & \lambda_2\gamma_{11}\phi_{11} & \phi_{11} + \mathrm{VAR}(\delta_1) & \\ \lambda_1\gamma_{11}\phi_{11} & \lambda_1\lambda_2\gamma_{11}\phi_{11} & \lambda_1\phi_{11} & \lambda_1^2\phi_{11} + \mathrm{VAR}(\delta_2) \end{bmatrix} \quad (8.24)$$

where

$$\theta' = [\lambda_1 \quad \lambda_2 \quad \gamma_{11} \quad \phi_{11} \quad \mathrm{VAR}(\epsilon_1) \quad \mathrm{VAR}(\epsilon_2) \quad \mathrm{VAR}(\delta_1) \quad \mathrm{VAR}(\delta_2) \quad \psi_{11}].$$

The covariance structure of $\Sigma = \Sigma(\theta)$ leads to ten $[= \frac{1}{2}(4)(5)]$ equations in nine unknowns. I can establish the identification of this model if I solve these equations for the nine unknowns in θ.

To illustrate this, two of the ten equations implied by the model are

$$\begin{aligned} \mathrm{COV}(x_2, y_1) &= \lambda_1\gamma_{11}\phi_{11} \\ \mathrm{COV}(x_2, x_1) &= \lambda_1\phi_{11} \end{aligned} \quad (8.25)$$

which leads to

$$\gamma_{11} = \frac{\mathrm{COV}(x_2, y_1)}{\mathrm{COV}(x_2, x_1)} \quad (8.26)$$

This establishes that γ_{11} is identified. If I divide the $\mathrm{COV}(x_2, y_2)$ by $\mathrm{COV}(x_1, y_2)$ and the $\mathrm{COV}(x_2, y_2)$ by $\mathrm{COV}(x_2, y_1)$, I can show that λ_1 and λ_2 are identified. By a series of additional algebraic manipulations I can show that all elements of θ are identified. Implicit in claiming identification is that the covariances of observed variables or the parameters do not lead to undefined or improper solutions. For instance, if $\mathrm{COV}(x_1, x_2) = 0$, γ_{11} in (8.26) is not defined, or if $\gamma_{11} = 0$, the model is underidentified.

In this example and other relatively simple models the algebraic means to establish identification may have some utility. But, as seen in Chapters 4 and 7 for more complex models, this algebraic approach is extremely tedious and prone to mistakes. Rules exist that aid in the identification of the general model. I review several of these next.

t-Rule

As in structural equations with observed variables and confirmatory factor analysis models, the t-rule is a *necessary* but not sufficient condition of identification:

$$t < \tfrac{1}{2}(p + q)(p + q + 1) \tag{8.27}$$

where t is the number of free and unconstrained elements in $\boldsymbol{\theta}$. The justification for this rule is the same as before. The nonredundant elements of $\boldsymbol{\Sigma} = \boldsymbol{\Sigma}(\boldsymbol{\theta})$ imply $\frac{1}{2}(p + q)(p + q + 1)$ equations. If the number of unknowns in $\boldsymbol{\theta}$ exceeds the number of equations, identification is not possible. The model in Figure 8.2 and equations (8.24) show that $\boldsymbol{\theta}$ has nine elements and $\frac{1}{2}(p + q)(p + q + 1)$ is ten. Thus the t-rule is met.

Two-Step Rule

As the name suggests, the two-step rule has two parts. In the first step the analyst treats the model as a confirmatory factor analysis. This entails viewing the original $\mathbf{x}$ and $\mathbf{y}$ as x variables and the original $\boldsymbol{\xi}$ and $\boldsymbol{\eta}$ as ξ variables. The only relationships between the latent variables that are of concern are their variances and covariances ($\boldsymbol{\Phi}$). That is, ignore the $\mathbf{B}$, $\boldsymbol{\Gamma}$, and $\boldsymbol{\Psi}$ elements. With the model reformulated as a confirmatory factor analysis, determine if it is identified. At this point any of the rules of identification presented in Chapter 7 are available. If identification can be established, move to the second step. If not, this identification rule is not applicable.

The second step examines the latent variable equation of the original model (i.e., $\boldsymbol{\eta} = \mathbf{B}\boldsymbol{\eta} + \boldsymbol{\Gamma}\boldsymbol{\xi} + \boldsymbol{\zeta}$) and treats it as though it were a structural equation in observed variables. That is, assume that each latent variable is an observed variable that is perfectly measured. Next determine whether $\mathbf{B}$, $\boldsymbol{\Gamma}$, and $\boldsymbol{\Psi}$ are identified, using the identification rules developed in Chapter 4 and ignoring the measurement parameters considered in the first step. *If the first step shows that the measurement parameters are identified and the second step shows that the latent variable model parameters also are identified, then this is sufficient to identify the whole model.*

To understand why this rule leads to identified models, consider each step separately. The first step establishes that all parameters in the measurement model are identified, including the covariance matrix of the latent variables. In the second step the latent variable covariance matrix plays a role analogous to the covariance matrix of observed variables in the structural equation models of Chapter 4. There the question is whether the unknown parameters in $\mathbf{B}$, $\boldsymbol{\Gamma}$, $\boldsymbol{\Phi}$, and $\boldsymbol{\Psi}$ are functions of the identified elements of the observed variables' covariance matrix. Here I ask the same question, but the identified covariance matrix is of latent rather than of observed variables. The rules of Chapter 4 help to show whether the parameters associated with the latent variable model are identified. Putting the two pieces together establishes the identification of the whole model. I illustrate this with the four-indicator, two latent variable model in Figure 8.2 and equations (8.22). In the first step η_1 is redefined as ξ_2, y_1 and y_2 are now x_3 and x_4, ϵ_1 and ϵ_2 are δ_3 and δ_4, and ζ_1 and γ_{11} are not considered. Instead, we now examine the variances and covariance of ξ_1 and the new $\xi_2(=\eta_1)$. This reformulation is represented in Figure 8.3(a).

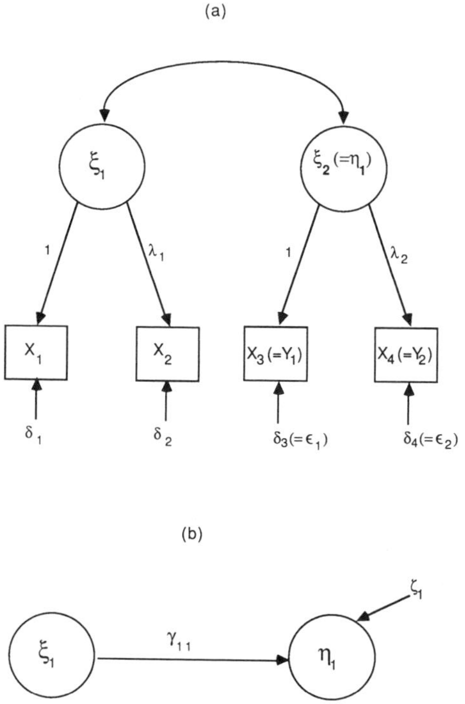

Figure 8.3 Illustration of the Two-Step Identification Rule

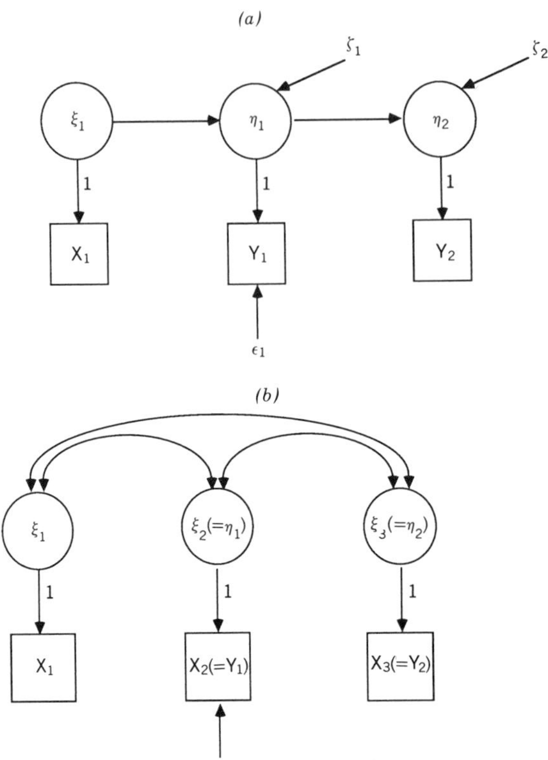

Figure 8.4 An Example of a Model That Fails the Two-Step Identification Rule but Is Identified

Using the two-indicator rule from Chapter 7, I can establish that the model is identified as long as ϕ_{12} is not zero. This completes the first step. The second step is represented in Figure 8.3(b). If I assume that η_1 and ξ_1 are perfectly measured, it is clear that the model is identified by the null **B** rule I presented in Chapter 4. Since both steps are satisfied, the model is identified.

Remember that this is a *sufficient* condition for model identification. A model could fail to meet it and still be identified. For instance, consider the model in Figure 8.4(a). If I take the first step of reformulating it into a confirmatory factor analysis model, the result is three latent variables with three indicators as represented in Figure 8.4(b). Even with no measurement error in x_1 and x_3 $(= y_2)$, the model in Figure 8.4(b) is not identified so the first step is not met. Yet in Chapter 5 I showed that the original model in Figure 8.4(a) is identified. This exemplifies the possibility that constraints in the latent variable relations can aid the identification of measurement parameters so that even if a model fails the two-step rule, it still may be possible to find unique solutions for the unknown parameters.

MIMIC Rule

Some researchers have given particular attention to one special case of the general model referred to as MIMIC models. These contain observed variables that are *M*ultiple *I*ndicators and *M*ult*I*ple *C*auses of a single latent variable. The equations for this model are

$$\begin{aligned} \eta_1 &= \boldsymbol{\Gamma}\mathbf{x} + \zeta_1 \\ \mathbf{y} &= \boldsymbol{\Lambda}_y \eta_1 + \boldsymbol{\epsilon} \\ \mathbf{x} &= \boldsymbol{\xi} \end{aligned} \tag{8.28}$$

Note that $\mathbf{x}$ is a perfect measure of $\boldsymbol{\xi}$ and that only one latent variable, η_1, is present. The η_1 is directly affected by one or more x variables, and it is indicated by one or more y variables. As an example, the latent variable might be the latent economic values of automobiles, the y's different estimates of their value, and the x's different characteristics of the cars that could affect their value (e.g., age, mileage, size, fuel economy).

Identification of MIMIC models that conform to (8.28) follows if p (the number of y's) is two or greater and q (the number of x's) is one or more, provided that η_1 is assigned a scale as discussed in Chapter 7.

The MIMIC rule that $p \geq 2$ and $q \geq 1$ is a *sufficient* condition for identification but not a necessary one. Also it applies only to models that conform to (8.28) so that it is useful for a relatively narrow range of applications. Stapleton (1977) provides an example of proving identification for a MIMIC model with multiple η's. Robinson (1974) presents identification for more general MIMIC models.

Summary of Identification Rules

Table 8.1 summarizes the identification rules described in this section. All the rules apply to the identification of the whole model. Whenever appropriate the researcher can apply one or more of these to help assess a model's identification. Unfortunately, none of these is a *necessary* and *sufficient* condition for model identification. Wald's rank rule, the test of singularity of the information matrix, varying starting values, and analyzing the fitted covariance matrix can provide information on local identification. I refer the reader to Chapter 7 for details.

Table 8.1 Identification Rules for General Model ($\eta = B\eta + \Gamma\xi + \zeta$, $y = \Lambda_y\eta + \epsilon$, $x = \Lambda_x\xi + \delta$)

Identification Rule	Requirements	Necessary Condition	Sufficient Condition
t-Rule	$t \leq \frac{1}{2}(p+q)(p+q+1)$	yes	no
Two-Step Rule	1. Reformulate original model as measurement model, eliminating **B**, **Γ**, and **Ψ**. Establish identification of measurement model. 2. Establish identification of latent variable model *as if* latent variables observed with no measurement error (i.e., treat latent variable model as structural equation with observed variables).	no	yes
MIMIC Rule	Model Form: $\eta_1 = \Gamma x + \zeta_1$, $p \geq 2$ $y = \Lambda_y \eta_1 + \epsilon$, $q \geq 1$ $x = \xi$, η_1 scaled	no	yes

Industrialization and Political Democracy Example

I illustrate how to establish identification with the industrialization and political democracy model. Figure 8.1 and equations (8.13) to (8.15) show its structure. A quick way to detect some underidentified models is by the t-rule. The number of free parameters is 28 and $\frac{1}{2}(p+q)(p+q+1)$ is 66. Thus the model *may* be identified. To further investigate its identification, I apply the two-step rule. In the first step I reformulate the model as a confirmatory factor analysis with three latent variables [see Figure 8.5(a)]. The model that results, ignoring ξ_1, x_1, x_2, x_3, ϕ_{13}, and ϕ_{12}, is the political democracy panel data model of Chapter 7. The identification of this part of the model was established there. What remains to complete this first step is to show that λ_6, λ_7, VAR(δ_1), VAR(δ_2), VAR(δ_3), ϕ_{11}, ϕ_{12}, and ϕ_{13} are identified. Considering ξ_1 and its three indicators in isolation, we know that λ_6, λ_7 VAR(δ_1), VAR(δ_2), VAR(δ_3), and ϕ_{11} are identified by the three-indicator rule of Chapter 7. The only remaining parameters are ϕ_{12} and ϕ_{13}. Using the measurement equations for x_1, x_4, and x_8 and the zero covariances of δ_1 with δ_4 and δ_8, the COV(x_1, x_4) is ϕ_{12} and COV(x_1, x_8) is ϕ_{13}. Thus the first step in the two-step rule is satisfied.

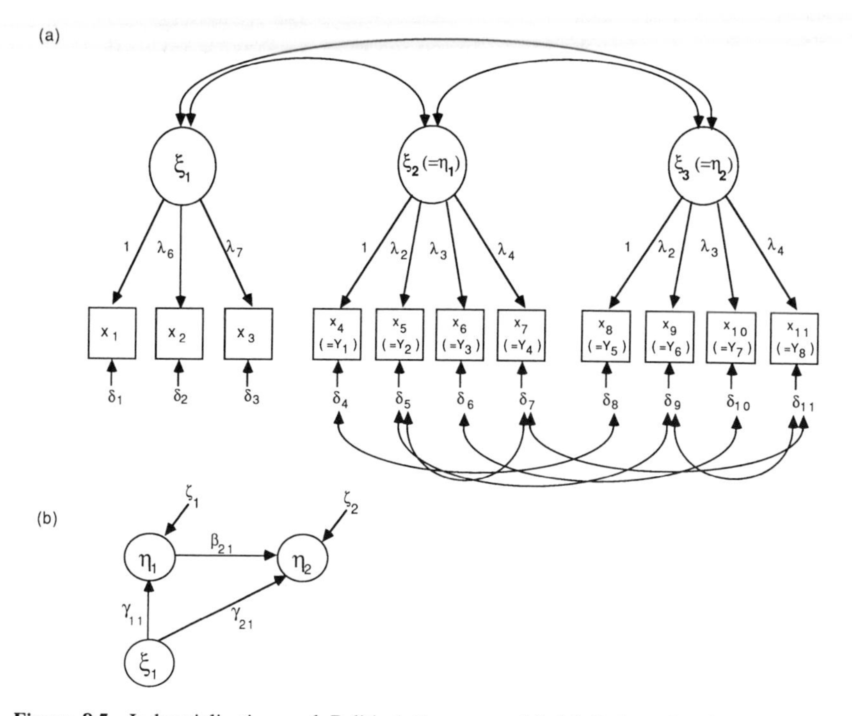

Figure 8.5 Industrialization and Political Democracy Model Reformulated for Two-Step Rule

The second step is to show that the latent variable model is identified if the latent variables are treated as perfectly measured variables. The latent variable model is shown in Figure 8.5(*b*). As is evident from the diagram, the lower triangular form of $\mathbf{B}$, and the diagonal form of $\boldsymbol{\Psi}$, this corresponds to a recursive model, which is identified by the recursive rule of Chapter 4.

ESTIMATION AND MODEL EVALUATION

The hypothesis of our model is $\boldsymbol{\Sigma} = \boldsymbol{\Sigma}(\boldsymbol{\theta})$. Though the parameters in $\boldsymbol{\theta}$ may differ, the estimation problem is the same as in Chapters 4 and 7. Given the sample covariance matrix of the observed variables, $\mathbf{S}$, how can we choose $\boldsymbol{\theta}$ so that $\boldsymbol{\Sigma}(\boldsymbol{\theta})$ is close to $\mathbf{S}$?

Three answers to this question are the ML, GLS, and ULS fitting functions introduced in Chapter 4. These are appropriate for the general model as well. I list these fitting functions in equations (8.33) to (8.35) for easy reference:

$$F_{\mathrm{ML}} = \log|\boldsymbol{\Sigma}(\boldsymbol{\theta})| + \mathrm{tr}\{\mathbf{S}\boldsymbol{\Sigma}^{-1}(\boldsymbol{\theta})\} - \log|\mathbf{S}| - (p+q) \quad (8.29)$$

$$F_{\mathrm{GLS}} = \left(\tfrac{1}{2}\right)\mathrm{tr}\left\{\left[\mathbf{I} - \boldsymbol{\Sigma}(\boldsymbol{\theta})\mathbf{S}^{-1}\right]^2\right\} \quad (8.30)$$

$$F_{\mathrm{ULS}} = \left(\tfrac{1}{2}\right)\mathrm{tr}\left\{\left[\mathbf{S} - \boldsymbol{\Sigma}(\boldsymbol{\theta})\right]^2\right\} \quad (8.31)$$

Each of these functions is minimized with respect to $\boldsymbol{\theta}$.

I illustrate the three different estimators with the industrialization-democracy panel data. The lower half of the sample covariance matrix, **S**, is in Table 8.2. Table 8.3 presents the estimates from the F_{GLS}, F_{ML}, and F_{ULS} fitting functions. The λ_1 to λ_4 coefficients are the factor loadings for the 1960 and 1965 political democracy measures where the 1960 coefficients are constrained to equal their 1965 counterparts. And λ_5 to λ_7 are the coefficients for the 1960 indicators of industrialization (x_1 to x_3). The estimates of γ_{11} and γ_{21} show the influence of industrialization (ξ_1) on political democracy in 1960 (η_1) and 1965 (η_2), whereas that for β_{21} reveals the effect of η_1 on η_2. The VAR(δ_1) to VAR(δ_3) are the error variances for the three measures of industrialization. To conserve space, I omit the

Table 8.2 Covariance Matrix and Means for Political Democracy and Industrialization Indicators (y_1 to y_8, x_1 to x_3) for 75 Developing Countries

	y_1	y_2	y_3	y_4	y_5	y_6	y_7	y_8	x_1	x_2	x_3
y_1	6.89										
y_2	6.25	15.58									
y_3	5.84	5.84	10.76								
y_4	6.09	9.51	6.69	11.22							
y_5	5.06	5.60	4.94	5.70	6.83						
y_6	5.75	9.39	4.73	7.44	4.98	11.38					
y_7	5.81	7.54	7.01	7.49	5.82	6.75	10.80				
y_8	5.67	7.76	5.64	8.01	5.34	8.25	7.59	10.53			
x_1	0.73	0.62	0.79	1.15	1.08	0.85	0.94	1.10	0.54		
x_2	1.27	1.49	1.55	2.24	2.06	1.81	2.00	2.23	0.99	2.28	
x_3	0.91	1.17	1.04	1.84	1.58	1.57	1.63	1.69	0.82	1.81	1.98
Means	5.46	4.26	6.56	4.45	5.14	2.98	6.20	4.04	5.05	4.79	3.56

Table 8.3 The GLS, ML, and ULS Estimates for the Industrialization and Political Democracy Panel Data for Developing Countries, 1960 to 1965 ($N = 75$)

	Estimates (standard errors)		
Parameter	GLS	ML	ULS
λ_1	1.00[c]	1.00[c]	1.00[c]
	(—)	(—)	
λ_2	1.18[e]	1.19[e]	1.21[e]
	(0.15)	(0.14)	
λ_3	1.22[e]	1.18[e]	1.14[e]
	(0.13)	(0.12)	
λ_4	1.23[e]	1.25[e]	1.29[e]
	(0.13)	(0.12)	
λ_5	1.00[c]	1.00[c]	1.00[c]
	(—)	(—)	
λ_6	2.24	2.18	2.06
	(0.16)	(0.14)	
λ_7	1.89	1.82	1.62
	(0.18)	(0.15)	
γ_{11}	1.81	1.47	1.32
	(0.45)	(0.40)	
γ_{21}	0.72	0.60	0.41
	(0.28)	(0.23)	
β_{21}	0.80	0.87	0.92
	(0.08)	(0.08)	
VAR(δ_1)	0.05	0.08	0.02
	(0.02)	(0.02)	
VAR(δ_2)	0.15	0.12	0.07
	(0.07)	(0.07)	
VAR(δ_3)	0.40	0.47	0.60
	(0.09)	(0.09)	

Note: c = constrained to equal 1. $e = \lambda_i$ for same 1960 and 1965 indicator set equal.

estimates for Θ_ϵ, these being very similar to the estimates reported in Chapter 7.

The estimates of the factor loadings (λ_1 to λ_7) are relatively close regardless of the fitting function, though the ULS estimates of the loadings on ξ_1 (λ_6 and λ_7) are lower than the GLS and ML ones. Larger differences occur for the coefficients for the latent variable model (e.g., GLS $\hat{\gamma}_{11}$ is 1.81 vs. 1.47 and 1.32 for ML and ULS) and for the error variances of x_1 to x_3. With few exceptions the GLS and ML estimates are closer to one another

Table 8.4 The Measures of Overall Model Fit in the Industrialization and Political Democracy Panel Data for Developing Countries, 1960 to 1965 ($N = 75$)

	Value		
Summary Fit Measure	GLS	ML	ULS
χ^2	38.6	39.6	—
df	38	38	38
prob.	0.44	0.40	—
χ^2/df	1.02	1.04	—
Δ_1	1.00	0.95	0.99
Δ_2	1.00	1.00	—
ρ_1	0.99	0.92	0.99
ρ_2	1.00	1.00	—
GFI	0.91	0.92	0.99
AGFI	0.85	0.86	0.99

than either is to the ULS estimates. It is not known whether this will occur in general.

The overall model fit is shown in Table 8.4. All of these measures were presented in Chapter 7. The overall fit measures show a good to excellent match of the data to the model. The chi-square values are low with high probability values; the normed and nonnormed fit measures, and the GFIs are extremely high. The AGFIs are generally lower. The baseline model chosen for the normed and nonnormed measures is that of zero correlations between measures.[4] We should view these results with some caution for two reasons. One is that part of this model was estimated in Chapter 7 so that these results are not independent of the earlier analysis. Ideally, the model should be estimated with a new data set to see if the fit is still as good. Second, these are measures of overall fit. The component fit indices remain to be examined.

The component measures are quite good. All the coefficients are positive as expected, and most of the GLS and ML estimates are at least twice their standard errors.[5] The R^2's for the indicators of η_1 and η_2 are moderate to high. For the GLS solution they are

	y_1	y_2	y_3	y_4	y_5	y_6	y_7	y_8
$R^2_{y_i}$	0.76	0.49	0.63	0.70	0.68	0.60	0.66	0.68

[4]Other baseline models could be selected. See Chapter 7 for further discussion.
[5]As in Chapter 7, some estimates from $\mathbf{\Theta}_\epsilon$ are less than twice their standard errors.

The R^2's from the ML and ULS estimates are very similar. The values show a fairly strong relation between the latent political democracy variables and their indicators. The strength of association is even stronger for the indicators of industrialization. The $R_{x_i}^2$ based on the GLS estimates are

	x_1	x_2	x_3
$R_{x_i}^2$	0.87	0.92	0.76

The $R_{x_i}^2$ for GNP per capita (x_1) and energy consumption per capita (x_2) are particularly impressive, showing about 90% of the variance in x_1 and x_2 accounted for by the latent industrialization variable (ξ_1).

The coefficient of determination for the y measurement model provides a summary of the joint fit of the y variables.[6] Calculated as $1 - |\hat{\Theta}_\epsilon|/|\hat{\Sigma}_{yy}|$ its value for the GLS results is 0.94. The analogous coefficient of determination for $\mathbf{x}$ is $1 - |\hat{\Theta}_\delta|/|\hat{\Sigma}_{xx}|$ which equals 0.96. Each measure reinforces the earlier findings of indicators closely related to their respective constructs.

The $R_{\eta_i}^2$ provides a measure of latent variable equation fit for each η_i. The $R_{\eta_1}^2$ for the 1960 political democracy variable is 0.27, whereas $R_{\eta_2}^2$ for 1965 democracy is 0.94. The considerable difference in these values largely results from using industrialization (ξ_1) and the lagged political democracy variable (η_1) in the 1965 political democracy (η_2) equation, while the 1960 democracy (η_1) equation only has 1960 industrialization (ξ_1) as an explanatory factor. The total coefficient of determination for the latent variable model is $1 - |\hat{\Psi}|/|\hat{\Sigma}_{\eta\eta}|$, which for this model is 0.55.

An examination of the unstandardized, correlation, and normalized residuals did not reveal any serious problems. However, the residuals do differ across fitting function. For instance, most (48 out of 66) of the residuals from GLS are positive, but less than half (29) of the ones from ML are positive. The largest GLS residuals are for the variances of the observed variables, but this is not true for the ML residuals. This imperfect association between residuals for the two fitting functions suggests that examining the pattern of residuals for model specification errors occasionally can mislead analysts. This follows since residuals depend not only on sampling variability and specification errors but also on the fitting function employed. One strategy that can minimize the confounding effects of the fitting function is to check the residuals from GLS, ML, or other procedures to ensure that the corresponding residuals are roughly the same magnitude and same sign regardless of the fitting function.

[6] The cautionary remarks about the coefficient of determination I made in Chapter 4 hold here as well.

In sum, both the overall and component fit measures suggest that the model adequately matches the data. Industrialization (ξ_1) has a positive contemporaneous and lagged effect on political democracy (η_1 and η_2) in developing countries. Industrialization in 1960 alone explained a modest amount of the variance (about 25%) of 1960 political democracy. The combined influence of 1960 industrialization and 1960 democracy led to about 95% explained variance in 1965 political democracy. In addition the indicators of political democracy were fairly good, with 49% to 76% of the variance explained by the latent democracy factors. The industrialization measures were even better with percentages of explained variances from 76% to 92%.

Power of Significance Test

Type I error in hypothesis testing is the rejection of a true null hypothesis, H_0.[7] Type II error is not rejecting a false H_0. The power of a test equals one minus the probability of a Type II error—so it is the probability of rejecting H_0 when it is incorrect, given that an alternative hypothesis, H_a, is true. So far for all the statistical tests (e.g., likelihood ratio, Wald) I have emphasized the probabilities of rejecting a true null hypothesis (Type I errors). The cost of neglecting statistical power is most obvious in the ambiguities of interpreting the chi-square tests of overall model fit. Large sample sizes have greater power to detect false H_0 than do smaller samples for the same model. For large samples we face the question of whether a statistically significant chi-square estimate of overall fit means that serious specification errors are present or whether the test has excessively high power. Nonsignificant chi-squares can occur in the face of substantial specification errors in small samples where the power is more likely to be low. In this section I provide procedures to estimate the power of statistical tests of not only overall model fit but also for tests of individual or groups of parameters.

I begin by illustrating the relation between Type I error probability and statistical power for a test of one parameter. Suppose H_0 is $\theta = 0$, whereas H_a is $\theta \neq 0$. Figure 8.6 contains three hypothetical distributions of the test statistic when testing H_0. Part (a) shows the distribution when H_0 is true. The probability of a Type I error, α, is the lined area to the right of the critical value. Assume that the true parameter is c_1 for the alternative

[7]A null hypothesis differs from the null or baseline model discussed in Chapter 7. The null hypothesis stands for the constraints imposed on parameters by the researcher's model. The null or baseline model is a highly restricted model. Researchers compare the fit of their chosen model to that of the null or baseline one.

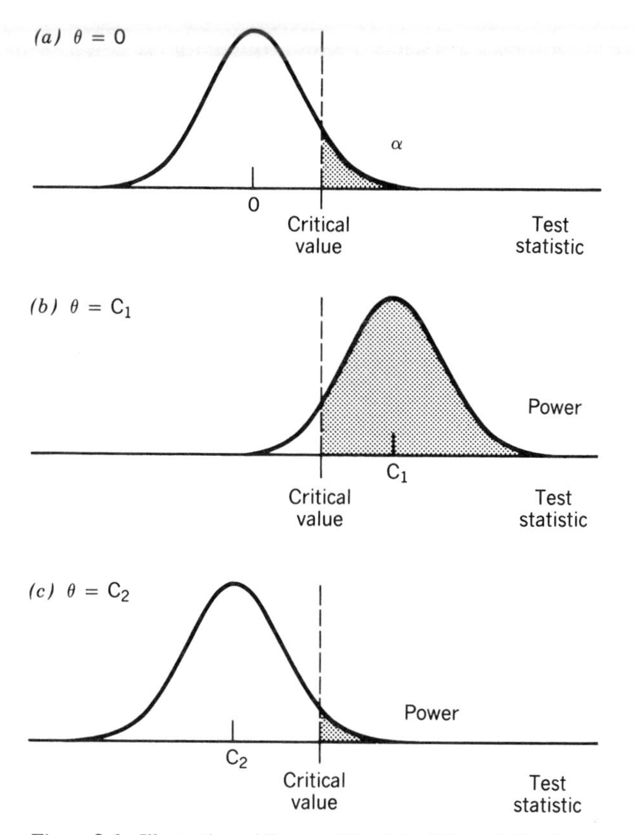

Figure 8.6 Illustration of Power of Statistical Test of H_0: $\theta = 0$

hypothesis, H_a [see Figure 8.6(b)]. The power of the test of H_0 when H_a is true is the lined area to the right of the critical value in Figure 8.6(b). It is the probability of correctly rejecting H_0 when $\theta = c_1$ is valid. As (b) shows, this probability is high so that the false H_0 has a good chance of being detected. In contrast Figure 8.6(c) illustrates the consequences of the true parameter being c_2 so that H_a' is correct, but H_0 is tested at significance level α. Here the power also is shown as the lined area to the right of the critical test value, but the power is low.

This simple case for one parameter illustrates basic principles that hold for tests of multiple parameters as well as for tests of overall model fit. First, the probability of a Type I error and the statistical power of a test are positively related. If α is made smaller (e.g., 0.05 to 0.01), the critical value in Figure 8.6 moves to the right, and the power of the test under the two alternative hypotheses also gets smaller, other things equal. But a larger α (e.g., 0.05 to 0.10) moves the critical value to the left and increases power.

A second basic principle illustrated by Figure 8.6 is that power depends on the specific value of a parameter under the alternative hypothesis. For instance, suppose that H_0 is $\lambda_{12} = 0$. The power of the statistical test of H_0 generally is lower if the true λ_{12} is 0.5 than if it is 1.2. Similarly, if a group of parameters in a structural equation model are constant values under H_0 (e.g., $\gamma_{21} = 0$, $\beta_{12} = 1$, $\phi_{21} = 0$), the power of the statistical test of H_0 depends on the specific values for these parameters under H_a (e.g., $\gamma_{21} = 0.8$, $\beta_{32} = -0.5$, $\phi_{21} = 1.3$). Thus the power of a test is always with respect to a specific alternative set of parameters.

It also is evident from Figure 8.6 that to know the power of a test, we must know the distribution of the test statistic under the alternative hypothesis. Under the standard assumptions the likelihood ratio (LR), Wald (W), and Lagrangian multiplier (LM) test statistics have an asymptotic chi-square distribution when H_0 is valid (see Chapters 4 and 7). However, when H_a is correct and H_0 is tested, these test statistics have an asymptotic noncentral chi-square distribution (Satorra and Saris, 1985). As with the central chi-square distributions, the noncentral one has a parameter for degrees of freedom (df), but it also has a noncentrality parameter. If we determine the df and the noncentrality parameter, then we can know the distribution of the test statistic when H_a is valid and H_0 is tested. And from tabled values of the noncentral chi-square, we can estimate the power of a test.

The df parameter is the difference in the number of free parameters in H_0 and H_a where the model for H_0 is nested in the model for H_a. In a statistical test of a single parameter, the df equals one. For the usual chi-square test of overall model fit where H_a is an exactly identified (though usually not fully described) alternative model, the df equals $\frac{1}{2}(p+q)(p+q+1) - t$, where t is the number of free parameters for H_0. So in a test of overall model fit, the df for the noncentral chi-square equals the df used in the standard chi-square test.

Any of three procedures provide estimates of the noncentrality parameter. They have their basis in the W, LR, and LM tests of Chapter 7. To explain them, some additional notation is needed. To begin with, partition $\boldsymbol{\theta}$, the vector of parameters, into two components so that $\boldsymbol{\theta} = [\boldsymbol{\theta}_a', \boldsymbol{\theta}_b']'$, where $\boldsymbol{\theta}_a$ is $a \times 1$, $\boldsymbol{\theta}_b$ is $b \times 1$. The null hypothesis is H_0: $\boldsymbol{\theta}_a = \boldsymbol{\theta}_0$, and for H_0, $\boldsymbol{\theta}_r = [\boldsymbol{\theta}_0', \boldsymbol{\theta}_b']'$, with $\boldsymbol{\theta}_b$ containing the only free parameters. The alternative hypothesis is H_a: $\boldsymbol{\theta}_a \neq \boldsymbol{\theta}_0$, and for H_a, $\boldsymbol{\theta} = [\boldsymbol{\theta}_a', \boldsymbol{\theta}_b']'$ so that both $\boldsymbol{\theta}_a$ and $\boldsymbol{\theta}_b$ are free. The $\hat{\boldsymbol{\theta}}$ is the ML estimator of $\boldsymbol{\theta}$ under H_a with its asymptotic covariance matrix represented by ACOV($\hat{\boldsymbol{\theta}}$). The asymptotic covariance matrix for $\hat{\boldsymbol{\theta}}_a$ is a submatrix of ACOV($\hat{\boldsymbol{\theta}}$) represented by ACOV($\hat{\boldsymbol{\theta}}_a$).

The first way to estimate the noncentrality parameter is based on the Wald statistic. From Chapter 7 we know that the W test of H_0: $\boldsymbol{\theta}_a = \boldsymbol{\theta}_0$ is

$$\mathrm{W} = (\hat{\boldsymbol{\theta}}_a - \boldsymbol{\theta}_0)'[\mathrm{ACOV}(\hat{\boldsymbol{\theta}}_a)]^{-1}(\hat{\boldsymbol{\theta}}_a - \boldsymbol{\theta}_0) \tag{8.32}$$

As mentioned above, W has an asymptotic chi-square distribution with df $= a$ (= row dimension of $\boldsymbol{\theta}_a$) when H_0 is correct. Otherwise, it has an asymptotic noncentral chi-square distribution with df $= a$ and a noncentrality parameter of (Kendall and Stuart 1979, 246–247)

$$(\boldsymbol{\theta}_a - \boldsymbol{\theta}_0)'[\mathrm{ACOV}(\hat{\boldsymbol{\theta}}_a)]^{-1}(\boldsymbol{\theta}_a - \boldsymbol{\theta}_0) \tag{8.33}$$

Satorra and Saris (1985) and Matsueda and Bielby (1986) use these results to propose a way of estimating the noncentrality parameter that we can use to estimate the statistical power of the test of H_0: $\boldsymbol{\theta}_a = \boldsymbol{\theta}_0$. The procedure assumes that the researcher has a null hypothesis and an alternative model complete with its parameter values. There are four steps:

1. Determine the specific values of $\boldsymbol{\theta}_a$ and $\boldsymbol{\theta}_b$ for the alternative model, say, $\boldsymbol{\theta} = (\boldsymbol{\theta}_a', \boldsymbol{\theta}_b')' = \mathbf{c}$, where $\mathbf{c}$ is an $(a + b) \times 1$ vector of known constants.
2. Generate the implied covariance matrix $\boldsymbol{\Sigma}(\boldsymbol{\theta})$ with $\boldsymbol{\theta} = \mathbf{c}$.
3. Analyze the implied covariance matrix from step 2 under H_a, the alternative (not the null) hypothesis, while keeping the sample size the same, and get $\mathrm{ACOV}(\hat{\boldsymbol{\theta}})$.
4. Take the submatrix $\mathrm{ACOV}(\hat{\boldsymbol{\theta}}_a)$ from the asymptotic covariance matrix of step 3, and substitute this and the values of $\boldsymbol{\theta}_a$ and $\boldsymbol{\theta}_0$ into (8.33) to estimate the noncentrality parameter.

The specification of the values of $\boldsymbol{\theta}$ for the alternative hypothesis plays a key role. By providing population values of $\boldsymbol{\theta}$, we know the population covariance matrix $\boldsymbol{\Sigma}(\boldsymbol{\theta})$, we know the asymptotic covariance matrix $\mathrm{ACOV}(\hat{\boldsymbol{\theta}}_a)$, and we know the difference $(\boldsymbol{\theta}_a - \boldsymbol{\theta}_0)$. The H_a model fitted to $\boldsymbol{\Sigma}(\boldsymbol{\theta})$ has a perfect fit and will exactly reproduce the values of $\boldsymbol{\theta}$ (= $\mathbf{c}$). A Wald-based statistic gives a value for the noncentrality parameter when we restrict $\boldsymbol{\theta}_a$ to $\boldsymbol{\theta}_0$. By comparing the noncentrality parameter to the tabled values of the noncentral chi-square distribution with df $= a$, we can estimate the power of a test of H_0 at a given α level. Tables are available to estimate the power of the chi-square tests for different df and α levels (see Saris and Stronkhorst 1984; Haynam, Govindarajulu, and Leone 1973).

When the preceding procedure is applied to a single parameter, θ_a is a scalar and the noncentrality parameter is $(\theta_a - \theta_0)^2/\mathrm{AVAR}(\hat{\theta}_a)$. The $\mathrm{AVAR}(\hat{\theta}_a)$ is the value of the asymptotic variance of θ_a from the analysis of $\Sigma(\boldsymbol{\theta})$ with $\boldsymbol{\theta} = \mathbf{c}$. Testing several parameters follows the same procedure except that $(\boldsymbol{\theta}_a - \boldsymbol{\theta}_0)$ is a vector and $\mathrm{ACOV}(\hat{\boldsymbol{\theta}}_a)$ is a matrix. If the alternative model is exactly identified so that $(a + b) = \frac{1}{2}(p + q)(p + q + 1)$, this method provides an estimate of the power of the chi-square test of overall model fit.

To illustrate this technique, consider the two-equation recursive model ($N = 200$):

$$y_2 = \beta_{21} y_1 + \gamma_{21} x_1 + \zeta_2 \tag{8.34}$$

$$y_1 = \gamma_{11} x_1 + \zeta_1 \tag{8.35}$$

where $\mathrm{COV}(\zeta_1, \zeta_2) = 0$ and $\mathrm{COV}(x_1, \zeta_i) = 0$, for $i = 1, 2$. Suppose that H_0 and H_a are

$$H_0 : \gamma_{21} = 0 \tag{8.36}$$

$$H_a : \gamma_{21} \neq 0 \tag{8.37}$$

with α selected at 0.05. Here $\theta_a = [\gamma_{21}]$, $\boldsymbol{\theta}_b' = [\gamma_{11}\ \beta_{21}\ \psi_{11}\ \psi_{22}\ \phi_{11}]$, $\boldsymbol{\theta} = [\theta_a, \boldsymbol{\theta}_b']'$, and $\boldsymbol{\theta}_r = [0, \boldsymbol{\theta}_b']'$. Following step 1, we specify the values of the parameters for the alternative model:

$$\begin{aligned} \gamma_{11} &= 1, & \psi_{11} &= 0.4 \\ \gamma_{21} &= 0.5, & \psi_{22} &= 0.4 \\ \beta_{21} &= 0.5, & \phi_{11} &= 1 \end{aligned} \tag{8.38}$$

This leads to $\Sigma(\boldsymbol{\theta})$ for step 2:

$$\Sigma(\boldsymbol{\theta}) = \begin{bmatrix} 1.40 & & \\ 1.20 & 1.50 & \\ 1.00 & 1.00 & 1.00 \end{bmatrix} \tag{8.39}$$

Analyzing (8.39) under H_a with γ_{21} free completes step 3. The asymptotic covariance matrix $\mathrm{ACOV}(\hat{\boldsymbol{\theta}}_a)$ simplifies to a scalar, the asymptotic variance of $\hat{\gamma}_{21}$ or $\mathrm{AVAR}(\hat{\gamma}_{21})$. For step 4, the noncentrality parameter is $\gamma_{21}^2/\mathrm{AVAR}(\hat{\gamma}_{21})$. The $\mathrm{AVAR}(\hat{\gamma}_{21})$ from the analysis of (8.39) under H_a is 0.007056. For $\gamma_{21} = 0.5$, the noncentrality parameter is 35.4, which with an α of 0.05 leads to a power estimate of 1.00. This means that if γ_{21} is 0.5 and

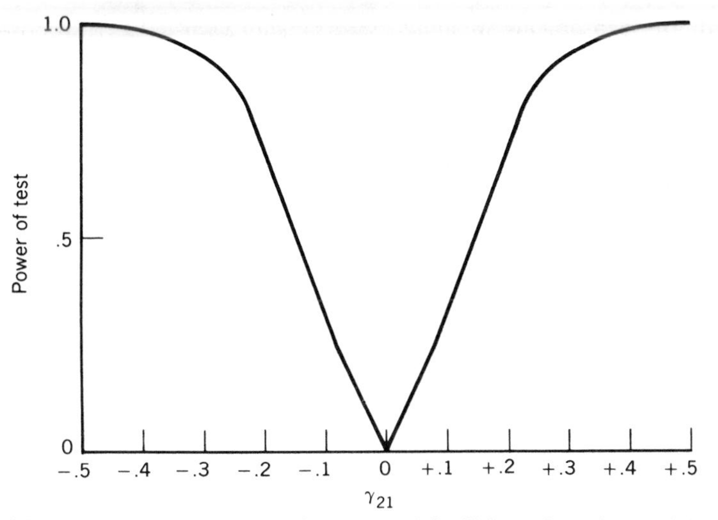

Figure 8.7 Power of Statistical Test of H_0: $\gamma_{21} = 0$ for Values of γ_{21} from -0.5 to $+0.5$ ($N = 200$, $\alpha = 0.05$)

we test H_0: $\gamma_{21} = 0$, we will, with near certainty, reject the false H_0. The power of the test depends on, among other things, the true value of γ_{21}. Figure 8.7 plots the power of the test of H_0 ($\alpha = 0.05$) for a range of true values for γ_{21} from -0.5 to $+0.5$. This *power curve* shows the sensitivity of the power of the test of H_0 to the value of γ_{21}. We are nearly certain to detect a false H_0 when the absolute value of the parameter deviates 0.5 or more from zero. The closer the values of γ_{21} are to zero, the lower will be the power of the test.

Since in the preceding example the alternative model is exactly identified, the power analysis is equivalent for the chi-square test of overall model fit of H_0. More typically, the formulation of an exactly identified alternative model is not as straightforward. Recall that the chi-square test of overall model fit compares the hypothesized model to an exactly identified alternative model that perfectly reproduces **S**. The procedure requires that we completely describe this usually unarticulated alternative model and the values for all of its parameters. In moderately complex models the determination of an exactly identified alternative model is not a trivial task. Often several or more exactly identified models are possible, each of which could lead to different estimates of power. In other situations a plausible alternative model may not be identified so that the asymptotic covariance matrix required for this method is unavailable (see Matsueda and Bielby 1986, 145–146; Satorra and Saris 1985, 85).

With these problems in mind, Satorra and Saris (1985) proposed an alternative LR-based technique to estimate statistical power. With H_0 chosen, their procedure has these four steps:

1. Determine the specific values for $\boldsymbol{\theta}_a$ and $\boldsymbol{\theta}_b$ for the alternative model, say, $\boldsymbol{\theta} = [\boldsymbol{\theta}_a', \boldsymbol{\theta}_b']' = \mathbf{c}$.
2. Generate the implied covariance matrix, $\boldsymbol{\Sigma}(\boldsymbol{\theta})$ with $\boldsymbol{\theta} = \mathbf{c}$.
3. Analyze the implied covariance matrix from step 2 under H_0 (not the alternative model) while keeping the sample size the same.
4. Take the chi-square value and degrees of freedom from step 3 as an approximation of the noncentrality parameter and its df.

The procedure is based on using the "wrong model" (H_0) on the correct implied covariance matrix for the alternative model. As with the usual LR test statistic for overall model fit, H_0 is compared to a perfectly fitting alternative model. The differences are that here we know that H_0 is incorrect, and we analyze a population covariance matrix $\boldsymbol{\Sigma}(\boldsymbol{\theta})$ where $\boldsymbol{\theta}$ is known. The chi-square for the alternative model is zero by construction, whereas that for H_0 is typically positive. The chi-square estimate reflects the difference in fit of H_0 and H_a, as in the usual LR test, except that the value estimates the noncentrality parameter. The df equals the number of parameters that distinguish H_0 from an exactly identified alternative model. Satorra and Saris (1985) provide a rigorous justification for this procedure.

Like the first procedure, the technique applies to tests of individual or groups of parameters, and not only to tests of overall fit. These power calculations are based on the chi-square difference LR test for comparing nested models where the parameter values for the least restrictive model usually generate the $\boldsymbol{\Sigma}(\boldsymbol{\theta})$ that is analyzed. The models can differ in a single or in many parameters. The preceding steps are followed except that both nested models are fitted to the implied covariance matrix of step 2. The noncentrality parameter estimate equals the difference in the noncentrality estimates of the two models, with df equal to the difference in the df's for the models (Saris, den Ronden, and Satorra 1984). Generally, there is no need to fit the least restrictive model since its noncentrality parameter estimate is zero by construction.

The final procedure is based on the LM test statistic (see Chapter 7). It has the same first three steps as the LR-based technique, but in the fourth step the LM test statistic for freeing the restrictions on $\boldsymbol{\theta}$ is the estimate of the noncentrality parameter. This too can be applied for one or multiple parameters as well as for overall model tests. In the special case where H_0 and H_a differ in only one parameter, the noncentrality parameter estimate equals the value of the univariate LM statistic (the "modification index") for a parameter when $\boldsymbol{\Sigma}(\boldsymbol{\theta})$ with $\boldsymbol{\theta} = \mathbf{c}$ is analyzed (see Saris, Satorra, and Sŏrbom 1987).

Comparisons of these procedures are at an early stage of research. Intuitively, the alternative procedures build upon the asymptotic equivalency of the W, LR, and LM tests discussed in Chapter 7. Computationally, the LR-based estimate of this section is easier when H_a is an exactly identified (or underidentified) alternative model whose structure is not fully specified. This is common in moderately complex models for power estimates of the overall fit test. With a program such as EQS where W- and LM-based tests are readily available, the W- and LM-based procedures are straightforward when testing one or a few parameters or when the exactly identified alternative model is fully specified. Asymptotically, the procedures are identical, but Saris, Satorra, and Sörbom (1987) suggest that the LR-based procedure gives the most accurate approximation of power for small samples. The LR procedure works best for large samples and for relatively small specification errors in the null hypothesis. Saris and Satorra (1985) and Matsueda and Bielby (1986) find that the noncentrality parameter estimates computed with either the W- or LR-based methods are generally within 10% to 15% of one another.

I illustrate all three procedures in determining the power of the chi-square test of overall model fit for the union sentiment example of Chapter 4. Figure 8.8 is a path diagram of the original model (see solid lines) and an exactly identified alternative model (see solid and dashed lines). Suppose that we plan to collect new data on which to test the original specification. But we want an estimate of the power of the test before proceeding with

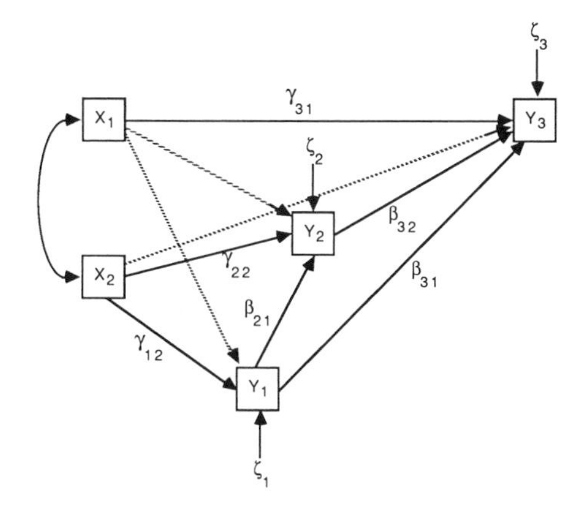

Figure 8.8 Union Sentiment Example for Original Model (Solid Lines) and Exactly Identified Alternative Model (Solid and Dashed Lines)

data collection. The original model (solid lines) provides H_0. Full determination of the alternative model requires provision of the exact values for all parameters. We generally would have no way of knowing these values, but reasonable choices for the free parameters of H_0 are the estimates from Chapter 4. For the remaining parameters that distinguish the alternative model from H_0, we can choose values that are large enough that the chi-square test should have the power to detect them if present. Though the choice is somewhat arbitrary, I set all these coefficients to values that, when standardized,[8] equal 0.1. So the unstandardized coefficients for the alternative model are:

$$\mathbf{B} = \begin{bmatrix} 0 & 0 & 0 \\ -0.285 & 0 & 0 \\ -0.218 & 0.850 & 0 \end{bmatrix}, \qquad \boldsymbol{\Gamma} = \begin{bmatrix} 0.378 & -0.087 \\ 0.328 & 0.058 \\ 0.860 & 0.039 \end{bmatrix}$$

$$\boldsymbol{\Psi} = \begin{bmatrix} 12.961 & 0 & 0 \\ 0 & 8.488 & 0 \\ 0 & 0 & 19.455 \end{bmatrix}, \qquad \boldsymbol{\Phi} = \begin{bmatrix} 1.021 & 7.139 \\ 7.139 & 215.662 \end{bmatrix}$$

The second step for all three procedures is to generate the implied covariance matrix from these parameter values. This can be accomplished in several ways. One is to use a matrix programming language and to substitute the parameter values into the formula for $\boldsymbol{\Sigma}(\boldsymbol{\theta})$ [see (8.21)]. Another is to use a program such as LISREL, start all parameter matrices at the values listed above, and fix all matrices so that no parameters are estimated. The "fitted moment matrix" from this run is $\boldsymbol{\Sigma}(\boldsymbol{\theta})$ at $\boldsymbol{\theta} = \mathbf{c}$.

The third step for the W-based procedure is to estimate the H_a model for the implied covariance matrix while keeping the sample size the same. In the fourth step Bentler's (1985) EQS program has a W test procedure that can be employed to generate the estimate of the noncentrality parameter when moving from H_a, the exactly identified model, to H_0. Otherwise, $\text{ACOV}(\hat{\boldsymbol{\theta}}_a)$ is calculable from the correlation matrix of parameter estimates and the standard errors that are available in LISREL or EQS. This $\text{ACOV}(\hat{\boldsymbol{\theta}}_a)$ is substituted into (8.33) along with $\boldsymbol{\theta}_a$ and $\boldsymbol{\theta}_0$ to calculate the noncentrality parameter.

For the LR- and LM-based estimates H_0 is fitted to $\boldsymbol{\Sigma}(\boldsymbol{\theta})$ $(\boldsymbol{\theta} = \mathbf{c})$ in the third step. The fourth step in the LR-based procedure is to take the chi-square from this run as the estimate of the noncentrality parameter. The

[8] I use the standard deviations from the original covariance matrix in calculating the unstandardized coefficient necessary to lead to standardized coefficients of 0.1.

fourth step in the LM method is to compute the LM test statistic for moving from H_0 to H_a as the estimate of the noncentrality parameter.

Following these procedures, the three estimates of the noncentrality parameter for this model are:

	W Test	LR Test	LM Test
Noncentrality Parameter	5.06	5.03	5.00

All three are virtually the same. The df = 3, $\alpha = 0.05$, $N = 173$, and with a noncentrality parameter of about five the power of the test of H_0 for overall model fit is 0.44. Under these conditions we have about a 0.44 probability of detecting a false H_0 when the alternative parameterization is true.

The power of a statistical test can help researchers to assess the quality of their chi-square test of model fit. Figure 8.9 illustrates several possible outcomes. In it the horizontal axis is the probability of a Type I error with the preselected α level marked. The preselected minimum power level is on the power axis. If the model H_0 has a Type I probability lower than the

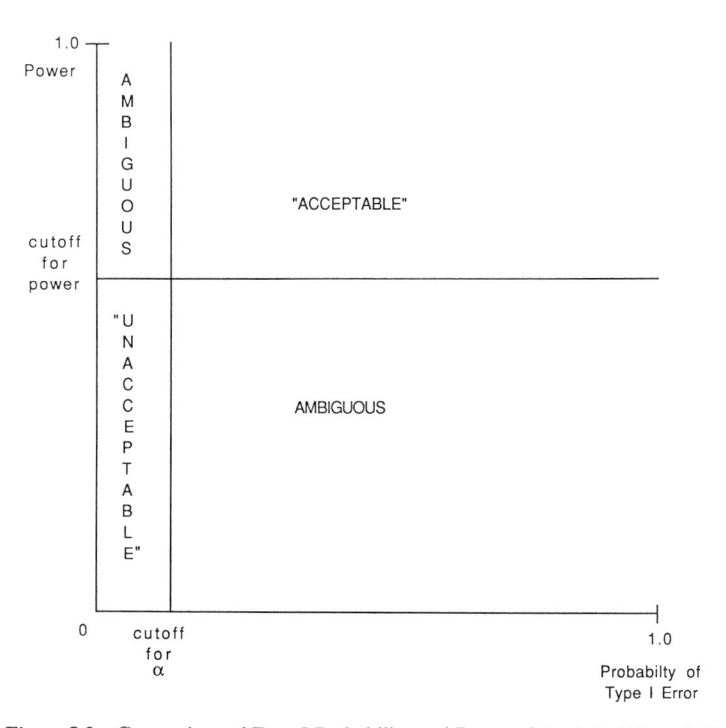

Figure 8.9 Comparison of Type I Probability and Power of Statistical Test of H_0

cutoff α and if the power of the test with respect to H_a is lower than the cutoff for power, this case falls in the lower left-hand quadrant of Figure 8.9 labeled "unacceptable." As the label suggests, H_0 would not be very acceptable since with low power and a significant chi-square value, it is likely that more than minor specification errors are in the H_0 model. In contrast, suppose that the Type I probability is larger than α and that the power of the test is high; a situation shown in the upper-right corner of Figure 8.9. The high power means a high probability of detecting a false H_0. The high Type I probability and high power suggest that the H_0 model is very consistent with the data since, if it were not so, these discrepancies would be detected.

The two remaining quadrants of Figure 8.9 represent more ambiguous situations. The lower-right one occurs when the Type I probability for H_0 is greater than α, but the power of the test is low. We cannot be certain whether H_0 is appropriate or whether the low power of the test prevents the detection of large errors. The upper-left quadrant presents a different problem; the Type I probability of H_0 is lower than α, but the power of the test is high. Here we do not know whether the H_0 model has major misspecifications or if the high power of the test reflects minor discrepancies.

As an illustration of the preceding assessment of test quality consider the union sentiment example given earlier. In Chapter 4 the chi-square estimate of H_0 was 1.26 with df = 3. The Type I probability was very high (= 0.74), indicating a nonsignificant test statistic. The power estimate of 0.44 is moderate and suggests that even with a nonsignificant chi-square estimate for the test of H_0, we cannot be sure whether important errors are present in the model. This ambiguous situation corresponds to the lower-right quadrant of Figure 8.9. This demonstrates the need to redesign the study so as to increase the power of the chi-square test.

Power is influenced by many factors. As mentioned earlier, if α is increased, the power of a test goes up. This is one possible means to increase power. Another is to increase the sample size, if this is possible, when collecting new data. In general, power also can be enhanced by using a model with more indicators, though the impact of doing so depends on the reliability of the measures. The greatest gains are likely when new measures are added to variables with low reliabilities. Similarly, increases of power are possible by using measures with higher reliabilities. In short, the relation between all the factors that affect power is complex, but the researcher has several options to increase power, including increasing the sample size, the cutoff α level, the number of indicators of the latent variables, and the reliability of indicators (see Matsueda and Bielby 1986). Too much power also can be a problem. Trivial deviations can lead to significant chi-square estimates because of very high power. A simple adjustment to reduce power is to reduce the chosen α level.

In sum, given the controversy surrounding the goodness of fit measures (GFI, Δ_1, ρ_2, etc.), the analysis of the power of the chi-square test can be a very useful aid in assessing model fit. Of course, it too involves judgment in the selection of the alternative model, H_a, and its parameter values, the choice of α, and the desired power level.

STANDARDIZED AND UNSTANDARDIZED COEFFICIENTS

An analysis of the covariance matrix of observed variables leads to unstandardized coefficients that depend upon the units in which the variables are scaled. As discussed in Chapters 4 and 7, many times these units are arbitrary. It can be difficult to compare the effects of two or more variables on the same dependent variable when they have different units of measurement. Standardized coefficients can be useful in assessing relative effects of different explanatory variables. They are defined as

$$\hat{\lambda}_{ij}^{s} = \hat{\lambda}_{ij}\left(\frac{\hat{\sigma}_{jj}}{\hat{\sigma}_{ii}}\right)^{1/2}$$

$$\hat{\beta}_{ij}^{s} = \hat{\beta}_{ij}\left(\frac{\hat{\sigma}_{jj}}{\hat{\sigma}_{ii}}\right)^{1/2}$$

$$\hat{\gamma}_{ij}^{s} = \hat{\gamma}_{ij}\left(\frac{\hat{\sigma}_{jj}}{\hat{\sigma}_{ii}}\right)^{1/2}$$

where the superscript s represents a standardized coefficient, i is the influenced ("dependent") variable, j is the explanatory ("independent") variable, and $\hat{\sigma}_{ii}$ and $\hat{\sigma}_{jj}$ are the model-predicted variances of the ith and jth variables. The standardized coefficient is the expected shift in standard deviation units of the dependent variable that is due to a one standard deviation shift in the independent variable when the other variables are held constant.[9]

[9]The elasticities at the means are another measure of relative importance defined in Chapter 4. Note that these elasticities are not defined if the means of the latent variables are zero, as is true with deviation scores. They are defined at values other than the means.

The limits of the standardized coefficients are the same as described in Chapter 4: the standard errors for the standardized coefficients obtained when the correlation matrix is analyzed generally are not correct, and in most cases the standardized coefficients should not be tested for equivalence across different groups. One source of confusion is that the "standardized" solution reported in LISREL VI (as well as V and IV) and EQS are different. In LISREL only the latent variables ξ and η are transformed to have a variance of one. The observed variables are not standardized so that LISREL's "standardized solution" does not provide the standardized coefficients as defined here. In the standardized solution of EQS all the variables are rescaled to have a variance of one, and the coefficients are equivalent to those defined in the previous paragraph.

The matrix expressions for the estimate of the standardized coefficient matrices are

$$\hat{\Lambda}_y^s = \left[\operatorname{diag}\left(\hat{\Sigma}_{yy}\right)\right]^{-1/2} \hat{\Lambda}_y \left[\operatorname{diag}(\hat{\mathbf{C}})\right]^{1/2}$$

$$\hat{\Lambda}_x^s = \left[\operatorname{diag}\left(\hat{\Sigma}_{xx}\right)\right]^{-1/2} \hat{\Lambda}_x \left[\operatorname{diag}(\Phi)\right]^{1/2}$$

$$\hat{\mathbf{B}}^s = \left[\operatorname{diag}(\hat{\mathbf{C}})\right]^{-1/2} \hat{\mathbf{B}} \left[\operatorname{diag}(\hat{\mathbf{C}})\right]^{1/2}$$

$$\hat{\Gamma}^s = \left[\operatorname{diag}(\hat{\mathbf{C}})\right]^{-1/2} \hat{\Gamma} \left[\operatorname{diag}(\hat{\Phi})\right]^{1/2}$$

where $\hat{\mathbf{C}}$ is the estimate of covariance matrix of

$$\eta \left[= (\mathbf{I} - \hat{\mathbf{B}})^{-1}(\hat{\Gamma}\hat{\Phi}\hat{\Gamma}' + \hat{\Psi})(\mathbf{I} - \hat{\mathbf{B}})^{-1'}\right].$$

MEANS AND EQUATION CONSTANTS

As was true in Chapter 4 for structural equations with observed variables and in Chapter 7 for CFA models, it is possible to estimate the general model with intercept terms for the measurement and latent variable equations and to estimate latent variable means. The extended notation to accomplish this was partially presented in Chapters 4 and 7. The complete notation is

$$\eta = \alpha + \mathbf{B}\eta + \Gamma\xi + \zeta \tag{8.40}$$

$$\mathbf{y} = \upsilon_y + \Lambda_y \eta + \epsilon \tag{8.41}$$

$$\mathbf{x} = \upsilon_x + \Lambda_x \xi + \delta \tag{8.42}$$

The means of the exogenous variables, $\boldsymbol{\xi}$, are in a $n \times 1$ vector, $\boldsymbol{\kappa}$. The expected values of the latent endogenous variables are

$$\begin{aligned} E(\boldsymbol{\eta}) &= E\left[(\mathbf{I} - \mathbf{B})^{-1}(\boldsymbol{\alpha} + \boldsymbol{\Gamma}\boldsymbol{\xi} + \boldsymbol{\zeta})\right] \\ &= (\mathbf{I} - \mathbf{B})^{-1}(\boldsymbol{\alpha} + \boldsymbol{\Gamma}\boldsymbol{\kappa}) \end{aligned} \tag{8.43}$$

The mean vectors of $\mathbf{x}$ and $\mathbf{y}$ are

$$E(\mathbf{x}) = \boldsymbol{\upsilon}_x + \boldsymbol{\Lambda}_x \boldsymbol{\kappa} \tag{8.44}$$

$$E(\mathbf{y}) = \boldsymbol{\upsilon}_y + \boldsymbol{\Lambda}_y (\mathbf{I} - \mathbf{B})^{-1}(\boldsymbol{\alpha} + \boldsymbol{\Gamma}\boldsymbol{\kappa}) \tag{8.45}$$

As (8.43) shows, the mean of $\boldsymbol{\eta}$ is not only a function of $\boldsymbol{\kappa}$, the mean of the exogenous variables, but also a function of the structural parameters in $\mathbf{B}$, $\boldsymbol{\Gamma}$, and $\boldsymbol{\alpha}$. Similarly, the mean of $\mathbf{y}$ is determined by these matrices as well as by $\boldsymbol{\upsilon}_y$ and $\boldsymbol{\Lambda}_y$. The expected value of $\mathbf{x}$ is influenced by $\boldsymbol{\upsilon}_x$, $\boldsymbol{\Lambda}_x$, and $\boldsymbol{\kappa}$.

The identification problem is complicated by the additional parameters. I illustrate this with a simple example of two latent variables, each measured with two indicators:

$$\begin{aligned} \eta_1 &= \alpha_1 + \gamma_{11}\xi_1 + \zeta_1 \\ X_1 &= \upsilon_{X_1} + \xi_1 + \delta_1 \\ X_2 &= \upsilon_{X_2} + \lambda_2 \xi_1 + \delta_2 \\ Y_1 &= \upsilon_{Y_1} + \eta_1 + \epsilon_1 \\ Y_2 &= \upsilon_{Y_2} + \lambda_4 \eta_1 + \epsilon_2 \end{aligned} \tag{8.46}$$

The scale of ξ_1 is set to $X_1(\lambda_1 = 1)$ and that for η_1 is set to $Y_1(\lambda_3 = 1)$. The implied covariance matrix for $\mathbf{y}$ and $\mathbf{x}$ is the same as when deviation scores were employed. This follows since adding or subtracting a constant (i.e., the mean) from a vector does not change its covariances. As can be easily shown, $\boldsymbol{\Lambda}_x$, $\boldsymbol{\Lambda}_y$, $\boldsymbol{\Gamma}$, $\boldsymbol{\Theta}_\epsilon$, $\boldsymbol{\Theta}_\delta$, $\boldsymbol{\Phi}$, and $\boldsymbol{\Psi}$ are identified by using the covariances and variances of the observed variables. The means of X_1, X_2, Y_1, and Y_2 are a new source of information, but new parameters also are introduced. The equations relating the means to the structural parameters are

$$\begin{aligned} E(\eta_1) &= \alpha_1 + \gamma_{11}\kappa_1 \\ E(X_1) &= \upsilon_{X_1} + \kappa_1 \\ E(X_2) &= \upsilon_{X_2} + \lambda_2 \kappa_1 \\ E(Y_1) &= \upsilon_{Y_1} + (\alpha_1 + \gamma_{11}\kappa_1) \\ E(Y_2) &= \upsilon_{Y_2} + \lambda_4(\alpha_1 + \gamma_{11}\kappa_1) \end{aligned} \tag{8.47}$$

Equation (8.47) reveals that six new parameters ($\alpha_1, \kappa_1, \upsilon_{X_1}, \upsilon_{X_2}, \upsilon_{Y_1}, \upsilon_{Y_2}$) appear, with only four mean values of the observed variables. Clearly, without further restrictions, the new parameters cannot be identified. It is necessary to assign an origin to ξ_1 and η_1 as discussed in Chapter 7. One way to do this is to make each latent variable's mean equal to zero. For ξ_1, this is accomplished by fixing κ_1 to zero. For η_1, the added constraint of α_1 equal to zero is required. An alternative strategy is to set the origin and scale of a latent variable so that it is the same as one of the observed variables. The ξ_1 variable already is scaled to X_1. Its origin can be made the same as X_1 by defining υ_{X_1} to be zero. Similarly, υ_{Y_1} set to zero provides an origin for η_1. With υ_{X_1} and υ_{Y_1} at zero, the remaining unknowns are identified from the observed variable means:

$$\begin{aligned}
\kappa_1 &= E(X_1) \\
E(\eta_1) &= E(Y_1) \\
\upsilon_{X_2} &= E(X_2) - \lambda_2 E(X_1) \\
\upsilon_{Y_2} &= E(Y_2) - \lambda_4 E(Y_1) \\
\alpha_1 &= E(Y_1) - \gamma_{11} E(X_1)
\end{aligned} \tag{8.48}$$

(The covariance matrix of the observed variables serves to identify the coefficients λ_2, λ_4, and γ_{11}.) Just as the sample variances and covariances of the observed variables are used in estimation in place of the population values, so are the sample means of the observed variables substituted for the expected values of the observed variables.

To program intercept and mean terms in LISREL VI (and earlier versions) requires an alternative notation. A new $\mathbf{y}^{\cdot}$ variable contains the original $\mathbf{y}$ and $\mathbf{x}$, and a new $\boldsymbol{\eta}^{\cdot}$ contains $\boldsymbol{\eta}$, $\boldsymbol{\xi}$, and a last row that equals one. The new $\xi_1^{\cdot}$ is a fixed constant equal to one. The representation in matrix terms is

$$\begin{array}{ccccccccc}
\boldsymbol{\eta}^{\cdot} & = & \mathbf{B}^{\cdot} & \boldsymbol{\eta}^{\cdot} & + & \boldsymbol{\Gamma}^{\cdot} & \xi_1^{\cdot} & + & \boldsymbol{\zeta}^{\cdot} \\
\begin{bmatrix} \boldsymbol{\eta} \\ \boldsymbol{\xi} \\ 1 \end{bmatrix} & = & \begin{bmatrix} \mathbf{B} & \boldsymbol{\Gamma} & 0 \\ \mathbf{0} & \mathbf{0} & 0 \\ \mathbf{0}' & \mathbf{0}' & 0 \end{bmatrix} & \begin{bmatrix} \boldsymbol{\eta} \\ \boldsymbol{\xi} \\ 1 \end{bmatrix} & + & \begin{bmatrix} \boldsymbol{\alpha} \\ \boldsymbol{\kappa} \\ 1 \end{bmatrix} & (1) & + & \begin{bmatrix} \boldsymbol{\zeta} \\ \boldsymbol{\xi} - \boldsymbol{\kappa} \\ 0 \end{bmatrix}
\end{array} \tag{8.49}$$

$$\begin{array}{ccccccc}
\mathbf{y}^{\cdot} & = & \boldsymbol{\Lambda}_y^{\cdot} & \boldsymbol{\eta}^{\cdot} & + & \boldsymbol{\epsilon}^{\cdot} \\
\begin{bmatrix} \mathbf{y} \\ \mathbf{x} \end{bmatrix} & = & \begin{bmatrix} \boldsymbol{\Lambda}_y & \mathbf{0} & \boldsymbol{\upsilon}_y \\ \mathbf{0} & \boldsymbol{\Lambda}_x & \boldsymbol{\upsilon}_x \end{bmatrix} & \begin{bmatrix} \boldsymbol{\eta} \\ \boldsymbol{\xi} \\ 1 \end{bmatrix} & + & \begin{bmatrix} \boldsymbol{\epsilon} \\ \boldsymbol{\delta} \end{bmatrix}
\end{array} \tag{8.50}$$

$$\begin{aligned}
\xi^{\cdot} &= [X_1] \\
X_1 &= (1) \quad (\text{fixed } X)
\end{aligned} \tag{8.51}$$

The intercept term, $\boldsymbol{\alpha}$, and the vector of $\boldsymbol{\xi}$ means, $\boldsymbol{\kappa}$, are located in the $\boldsymbol{\Gamma}^{\cdot}$ matrix. The intercepts for the measurement equations, $\boldsymbol{\upsilon}_y$ and $\boldsymbol{\upsilon}_x$, are in the last column of $\boldsymbol{\Lambda}_y^{\cdot}$. In addition the $\boldsymbol{\Psi}^{\cdot}$ matrix has fixed zeros in its last row. The $\boldsymbol{\Theta}_\epsilon$ matrix is designed to represent the pattern of variances and covariances for the errors of measurement.

I illustrate this procedure with the industrialization–political democracy panel data. To estimate the intercept and means, the latent variable equation in the modified notation is

$$\begin{array}{ccccccc} \boldsymbol{\eta}^{\cdot} & = & \mathbf{B}^{\cdot} & \boldsymbol{\eta}^{\cdot} & + \ \boldsymbol{\Gamma}^{\cdot} & \boldsymbol{\xi}^{\cdot} + & \boldsymbol{\zeta}^{\cdot} \\ \begin{bmatrix} \eta_1 \\ \eta_2 \\ \xi_1 \\ 1 \end{bmatrix} & = & \begin{bmatrix} 0 & 0 & \gamma_{11} & 0 \\ \beta_{21} & 0 & \gamma_{21} & 0 \\ 0 & 0 & 0 & 0 \\ 0 & 0 & 0 & 0 \end{bmatrix} & \begin{bmatrix} \eta_1 \\ \eta_2 \\ \xi_1 \\ 1 \end{bmatrix} & + \begin{bmatrix} \alpha_1 \\ \alpha_2 \\ \kappa_1 \\ 1 \end{bmatrix} & (1) + & \begin{bmatrix} \zeta_1 \\ \zeta_2 \\ \xi_1 - \kappa_1 \\ 0 \end{bmatrix} \end{array} \tag{8.52}$$

The $\boldsymbol{\Psi}^{\cdot}$ matrix is

$$\boldsymbol{\Psi}^{\cdot} = \operatorname{diag}[\psi_{11} \ \psi_{22} \ \phi_{11} \ 0] \tag{8.53}$$

I assign a scale and origin to η_1 and η_2 by setting the λ's for Y_1 and Y_4 to one and their intercepts, υ_{Y_1} and υ_{Y_5}, to zero. I assume that the intercepts for the $\mathbf{y}$ variables are the same in 1965 as in 1960. Industrialization is scaled to X_1 by setting the appropriate λ to one, and υ_{X_1} is constrained to zero.

$$\boldsymbol{\Lambda}_y^{\cdot} = \begin{bmatrix} 1 & 0 & 0 & 0 \\ \lambda_2 & 0 & 0 & \upsilon_{Y_2} \\ \lambda_3 & 0 & 0 & \upsilon_{Y_3} \\ \lambda_4 & 0 & 0 & \upsilon_{Y_4} \\ 0 & 1 & 0 & 0 \\ 0 & \lambda_2 & 0 & \upsilon_{Y_2} \\ 0 & \lambda_3 & 0 & \upsilon_{Y_3} \\ 0 & \lambda_4 & 0 & \upsilon_{Y_4} \\ 0 & 0 & 1 & 0 \\ 0 & 0 & \lambda_6 & \upsilon_{X_2} \\ 0 & 0 & \lambda_7 & \upsilon_{X_3} \end{bmatrix} \tag{8.54}$$

The moments around zero are analyzed. These can be calculated with the covariances and the means for the observed variables (see Table 8.2). The estimates and standard errors (in parentheses) of the intercept and mean terms are

υ_{Y_2}	υ_{Y_3}	υ_{Y_4}	υ_{X_2}	υ_{X_3}	α_1	α_2	κ_1
− 3.07	0.18	−2.38	−6.23	−5.63	−1.91	−2.67	5.05
(0.79)	(0.68)	(0.66)	(0.70)	(0.77)	(1.99)	(1.02)	(0.08)

Except for the estimates of υ_{Y_3} and α_1, all estimates are at least twice their standard errors. The estimate of κ_1 is the mean of the latent industrialization variable in 1960. It is assigned the same scale and origin as X_1, which is the logarithm of GNP per capita. Exponentiating $\hat{\kappa}_1$ leads to 156.02 dollars per capita as the mean level of industrialization in dollar units, which is easier to understand than the log transformation of this value. If I had set the scale and origin to X_2 instead of X_1, the mean of ξ_1 would be in units of the log of energy consumption per capita. Since there is no consensus on the appropriate unit for industrialization, the choice is arbitrary. The intercept terms can be interpreted the same as the constant terms in regression equations. The υ_{Y_i} and υ_{X_j} are intercepts in the equations relating the latent variable to the corresponding observed variable. The α_1 and α_2 are intercepts in the equations relating the latent variables.

An issue of interest in panel data such as these is whether the mean level of the latent variable has stayed the same over the time interval. To determine this, I can compute the mean level of the latent endogenous variables, using equation (8.43), and substitute in the estimates for **B**, **α**, Γ, and **κ**. Obtaining a standard error for the difference in the means over time is more difficult. There is, however, an easier alternative. Reestimate the industrialization—political democracy panel data, treating all three latent variables as freely correlating ξ_i variables. This can be accomplished by keeping all the parameter matrices as before except $\mathbf{B}^{\cdot}$, which is set to zero, and then using the new definitions for the following matrices:

$$\boldsymbol{\eta}^{\cdot}=\begin{bmatrix}\xi_1\\ \xi_2\\ \xi_3\\ 1\end{bmatrix},\quad \boldsymbol{\zeta}^{\cdot}=\begin{bmatrix}\xi_1-\kappa_1\\ \xi_2-\kappa_2\\ \xi_3-\kappa_3\\ 0\end{bmatrix},\quad \boldsymbol{\Gamma}^{\cdot}=\begin{bmatrix}\kappa_1\\ \kappa_2\\ \kappa_3\\ 1\end{bmatrix} \tag{8.55}$$

$$\boldsymbol{\Psi}^{\cdot}=\begin{bmatrix}\phi_{11} & & & \\ \phi_{21} & \phi_{22} & & \\ \phi_{31} & \phi_{32} & \phi_{33} & \\ 0 & 0 & 0 & 0\end{bmatrix} \tag{8.56}$$

The ξ_1 and ξ_2 in (8.55) correspond to 1960 and 1965 political democracy, and ξ_3 is industrialization. The first two elements of $\Gamma^{\cdot}$ are the means of 1960 and 1965 political democracy. The estimated standard error of the difference of $\hat{\kappa}_1 - \hat{\kappa}_2$ is computed as the square root of $\text{var}(\hat{\kappa}_1) + \text{var}(\hat{\kappa}_2) - 2\,\text{cov}(\hat{\kappa}_1, \hat{\kappa}_2)$. For these data, $\hat{\kappa}_1 = 5.50$ and $\hat{\kappa}_2 = 5.09$, so the difference in mean level of political democracy from 1965 to 1960 is 0.41. The estimated standard errors of $\hat{\kappa}_1$ and $\hat{\kappa}_2$ are 0.291 and 0.290, and their covariance is 0.0743, so the standard error of their difference is 0.142 $[= ((0.291)^2 + (0.290)^2 - 2(0.074))^{1/2}]$. The difference in means, 0.41, is more than twice its standard error, 0.142, which suggests a statistically significant decline in political democracy. (Alternatively, we could use the LR or W test to determine whether a significant decline in fit occurs by imposing the restriction that $\kappa_1 = \kappa_2$.) The substantive significance of the decline is a matter of judgment. Relative to the 1960 mean democracy level of 5.50, the 0.41 decline is about a 7% drop, which I would evaluate as a weak to moderate decline. All of the preceding interpretation is conditional upon the assumption that the relation of the 1960 latent political democracy variable to Y_1 (freedom of the press, 1960) is the same as that for the 1965 latent democracy to Y_5 (freedom of the press, 1965). This is so because I have scaled and set the origin of the latent democracy variables to these measures. If the intercept for the Y_5 measurement equation is lower in 1965 than for Y_1 in 1960 and it is assumed that both equations have a zero intercept, then this could lead to differences in $\hat{\kappa}_1$ and $\hat{\kappa}_2$ when there are none. I have no reason to suspect this, but you should be alert to the possibility in this and other applications.

COMPARING GROUPS

There are many situations where we want to know if a measurement or latent variable model for one group has the same parameter values as that in another group. For instance, do those experiencing job-training programs have a steeper positive slope between years on the job and performance than those without training? Or, are the mean levels of job satisfaction in a corporation equal for males and females? If parameter values differ across groups, we run the risk of making serious errors. As an illustration, suppose that a measure of job satisfaction for males is related to a latent satisfaction variable such that $X = \xi + \delta$, whereas for females, the measurement equation is $X = 0.8\xi + \delta$. Even if males and females have the same nonzero mean for unobserved job satisfaction, the means of the satisfaction measures would tend to suggest that females' average satisfaction is lower. Or, we might examine the relation of job satisfaction to education and find significant differences in the slopes relating the latent variables when this is due to assuming that the satisfaction measures are scaled the same in both groups. The situation is analogous to that of studying food weight and its cost in two different countries without considering the differences in currencies and the units of weight. We would expect to find different coefficients relating cost to weight even if the slopes were equivalent after the proper conversion of the measures.

In this section I present methods of comparing structural equations, with and without latent variables, in different populations. These procedures assume that independent, random samples are available from each population. If selectivity factors affect whether an individual falls into a group, this may call for the addition of equations to model the selectivity process (see Muthén and Jöreskog 1983). I assume that problems with selectivity are not present. I first treat models in which all variables are in deviation form. Then I turn to cases where the intercepts and means are compared.

The notation is largely the same as before, with the exceptions that the model matrices and parameters have superscripts of "(g)" [e.g., $\Lambda_x^{(g)}$, $\Phi^{(g)}$, $\beta_{21}^{(g)}$]; the g provides the group number. The g runs from $1, 2, \ldots, G$, where G is the total number of groups. So $\Lambda_x^{(1)}$ refers to Λ_x for group 1, and $\gamma_{12}^{(3)}$ is γ_{12} for group 3.

Comparability or invariance in models represents a continuum. I divide comparability into two overlapping dimensions: one is model form, and the other is similarity in parameter values. Models with different forms mostly represent the lower range of the invariance continuum. Two models have the same form if the model for each group has the same parameter matrices with the same dimensions and the same location of fixed, free, and constrained parameters. Figure 8.10(a) illustrates two models in different groups that have different forms. Each model has the same indicators (x_1, x_2, and x_3). But in group 1 all three load on ξ_1, whereas in group 2, x_1 and x_2 load on ξ_1 and x_3 loads on ξ_2. The similarity of form can range from a low point of a very different number of latent variables and pattern of loadings to a relatively minor difference in form where, say, the models are identical except for the pattern of correlated measurement errors. Figure 8.10(b) shows models in two groups that have the same form.

In most applications to date, researchers assume that the form of two models is the same, and they concentrate on the similarity of parameter values within a given form. Similarity, like form, is a matter of degree. Researchers must decide which elements or matrices of parameters should be tested for equality across groups and in what order these tests should be made. Testing comparability is possible for all of the types of structural equation models treated in the book. I start with structural equations with

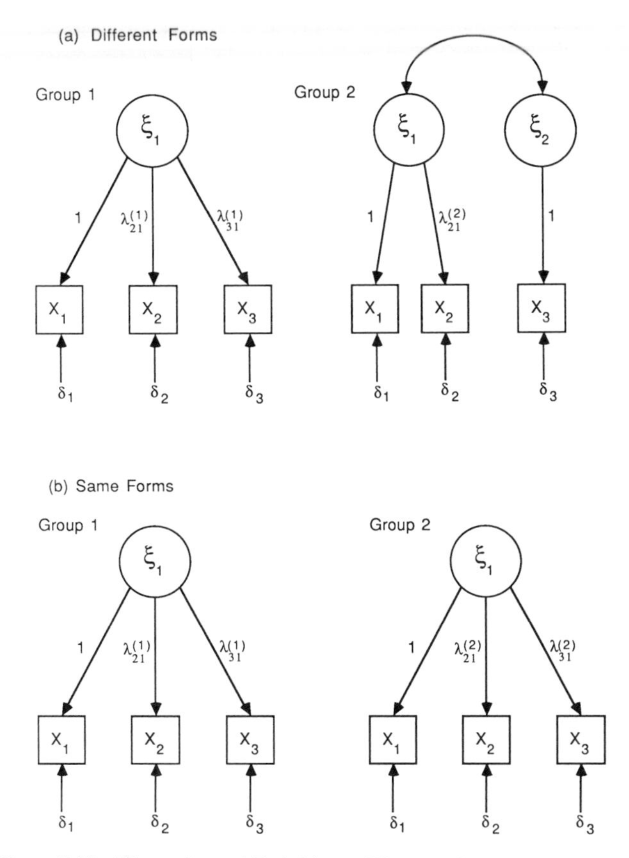

Figure 8.10 Illustrations of Definitions of Form and Structure of Models

observed variables. Figure 8.11 shows a hypothetical model relating party loyalty (y_2) to income (y_1) and education (x_1) for Democrats and Republicans. One possible order in which to test for similarity is represented as

H_{form}: same form (same dimensions and same patterns of fixed, free, and constrained values in $\mathbf{B}$, $\boldsymbol{\Gamma}$, $\boldsymbol{\Psi}$, and $\boldsymbol{\Phi}$)

$H_{\mathbf{B}\boldsymbol{\Gamma}}$: $\mathbf{B}^{(1)} = \mathbf{B}^{(2)}$, $\quad \boldsymbol{\Gamma}^{(1)} = \boldsymbol{\Gamma}^{(2)}$

$H_{\mathbf{B}\boldsymbol{\Gamma}\boldsymbol{\Psi}}$: $\mathbf{B}^{(1)} = \mathbf{B}^{(2)}$, $\quad \boldsymbol{\Gamma}^{(1)} = \boldsymbol{\Gamma}^{(2)}$, $\quad \boldsymbol{\Psi}^{(1)} = \boldsymbol{\Psi}^{(2)}$

$H_{\mathbf{B}\boldsymbol{\Gamma}\boldsymbol{\Psi}\boldsymbol{\Phi}}$: $\mathbf{B}^{(1)} = \mathbf{B}^{(2)}$, $\quad \boldsymbol{\Gamma}^{(1)} = \boldsymbol{\Gamma}^{(2)}$, $\quad \boldsymbol{\Psi}^{(1)} = \boldsymbol{\Psi}^{(2)}$, $\quad \boldsymbol{\Phi}^{(1)} = \boldsymbol{\Phi}^{(2)}$

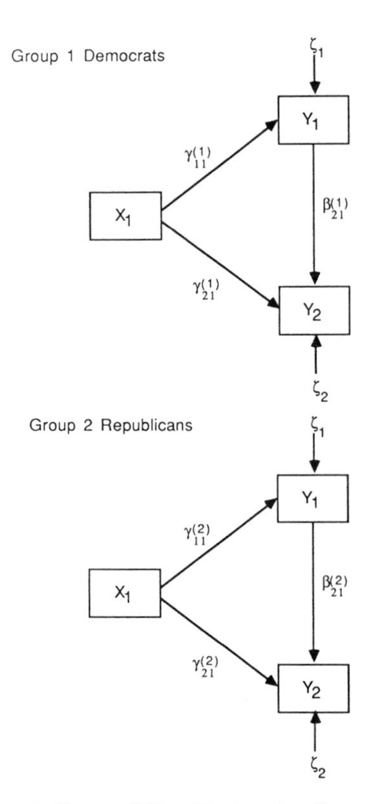

Figure 8.11 Illustration of Comparability Testing for Structural Equations with Observed Variables

The least demanding test of similarity is whether the two groups have the same form, without restricting any of the nonfixed parameters to have the same value across groups (H_{form}). Since this is the least restrictive test in the hierarchy, a poor fit means that it makes little sense to move to the more restrictive hypotheses. If, however, the fit is satisfactory, a move to $H_{\mathbf{B}\Gamma}$ is appropriate. This hypothesis constrains the elements of $\mathbf{B}$ and Γ to be the same in all groups. For Figure 8.11, this translates into $\gamma_{11}^{(1)} = \gamma_{11}^{(2)}$, $\gamma_{21}^{(1)} = \gamma_{21}^{(2)}$, and $\beta_{21}^{(1)} = \beta_{21}^{(2)}$. If we had special interest in $\mathbf{B}$ or Γ separately, then $H_{\mathbf{B}}$ or H_{Γ} could be tested instead of the combination hypothesis $H_{\mathbf{B}\Gamma}$. Otherwise, $H_{\mathbf{B}\Gamma}$ provides a simultaneous test of all coefficients across groups. Assuming that this hypothesis shows an adequate fit, we can move to $H_{\mathbf{B}\Gamma\Psi}$. This hypothesis maintains the equality constraints between $\mathbf{B}^{(1)}$ and $\mathbf{B}^{(2)}$, and $\Gamma^{(1)}$ and $\Gamma^{(2)}$ but adds an equality for the covariance matrices of the disturbance terms. The final most restrictive hypothesis in this hierarchy is $H_{\mathbf{B}\Gamma\Psi\Phi}$. Here all parameter matrices are constrained to be equal

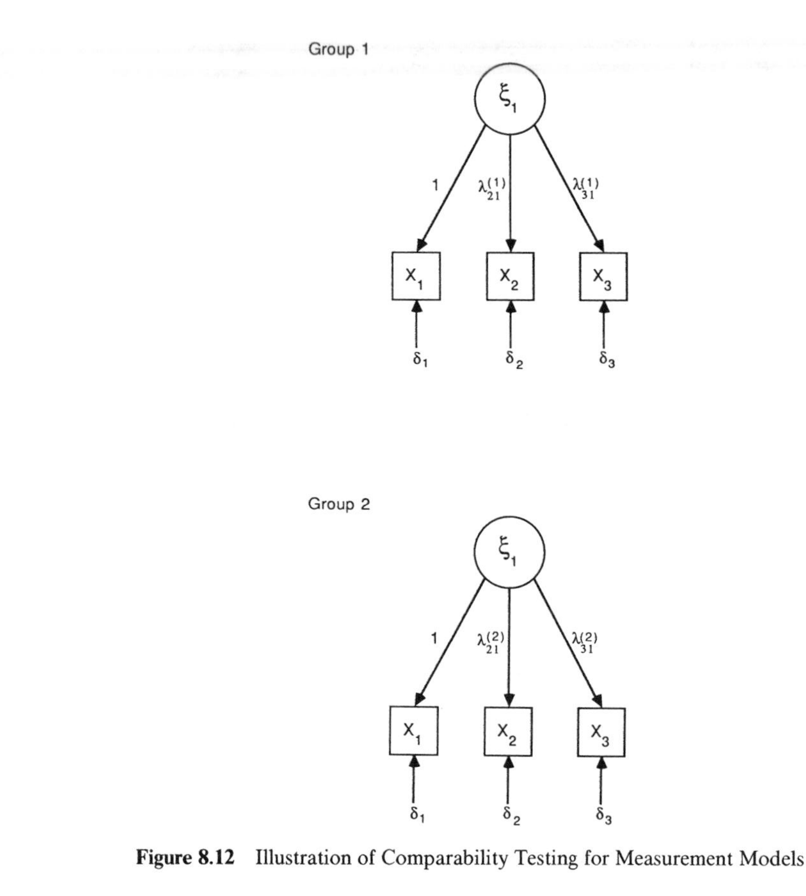

Figure 8.12 Illustration of Comparability Testing for Measurement Models

in the different groups. If a structural equation model with observed variables still matches the data under this highly restrictive hypothesis, the results are consistent with the assumption that the same model operates in both groups. In terms of the hypothetical example in Figure 8.11, such a result would suggest that the effects of education on income, of education on party loyalty, and of income on partly loyalty are the same for Democrats and Republicans. In addition, the variances of education and the unexplained variances in income and party loyalty are equal for both political parties. The model is invariant across groups.

A similar hierarchy of invariance is applicable to measurement models, an example of which is in Figure 8.12. The x_1 to x_3 variables could be indicators of the latent variable inflationary expectations (ξ_1), with the two groups being corporate executives in two different organizations. The goal is to determine if the same relation of indicators to latent variable holds in both groups. One possible testing hierarchy is

H_{form}: same form (same dimensions and same patterns of fixed, free, and constrained elements in Λ_x, Θ_δ, and Φ)

$$H_{\Lambda_x}: \quad \Lambda_x^{(1)} = \Lambda_x^{(2)}$$

$$H_{\Lambda_x \Theta_\delta}: \quad \Lambda_x^{(1)} = \Lambda_x^{(2)}, \quad \Theta_\delta^{(1)} = \Theta_\delta^{(2)}$$

$$H_{\Lambda_x \Theta_\delta \Phi}: \quad \Lambda_x^{(1)} = \Lambda_x^{(2)}, \quad \Theta_\delta^{(1)} = \Theta_\delta^{(2)}, \quad \Phi^{(1)} = \Phi^{(2)}$$

This hierarchy is similar to the last one. It begins with the hypothesis that the form of the model is the same. If this does not hold, then it makes little sense to go further. Assuming this hypothesis fits, the next step is to assess whether the coefficients linking the latent to the observed variables are the same in both groups. The equality of scaling is generally of a higher priority than the equality of measurement error variances or the equality of the covariance matrices in different groups, so H_{Λ_x} precedes the last two hypotheses of $H_{\Lambda_x \Theta_\delta}$ and $H_{\Lambda_x \Theta_\delta \Phi}$. The selection of $H_{\Lambda_x \Theta_\delta}$ instead of $H_{\Lambda_x \Phi}$ is somewhat arbitrary, but in many cases greater interest lies in whether the degree of measurement error is the same across groups instead of whether the covariances of the latent variables are equivalent. The highest step in this hierarchy is $H_{\Lambda_x \Theta_\delta \Phi}$ where all three parameter matrices are simultaneously tested for equality.

Finally, a hierarchy of hypotheses helps to assess invariance for the general structural equation model with latent variables. From first to last this is H_{form}, H_{Λ}, $H_{\Lambda \mathbf{B} \Gamma}$, $H_{\Lambda \mathbf{B} \Gamma \Theta}$, $H_{\Lambda \mathbf{B} \Gamma \Theta \Psi}$, and $H_{\Lambda \mathbf{B} \Gamma \Theta \Psi \Phi}$, where Λ represents Λ_x and Λ_y and Θ stands for Θ_ϵ and Θ_δ. As before, the subscripts of H show the matrices that are constrained to be equal. The order in which parameter equalities are tested can be altered in accordance with substantive interest. For example, if a researcher's work places greater emphasis on the equality of the parameters linking the latent variables ($\mathbf{B}$ and Γ) than the coefficients from the measurement model (Λ_x and Λ_y), then a test for invariance of $\mathbf{B}$ and Γ could precede that for Λ_x and Λ_y. Once a hierarchy is established, we can test the hypotheses and assess which degree of invariance best matches the data.

Each group's covariance matrix ($\mathbf{S}_g$) is the object of analysis. As before, the hypothesized structure implies a covariance matrix $\Sigma_g(\theta_g)$ for each group. The "closer" $\Sigma_g(\theta_g)$ is to $\mathbf{S}_g$ for all groups, the better the model fits. The fit function is a weighted combination of the fit for all groups:

$$F = \sum_{g=1}^{G} \left(\frac{N_g}{N} \right) F_g\big(\mathbf{S}_g, \Sigma_g(\theta_g)\big) \tag{8.57}$$

where F is a general fit function, N_g is the sample size in the gth group, $N = N_1 + N_2 + \cdots + N_G$, and $F_g(\mathbf{S}_g, \mathbf{\Sigma}_g(\boldsymbol{\theta}_g))$ is the fit function for the gth group. Equation (8.57) shows that the largest groups receive the greatest weights (N_g/N) in minimizing F. The F_g for ML, ULS, and GLS are

$$F_{g\text{ML}} = \log|\mathbf{\Sigma}_g| + \text{tr}\left(\mathbf{S}_g\mathbf{\Sigma}_g^{-1}\right) - \log|\mathbf{S}_g| - (p + q) \tag{8.58}$$

$$F_{g\text{ULS}} = \left(\tfrac{1}{2}\right)\text{tr}\left[\left(\mathbf{S}_g - \mathbf{\Sigma}_g\right)^2\right] \tag{8.59}$$

$$F_{g\text{GLS}} = \left(\tfrac{1}{2}\right)\text{tr}\left[\left(\mathbf{I} - \mathbf{\Sigma}_g\mathbf{S}_g^{-1}\right)^2\right] \tag{8.60}$$

where $\mathbf{\Sigma}_g = \mathbf{\Sigma}_g(\boldsymbol{\theta}_g)$. Equations (8.58) to (8.60) show that the fitting functions for each group are the same as previously. The major differences are the inclusion of parameter restrictions across groups and the simultaneous minimization of a composite fit function in two or more groups rather than one. Measures of model fit are analogous to those presented in single group analyses. A χ^2 variate is approximated by $(N - 1)$ times the minimum of F for $F_{g\text{ML}}$ and $F_{g\text{GLS}}$. There is only one chi-square estimate for all the groups. The null hypothesis is that the constraints of the model in all groups are correct. The degrees of freedom equal $\frac{1}{2}(G)(p + q)(p + q + 1) - t$, where t is the number of independent parameters estimated in all groups. Another important property holds for nested models. As before, the difference in chi-squares for nested models is distributed as a new chi-square variable with df equal to the difference in df for the two models. Since the hierarchies of invariances contain nested models (e.g., $H_{\Lambda_x\Theta_\delta}$ is nested in H_{Λ_x}), this provides a means of assessing relative fit as we move up the hierarchy.

In addition the other measures of overall fit (e.g., GFI, Δ_1, Δ_2, ρ_1, and ρ_2) are available. The GFI differs from the others in that it is calculable for each group. The selection of the baseline model for Δ_1, Δ_2, ρ_1, and ρ_2 presents some ambiguity. A model might restrict the covariances of observed variables to be zero in all groups, but should the variances be equal or unconstrained across groups? Similar questions will arise for other models. Also the discussion on the power of significance tests is relevant here.

To illustrate these procedures, I use the objective-subjective socioeconomic status model from Chapter 5. The path diagram is in Figure 8.13. The covariance matrices for blacks and whites are in Table 8.5. The observed variables are income (x_1), occupational prestige (x_2), subjective income (y_1), subjective occupational prestige (y_2), and subjective overall status (y_3). The latent variables are unobserved subjective income (η_1),

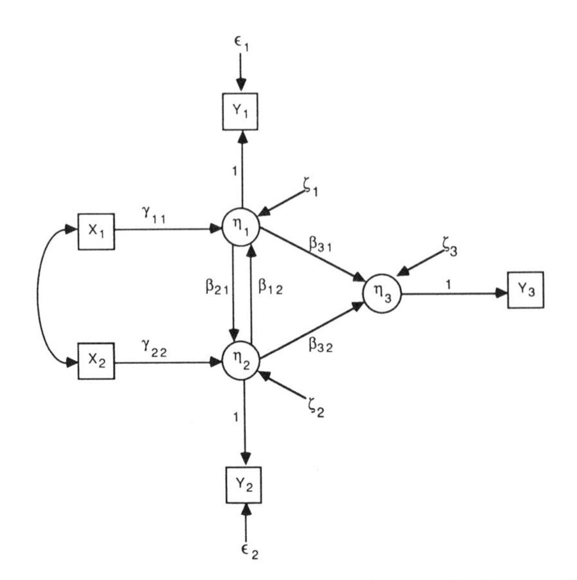

Figure 8.13 Model of Objective and Subjective Socioeconomic Status with Measurement Error in Y_1 and Y_2

Table 8.5 Covariance Matrices for Five Variables on Objective and Subjective Socioeconomic Status for Blacks (N = 368) and Whites (N = 432)

	y_1	y_2	y_3	x_1	x_2
y_1	0.663 / 0.449	0.334	0.301	0.724	2.417
y_2	0.166	0.558 / 0.410	0.282	0.420	3.357
y_3	0.226	0.173	0.615 / 0.393	0.391	1.806
x_1	0.564	0.259	0.382	4.397 / 4.381	7.485
x_2	2.366	3.840	3.082	13.656	263.218 / 452.77

Note: Covariances for blacks in top half; those for whites in bottom half.

subjective occupational prestige (η_2), and subjective overall status (η_3). Up to now I have only treated the sample of whites. The same observed variables are available for a sample of blacks. My goal is to determine the degree of invariance of the model for whites and blacks. The hierarchy of invariance I choose is H_{form}, H_{Γ}, $H_{\Gamma B}$, $H_{\Gamma B \Theta_\epsilon}$, and $H_{\Gamma B \Theta_\epsilon \Psi}$. The testing begins with determining whether the same model form is appropriate for blacks and whites. Suppose that I am most interested in whether the relation between the objective and subjective income and occupational variables is the same for both groups. The equality of the coefficients relating the subjective variables to each other is of secondary concern. In this case H_{Γ} comes after H_{form} and $H_{\Gamma B}$ follows H_{Γ}. If my next priorities are testing whether the groups' measurement error variances are the same and, finally, whether their equation error variances are equal, then the next two hypotheses are $H_{\Gamma B \Theta_\epsilon}$ and $H_{\Gamma B \Theta_\epsilon \Psi}$. I have made the assumption that the scale coefficients relating the latent (e.g., η_1) to observed variables (e.g., y_1) are equal to one for blacks and whites. The same is implicitly done when any structural equation with observed variables is compared across groups. Unfortunately, without additional indicators or other information investigators cannot explore the validity of this constraint.

Table 8.6 reports the results of testing the invariance hypotheses. The H_{form} hypothesis shows an excellent match to the data with a χ^2 less than one, and the GFI's, Δ_1, Δ_2, and other measures near perfect. The baseline model for these latter three fit measures sets the covariances of all observed variables to zero except for that of x_1 and x_2, which can vary across groups, and it allows the variances of all observed variables to differ for blacks and whites. For this baseline, the chi-square estimate is 709.80, with 18 degrees of freedom. All the fit measures provide strong evidence that the same model form holds for both groups. The next hypothesis H_{Γ} also has an excellent fit. The chi-square estimate of 2.10 (4 df) has a probability of 0.717. The chi-square difference of H_{Γ} and H_{form} is 1.15 ($= 2.10 - 0.95$), with 2 df which is not statistically significant ($\alpha = 0.05$ or 0.10). The other

Table 8.6 Tests of Invariance of Path Model in Figure 8.13 for Blacks and Whites

				GFI			
Hypothesis	χ^2	df	prob	Whites	Blacks	Δ_1	Δ_2
H_{form}	0.95	2	0.621	1.000	0.999	0.999	1.001
H_{Γ}	2.10	4	0.717	1.000	0.998	0.997	1.003
$H_{\Gamma B}$	13.85	8	0.086	0.995	0.991	0.980	0.992
$H_{\Gamma B \Theta_\epsilon}$	16.28	10	0.092	0.994	0.989	0.977	0.991
$H_{\Gamma B \Theta_\epsilon \Psi}$	60.66	13	0.000	0.970	0.971	0.915	0.932

measures also show little deterioration in fit by imposing H_Γ. In substantive terms this means that income's influence on subjective income and occupational prestige's influence on subjective occupational prestige appear equal for blacks and whites (i.e., $\gamma_{11}^{(1)} = \gamma_{11}^{(2)}$, $\gamma_{22}^{(1)} = \gamma_{22}^{(2)}$).

Adding the restriction on **B** leads to $H_{\Gamma \mathbf{B}}$ with a chi-square estimate of 13.85 (8 df) and a probability level of 0.086. The decline in chi-squares is 11.75 (= 13.85 − 2.10) with 4 df. This is statistically significant at the 0.05 level though not at the 0.01 level. A comparison of the other overall fit measures for H_Γ and $H_{\Gamma \mathbf{B}}$ are less suggestive of a sharp decline in fit (e.g., compare the GFI's and Δ's). An examination of the components of each model may prove helpful. The $\mathbf{B}^{(1)}$ and $\mathbf{B}^{(2)}$ for whites (1) and blacks (2) for H_Γ are

$$\hat{\mathbf{B}}^{(1)} = \begin{bmatrix} 0 & 0.241 & 0 \\ 0.325 & 0 & 0 \\ 0.388 & 0.529 & 0 \end{bmatrix}$$
$$\hat{\mathbf{B}}^{(2)} = \begin{bmatrix} 0 & 0.521 & 0 \\ 0.518 & 0 & 0 \\ 0.339 & 0.442 & 0 \end{bmatrix} \tag{8.61}$$

The chi-square difference between H_Γ and $H_{\Gamma \mathbf{B}}$ is a joint significant test of the equality of all coefficients in **B**. Equation (8.61) reveals the location and magnitude of these differences. The largest differences in $\hat{\mathbf{B}}^{(1)}$ for whites and $\hat{\mathbf{B}}^{(2)}$ for blacks are for the coefficients of subjective occupational prestige and subjective income's reciprocal relation. The difference in $\hat{\beta}_{12}$ is about 0.3, and that for $\hat{\beta}_{21}$ is about 0.2. The remaining coefficients $\hat{\beta}_{31}$ and $\hat{\beta}_{32}$ are smaller for blacks than for whites. Compared to whites, blacks tend to have a stronger reciprocal relation between subjective income and subjective occupational prestige and a slightly weaker effect for these two variables on subjective overall status.

The residuals, $(s_{ij} - \hat{\sigma}_{ij})$, from $H_{\Gamma \mathbf{B}}$ exhibit a disturbing pattern. For the whites, only two out of the 15 residuals are positive, three are zero, and the remaining 10 are negative. For blacks, a reverse pattern holds, and there is only one negative, three zero, and 11 positive residuals. Thus with $H_{\Gamma \mathbf{B}}$ there is a tendency to overpredict the covariances of the observed variables for whites and to underpredict the covariances for blacks.[10]

Should $H_{\Gamma \mathbf{B}}$ be rejected? Favoring its rejection is the statistically significant chi-square difference compared to H_Γ, the unusual pattern of residuals,

[10] This tendency was partially visible even with H_{form}. However, in H_{form} and H_Γ it was not as pronounced for both groups and the magnitudes of the residuals were considerably smaller.

and the magnitude of the differences for some of the $\hat{\beta}_{ij}$'s. The decision to maintain the hypothesis is supported by the large sample size ($N = 800$) and its potential influence on the chi-square LR tests, the nonsignificance of $H_{\boldsymbol{\Gamma}\mathbf{B}}$ at the 0.05 level, and the relatively large magnitude of the GFIs and Δ's. Resolution of this decision has to await a replication of the analysis on a fresh sample. If the same pattern is observed this would lend support to a decision to reject $H_{\boldsymbol{\Gamma}\mathbf{B}}$. In this event there would be no need to take the comparisons further. But, since the present evidence is ambiguous, I continue to examine the more restrictive models.

The next hypothesis $H_{\boldsymbol{\Gamma}\mathbf{B}\boldsymbol{\Theta}_\epsilon}$ adds the constraint that the measurement error variances are equal for blacks and whites. The chi-square for it is 16.28 with 10 df which has a probability level of 0.092. The chi-square difference of $H_{\boldsymbol{\Gamma}\mathbf{B}\boldsymbol{\Theta}_\epsilon}$ and $H_{\boldsymbol{\Gamma}\mathbf{B}}$ is a nonsignificant 2.43 with 2 df. Moreover, at worst, there is a slight decline in the measures of overall fit. This suggest that there is little change in model fit when an equality constraint for $\boldsymbol{\Theta}_\epsilon$ is added to $H_{\boldsymbol{\Gamma}\mathbf{B}}$.

The $H_{\boldsymbol{\Gamma}\mathbf{B}\boldsymbol{\Theta}_\epsilon\boldsymbol{\Psi}}$ leads to a chi-square estimate of 60.66 (13 df) and a chi-square difference of 44.38 (3 df) compared to $H_{\boldsymbol{\Gamma}\mathbf{B}\boldsymbol{\Theta}_\epsilon}$, both of which are statistically significant ($p < 0.001$). In addition the residuals (normalized and unstandardized) are much larger for both groups. Furthermore, for whites, all residuals are zero or negative; for blacks all residuals are zero or positive. The changes in the GFIs and Δ's do not reveal as great a difference, though their largest absolute decline occurs in this step. The assumption that the equation error variances are equal for blacks and whites is not tenable.

In sum, this test of invariance favors the hypotheses that both groups have the same model form and that the influences of the objective variables on their subjective counterparts are the same. When adding the equality of $\mathbf{B}$'s to the model, some decline in fit does occur. Whether the decline is sufficient to reject the equality of $\mathbf{B}$ is subject to debate. Not much additional change occurs if $H_{\boldsymbol{\Gamma}\mathbf{B}\boldsymbol{\Theta}_\epsilon}$ is compared to $H_{\boldsymbol{\Gamma}\mathbf{B}}$. The sharpest decline in fit occurs with $H_{\boldsymbol{\Gamma}\mathbf{B}\boldsymbol{\Theta}_\epsilon\boldsymbol{\Psi}}$. Thus the invariance of the models seems safe at least for the form and $\boldsymbol{\Gamma}^{(1)} = \boldsymbol{\Gamma}^{(2)}$. It is a little less certain adding $\mathbf{B}^{(1)} = \mathbf{B}^{(2)}$ and $\boldsymbol{\Theta}_\epsilon^{(1)} = \boldsymbol{\Theta}_\epsilon^{(2)}$, but it is unacceptable if we assume $\boldsymbol{\Psi}^{(1)} = \boldsymbol{\Psi}^{(2)}$. Though not done here, the power of the chi-square tests could help to assess the quality of these tests.

Invariance of Intercepts and Means

Earlier in this chapter I presented the methods of estimating structural equation models with intercepts and the means of the latent variables. These procedures also prove useful when testing invariance of means and intercepts in groups. For each group g, the latent variable and measurement equations are

$$\boldsymbol{\eta}^{(g)} = \boldsymbol{\alpha}^{(g)} + \mathbf{B}^{(g)}\boldsymbol{\eta}^{(g)} + \boldsymbol{\Gamma}^{(g)}\boldsymbol{\xi}^{(g)} + \boldsymbol{\zeta}^{(g)} \tag{8.62}$$

$$\mathbf{y}^{(g)} = \boldsymbol{\upsilon}_y^{(g)} + \boldsymbol{\Lambda}_y^{(g)}\boldsymbol{\eta}^{(g)} + \boldsymbol{\epsilon}^{(g)} \tag{8.63}$$

$$\mathbf{x}^{(g)} = \boldsymbol{\upsilon}_x^{(g)} + \boldsymbol{\Lambda}_x^{(g)}\boldsymbol{\xi}^{(g)} + \boldsymbol{\delta}^{(g)} \tag{8.64}$$

where $\boldsymbol{\zeta}^{(g)}$, $\boldsymbol{\epsilon}^{(g)}$, and $\boldsymbol{\delta}^{(g)}$ have expected values of zero, $\boldsymbol{\xi}^{(g)}$ and $\boldsymbol{\eta}^{(g)}$ no longer are deviation scores, $\boldsymbol{\zeta}^{(g)}$, $\boldsymbol{\delta}^{(g)}$, and $\boldsymbol{\epsilon}^{(g)}$ are uncorrelated with $\boldsymbol{\xi}^{(g)}$, and $\boldsymbol{\epsilon}^{(g)}$ and $\boldsymbol{\delta}^{(g)}$ are uncorrelated with $\boldsymbol{\zeta}^{(g)}$.

As in the single population analysis the latent variables must be assigned a scale and an origin. There are several ways to accomplish this, but a straightforward strategy is that followed earlier in this chapter. Each latent variable has its scale and origin matched to one of the observed variables. The scale is assigned by setting the λ_{ij} of the ith observed variable to one and the intercept (e.g., υ_{X_i}) for the same variable to zero. For the ith observed variable this leads to $E(X_i) = \kappa_j$, where κ_j is a single latent variable that underlies the observed X_i. Thus the ξ_j variable is given the same mean and units as X_i.

The hierarchy of hypotheses for the invariance of models between groups is flexible as was true earlier. The new issue is the placement of the hypotheses about the latent variable intercepts ($\boldsymbol{\alpha}^{(g)}$), the measurement equation intercepts ($\boldsymbol{\upsilon}_x^{(g)}$ and $\boldsymbol{\upsilon}_y^{(g)}$), and the means of the latent exogenous variables ($\boldsymbol{\kappa}^{(g)}$). Applications to date require that at a minimum the invariance of form and the invariance of factor loadings should hold before testing restrictions on means and intercepts. I follow this convention.

I illustrate these ideas with data on "sense of belonging" and "morale." The covariance matrices and means are in Table 8.7. The sample consists of a random sample of 106 students who were undergraduates in 1984 at a private liberal arts college. There are 49 females and 57 males. The students' morale (ξ_1) was measured with four indicators (X_1 to X_4) and their sense of belonging (ξ_2) was gauged by three measures (X_5 to X_7).[11] The measurement equations are of the form $X_i^{(g)} = \upsilon_{X_i}^{(g)} + \lambda_{ij}^{(g)}\xi_j^{(g)} + \delta_i^{(g)}$. For morale ($\xi_1$), $j = 1$ and $i = 1, 2, 3, 4$, whereas for sense of belonging (ξ_2), $j = 2$

[11] Students were asked to indicate their extent of agreement on a 0 to 10 scale with four statements on morale: (1) "I am enthusiastic about ____." (2) "My ____ school spirit is low." (3) "I am happy to be at ____." (4) "____ is one of the best schools in the nation." There were three statements on belonging: (1) "I feel a sense of belonging to ____." (2) "I feel that I am a member of the ____ community." (3) "I see myself as part of the ____ community."

Table 8.7 Covariance Matrices and Means for Four Measures of Morale (x_1 to x_4) and Three Measures of Belonging (x_5 to x_7) for Females ($N = 47$) and Males ($N = 59$)

	x_1	x_2	x_3	x_4	x_5	x_6	x_7	Means
	4.023	−2.954	2.256	1.808	2.756	2.132	1.978	8.347
x_1	4.096							
		7.223	−2.340	−2.598	−3.163	−1.997	−2.614	2.163
x_2	−2.663	5.446						
			2.665	1.965	2.517	1.772	1.810	9.041
x_3	2.762	−2.334	2.848					
				2.473	2.121	1.455	1.677	8.837
x_4	2.917	−2.514	2.471	3.429				
					3.969	2.690	2.410	8.102
x_5	3.553	−3.479	2.919	3.017	4.707			
						4.604	1.835	7.980
x_6	3.652	−3.403	2.773	3.420	4.039	6.070		
							4.267	8.061
x_7	3.522	−3.142	2.811	3.331	3.923	4.146	4.427	
Means	8.105	2.018	8.719	8.772	7.842	7.579	7.965	

Note: Covariances for females in top half; those for males in bottom half

and $i = 5, 6, 7$. The superscript of g runs from 1 for females to 2 for males. To assign a scale to morale (ξ_1), I set υ_{X_1} to zero and λ_{11} to one, and to scale belonging (ξ_2), I set υ_{X_5} to zero and λ_{52} to one.

Because the school was formerly all male and has a recent history of coeducation, an interesting question might be whether the males or females have a higher school morale and sense of belonging. The long tradition as a male school might lead males to feel more comfortable and to have more support networks to draw upon than females. This could lead to a higher morale and sense of belonging for them. On the other hand, the transition to a coeducational institution might have diminished the dominance of males and lowered their morale and sense of belonging. Both hypotheses can be explored by comparing $\kappa_1^{(1)}$ and $\kappa_1^{(2)}$, for morale, and $\kappa_2^{(1)}$ and $\kappa_2^{(2)}$ for sense of belonging.

Another set of hypotheses concerns the intercepts for the measurement equations. It could be that one group tends to respond systematically higher or lower to the questions even if both groups have the same slope relating the measures to the latent variables. This would be revealed in a difference in the intercepts for the same indicator for males versus females.

Table 8.8 Tests of Invariance of Morale (ξ_1) and Sense of Belonging (ξ_2) for Females and Males

				GFI			
Hypothesis	χ^2	df	prob	Females	Males	Δ_1	Δ_2
H_{form}	24.88	26	0.526	0.957	0.938	0.959	1.002
H_{Λ_x}	32.47	31	0.394	0.935	0.927	0.946	0.997
$H_{\Lambda_x \upsilon_x}$	35.37	36	0.499	0.926	0.923	0.942	1.001
$H_{\Lambda_x \upsilon_x \kappa}$	35.78	38	0.572	0.925	0.922	0.941	1.004

I explore these and other issues by estimating a confirmatory factor analysis of the seven indicators for the two groups. I program this problem in the same fashion as discussed earlier for models that allow means and intercepts. The major difference is that the estimation is for two groups.

I choose the following hierarchy of hypotheses: H_{form}, H_{Λ_x}, $H_{\Lambda_x \upsilon_x}$, and $H_{\Lambda_x \upsilon_x \kappa}$. Each hypothesis is nested in the one that precedes it. Table 8.8 reports a summary of the fit of each hypothesis. Beginning with H_{form}, the form of the model leads to an excellent fit for both groups. Two factors—morale and belonging—without any correlated errors of measurement are quite consistent with the data, as shown by the nonsignificant chi-square estimate of 24.88 with 26 df ($p = 0.526$) and the high values of the GFIs, and Δ's. The H_{Λ_x} restricts the slopes relating the measures to the latent variables to be equal for males and females. A one-unit increase in morale or belonging leads to the same expected change in the same measure for both groups. This hypothesis also is a good match to the data. The χ^2 of 32.47 (31 df) is not statistically significant, and the other fit measures change only slightly. The chi-square difference of 7.59 (5 df) between H_{form} and H_{Λ_x} is nonsignificant ($p > 0.10$).

The next step is to test if the intercept term υ_x also are equivalent. The hypothesis $H_{\Lambda_x \upsilon_x}$ has a nonsignificant chi-square estimate of 35.37 with 36 df ($p = 0.499$), and there is hardly any change in the fit measures. The chi-square difference for H_{Λ_x} and $H_{\Lambda_x \upsilon_x}$ clearly is not significant. This provides strong support for the hypothesis that neither group has a constant tendency to answer higher or lower than the other group.

The final step is testing $H_{\Lambda_x \upsilon_x \kappa}$, which reveals whether the mean values of the two latent variables differ. This added restriction does not lead to a statistically significant chi-square (= 39.78, 38 df, $p = 0.572$). There is barely any decline in the other fit measures. These results suggest that males and females have the same mean values of morale and sense of belonging. In addition the evidence supports the view that the slopes and intercepts for these observed variables do not vary for the groups.

The component fit measures did reveal some characteristics that bear mentioning. First, the intercepts υ_x were all close to zero for all measures except X_2. This variable is a negatively worded question, whereas the others are positive, so its different intercept is not surprising. These results suggest that not only are the intercepts for each variable equal for males and females but most of the intercepts (except those for X_2) are zero. Second, starting with $H_{\Lambda_x \upsilon_x}$, an unusual pattern of the residuals becomes evident. The residuals $(s_{ij} - \hat{\sigma}_{ij})$ for covariances involving X_2 in the female group are all positive, while the corresponding ones for the males are all negative. Though all these residuals in both groups are small (absolute value of normalized residuals less than one), they are the largest normalized residuals for both groups. For $H_{\Lambda_x \upsilon_x \kappa}$, the pattern grows more extreme, with all residuals but three being positive for females and all residuals negative for males. If such a pattern replicates in independent samples, its origin should be investigated despite the other favorable overall fit measures.

An aspect of assessing invariance that I have not mentioned is the impact of sequential testing on statistical significance levels. With repeated testing on the same data with just varying the equality constraints, the chi-square test statistic is correlated with the ones that come before and after it. The usual probability levels do not reflect this dependency. However, the chi-square difference tests are asymptotically uncorrelated (see Steiger, Shapiro, and Browne 1985), and in this sense they are preferable to the individual chi-square tests.[12]

MISSING VALUES

Throughout the book I have implicitly assumed that the full random sample of cases is available for the observed variables. Typically, information is missing for some cases on some variables. This section gives an overview of the problem of missing values in covariance structure models. Treatments of missing values for related techniques can be found in Afifi and Elashoff (1966), Haitovsky (1968), Timm (1970), Rubin (1976), and Brown (1983), among others.

Missing values may follow completely random or other patterns. By completely random missing values, I refer to the situation where the observations with any missing values are a simple random subsample of the full sample (Rubin 1976). I concentrate on completely random missing

[12] The problem of testing multiple hypotheses remains but can be treated by using Bonferroni adjustments to the α levels. For instance, with five invariance chi-square difference tests, the critical probability level should be set to $0.05/5 = 0.01$ to maintain an overall α of 0.05.

values, though I will mention relevant work for patterned missing values at the end of this section.

The two approaches to random missing values are alternative estimators of covariance matrices and explicit consideration of the missing values in the covariance structure estimation routine. I review each of these next.

Alternative Estimators of Covariance Matrix

The common characteristics of the covariance matrix estimator approach to missing values is that a method is selected to estimate the covariance matrix and then this matrix is analyzed in the usual fashion. It is an indirect technique since only the covariance matrix is treated and the estimators are not modified. The three major methods are (1) listwise deletion, (2) pairwise deletion, and (3) missing values replacements. Less widespread are methods to estimate the covariance matrix of the observed variables with pairwise ML and full ML procedures, such as discussed in Brown (1983).

The listwise deletion method removes any observations that have missing information for any of the variables. The covariance matrix for the remaining cases, say, $\mathbf{S}_L$, is the input to analysis with the number of cases set to the number with complete information, say, N_L. Several properties of this strategy follow. First, the sample size based on the listwise deletion is always less than N when any cases have missing information. In some situations the loss of cases can be quite severe. Even if each variable has only 5% missing values, the percent of missing cases using listwise deletion can range from 5% to $(p + q)5\%$ of the original N. Also the nonmissing values of variables for the dropped cases are not utilized. Relative to the full sample, the analysis of $\mathbf{S}_L$ leads to less efficient estimators than would be available with $\mathbf{S}$ from the full sample.

On the positive side, listwise deletion leads to a consistent estimator of $\boldsymbol{\theta}$, using F_{ML}, F_{GLS}, or F_{ULS} as long as $N_L \to \infty$ as $N \to \infty$. When $N_L > (p + q)$, the $\mathbf{S}_L$ matrix is positive-definite, provided $\boldsymbol{\Sigma}$ is (Dijkstra 1981, 18). And if the multinormal distribution assumption is satisfied for the full sample, it also is met for the listwise deletion sample. This means that the usual test statistics are appropriate when $\mathbf{S}_L$ replaces $\mathbf{S}$ and N_L replaces N.

Pairwise deletion forms a sample covariance matrix, $\mathbf{S}_P$, by using all cases with nonmissing values to compute each covariance or variance. The elements of $\mathbf{S}_P$ typically are based on different sets of cases. A desirable feature of $\mathbf{S}_P$ is that it uses all available observations in developing estimates and does not "throw out" cases that lack only a subset of values. Pairwise deletion does have its drawbacks, however. First, $\mathbf{S}_P$ may not be positive-definite, and the fitting functions no longer need have a minimum of zero (Browne 1982, 88).[13] Second, the choice of N_P in a covariance structure analysis of $\mathbf{S}_P$ is ambiguous since the elements of $\mathbf{S}_P$ are determined with different numbers of cases. In general, this is not a trivial problem since the researcher's choice of N_P plays a role in the chi-square tests of model fit and the estimated asymptotic standard errors of the parameter estimates.

Assuming that the variables for the complete data are multinormally distributed, Brown (1983, 288) lists the asymptotic covariance of s_{ij} and s_{gh} for pairwise covariances as approximately:

$$\left\{\frac{N_{ijgh}}{N_{ij}N_{gh}}\right\}(\sigma_{ig}\sigma_{jh} + \sigma_{ih}\sigma_{jg})$$

where N_{ijgh} is the number of observations with complete information on the i, j, g, and h variables and N_{ij} and N_{gh} have analogous definitions. This differs from the usual assumption for the asymptotic covariances which underlie F_{GLS} and F_{ML} in that $N_{ijgh}/(N_{ij}N_{gh})$ appears instead of $1/N$. Thus to some extent, the distributional assumptions associated with the F_{ML} and F_{GLS} fitting functions are violated, and it raises questions about the appropriateness of using the uncorrected standard errors and chi-square test that result from analyzing $\mathbf{S}_P$.

Perhaps the third most common means to handle missing data is to estimate missing values and then to compute the covariance matrix, $\mathbf{S}_M$, for all N observations. The missing value may be replaced by the sample mean of the observed variables, a regression estimate of its values, or by some other number. The major advantage of this procedure is that all N observations are available for estimating the covariance matrix. However, there are drawbacks. Some of these can be illustrated by means of an example with a single indicator:

$$x^* = \lambda\xi + \delta \tag{8.65}$$

In (8.65) x^* represents the ideal situation where the indicator is available for all N observations. As usual, I assume that $\mathrm{COV}(\xi, \delta) = 0$ and $E(\delta) = 0$. Suppose that when x^* is missing, we estimate it, and when present, we use it. The observed x variable for all N observations is

$$x = x^* + e \tag{8.66}$$

[13]Procedures to "smooth" indefinite covariance matrices are available (Schwertman and Allen 1979).

where e is a random variable that is the discrepancy between the estimated value, x, and the missing x^* variable, and $e = 0$ when x^* is present. I assume that the $\text{COV}(x^*, e)$ is zero and $E(e) = 0$.

Consider the $\text{VAR}(x)$ for those observations with complete information (i.e., when $x = x^*$):

$$\text{VAR}(x) = \lambda^2\phi + \text{VAR}(\delta) \tag{8.67}$$

Compare (8.67) to the observations that are estimated (i.e., when $x = x^* + e$):

$$\text{VAR}(x) = \lambda^2\phi + \text{VAR}(\delta) + \text{VAR}(e) \tag{8.68}$$

The $\text{VAR}(x)$ is greater for the observations with estimated values of x^* than it is for the cases where x equals x^*. Ignoring this difference creates heteroscedasticity for the error in the equation for x such that the error variance is greater for those cases with estimated values.

Another problem is that the distribution of x is unlikely to be normal even if the distribution of x^* is normal. The x variable is the sum of x^* and e, so its distribution depends on the distribution of these variables. The variable e for the full sample is unlikely to have a normal distribution since its value is zero whenever there is a nonmissing value. Usually, the missing values of x^* are a small percentage of all cases, and only a small percentage of the values of e are nonzero. This makes it highly unlikely that e has a normal distribution across all N cases. Since x is a sum of a normal and nonnormal random variable, it will have a nonnormal distribution. I can make the same argument for other indicators with estimated values. Furthermore the GLS assumptions for the distribution of the sample covariance elements generally will fail to hold. It follows that the distributional assumptions which underlie F_{GLS} and F_{ML} are violated and that the usual formulas for asymptotic standard errors and the chi-square tests are likely to be inaccurate.

For structural equations with observed variable models, the consequences of estimating missing values for y are similar to that described for indicators: the error variance is heteroscedastic, being too high for the cases with missing values and the distribution of y is unlikely to be normal even if y^* is normal. If any explanatory variables in $\mathbf{x}$ have estimated values, the situation parallels the error in the variable case and the consequences presented in Chapter 5 follow (e.g., inconsistent estimators of $\mathbf{B}$ and $\mathbf{\Gamma}$).

In short, using $\mathbf{S}_L$, $\mathbf{S}_P$, or $\mathbf{S}_M$ in place of $\mathbf{S}$ can have consequences for the analysis of covariance structure models. It is difficult to make general statements about the seriousness and magnitude of the effects, but it is reasonable to expect that the smaller the original sample and the larger the percent of missing data, the more grave the consequences. Simulation studies are sometimes conflicting in their evaluations of these methods. For instance, Haitovsky (1968) finds that the efficiency of the OLS estimator in regression analysis using the pairwise deletion covariance matrix inferior to that for the listwise deletion covariance matrix, whereas Kim and Curry (1977) reach the opposite conclusion. The covariance structure models of primary interest here have been neglected.

Explicit Estimation

A second approach is to consider the missing and complete data as part of the ML or GLS estimator, rather than constructing a covariance matrix that is then analyzed without further adjustments. Werts, Rock, and Grandy (1979, 201–204), Baker and Fulker (1983), and Allison (1987) suggest a way to use multiple group analysis to treat missing values. I present an observed variable model to illustrate their method.

Suppose that we have a two-equation recursive observed variable model where the usual assumptions hold:

$$\begin{aligned} y_1 &= \gamma_{11} x_1 + \zeta_1 \\ y_2 &= \beta_{21} y_1 + \gamma_{21} x_1 + \zeta_2 \end{aligned} \tag{8.69}$$

Out of a sample of 500, we are missing 100 cases for x_1. The procedure modifies the standard notation, uses the means of the observed variables, and analyzes the moments around zero instead of the covariances. It begins by forming two groups: one for the 400 complete observations and the second for the remaining 100 cases. The moment matrix can be created from the usual covariance matrix and a vector of means, which for the first group ($N = 400$) might look like

$$\mathbf{S}^{(1)} = \begin{bmatrix} 10.5 & & \\ 6.9 & 12.3 & \\ 7.2 & 6.2 & 9.8 \end{bmatrix}$$

$$\text{Means} = \begin{bmatrix} 5.1 & 6.7 & 6.1 \end{bmatrix}$$

The second group ($N = 100$) has no values for X_1. To keep $\mathbf{S}^{(2)}$ a 3×3 matrix and the means a 1×3 vector, we replace the missing information for X_1 with pseudo-values:

$$\mathbf{S}^{(2)} = \begin{bmatrix} 9.8 & & \\ 7.1 & 10.9 & \\ 0 & 0 & 1 \end{bmatrix}$$

$$\text{Means} = [4.9 \quad 6.2 \quad 0]$$

The missing covariances of X_1 with Y_1 and Y_2 and the mean of X_1 are replaced by zeros. The variance of X_1 is set to one. This is a programming "trick" to allow estimation of the model in groups with different numbers of observed variables.

Next we construct a measurement equation:

$$\mathbf{y} = \boldsymbol{\Lambda}_y \boldsymbol{\eta} + \boldsymbol{\epsilon} \tag{8.70}$$

$$\begin{bmatrix} Y_1 \\ Y_2 \\ X_1 \end{bmatrix} = \begin{bmatrix} 1 & 0 & 0 & \lambda_{14} \\ 0 & 1 & 0 & \lambda_{24} \\ 0 & 0 & \lambda_{33} & \lambda_{34} \end{bmatrix} \begin{bmatrix} \eta_1 \\ \eta_2 \\ \eta_3 \\ \eta_4 \end{bmatrix} + \begin{bmatrix} \epsilon_1 \\ \epsilon_2 \\ \epsilon_3 \end{bmatrix} \tag{8.71}$$

where I use Y_1, Y_2, and X_1 to represent the observed variables in their nondeviation form. Here we combine all of the observed variables (Y_1, Y_2, and X_1) into a new vector $\mathbf{y}$. Each observed variable corresponds to a latent variable (η_1, η_2, or η_3), and η_4 is a constant with its coefficients (λ_{i4}, $i = 1, 2, 3$) corresponding to the means of the three observed variables. In group 1 with complete information, $\lambda_{33} = 1$ and $\text{VAR}(\epsilon_i) = 0$ for $i = 1, 2, 3$. In group 2, where X_1 is missing, we use the same specification except that $\lambda_{33} = \lambda_{34} = 0$ and $\text{VAR}(\epsilon_3) = 1$. These latter constraints enable us to reproduce the pseudo-values of zero covariances, zero mean, and a variance of one for X_1 in group 2, (see $\mathbf{S}^{(2)}$ and the mean vector for group 2).

The latent variable model is

$$\boldsymbol{\eta} = \mathbf{B} \boldsymbol{\eta} + \boldsymbol{\Gamma} \boldsymbol{\xi} + \boldsymbol{\zeta} \tag{8.72}$$

$$\begin{bmatrix} \eta_1 \\ \eta_2 \\ \eta_3 \\ \eta_4 \end{bmatrix} = \begin{bmatrix} 0 & 0 & \beta_{13} & 0 \\ \beta_{21} & 0 & \beta_{23} & 0 \\ 0 & 0 & 0 & 0 \\ 0 & 0 & 0 & 0 \end{bmatrix} \begin{bmatrix} \eta_1 \\ \eta_2 \\ \eta_3 \\ \eta_4 \end{bmatrix} + \begin{bmatrix} 0 \\ 0 \\ 0 \\ 1 \end{bmatrix} [\xi_1] + \begin{bmatrix} \zeta_1 \\ \zeta_2 \\ \zeta_3 \\ \zeta_4 \end{bmatrix} \tag{8.73}$$

Equation (8.73) is the same for both groups. The $\text{VAR}(\zeta_4) = 0$, $\xi_1 = 1$, and

$\eta_4 = \xi_1 = 1$. The "x_1" ($= \xi_1$) is treated as "fixed," or nonstochastic. The γ_{11} and γ_{21} of the original model in (8.69) are β_{13} and β_{23}, while β_{21} is called the same in the preceding **B**. During estimation we restrict the **B** and $\boldsymbol{\Psi}$ matrices, λ_{14} and λ_{24} to be equal across groups.

We estimate this like any other multigroup analysis. The one exception is that the degrees of freedom associated with the chi-square estimate is too high. To adjust it, we must count the number of pseudo-values of zeros and ones in the covariance matrix and mean vector for the group with missing variables and then subtract this number from the reported degrees of freedom. In the preceding illustration, we have 9 df before adjustment and four pseudo-values. So the correct degrees of freedom is 5. The preceding hypothetical example leads to a chi-square estimate of 6.0 with $p > 0.3$. Appendix 8B lists the LISREL VI program for this example.

The same general procedure works for confirmatory factor analysis or other structural equation models. See Allison (1987), Baker and Fulker (1983), and Werts et al. (1979) for additional illustrations.

A desirable characteristic of this strategy is that the estimator uses all available information. In addition the equality of the parameters for the groups containing missing and nonmissing data can be tested rather than only assumed. One practical difficulty with the procedure arises when many observed variables have missing values that do not follow simple patterns. For instance, suppose we have six variables and each variable has some missing values. Following the preceding procedures we would form many groups to cover all combinations of cases missing one variable at a time, two at a time, etc. Lee (1986) provides a way to handle this more general case for GLS and ML estimators and demonstrates its implementation for a confirmatory factor analysis example.

Systematic Missing Values

The observations with complete information are not always a simple random sample of the original sample as I just assumed. When the data are *not* missing completely at random, then use of listwise, pairwise, or alternative estimators of the covariance matrix can lead to inconsistent parameter estimators. The multigroup approach of the last section maintains consistency and other desirable properties under a less restrictive assumption described by Rubin (1976) as data missing at random. Data are *missing at random* if, given the values of the variables that are observed, the probability of a value being missing does not depend on the values of the missing variable. To illustrate this, suppose we have two variables, salary and seniority, for a random sample of employees at a large corporation. We have complete information on seniority but missing data on salary. If, for all cases with the same seniority, the probability of their salary value being missing is unrelated to their salary, then the data are missing at random. If, on the other hand, at the same seniority level those with high salary levels are more likely to be missing than others, the data are not missing at random.

The failure to have data missing at random calls for models of the possible selectivity bias (e.g., see Heckman 1979). This approach introduces a new "selection equation" to model which cases are included or excluded from the sample. The equation is considered simultaneously with the structural equations. Muthén and Jöreskog (1983) provide an overview of this problem that is consistent with the covariance structure models of this book. However, work in this area has not provided a fully inclusive approach to systematic missing values on many variables combined with a general latent variable and measurement model.

TOTAL, DIRECT, AND INDIRECT EFFECTS

As I mentioned in Chapter 2, path analysis allows researchers to decompose the effects of one variable on another into direct, indirect, and total effects. Direct effects are the influences of one variable on another that are not mediated by any other variable. In Figure 8.1 the direct effects of 1960 industrialization (ξ_1) on 1965 political democracy (η_2) is γ_{21}. The direct effect of industrialization (ξ_1) on energy consumption per capita (x_2) is λ_6. More generally, the direct effects are in the $\mathbf{B}$, $\boldsymbol{\Gamma}$, $\boldsymbol{\Lambda}_x$, and $\boldsymbol{\Lambda}_y$ matrices, where, for example, the direct effects of $\boldsymbol{\xi}$ on $\boldsymbol{\eta}$ are in $\boldsymbol{\Gamma}$ and the direct effects of $\boldsymbol{\eta}$ on $\mathbf{y}$ are in $\boldsymbol{\Lambda}_y$.

Indirect effects are ones that are mediated by at least one other variable, and the total effects are the sum of direct and indirect effects. Figure 8.1 shows the indirect effect of 1960 industrialization (ξ_1) on 1965 political democracy (η_2) to be $\gamma_{11}\beta_{21}$, while the indirect effect of 1960 political democracy (η_1) on the 1965 freedom of group opposition measure (y_6) is $\beta_{21}\lambda_2$. The total effects of 1960 industrialization (ξ_1) on 1965 political democracy (η_2) is $(\gamma_{21} + \gamma_{11}\beta_{21})$. I provide more general expressions for indirect and total effects in this section.

The indirect and total effects can help to answer important questions that are not addressed by examining the direct effects. For instance, suppose that participation in a social welfare program has a positive direct effect on household income but a negative indirect effect because it reduces the number of hours of part-time work, which in turn reduces other sources of income. The direct effect provides a misleading impression of the influence of program participation on income. It is the total effect that is

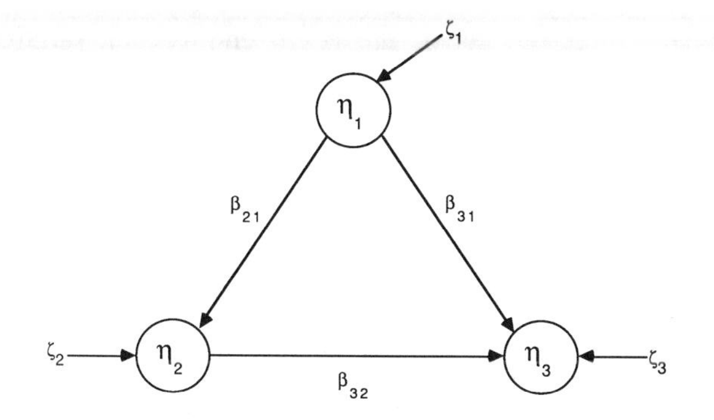

Figure 8.14 A Simple Recursive Model for Three Endogenous Variables

more relevant. Or, suppose that you have a second-order factor analysis (see Chapter 7) and wish to know which indicator is most strongly affected by the second-order factor. The total effects of the higher-order factor on the measures provides such information.

In this section I define the total and indirect effects and introduce the idea of specific effects and how to compute them.[14] Appendix 8A shows the computation of estimated asymptotic variances for these effects.

Total effects are defined in two ways. The first sets them equal to a sum of powers of coefficient matrices. The other assigns meaning to total effects by using reduced-form coefficients. Both definitions lead to the same results. I present the "infinite sum" definition and refer the reader to Alwin and Hauser (1975), Graff and Schmidt (1982), or Bollen (1987) for the reduced-form definition. The total effects of $\boldsymbol{\eta}$ on $\boldsymbol{\eta}$ or $\mathbf{T}_{\eta\eta}$ are

$$\mathbf{T}_{\eta\eta} = \sum_{k=1}^{\infty} \mathbf{B}^k \tag{8.74}$$

$\mathbf{T}_{\eta\eta}$ is only defined if the infinite sum in (8.74) converges to a matrix with finite elements. To better understand this definition, consider the relation between three latent endogenous variables, as drawn in the path diagram in Figure 8.14. The $\mathbf{B}$ matrix is

$$\mathbf{B} = \begin{bmatrix} 0 & 0 & 0 \\ \beta_{21} & 0 & 0 \\ \beta_{31} & \beta_{32} & 0 \end{bmatrix} \tag{8.75}$$

[14]Most of this section is based on Bollen (1987).

B provides the direct effects of the latent endogenous variables on one another. The first few terms of the infinite series defining the total effects of $\boldsymbol{\eta}$ on $\boldsymbol{\eta}$ are

$$\begin{aligned} \mathbf{T}_{\eta\eta} &= \mathbf{B} + \mathbf{B}^2 + \mathbf{B}^3 + \cdots \\ &= \begin{bmatrix} 0 & 0 & 0 \\ \beta_{21} & 0 & 0 \\ \beta_{31} & \beta_{32} & 0 \end{bmatrix} + \begin{bmatrix} 0 & 0 & 0 \\ 0 & 0 & 0 \\ \beta_{21}\beta_{32} & 0 & 0 \end{bmatrix} + \mathbf{O} + \cdots \end{aligned} \tag{8.76}$$

Clearly, since all $\mathbf{B}^k$ for $k \geq 3$ equal zero, the series converges and the total effects are defined. The first term in the series is the direct effects of $\boldsymbol{\eta}$ on $\boldsymbol{\eta}$. The second and higher power terms define indirect effects of varying lengths. For Figure 8.14, the only indirect effect of length 2 is $\beta_{21}\beta_{32}$, the influence of η_1 on η_3 mediated by η_2. The zero values for $\mathbf{B}^3$ and the higher-power terms indicate that all indirect effects of length 3 or greater equal zero. Summing the right-hand side of (8.76) leads to

$$\mathbf{T}_{\eta\eta} = \begin{bmatrix} 0 & 0 & 0 \\ \beta_{21} & 0 & 0 \\ \beta_{31} + \beta_{21}\beta_{32} & \beta_{32} & 0 \end{bmatrix} \tag{8.77}$$

In general, the indirect effects are the differences between the total and direct effects. Subtracting $\mathbf{B}$ from $\mathbf{T}_{\eta\eta}$ yields $\mathbf{B}^2$, the indirect effects ($\mathbf{I}_{\eta\eta}$) for this model.

This result generalizes. For recursive systems where $\mathbf{B}$ can be written as a lower triangular matrix, $\mathbf{B}^k$ equals zero for $k \geq m$ (where m is the number of η's). Thus $\mathbf{B}$ converges, and the total effects are defined. These total effects equal $\Sigma_{k=1}^{m-1}\mathbf{B}^k$. It then follows that the indirect effects are $\mathbf{T}_{\eta\eta} - \mathbf{B}$ or $\Sigma_{k=2}^{m-1}\mathbf{B}^k$.

For nonrecursive models, the situation is more complicated. Consider the example in Figure 8.15.

$$\mathbf{B} = \begin{bmatrix} 0 & \beta_{12} \\ \beta_{21} & 0 \end{bmatrix} \tag{8.78}$$

where β_{21} is the direct effect of η_1 on η_2 and β_{12} that for η_2 on η_1. The indirect effects of $\boldsymbol{\eta}$ on $\boldsymbol{\eta}$ of particular lengths can be determined by raising

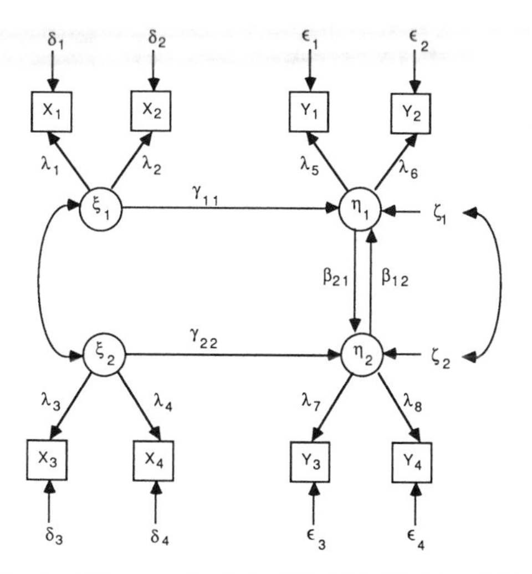

Figure 8.15 Path Diagram of Nonrecursive Latent Variable Model and Measurement Model

B to the appropriate powers. The indirect effects for lengths 2, 3, and 4 are

$$\mathbf{B}^2 = \begin{bmatrix} \beta_{21}\beta_{12} & 0 \\ 0 & \beta_{21}\beta_{12} \end{bmatrix}, \qquad \mathbf{B}^3 = \begin{bmatrix} 0 & \beta_{21}\beta_{12}^2 \\ \beta_{21}^2\beta_{12} & 0 \end{bmatrix}$$
$$\mathbf{B}^4 = \begin{bmatrix} \beta_{21}^2\beta_{12}^2 & 0 \\ 0 & \beta_{21}^2\beta_{12}^2 \end{bmatrix} \tag{8.79}$$

Unlike recursive models, $\mathbf{B}^k$ is not necessarily zero for $k \geq m$. Also, as $\mathbf{B}^2$ and $\mathbf{B}^4$ illustrate, an endogenous variable can have an indirect effect on itself. For instance, η_1 changes η_2, β_{21} units, but the β_{21} change in η_2 leads to a $\beta_{21}\beta_{12}$ shift in η_1. It is evident that the sum of $\mathbf{B}^k$ as $k \to \infty$ leads each element of $\mathbf{T}_{\eta\eta}$ to be an infinite sum.

In general, for the total effects $\mathbf{T}_{\eta\eta}$ $(= \Sigma_{k=1}^{\infty}\mathbf{B}^k)$ to be defined, $\mathbf{B}^k$ must converge to zero as $k \to \infty$ (Ben-Israel and Greville 1974). This occurs if and only if the modulus or absolute value of the largest eigenvalue of $\mathbf{B}$ is less than one (Bentler and Freeman 1983, 144).[15] To find the value to which $\mathbf{T}_{\eta\eta}$ converges, first add $\mathbf{I}$ $(= \mathbf{B}^0)$ to $\mathbf{T}_{\eta\eta}$, and then premultiply the sum

[15]The modulus of a complex number $z = a + ib$ is $(a^2 + b^2)^{1/2}$, where a and b are real constants and i is the imaginary number of $\sqrt{-1}$. The similar canonical form of $\mathbf{B}$ is useful in studying $\mathbf{B}^k$ and in understanding the justification for placing this condition on the eigenvalues of $\mathbf{B}$. See Searle (1982, 282–289).

$\mathbf{I} + \mathbf{B} + \mathbf{B}^2 + \cdots + \mathbf{B}^k$ by $(\mathbf{I} - \mathbf{B})$:

$$(\mathbf{I} - \mathbf{B})(\mathbf{I} + \mathbf{B} + \mathbf{B}^2 + \cdots + \mathbf{B}^k) = \mathbf{I} - \mathbf{B}^{k+1}. \tag{8.80}$$

Since $\mathbf{B}^{k+1} \to 0$ as $k \to \infty$, the last term in (8.80) approaches zero, leaving $\mathbf{I}$. For the product of the two left-hand terms in (8.80) to equal $\mathbf{I}$, $(\mathbf{I} + \mathbf{B} + \mathbf{B}^2 + \cdots + \mathbf{B}^k)$ must converge to $(\mathbf{I} - \mathbf{B})^{-1}$ as $k \to \infty$. If $\mathbf{I}$ is subtracted from this value, $\mathbf{T}_{\eta\eta}$ results

$$\mathbf{T}_{\eta\eta} = (\mathbf{I} - \mathbf{B})^{-1} - \mathbf{I}. \tag{8.81}$$

Indirect effects follow when $\mathbf{B}$ is subtracted from $\mathbf{T}_{\eta\eta}$:

$$\mathbf{I}_{\eta\eta} = (\mathbf{I} - \mathbf{B})^{-1} - \mathbf{I} - \mathbf{B}. \tag{8.82}$$

The preceding analysis shows the decomposition of effects for the latent endogenous variables on one another. Decompositions of the latent exogenous variables on the latent endogenous ones are closely related. The total effects of $\boldsymbol{\xi}$ on $\boldsymbol{\eta}$ can be determined by substituting the equation $\boldsymbol{\eta} = \mathbf{B}\boldsymbol{\eta} + \boldsymbol{\Gamma}\boldsymbol{\xi} + \boldsymbol{\zeta}$ for $\boldsymbol{\eta}$ into the right-hand side of the same equation (Bentler and Freedman 1988, 143) repeatedly:

$$\begin{aligned}
\boldsymbol{\eta} &= \mathbf{B}\boldsymbol{\eta} + \boldsymbol{\Gamma}\boldsymbol{\xi} + \boldsymbol{\zeta} \\
&= \mathbf{B}(\mathbf{B}\boldsymbol{\eta} + \boldsymbol{\Gamma}\boldsymbol{\xi} + \boldsymbol{\zeta}) + \boldsymbol{\Gamma}\boldsymbol{\xi} + \boldsymbol{\zeta} \\
&= \mathbf{B}^2\boldsymbol{\eta} + (\mathbf{I} + \mathbf{B})(\boldsymbol{\Gamma}\boldsymbol{\xi} + \boldsymbol{\zeta}) \\
&= \mathbf{B}^2(\mathbf{B}\boldsymbol{\eta} + \boldsymbol{\Gamma}\boldsymbol{\xi} + \boldsymbol{\zeta}) + (\mathbf{I} + \mathbf{B})(\boldsymbol{\Gamma}\boldsymbol{\xi} + \boldsymbol{\zeta}) \\
&= \mathbf{B}^3\boldsymbol{\eta} + (\mathbf{I} + \mathbf{B} + \mathbf{B}^2)(\boldsymbol{\Gamma}\boldsymbol{\xi} + \boldsymbol{\zeta}) \\
&\quad \vdots \qquad \vdots \qquad \vdots \\
&= \mathbf{B}^k\boldsymbol{\eta} + (\mathbf{I} + \mathbf{B} + \mathbf{B}^2 + \cdots + \mathbf{B}^{k-1})(\boldsymbol{\Gamma}\boldsymbol{\xi} + \boldsymbol{\zeta}).
\end{aligned} \tag{8.83}$$

The total effects of $\boldsymbol{\xi}$ on $\boldsymbol{\eta}$ are in the coefficient matrix for $\boldsymbol{\xi}$ in the last line of (8.83): $(\mathbf{I} + \mathbf{B} + \mathbf{B}^2 + \cdots + \mathbf{B}^{k-1})\boldsymbol{\Gamma}$. The $(\mathbf{I} + \mathbf{B} + \mathbf{B}^2 + \cdots + \mathbf{B}^{k-1})$ term is an infinite sum that converges to $(\mathbf{I} - \mathbf{B})^{-1}$ under the same conditions stated earlier. That is, the absolute value or modulus of the largest eigenvalue of $\mathbf{B}$ must be less than one. The $\mathbf{T}_{\eta\xi}$ matrix under this condition is

$$\mathbf{T}_{\eta\xi} = (\mathbf{I} - \mathbf{B})^{-1}\boldsymbol{\Gamma}. \tag{8.84}$$

Since the direct effects of $\boldsymbol{\xi}$ on $\boldsymbol{\eta}$ are in $\boldsymbol{\Gamma}$, the indirect effects of $\boldsymbol{\xi}$ on $\boldsymbol{\eta}$ are

$$\mathbf{I}_{\eta\xi} = (\mathbf{I} - \mathbf{B})^{-1}\boldsymbol{\Gamma} - \boldsymbol{\Gamma} = \left[(\mathbf{I} - \mathbf{B})^{-1} - \mathbf{I}\right]\boldsymbol{\Gamma}. \tag{8.85}$$

Equation (8.85) shows the indirect effects of $\boldsymbol{\xi}$ on $\boldsymbol{\eta}$ to equal the product of the total effects of $\boldsymbol{\eta}$ on $\boldsymbol{\eta}$ times the direct effects of $\boldsymbol{\xi}$ on $\boldsymbol{\eta}$.

The derivation of the total and indirect effects of $\boldsymbol{\xi}$ on $\mathbf{y}$ follows a similar strategy. Repeated substitution of $\mathbf{B}\boldsymbol{\eta} + \boldsymbol{\Gamma}\boldsymbol{\xi} + \boldsymbol{\zeta}$ for $\boldsymbol{\eta}$ in the measurement equation for $\mathbf{y}$ leads to

$$\begin{aligned}\mathbf{y} &= \boldsymbol{\Lambda}_y\mathbf{B}^k\boldsymbol{\eta} + \boldsymbol{\Lambda}_y(\mathbf{I} + \mathbf{B} + \mathbf{B}^2 + \cdots + \mathbf{B}^{k-1})\boldsymbol{\Gamma}\boldsymbol{\xi} \\ &\quad + \boldsymbol{\Lambda}_y(\mathbf{I} + \mathbf{B} + \mathbf{B}^2 + \cdots + \mathbf{B}^{k-1})\boldsymbol{\zeta} + \boldsymbol{\epsilon}.\end{aligned} \tag{8.86}$$

For convergent $\mathbf{B}$, the total effects $\mathbf{T}_{y\xi}$ are

$$\mathbf{T}_{y\xi} = \boldsymbol{\Lambda}_y(\mathbf{I} - \mathbf{B})^{-1}\boldsymbol{\Gamma}. \tag{8.87}$$

while the indirect effects also equal (8.87) since $\boldsymbol{\xi}$ has no direct effects on $\mathbf{y}$. By the same logic, the total effects of $\boldsymbol{\eta}$ on $\mathbf{y}$ are

$$\begin{aligned}\mathbf{T}_{y\eta} &= \boldsymbol{\Lambda}_y(\mathbf{I} + \mathbf{B} + \mathbf{B}^2 + \cdots + \mathbf{B}^k) \\ &= \boldsymbol{\Lambda}_y(\mathbf{I} - \mathbf{B})^{-1},\end{aligned} \tag{8.88}$$

with indirect effects of

$$\mathbf{I}_{y\eta} = \boldsymbol{\Lambda}_y(\mathbf{I} - \mathbf{B})^{-1} - \boldsymbol{\Lambda}_y = \boldsymbol{\Lambda}_y\left[(\mathbf{I} - \mathbf{B})^{-1} - \mathbf{I}\right]. \tag{8.89}$$

As stated earlier, the total and indirect effects are only defined under certain conditions. A sufficient condition for all total effects to exist is that the absolute value or modulus of the eigenvalues of $\mathbf{B}$ must be less than one. The eigenvalues are not always readily available. However, two shortcuts can identify convergent $\mathbf{B}$. One has already been discussed. If $\mathbf{B}$ is a lower triangular matrix, then $\mathbf{B}^k$ is zero for $k \geq m$. A second check is applicable to nonrecursive models. If the elements of $\mathbf{B}$ are positive and the sum of the elements in each column is less than one, the absolute values of the eigenvalues are less than one (Goldberg 1958, 237–238). Note that this check is a sufficient but not necessary condition for stability. Given its ease of calculation, it should prove useful in many situations.

To illustrate the decompositions, I refer to Figure 8.15. The total effects of η on η are

$$\mathbf{T}_{\eta\eta} = (1 - \beta_{21}\beta_{12})^{-1}\begin{bmatrix} \beta_{21}\beta_{12} & \beta_{12} \\ \beta_{21} & \beta_{21}\beta_{12} \end{bmatrix} \tag{8.90}$$

with indirect effects of

$$\begin{aligned} \mathbf{I}_{\eta\eta} &= \mathbf{T}_{\eta\eta} - \mathbf{B} \\ &= (1 - \beta_{21}\beta_{12})^{-1}\begin{bmatrix} \beta_{21}\beta_{12} & \beta_{21}\beta_{12}^{2} \\ \beta_{21}^{2}\beta_{12} & \beta_{21}\beta_{12} \end{bmatrix} \end{aligned} \tag{8.91}$$

In a similar fashion the total and indirect effects for the remaining variables can be determined by substituting the $\mathbf{B}$, $\boldsymbol{\Gamma}$, and $\boldsymbol{\Lambda}_y$ for the model in Figure 8.15 into the appropriate decomposition formula derived earlier.

Table 8.9 summarizes the decomposition of effects for the general structural equation model with latent variables. These decompositions can be specialized to treat any of the models in the book. For example, if **x** were substituted for $\boldsymbol{\xi}$, and **y** for $\boldsymbol{\eta}$, the decomposition of effects in Table 8.9 are for structural equations with observed variables.

Table 8.9 Direct, Indirect, and Total Effects of $\boldsymbol{\xi}$ and $\boldsymbol{\eta}$ on $\boldsymbol{\eta}$, y, and x

	Effects on:		
	$\boldsymbol{\eta}$	y	x
Effects of ξ:			
Direct Effect	$\boldsymbol{\Gamma}$	$\mathbf{0}$	$\boldsymbol{\Lambda}_x$
Indirect Effect	$(\mathbf{I} - \mathbf{B})^{-1}\boldsymbol{\Gamma} - \boldsymbol{\Gamma}$	$\boldsymbol{\Lambda}_y(\mathbf{I} - \mathbf{B})^{-1}\boldsymbol{\Gamma}$	$\mathbf{0}$
Total Effect	$(\mathbf{I} - \mathbf{B})^{-1}\boldsymbol{\Gamma}$	$\boldsymbol{\Lambda}_y(\mathbf{I} - \mathbf{B})^{-1}\boldsymbol{\Gamma}$	$\boldsymbol{\Lambda}_x$
Effects of η:			
Direct Effect	$\mathbf{B}$	$\boldsymbol{\Lambda}_y$	$\mathbf{0}$
Indirect Effect	$(\mathbf{I} - \mathbf{B})^{-1} - \mathbf{I} - \mathbf{B}$	$\boldsymbol{\Lambda}_y(\mathbf{I} - \mathbf{B})^{-1} - \boldsymbol{\Lambda}_y$	$\mathbf{0}$
Total Effect	$(\mathbf{I} - \mathbf{B})^{-1} - \mathbf{I}$	$\boldsymbol{\Lambda}_y(\mathbf{I} - \mathbf{B})^{-1}$	$\mathbf{0}$

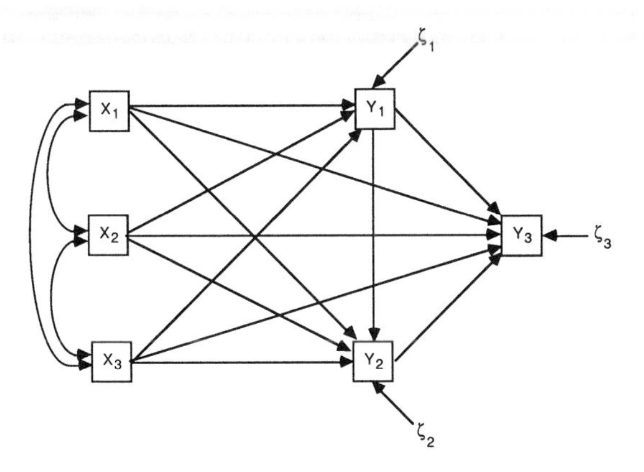

Figure 8.16 Recursive Model to Illustrate Specific Effects

Specific Effects

The indirect effects comprise all of the indirect paths from one variable to another. As a consequence the contribution of particular mediating variables(s) can be obscured. It is possible to analyze *specific indirect effects*, that is, those effects transmitted by a particular variable or group of variables. In this subsection I present a technique proposed in Bollen (1987) that can determine the specific effects transmitted across any path or combination of paths in a model. Once the analyst designates a specific effect for study, the method has several steps: (1) identify the changes needed in the coefficient matrices, (2) modify the coefficient matrices, (3) if **B** is changed, check the modulus or absolute value of the largest eigenvalue of the new **B** to ensure that it is less than one, and, (4) calculate the direct, indirect, or total effects with the modified matrices. An optional fifth step is (5) subtract the new decompositions from the old.

I demonstrate this procedure with the recursive model in Figure 8.16. The **B** and **Γ** for this model are

$$\mathbf{B} = \begin{bmatrix} 0 & 0 & 0 \\ \beta_{21} & 0 & 0 \\ \beta_{31} & \beta_{32} & 0 \end{bmatrix}, \quad \boldsymbol{\Gamma} = \begin{bmatrix} \gamma_{11} & \gamma_{12} & \gamma_{13} \\ \gamma_{21} & \gamma_{22} & \gamma_{23} \\ \gamma_{31} & \gamma_{32} & \gamma_{33} \end{bmatrix} \tag{8.92}$$

Suppose that I want to estimate all of the specific indirect effects of **x** on **y** through y_1. The standard indirect effects provide the effects through y_1, as well as the other variables in the model. If paths through y_1 were eliminated and if we calculated the resulting decomposition, we would know the decomposition of effects *not* due to y_1 but due only to the remaining variables. Subtracting the second from the first gives only those specific indirect effects through y_1, the quantity desired. For Figure 8.16, the total indirect effects of $\mathbf{x}$ on $\mathbf{y}$ are $(\mathbf{I} - \mathbf{B})^{-1}\mathbf{\Gamma} - \mathbf{\Gamma}$. To remove the influence of y_1, all paths coming into or leaving y_1 are set to zero. Thus the $(2,1)$ and $(3,1)$ elements of $\mathbf{B}$ and the $(1,1)$, $(1,2)$, and $(1,3)$ elements of $\mathbf{\Gamma}$ are set to zero [see (8.92) for $\mathbf{B}$ and $\mathbf{\Gamma}$]. The altered $\mathbf{B}$ and $\mathbf{\Gamma}$ matrices are represented as $\mathbf{B}_{(l_1)}$ and $\mathbf{\Gamma}_{(l_2)}$:

$$\mathbf{B}_{(l_1)} = \begin{bmatrix} 0 & 0 & 0 \\ 0 & 0 & 0 \\ 0 & \beta_{32} & 0 \end{bmatrix}, \qquad \mathbf{\Gamma}_{(l_2)} = \begin{bmatrix} 0 & 0 & 0 \\ \gamma_{21} & \gamma_{22} & \gamma_{23} \\ \gamma_{31} & \gamma_{32} & \gamma_{33} \end{bmatrix} \tag{8.93}$$

where

$$l_1 = \{(.,1) = 0\}$$
$$l_2 = \{(1,.) = 0\}$$

The (l_1) subscript indicates a set of modifications given in the statement defining l_1. For (8.93), the first column of $\mathbf{B}$ is set to zero, which is represented by $(.,1) = 0$. Similarly, the first row of $\mathbf{\Gamma}$ is set to zero, and this is symbolized as $\mathbf{\Gamma}_{(l_2)}$ with l_2 representing $(1,.) = 0$. In this notation the "." in the row position indicates all rows in the stated column are zero. The "." in the column position defines all columns in the stated row to zero.

Before calculating indirect effects with $\mathbf{B}_{(l_1)}$, we need to check whether it satisfies the stability conditions presented earlier. That is, the modulus or absolute value of the largest eigenvalue must be less than one. In general, this is done because $\mathbf{B}_{(l_1)}$ can fail this condition even if $\mathbf{B}$ does not (Fisher 1970; Sobel 1986). In this example $\mathbf{B}$ and $\mathbf{B}_{(l_1)}$ are lower triangular, and the condition is satisfied.

The indirect effects $\mathbf{I}_{yx}$ based on the original $\mathbf{B}$ and $\mathbf{\Gamma}$ [see (8.92)] are

$$(\mathbf{I} - \mathbf{B})^{-1}\mathbf{\Gamma} - \mathbf{\Gamma} = \begin{bmatrix} 0 & 0 & 0 \\ \beta_{21}\gamma_{11} & \beta_{21}\gamma_{12} & \beta_{21}\gamma_{13} \\ (\beta_{31} + \beta_{21}\beta_{32})\gamma_{11} & (\beta_{31} + \beta_{21}\beta_{32})\gamma_{12} & (\beta_{31} + \beta_{21}\beta_{32})\gamma_{13} \\ +\beta_{32}\gamma_{21} & +\beta_{32}\gamma_{22} & +\beta_{32}\gamma_{23} \end{bmatrix} \tag{8.94}$$

The indirect effects based on the $\mathbf{B}_{(l_1)}$ and $\boldsymbol{\Gamma}_{(l_2)}$ of (8.93), in which the paths into or out of y_1 have been eliminated are

$$(\mathbf{I}-\mathbf{B}_{(l_1)})^{-1}\boldsymbol{\Gamma}_{(l_2)}-\boldsymbol{\Gamma}_{(l_2)}=\begin{bmatrix}0 & 0 & 0\\ 0 & 0 & 0\\ \beta_{32}\gamma_{21} & \beta_{32}\gamma_{22} & \beta_{32}\gamma_{23}\end{bmatrix} \tag{8.95}$$

Equation (8.95) reveals that the indirect effects of $\mathbf{x}$ on $\mathbf{y}$ that remain after those passing through y_1 have been eliminated are limited to indirect effects of $\mathbf{x}$ on y_3 operating through y_2. Subtracting (8.95) from (8.94) produces the specific effects of y_1:

$$\begin{bmatrix}0 & 0 & 0\\ \beta_{21}\gamma_{11} & \beta_{21}\gamma_{12} & \beta_{21}\gamma_{13}\\ (\beta_{31}+\beta_{21}\beta_{32})\gamma_{11} & (\beta_{31}+\beta_{21}\beta_{32})\gamma_{12} & (\beta_{31}+\beta_{21}\beta_{32})\gamma_{13}\end{bmatrix} \tag{8.96}$$

In general, the procedure for calculating all specific effects through one or more variables is to first form the indirect effects with the original matrices. Second, alter the $\mathbf{B}$ and $\boldsymbol{\Gamma}$ matrices so that all paths leading to or from the variable or variables of interest are set to zero. Third, assuming that the new $\mathbf{B}$ meets the stability condition, recalculate the indirect effects with these modified matrices. Finally, subtract the modified indirect effects from the original ones to obtain the specific effects through these variables.

Suppose that interests lie in the specific effects running through a path rather than through a variable. Analysts can determine the specific effects that result from any individual path, or group of paths, by first setting these paths to zero in the coefficient matrices. Then, they would compute the decompositions with the modified matrices and, finally, subtract the results from those obtained with the original matrices. The difference is attributable to the specific path(s) identified. I return to Figure 8.15 to illustrate this procedure.

I want to estimate that part of the total effects of $\boldsymbol{\xi}$ on $\boldsymbol{\eta}$ transmitted through the path from η_2 to η_1. All effects operating through this link would be eliminated if β_{12} in $\mathbf{B}$ were set to zero. Since the question at issue only involves the direct effect of η_2 on η_1, only the $\mathbf{B}$ matrix requires modification:

$$\mathbf{B}_{(l)}=\begin{bmatrix}0 & 0\\ \beta_{21} & 0\end{bmatrix}$$

where

$$l = \{\beta_{12} = 0\} \tag{8.97}$$

Next check the stability condition for $\mathbf{B}_{(l)}$. Since $\mathbf{B}_{(l)}$ is a lower triangular matrix, the condition is satisfied. The next step is to substitute (8.97) for $\mathbf{B}$ in the total effects formula for $\boldsymbol{\xi}$ on $\boldsymbol{\eta}$:

$$(\mathbf{I} - \mathbf{B}_{(l)})^{-1}\boldsymbol{\Gamma} = \begin{bmatrix} \gamma_{11} & 0 \\ \beta_{21}\gamma_{11} & \gamma_{22} \end{bmatrix} \tag{8.98}$$

The total effects of $\boldsymbol{\xi}$ on $\boldsymbol{\eta}$ are

$$(\mathbf{I} - \mathbf{B})^{-1}\boldsymbol{\Gamma} = (1 - \beta_{21}\beta_{12})^{-1}\begin{bmatrix} \gamma_{11} & \beta_{12}\gamma_{22} \\ \beta_{21}\gamma_{11} & \gamma_{22} \end{bmatrix} \tag{8.99}$$

Subtracting equation (8.98) from (8.99) gives that part of the total effects attributable to the direct influence of η_2 on η_1:

$$(\mathbf{I} - \mathbf{B})^{-1}\boldsymbol{\Gamma} - (\mathbf{I} - \mathbf{B}_{(l)})^{-1}\boldsymbol{\Gamma} = (1 - \beta_{21}\beta_{12})^{-1}\begin{bmatrix} \beta_{21}\beta_{12}\gamma_{11} & \beta_{12}\gamma_{22} \\ \beta_{21}^2\beta_{12}\gamma_{11} & \beta_{21}\beta_{12}\gamma_{22} \end{bmatrix} \tag{8.100}$$

Equation (8.100) shows, for instance, that $(1 - \beta_{21}\beta_{12})^{-1}\beta_{21}\beta_{12}\gamma_{11}$ of the total effects of ξ_1 on η_1 results from the η_2 to η_1 influence.

The decompositions I have discussed can be calculated in several ways. The Table 8.9 formulas may be programmed into software that has matrix capabilities. For instance, APL, GAUSS, the "PROC MATRIX" procedure in SAS, or SAS/IML can perform the required matrix algebra. An alternative procedure is to use LISREL V or VI. Each of these versions calculates the estimated total effects of $\boldsymbol{\xi}$ and $\boldsymbol{\eta}$. The estimated direct effects are the structural coefficients in $\hat{\boldsymbol{\Lambda}}_x$, $\hat{\boldsymbol{\Lambda}}_y$, $\hat{\boldsymbol{\Gamma}}$, and $\hat{\mathbf{B}}$. The indirect effects can be obtained by subtracting the direct effects from the total effects. Finally, for simple models, the decomposition of effects may be calculated with a hand calculator when the structural coefficient estimates are available. The more complicated the model, the less practical is this option. In Appendix 8A, I describe a method of obtaining estimated asymptotic variances and tests of significance for all types of effects.

To illustrate the preceding methods, I utilize the model relating objective and subjective socioeconomic status taken from Chapter 5. Figure 8.13 depicts this model. Define x_1 as income, x_2 as occupational prestige, η_1 as subjective income, η_2 as subjective occupational prestige, and η_3 as subjective overall status ranking. Let y_1, y_2, and y_3 be measures of the η_1, η_2, and η_3 latent variables. The model represents subjective income (η_1) and subjective occupational prestige (η_2) as dependent upon their objective counterparts, though reciprocally related to each other. In turn, the subjective income and occupation variables (η_1 and η_2) have direct effects on overall subjective status (η_3). I have used this model in Chapter 5 and in this chapter. The sample is for 432 whites.

The ML estimates based on the covariance matrix of these five variables leads to a $\hat{\mathbf{B}}$ and $\hat{\mathbf{\Gamma}}$ of

$$\hat{\mathbf{B}} = \begin{bmatrix} 0 & 0.288 & 0 \\ & (0.168) & \\ 0.330 & 0 & 0 \\ (0.090) & & \\ 0.399 & 0.535 & 0 \\ (0.139) & (0.201) & \end{bmatrix}, \quad \hat{\mathbf{\Gamma}} = \begin{bmatrix} 0.101 & 0 \\ (0.016) & \\ 0 & 0.007 \\ & (0.001) \\ 0 & 0 \end{bmatrix} \tag{8.101}$$

The standard errors are in parentheses below the coefficient estimates. All coefficient estimates carry the expected positive signs and all are more than twice their standard errors, except for $\hat{\beta}_{12}$ which has a critical ratio of 1.7. The goodness of fit of the overall model is excellent (χ^2 equals 0.27 with 1 df, GFI = 1.00, AFGI = 0.996).

The estimated indirect effects of $\boldsymbol{\eta}$ on $\boldsymbol{\eta}$ and $\mathbf{x}$ on $\boldsymbol{\eta}$ are

$$\hat{\mathbf{I}}_{\eta\eta} = \begin{bmatrix} 0.105 & 0.030 & 0 \\ (0.067) & (0.036) & \\ 0.035 & 0.105 & 0 \\ (0.026) & (0.067) & \\ 0.237 & 0.183 & 0 \\ (0.105) & (0.127) & \end{bmatrix}, \quad \hat{\mathbf{I}}_{\eta x} = \begin{bmatrix} 0.011 & 0.0022 \\ (0.006) & (0.0014) \\ 0.037 & 0.0007 \\ (0.011) & (0.0004) \\ 0.064 & 0.0048 \\ (0.013) & (0.0014) \end{bmatrix} \tag{8.102}$$

Considering $\hat{\mathbf{I}}_{\eta x}$ first, the ratios of estimates to standard errors are greater than three for half the elements and between 1.5 to 1.8 for the remaining ones. In contrast, none of the critical ratios of indirect effects of $\boldsymbol{\eta}$ on $\boldsymbol{\eta}$ are larger than 1.6 with the exception of subjective income's (η_1) effect on subjective overall status (η_3). This suggests that with one exception, the most significant effects of the subjective variables on one another are their direct not indirect effects. Relatively large estimate-to-standard-error ratios for direct effects need not be accompanied by such ratios for the indirect effects.

To explain the small critical ratios for the indirect effects of $\boldsymbol{\eta}$ on $\boldsymbol{\eta}$, one place to begin is with $\hat{\beta}_{12}$, the coefficient of the path from subjective occupational prestige (η_2) to subjective income (η_1). Relative to $\hat{\beta}_{21}$, the influence of subjective income on prestige, $\hat{\beta}_{12}$, is somewhat weak. In substantive terms, people who perceive their income as high are likely to perceive their occupational prestige as higher than those who view their income as low. Yet individuals with positions they perceive as prestigious have a less strong tendency to believe that their incomes are high. Since the $\hat{\beta}_{12}$ path channels a portion of the indirect effects, this may partially explain the relatively small magnitude of $\hat{\mathbf{I}}_{\eta\eta}$ compared to its estimated asymptotic variance. I investigate this possibility with the specific effects methodology. First, I estimate the indirect effects transmitted through the path from η_2 to η_1 (i.e., β_{12}). If the foregoing ideas are correct, these new specific indirect effects from $\boldsymbol{\eta}$ to $\boldsymbol{\eta}$ should have small critical ratios. Second, I estimate the indirect effects that do not cross the β_{12} path. If these have relatively large critical ratios, the idea that the small critical ratios for $\hat{\mathbf{I}}_{\eta\eta}$ are due at least partially to the weak link from η_2 to η_1 would receive further support. With $\hat{\mathbf{B}}_{(l)}$ defined so that $l = \{\hat{\beta}_{12} = 0\}$, the estimated indirect effects transmitted by the β_{12} path are

$$\hat{\mathbf{I}}^*_{\eta\eta} = \begin{bmatrix} \begin{matrix} 0.105 \\ (0.067) \end{matrix} & \begin{matrix} 0.030 \\ (0.036) \end{matrix} & 0 \\ \begin{matrix} 0.035 \\ (0.026) \end{matrix} & \begin{matrix} 0.105 \\ (0.067) \end{matrix} & 0 \\ \begin{matrix} 0.061 \\ (0.044) \end{matrix} & \begin{matrix} 0.183 \\ (0.127) \end{matrix} & 0 \end{bmatrix}, \quad \hat{\mathbf{I}}^*_{\eta x} = \begin{bmatrix} \begin{matrix} 0.011 \\ (0.006) \end{matrix} & \begin{matrix} 0.0022 \\ (0.0014) \end{matrix} \\ \begin{matrix} 0.003 \\ (0.002) \end{matrix} & \begin{matrix} 0.0007 \\ (0.0004) \end{matrix} \\ \begin{matrix} 0.006 \\ (0.004) \end{matrix} & \begin{matrix} 0.0012 \\ (0.0008) \end{matrix} \end{bmatrix} \tag{8.103}$$

In (8.103) the * indicates that these are indirect effects sent over particular paths, in this case that for β_{12}. Comparing $\hat{\mathbf{I}}^*_{\eta\eta}$ of (8.103) to $\hat{\mathbf{I}}_{\eta\eta}$ of (8.102) suggests that *all* of the indirect effects of $\boldsymbol{\eta}$ on $\boldsymbol{\eta}$ are transmitted through the β_{12} path except the indirect effects of subjective income (η_1) on subjective overall status (η_3). The indirect effect of η_1 on η_3 passed through the β_{12} path has a critical ratio of 1.4, although as shown in (8.102), the *total* indirect effect of η_1 on η_3 has a critical ratio of about 2.3.

The indirect effects of $\mathbf{x}$ on $\boldsymbol{\eta}$ through the path β_{12} are even more revealing. As $\hat{\mathbf{I}}^*_{\eta x}$ shows, none of the estimates of indirect effects through the β_{12} path are more than 1.8 times their standard errors. When contrasted with $\hat{\mathbf{I}}_{\eta x}$ in (8.102), this highlights the secondary role that the influence of subjective occupational prestige on subjective income plays in generating total indirect effects of $\mathbf{x}$ on $\boldsymbol{\eta}$.

Finally, the indirect effects form $\boldsymbol{\eta}$ to $\boldsymbol{\eta}$ and $\mathbf{x}$ to $\boldsymbol{\eta}$ *not* crossing the path from η_2 to η_1 (i.e., β_{12}) are

$$\hat{\mathbf{I}}^{**}_{\eta\eta} = \begin{bmatrix} 0 & 0 & 0 \\ 0 & 0 & 0 \\ \begin{matrix} 0.177 \\ (0.089) \end{matrix} & 0 & 0 \end{bmatrix}, \quad \hat{\mathbf{I}}^{**}_{\eta x} = \begin{bmatrix} 0 & 0 \\ \begin{matrix} 0.033 \\ (0.010) \end{matrix} & 0 \\ \begin{matrix} 0.058 \\ (0.012) \end{matrix} & \begin{matrix} 0.0036 \\ (0.0013) \end{matrix} \end{bmatrix} \tag{8.104}$$

The only indirect effect of $\boldsymbol{\eta}$ on $\boldsymbol{\eta}$ is the one from η_1 to η_3 through η_2. This effect is just short of twice its standard error. Turning to $\hat{\mathbf{I}}^{*}_{\eta x}$, all of the indirect effects of $\mathbf{x}$ on $\boldsymbol{\eta}$ not mediated by the path associated with β_{12} have critical ratios of more than 2.7. Thus the indirect effects that cross β_{12} path tend to have small critical ratios, while the indirect effects that do not cross the path are larger. Exact p-values are not assigned to each critical ratio because of the exploratory nature of the analysis and the multiple tests involved. Based on these exploratory results, hypotheses could be specified and tested on a new, independent random sample.

The example illustrates that not all specific indirect effects that comprise the total indirect effects contribute equally. Examining the critical ratios of only the total indirect effects can give a misleading portrayal of the critical ratios of the individual indirect paths. In other cases, if all the components of an effect are essentially zero, then the approximation for the standard error of the effect no longer holds (see Appendix 8A). For instance, if $\hat{\gamma}_{11}$ and $\hat{\beta}_{12}$ are small relative to their standard errors, then the method for estimating standard errors should not be applied to find the standard error of $\hat{\gamma}_{11}\hat{\beta}_{12}$.

SUMMARY

This chapter has presented the synthesis of the latent variable and measurement models. The implied covariance matrix, identification, model evaluation, and other topics that emerged for observed variable and measurement models arose for the general structural equation system. The fundamental hypothesis remains $\boldsymbol{\Sigma} = \boldsymbol{\Sigma}(\boldsymbol{\theta})$. I illustrated methods to estimate standardized coefficients, constant terms in equations, and the means of latent variables. Methods of comparing parameters across independent groups and of handling missing values also were covered. Finally, I demonstrated a procedure to decompose the total, direct, indirect, and specific effects that applies to any type of structural equation model.

With this chapter as a foundation, Chapter 9 considers more advanced topics.

APPENDIX 8A ASYMPTOTIC VARIANCES OF EFFECTS

This appendix briefly describes the means of obtaining estimated asymptotic variances for all types of effects. See Appendix B at the end of the book for a review of asymptotic distribution theory. The asymptotic variances of the direct effects are readily available, provided the assumptions for the ML and GLS estimators are satisfied. Significance testing of the indirect, total, and specific effects is more complicated, since it typically involves obtaining the variances of products of coefficient estimates. For instance, we can estimate the asymptotic variance of a ML estimator of a direct effect (e.g., $\hat{\beta}_{12}$), but the asymptotic variance of an indirect effect (e.g., $\hat{\beta}_{12}\hat{\gamma}_{11}$) is less obvious. The multivariate delta method (Bishop, Fienberg, and Holland 1975, 486–500; Rao 1973, 385–389) proves helpful in this situation. The delta method begins with the assumption that a parameter estimator has an asymptotically normal distribution with a mean of the parameter and an asymptotic covariance matrix. It then provides a method of estimating the asymptotic covariance matrix of functions of the parameter.

Folmer (1981, 1440–1442) and Sobel (1982; 1986) suggest applying the delta method to estimate the asymptotic variances of total, indirect, and other types of effects. The procedure is as follows. First, define $\boldsymbol{\theta}$ to be an s-dimensional vector of the unknown elements in $\mathbf{B}$, $\boldsymbol{\Gamma}$, and $\boldsymbol{\Lambda}_y$, while $\hat{\boldsymbol{\theta}}_N$, is the corresponding sample estimator of $\boldsymbol{\theta}$ for a sample of size N. Choose an estimator so that $\hat{\boldsymbol{\theta}}_N$ has an asymptotically normal distribution with a mean of $\boldsymbol{\theta}$ and an asymptotic covariance matrix, $\text{ACOV}(\hat{\boldsymbol{\theta}}_N) = N^{-1}\mathbf{V}$ where $\mathbf{V}$ is the covariance matrix of the limiting distribution of $\sqrt{N}(\hat{\boldsymbol{\theta}}_N - \boldsymbol{\theta})$ [see Appendix B]. Under appropriate assumptions the ML and GLS estimators of $\boldsymbol{\theta}$ satisfy these conditions.

Next, define an r-dimensional vector $\mathbf{f}(\boldsymbol{\theta})$ that is a differentiable function of $\boldsymbol{\theta}$. In this case $\mathbf{f}(\boldsymbol{\theta})$ contains the indirect (or total) effects that are functions of the direct effects. Under these conditions[16] the multivariate delta method states that the asymptotic distribution of $\mathbf{f}(\hat{\boldsymbol{\theta}}_N)$ is normal with a mean of $\mathbf{f}(\boldsymbol{\theta})$ and an asymptotic covariance matrix of

[16]See Bishop, Fienberg, and Holland's (1975) discussion of the delta method.

$(\partial\mathbf{f}/\partial\boldsymbol{\theta})'\text{ACOV}(\hat{\boldsymbol{\theta}}_N)(\partial\mathbf{f}/\partial\boldsymbol{\theta})$. The first row of $(\partial\mathbf{f}/\partial\boldsymbol{\theta})$ is $\partial f_1/\partial\theta_1$, $\partial f_2/\partial\theta_1, \ldots, \partial f_r/\partial\theta_1$, where f_i is the ith element of $\mathbf{f}(\boldsymbol{\theta})$. The second row is $\partial f_1/\partial\theta_2$, $\partial f_2/\partial\theta_2, \ldots, \partial f_r/\partial\theta_2$, and so on, so that $\partial\mathbf{f}/\partial\boldsymbol{\theta}$ is an $s \times r$ matrix. For large samples, $\hat{\boldsymbol{\theta}}_N$ is substituted for $\boldsymbol{\theta}$ to obtain an estimate of the asymptotic covariance matrix for $\mathbf{f}(\hat{\boldsymbol{\theta}}_N)$ that equals

$$\left(\frac{\partial\mathbf{f}}{\partial\hat{\boldsymbol{\theta}}_N}\right)'\text{acov}\left(\hat{\boldsymbol{\theta}}_N\right)\left(\frac{\partial\mathbf{f}}{\partial\hat{\boldsymbol{\theta}}_N}\right) \tag{8A.1}$$

To illustrate this procedure, consider the simple causal chain model:

$$\begin{aligned} \eta_1 &= \gamma_{11}\xi_1 + \zeta_1 \\ \eta_2 &= \beta_{21}\eta_1 + \zeta_2 \end{aligned} \tag{8A.2}$$

In (8A.2) assume that ζ_1 and ζ_2 are uncorrelated with each other and with ξ_1, $E(\zeta_i)$ is zero, and η_1, η_2, and ξ_1 are deviated from their means. Define $\gamma_{11}\beta_{21}$, the indirect effect of ξ_1 on η_2, as the single element of $\mathbf{f}(\boldsymbol{\theta})$, with $\boldsymbol{\theta}$ containing γ_{11} and β_{21}. The $\partial\mathbf{f}/\partial\boldsymbol{\theta}$ is $[\beta_{21} \quad \gamma_{11}]'$ and the asymptotic covariance matrix of $\hat{\boldsymbol{\theta}}_N$ is

$$\text{ACOV}\left(\hat{\boldsymbol{\theta}}_N\right) = \begin{bmatrix} N^{-1}V_{11} & 0 \\ 0 & N^{-1}V_{22} \end{bmatrix} \tag{8A.3}$$

The main diagonal of (8A.3) contains the asymptotic variances of $\hat{\gamma}_{11N}$ and $\hat{\beta}_{21N}$. The off-diagonal elements are zero since in a recursive system these two coefficients are uncorrelated. Combining these matrices with the multivariate delta method, the asymptotic variance of $\hat{\gamma}_{11}\hat{\beta}_{21}$ is the scalar:

$$N^{-1}\left[\beta_{21}^2V_{11} + \gamma_{11}^2V_{22}\right] \tag{8A.4}$$

If β_{21} and γ_{11} are zero, the delta method cannot be applied. Otherwise, substituting into (8A.4), the sample estimates, provides an estimate of the asymptotic variance of $\hat{\gamma}_{11N}\hat{\beta}_{21N}$ for large samples.

Considering each element of the indirect (or total) effects in the preceding fashion for more complicated models is extremely tedious. Sobel (1986) proposes a matrix formulation that is a far more efficient means of finding $\partial\mathbf{f}(\boldsymbol{\theta})/\partial\boldsymbol{\theta}$, which is required for the covariance matrix of $\mathbf{f}(\boldsymbol{\theta})$.

To simplify the results for the indirect effects, I define $\boldsymbol{\theta}$ to include only those unrestricted elements in $\mathbf{B}$, $\boldsymbol{\Gamma}$, and $\boldsymbol{\Lambda}_y$. Each indirect effect is treated separately. The partial derivatives for the indirect effects are

$$\frac{\partial \text{ vec } \mathbf{I}_{\eta\eta}}{\partial \boldsymbol{\theta}} = \mathbf{V}_B'\left((\mathbf{I} - \mathbf{B})^{-1} \otimes \left[(\mathbf{I} - \mathbf{B})^{-1}\right]' - \mathbf{I}_m \otimes \mathbf{I}_m\right) \quad (8A.5)$$

$$\frac{\partial \text{ vec } \mathbf{I}_{\eta\xi}}{\partial \boldsymbol{\theta}} = \mathbf{V}_B'\left\{(\mathbf{I} - \mathbf{B})^{-1}\boldsymbol{\Gamma} \otimes \left[(\mathbf{I} - \mathbf{B})^{-1}\right]'\right\}$$

$$+\mathbf{V}_\Gamma'\left\{\mathbf{I}_n \otimes \left[(\mathbf{I} - \mathbf{B})^{-1} - \mathbf{I}\right]'\right\} \quad (8A.6)$$

$$\frac{\partial \text{ vec } \mathbf{I}_{y\eta}}{\partial \boldsymbol{\theta}} = \mathbf{V}_{\Lambda_y}'\left\{\left[(\mathbf{I} - \mathbf{B})^{-1} - \mathbf{I}\right] \otimes \mathbf{I}_p\right\}$$

$$+\mathbf{V}_B'\left\{(\mathbf{I} - \mathbf{B})^{-1} \otimes \left[\boldsymbol{\Lambda}_y(\mathbf{I} - \mathbf{B})^{-1}\right]'\right\} \quad (8A.7)$$

$$\frac{\partial \text{ vec } \mathbf{I}_{y\xi}}{\partial \boldsymbol{\theta}} = \mathbf{V}_B'\left\{(\mathbf{I} - \mathbf{B})^{-1}\boldsymbol{\Gamma} \otimes \left[\boldsymbol{\Lambda}_y(\mathbf{I} - \mathbf{B})^{-1}\right]'\right\}$$

$$+\mathbf{V}_\Gamma'\left\{\mathbf{I}_n \otimes \left[\boldsymbol{\Lambda}_y(\mathbf{I} - \mathbf{B})^{-1}\right]'\right\}$$

$$+\mathbf{V}_{\Lambda_y}'\left[(\mathbf{I} - \mathbf{B})^{-1}\boldsymbol{\Gamma} \otimes \mathbf{I}_p\right] \quad (8A.8)$$

where

$$\mathbf{V}_B = \left[\text{vec } \frac{\partial \mathbf{B}}{\partial \boldsymbol{\theta}_1}, \text{vec } \frac{\partial \mathbf{B}}{\partial \boldsymbol{\theta}_2}, \ldots, \text{vec } \frac{\partial \mathbf{B}}{\partial \boldsymbol{\theta}_s}\right]$$

$$\mathbf{V}_\Gamma = \left[\text{vec } \frac{\partial \boldsymbol{\Gamma}}{\partial \boldsymbol{\theta}_1}, \text{vec } \frac{\partial \boldsymbol{\Gamma}}{\partial \boldsymbol{\theta}_2}, \ldots, \text{vec } \frac{\partial \boldsymbol{\Gamma}}{\partial \boldsymbol{\theta}_s}\right]$$

$$\mathbf{V}_{\Lambda_y} = \left[\text{vec } \frac{\partial \boldsymbol{\Lambda}_y}{\partial \boldsymbol{\theta}_1}, \text{vec } \frac{\partial \boldsymbol{\Lambda}_y}{\partial \boldsymbol{\theta}_2}, \ldots, \text{vec } \frac{\partial \boldsymbol{\Lambda}_y}{\partial \boldsymbol{\theta}_s}\right]$$

vec = vector operator

$\otimes$ = Kronecker's product

For the empirical illustration in Figure 8.14, $\boldsymbol{\theta} = [\beta_{12} \ \ \beta_{21} \ \ \beta_{31} \ \ \beta_{32} \ \ \gamma_{11} \ \ \gamma_{22}]'$.

The $\mathbf{V}_B$ and $\mathbf{V}_\Gamma$ matrices for $\mathbf{I}_{\eta\eta}$ and $\mathbf{I}_{\eta x}$ are

$$\mathbf{V}_B = \begin{bmatrix} 0 & 0 & 0 & 0 & 0 & 0 \\ 0 & 1 & 0 & 0 & 0 & 0 \\ 0 & 0 & 1 & 0 & 0 & 0 \\ 1 & 0 & 0 & 0 & 0 & 0 \\ 0 & 0 & 0 & 0 & 0 & 0 \\ 0 & 0 & 0 & 1 & 0 & 0 \\ 0 & 0 & 0 & 0 & 0 & 0 \\ 0 & 0 & 0 & 0 & 0 & 0 \\ 0 & 0 & 0 & 0 & 0 & 0 \end{bmatrix}, \qquad \mathbf{V}_\Gamma = \begin{bmatrix} 0 & 0 & 0 & 0 & 1 & 0 \\ 0 & 0 & 0 & 0 & 0 & 0 \\ 0 & 0 & 0 & 0 & 0 & 0 \\ 0 & 0 & 0 & 0 & 0 & 0 \\ 0 & 0 & 0 & 0 & 0 & 1 \\ 0 & 0 & 0 & 0 & 0 & 0 \end{bmatrix}.$$

These are utilized to estimate the standard errors for $\hat{\mathbf{I}}_{\eta\eta}$ and $\hat{\mathbf{I}}_{\eta x}$ reported in the text. With appropriate modifications the standard errors of the specific effects were estimated.

APPENDIX 8B LISTING OF THE LISREL VI PROGRAM FOR MISSING VALUE EXAMPLE

```
MISSING VALUES ILLUSTRATION: COMPLETE DATA
DATA NI=3 NOBS=400 MA=AM NG=2
CM
10.5
6.9 12.3
7.2 6.2 9.8
ME
5.1 6.7 6.1
MO NY=3 NX=1 NE=4 FI TE=FI BE=FI,FU GA=FI PS=DI
FR LY 1 4 LY 2 4 LY 3 4 BE 1 3 BE 2 1 BE 2 3
FI PS 4 4
VA 1.0 LY 1 1 LY 2 2 LY 3 3 GA 4 1
ST 6 LY 1 4 LY 2 4 LY 3 4
ST .5 BE 1 3 BE 2 1 BE 2 3
ST 3 PS 1 1 PS 2 2 PS 3 3
OU TO NS SE
MISSING VALUES ILLUSTRATION: INCOMPLETE DATA
DA NOBS=100
CM
9.8
7.1 10.9
```

```
0 0 1
ME
4.9 6.2 0
MO GA=IN BE=IN PS=IN LY=FU,FI
FR LY 1 4 LY 2 4
VA 1 LY 1 1 LY 2 2 TE 3 3
EQ LY 1 1 4 LY 2 1 4
EQ LY 1 2 4 LY 2 2 4
OU
```

CHAPTER NINE

The General Model, Part II: Extensions

In Chapter 8, I presented the "basics" of the general structural equation model. In this chapter I introduce extensions of it. The first section treats alternative model representations that enable some of the implicit constraints of the usual model to be removed. The next sections are about equality and inequality constraints, and interaction and quadratic terms. A section on instrumental variable estimators follows. Distributional assumptions and categorical variables are the final topics.

ALTERNATIVE NOTATIONS / REPRESENTATIONS

For most empirical applications the standard model and notation serves the analyst well. However, there are constraints in the scheme which on the surface appear to rule out some models. The usual representation does not allow: $\mathbf{x}$ to influence $\mathbf{y}$, $\boldsymbol{\eta}$, or $\boldsymbol{\xi}$; $\mathbf{y}$ to affect $\mathbf{x}$, $\boldsymbol{\eta}$, or $\boldsymbol{\xi}$; or a direct effect from $\boldsymbol{\xi}$ to $\mathbf{y}$. In addition the model imposes zero covariances between $\boldsymbol{\delta}$ and $\boldsymbol{\epsilon}$, $\boldsymbol{\delta}$ and $\boldsymbol{\zeta}$, and $\boldsymbol{\zeta}$ and $\boldsymbol{\epsilon}$. Some of these limitations were bypassed at several points in the previous chapters. For example, causal indicators (see Chapters 3 and 7) and MIMIC models (see Chapter 8) both require $\mathbf{x}$ to influence $\boldsymbol{\eta}$. This could be done by using the measurement model, $\mathbf{x} = \boldsymbol{\Lambda}_x\boldsymbol{\xi} + \boldsymbol{\Theta}_\delta$, with the appropriate λ_{ij}'s set to one and the corresponding $\boldsymbol{\Theta}_\delta$ element(s) set to zero so that $x_i = \xi_j$ for the cause indicators. Incorporating intercept constants in structural equations with observed variables (Chapter 4), in measurement models (Chapter 7), and in the general structural equation model (Chapter 8) required a modification of the standard notation as did higher-order factor analysis (Chapter 7).

Graff (1979), McDonald (1978; 1980), McArdle and McDonald (1984), and Bentler and Weeks (1980) have proposed alternative representations to overcome the restrictions of the usual model. For example, Graff (1979) suggests a two-equation model of

$$\boldsymbol{\eta}^{+} = \mathbf{B}^{+}\boldsymbol{\eta}^{+} + \boldsymbol{\zeta}^{+} \tag{9.1}$$

$$\mathbf{y}^{+} = \boldsymbol{\Lambda}_y^{+}\boldsymbol{\eta}^{+} \tag{9.2}$$

The "+" superscript signifies that $\boldsymbol{\eta}^{+}$, $\boldsymbol{\zeta}^{+}$, $\mathbf{B}^{+}$, and $\mathbf{y}^{+}$ are different from $\boldsymbol{\eta}$, $\boldsymbol{\zeta}$, $\mathbf{B}$, and $\mathbf{y}$. The relations between these new symbols and the old ones are given in (9.3) and (9.4):

$$\boldsymbol{\eta}^{+} = \begin{bmatrix} \mathbf{y} \\ \mathbf{x} \\ \boldsymbol{\eta} \\ \boldsymbol{\xi} \end{bmatrix}, \qquad \boldsymbol{\zeta}^{+} = \begin{bmatrix} \boldsymbol{\epsilon} \\ \boldsymbol{\delta} \\ \boldsymbol{\zeta} \\ \boldsymbol{\xi} \end{bmatrix}, \qquad \mathbf{y}^{+} = \begin{bmatrix} \mathbf{y} \\ \mathbf{x} \end{bmatrix} \tag{9.3}$$

$$\mathbf{B}^{+} = \begin{bmatrix} 0 & 0 & \boldsymbol{\Lambda}_y & 0 \\ 0 & 0 & 0 & \boldsymbol{\Lambda}_x \\ 0 & 0 & \mathbf{B} & \boldsymbol{\Gamma} \\ 0 & 0 & 0 & 0 \end{bmatrix} \tag{9.4}$$

The $\boldsymbol{\Lambda}_y^{+}$ matrix contains only zeros and ones and selects the observed variables from $\boldsymbol{\eta}^{+}$. Each row contains all zeros except a one that matches an observed variable to its counterpart in $\boldsymbol{\eta}^{+}$. The $\boldsymbol{\Lambda}_y^{+}$ matrix is $(p + q) \times (p + q + m + n)$, $\mathbf{y}^{+}$ is $(p + q) \times 1$, $\boldsymbol{\eta}^{+}$ and $\boldsymbol{\zeta}^{+}$ are $(p + q + m + n) \times 1$, and $\mathbf{B}^{+}$ is $(p + q + m + n) \times (p + q + m + n)$.

Equation (9.4) makes explicit the implicit constraints of the standard model mentioned above. For instance, the first two sets of columns of $\mathbf{B}^{+}$ contain the effects of $\mathbf{y}$ and $\mathbf{x}$ on the variables in $\boldsymbol{\eta}^{+}$. The zeros in these columns show that these variables are constrained to have no effects on the other variables. Similarly, the zero submatrix in the upper-right corner of $\mathbf{B}^+$ shows that the usual model prohibits $\boldsymbol{\xi}$ from directly influencing $\mathbf{y}$.

The final matrix for this alternative representation is the covariance matrix for $\boldsymbol{\zeta}^+$ called $\boldsymbol{\Psi}^+$. Its relation to the standard parameters is

$$\boldsymbol{\Psi}^+ = \begin{bmatrix} \boldsymbol{\Theta}_\epsilon & & & \\ \mathbf{0} & \boldsymbol{\Theta}_\delta & & \\ \mathbf{0} & \mathbf{0} & \boldsymbol{\Psi} & \\ \mathbf{0} & \mathbf{0} & \mathbf{0} & \boldsymbol{\Phi} \end{bmatrix} \tag{9.5}$$

The $\boldsymbol{\Psi}^+$ is a $(p+q+m+n)$ square and symmetric matrix. Equation (9.5) makes clear the zero restrictions for the covariances between the different types of disturbances and measurement errors and between $\boldsymbol{\xi}$ and all the disturbance terms.

The beauty of the alternative representation is that the restrictions of the standard model shown in (9.4) and (9.5) no longer need be imposed. For instance, a covariance between an ϵ_i and δ_j is estimable by allowing a free parameter for this in the $\boldsymbol{\Psi}^+$ matrix. Or, ξ_i can directly affect y_j by changing the fixed zero element in $\mathbf{B}^+$ to a free parameter. Indicators can directly affect one another as well as the latent variables. Many relations are made possible with this general model, provided, of course, that it is identified.

We could view Graff's (1979) model as more general than the standard one since we can write the usual model as a restrictive case of his model [see (9.1) to (9.5)]. However, we also could see Graff's model as a special case of the standard one where $\mathbf{x}$, $\boldsymbol{\xi}$, $\boldsymbol{\delta}$, and $\boldsymbol{\epsilon}$ are absent and $\boldsymbol{\Lambda}_y$ consists only of zeros and ones. Thus the generality of the models depends on which perspective we take. Analogous results hold for the other representations mentioned. What Graff (1979), McArdle and McDonald (1984), McDonald (1978; 1980), Bentler and Weeks (1980), and others have shown is that the usual restrictions on the relations between variables are in many cases more apparent than real.

I return to the objective-subjective social status data to illustrate the alternative notation. The five observed variables are income (x_1), occupational prestige (x_2), subjective income (y_1), subjective occupational prestige (y_2), and subjective general status (y_3). Another possible model for these variables is in Figure 9.1(a). This is a MIMIC model where income and occupational prestige have direct effects on a latent variable, overall subjective social status (η_1) which has y_1 to y_3 as indicators. In the standard

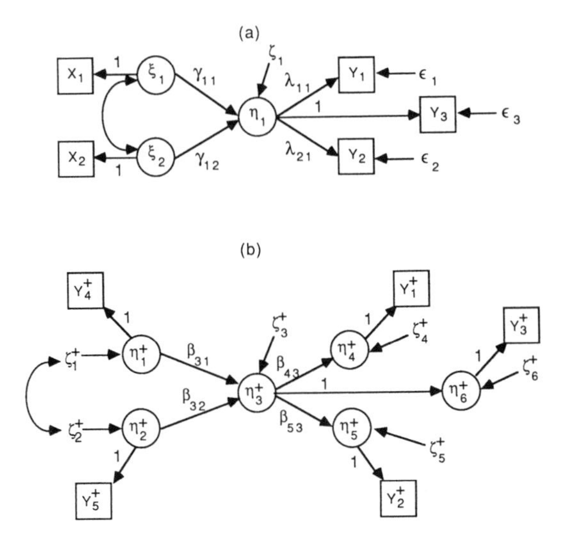

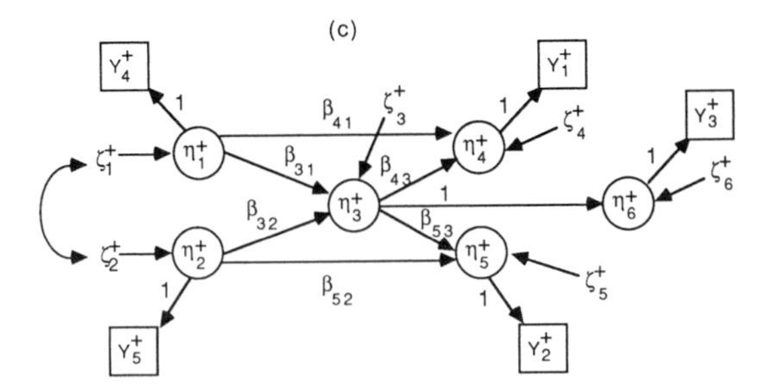

Figure 9.1 Objective and Subjective Social Status Models in Standard and Alternative Notations

model the matrices are

$$\mathbf{B} = [0]$$

$$\boldsymbol{\Gamma} = [\gamma_{11} \quad \gamma_{12}]$$

$$\boldsymbol{\Lambda}_x = \begin{bmatrix} 1 & 0 \\ 0 & 1 \end{bmatrix} \tag{9.6}$$

$$\boldsymbol{\Lambda}_y = \begin{bmatrix} \lambda_{11} \\ \lambda_{21} \\ 1 \end{bmatrix}$$

$$\Theta_{\delta} = \text{diag}[0 \quad 0]$$

$$\Theta_{\epsilon} = \text{diag}[\text{VAR}(\epsilon_1) \quad \text{VAR}(\epsilon_2) \quad \text{VAR}(\epsilon_3)] \tag{9.7}$$

$$\mathbf{\Phi} = \begin{bmatrix} \phi_{11} & \\ \phi_{21} & \phi_{22} \end{bmatrix}$$

$$\mathbf{\Psi} = [\psi_{11}] \tag{9.8}$$

Figure 9.1(b) shows the same MIMIC model cast in the alternative notation with $\boldsymbol{\eta}^{+} = [\xi_1 \quad \xi_2 \quad \eta_1 \quad y_1 \quad y_2 \quad y_3]$, $\mathbf{y}^{+} = [y_1 \quad y_2 \quad y_3 \quad x_1 \quad x_2]$, and $\boldsymbol{\zeta}^{+} = [\xi_1 \quad \xi_2 \quad \zeta_1 \quad \epsilon_1 \quad \epsilon_2 \quad \epsilon_3]$. I place the variables in $\boldsymbol{\eta}^{+}$ and $\boldsymbol{\zeta}^{+}$ in a different sequence than that in (9.3). I do this so that the subscripts of η_i and ζ_i increase as you move from left to right in Figure 9.1(b). Any ordering will do as long as it is consistent. I excluded x_1 and x_2 from $\boldsymbol{\eta}^{+}$ and excluded δ_1 and δ_2 from $\boldsymbol{\zeta}^{+}$ since the model assumes that $x_1 = \xi_1$ and $x_2 = \xi_2$. These equalities are set using $\mathbf{\Lambda}_y^{+}$ and (9.2). The other matrices are

$$\mathbf{B}^{+} = \begin{bmatrix} 0 & 0 & 0 & 0 & 0 & 0 \\ 0 & 0 & 0 & 0 & 0 & 0 \\ \beta_{31} & \beta_{32} & 0 & 0 & 0 & 0 \\ 0 & 0 & \beta_{43} & 0 & 0 & 0 \\ 0 & 0 & \beta_{53} & 0 & 0 & 0 \\ 0 & 0 & 1 & 0 & 0 & 0 \end{bmatrix}$$

$$\mathbf{\Psi}^{+} = \begin{bmatrix} \psi_{11} & & & & & \\ \psi_{21} & \psi_{22} & & & & \\ 0 & 0 & \psi_{33} & & & \\ 0 & 0 & 0 & \psi_{44} & & \\ 0 & 0 & 0 & 0 & \psi_{55} & \\ 0 & 0 & 0 & 0 & 0 & \psi_{66} \end{bmatrix} \tag{9.9}$$

$$\mathbf{\Lambda}_y^{+} = \begin{bmatrix} 0 & 0 & 0 & 1 & 0 & 0 \\ 0 & 0 & 0 & 0 & 1 & 0 \\ 0 & 0 & 0 & 0 & 0 & 1 \\ 1 & 0 & 0 & 0 & 0 & 0 \\ 0 & 1 & 0 & 0 & 0 & 0 \end{bmatrix} \tag{9.10}$$

Comparing (9.6)–(9.8) to (9.9)–(9.10) shows that the alternative notation requires fewer matrices but the matrices have larger dimensions and more zero elements than the standard one.

When these models are fitted to the data, the parameter estimates are identical as is expected. In addition their chi-square statistics are 26.6 with 4 df ($p < 0.001$). All estimates are at least twice their standard errors and are positive as expected. The overall fit measures are mixed (GFI = 0.98, AGFI = 0.91, $\Delta_1 = 0.93$, $\Delta_2 = 0.94$, $\rho_1 = 0.83$, $\rho_2 = 0.85$). The baseline model for the incremental fit measures has free parameters for the variances of the observed variables and the covariance between the exogenous variables, but all other covariances are forced to zero ($\chi^2 = 355.59$, df = 9).

With these results it seems worthwhile to pursue modifications that might improve the fit. One possibility is to allow additional paths from income to subjective income and from occupational prestige to subjective occupational prestige. This means that the indicators of subjective income and subjective occupational prestige are directly influenced not only by the latent overall subjective status but also by their objective counterparts. Unfortunately, this new model cannot be estimated with the standard notation of Figure 9.1(a); the ξ variables cannot have direct effects on y. The alternative notation easily can incorporate these changes, however. Figure 9.1(c) shows this model; it has the same matrices as (9.9) and (9.10), with the exception that β_{41} and β_{52} of $\mathbf{B}^+$ are free parameters rather than set to zero.

The chi-square estimate for this model is 4.21 with 2 df. Since the previous model is nested within this one, the difference in chi-square estimates approximates a chi-square variable that tests whether there is a significant improvement in fit for the less restrictive model. The difference is 22.38 (26.59 − 4.21) with 2 df (= 4 − 2), which is highly significant ($p < 0.01$). The GFI and AGFI of 1.00 and 0.97 are extremely high, as are the incremental fit measures (e.g., $\Delta_1 = 0.99$, $\Delta_2 = 1.0$). All parameter estimates are positive, greater than two times their standard errors, and all normalized residuals are less than two. Overall then, this new model is very consistent with the data, yet it is not one that could be treated with the standard notation. Appendix 9A provides a listing of the LISREL program to obtain these results.

I am not recommending that the alternative notation should replace the standard one for all models. Indeed, the standard notation encompasses the majority of applications for most researchers. Rather, my point is that the researcher's thinking about a problem should not be imprisoned in a single representation. The first priority should be the formulation of the models that best capture the relation between variables. The model notation and representation should follow, not precede, the substantive ideas.

EQUALITY AND INEQUALITY CONSTRAINTS

In most models one or more parameters are constrained. The most common case is when a parameter is set to zero, one, or some other constant. Another is setting two or more parameters equal, as I did with the factor loadings for the 1960 and 1965 indicators of political democracy (see Chapters 7 or 8). More complex constraints include general polynomial constraints (e.g., $\beta_{21} = 3\beta_{12}^2 + 2\gamma_{11}\gamma_{21} + 1$) and inequality restrictions [e.g., $\text{VAR}(\epsilon_1) > 0$].

LISREL, EQS, and most other structural equation programs allow simple equality constraints. Bentler's (1985) EQS also allows inequality restrictions and general linear equality constraints where a weighted sum of parameters is set equal to a constant ($\omega_i\theta_i + \omega_j\theta_j + \cdots + \omega_z\theta_z = c$). McDonald's (1978, 1980) COSAN and Schoenberg's (1987) LINCS programs allow these as well as more complex restrictions.

Technical discussions of direct methods to introduce complex restrictions into the estimation procedures are in McDonald (1980), Lee (1980), and Bentler and Lee (1983). Rindskopf (1983, 1984b) has suggested ways to "trick" programs with only equality constraints into incorporating a broader selection of restrictions. To illustrate his procedure, consider the causal chain model in Figure 9.2(*a*). The left-hand side contains the path diagram and the imposed constraint that $\beta_{42} = \gamma_{11}\beta_{21}$. This represents the direct method of implementing restrictions. The right-hand side has the indirect method. It shows a modified diagram with a "phantom" latent variable η_3 and the equality constraints required to obtain the same restriction as the example to its left. The phantom variable has no disturbance term since it is completely explained by the variables that precede it. The equations for this model are

$$\eta_1 = \gamma_{11}\xi_1 + \zeta_1 \tag{9.11}$$

$$\eta_2 = \beta_{21}\eta_1 + \zeta_2 \tag{9.12}$$

$$\eta_3 = \beta_{32}\eta_2 \tag{9.13}$$

$$\eta_4 = \beta_{43}\eta_3 + \zeta_4 \tag{9.14}$$

Substituting (9.13) in for η_3 in (9.14) and imposing the equality constraints of $\beta_{43} = \beta_{21}$ and $\beta_{32} = \gamma_{11}$ leads to the desired restriction:

$$\begin{aligned} \eta_4 &= \beta_{43}\beta_{32}\eta_2 + \zeta_4 \\ &= \beta_{21}\gamma_{11}\eta_2 + \zeta_4 \end{aligned} \tag{9.15}$$

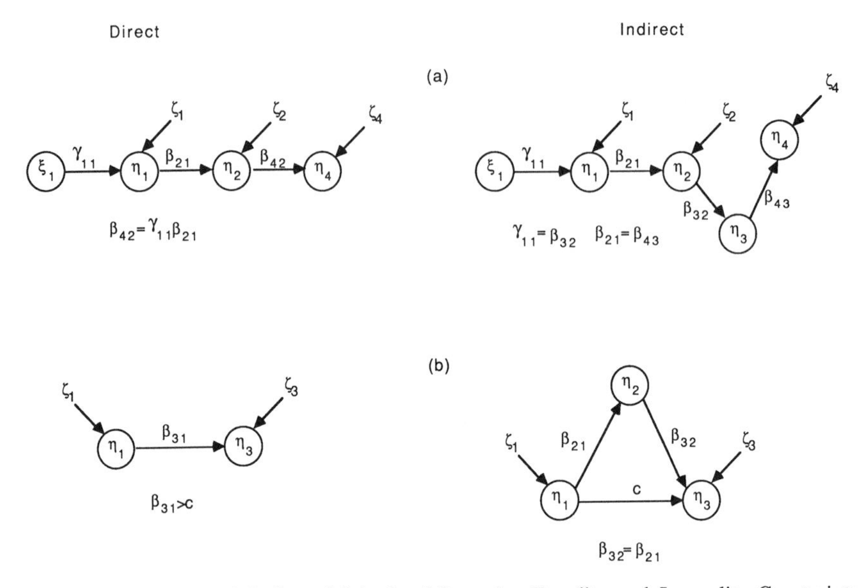

Figure 9.2 Direct and Indirect Methods of Imposing Equality and Inequality Constraints on Parameters

Alternatively, we can see the operation of the phantom variable and the equality constraints in creating the restriction $\beta_{42} = \gamma_{11}\beta_{21}$ by noting that in the path diagram, the indirect effect of η_2 on η_4 is the product of the direct effects of ξ_1 on η_1 and η_1 on η_2.

Inequality constraints also are feasible. Figure 9.2(*b*) demonstrates the restriction that a coefficient be greater than some constant c. The indirect method accomplishes this by forming

$$\eta_2 = \beta_{21}\eta_1 \tag{9.16}$$

$$\eta_3 = \beta_{32}\eta_2 + c\eta_1 + \zeta_3 \tag{9.17}$$

where $\beta_{32} = \beta_{21}$. Substituting (9.16) in for η_2 in (9.17) leads to

$$\eta_3 = \left(\beta_{21}^2 + c\right)\eta_1 + \zeta_3 \tag{9.18}$$

The β_{21}^2 must be a nonnegative number with its smallest value zero. The smallest coefficient for η_1 is c as required by the inequality constraint.

Rindskopf (1983, 1984b) provides other examples of restrictions introduced indirectly with only equality constraints. He does not propose a means to develop the asymptotic standard errors for the restricted parameters. However, the delta method (see Chapter 8) is one way of estimating these. Thus, when direct means of introducing complex restrictions are not available, this section illustrates that alternative representations of models and simple equality constraints can provide an indirect means of doing so.

QUADRATIC AND INTERACTION TERMS

Throughout Chapter 8 and this chapter I have assumed that the variables in the latent variable equation are linearly related. In some situations theory may suggest otherwise. For instance, a curvilinear relation between latent variables might be appropriately modeled by including a linear and a squared term of a latent exogenous variable. Or, the product of two latent variables can have an influence in addition to the linear effects of these variables considered individually.

In Chapter 4, I suggested methods to incorporate such relations in structural equation models with observed variables. In this section I address the more complicated task of doing the same for interaction or squared terms that involve latent variables. The two major approaches to this problem are the "product indicators" and the "corrected covariance matrix" methods. First I present the product indicators technique developed by Kenny and Judd (1984). A useful starting point is to consider an observed variable model (as in Chapter 4) that includes a squared term:

$$y_1 = \gamma_{11}x_1 + \gamma_{12}x_1^2 + \zeta_1 \tag{9.19}$$

Rewrite (9.19) as follows:

$$\eta_1 = \gamma_{11}\xi_1 + \gamma_{12}\xi_1^2 + \zeta_1 \tag{9.20}$$

$$\begin{aligned} x_1 &= \xi_1 \\ x_1^2 &= \xi_1^2 \\ y_1 &= \eta_1 \end{aligned} \tag{9.21}$$

All variables are in deviation form, and ζ_1 has an expected value of zero and is independent of ξ_1 and ξ_1^2. Equations (9.20) and (9.21) make explicit the implicit "latent" variable model and measurement model. The measurement equation for x_1 follows upon squaring both sides of the $x_1 = \xi_1$ equation.

Now suppose that (9.20) still describes the latent variable model but ξ_1 is measured with error and that it has two indicators, x_1 and x_2. The new measurement model is

$$\begin{aligned} x_1 &= \xi_1 + \delta_1 \\ x_2 &= \lambda_{21}\xi_1 + \delta_2 \\ y_1 &= \eta_1 \end{aligned} \tag{9.22}$$

where δ_1 and δ_2 have expected values of zero and are independent of each other and of ξ_1 and ζ_1. In equation (9.21) an indicator of ξ_1^2 is formed by squaring x_1. With (9.22) we can form x_1^2 plus x_2^2 and x_1x_2 as indicators of ξ_1^2. These squared and product terms equal:

$$\begin{aligned} x_1^2 &= \xi_1^2 + 2\xi_1\delta_1 + \delta_1^2 \\ x_2^2 &= \lambda_{21}^2\xi_1^2 + 2\lambda_{21}\xi_1\delta_2 + \delta_2^2 \\ x_1x_2 &= \lambda_{21}\xi_1^2 + \lambda_{21}\xi_1\delta_1 + \xi_1\delta_2 + \delta_1\delta_2 \end{aligned} \tag{9.23}$$

Equation (9.23) shows that x_1^2, x_2^2, and x_1x_2 measure ξ_1^2 as well as a number of new terms. The list of latent exogenous variables and error terms in the measurement model (9.22) and (9.23) now includes ξ_1, ξ_1^2, $\xi_1\delta_1$, $\xi_1\delta_2$, δ_1, δ_2, δ_1^2, δ_2^2, and $\delta_1\delta_2$. The list of indicators for these variables is x_1, x_2, x_1^2, x_2^2, and x_1x_2. The factor loadings are either constants (1 or 2) or functions of λ_{21}, so no new loadings are required. The effect of $\xi_1\delta_2$ on x_2^2, for instance, is $2\lambda_{21}$; $\xi_1\delta_2$'s effect on x_1x_2 is an implicit 1. The covariance matrix of the nine latent exogenous and error variables introduces additional parameters. Kenny and Judd (1984) suggest that the assumption that ξ_1, δ_1, and δ_2 are normally distributed can simplify matters greatly. With this assumption all the covariances between the nine latent exogenous variables are zero, and the variances are

$$\begin{array}{lll} \mathrm{VAR}(\xi_1) = \phi_{11}, & \mathrm{VAR}(\xi_1\delta_2) = \phi_{11}\mathrm{VAR}(\delta_2), & \mathrm{VAR}(\delta_1^2) = 2[\mathrm{VAR}(\delta_1)]^2 \\ \mathrm{VAR}(\xi_1^2) = 2\phi_{11}^2, & \mathrm{VAR}(\delta_1), & \mathrm{VAR}(\delta_2^2) = 2[\mathrm{VAR}(\delta_2)]^2 \\ \mathrm{VAR}(\xi_1\delta_1) = \phi_{11}\mathrm{VAR}(\delta_1), & \mathrm{VAR}(\delta_2), & \mathrm{VAR}(\delta_1\delta_2) = \mathrm{VAR}(\delta_1)\mathrm{VAR}(\delta_2) \end{array} \tag{9.24}$$

Equation (9.24) reveals that the variances are functions of only three parameters—ϕ_{11}, $\mathrm{VAR}(\delta_1)$, and $\mathrm{VAR}(\delta_2)$—when ξ_1, δ_1, and δ_2 are normally distributed with means of zero. Considering (9.23) and (9.24), the only new parameter in (9.20) compared to a model without a squared term is γ_{12}, the coefficient for ξ_1^2, and this parameter is identified.

An interaction term can be handled in a fashion analogous to that for the squared term. Consider the following latent variable model:

$$\eta_1 = \gamma_{11}\xi_1 + \gamma_{12}\xi_2 + \gamma_{13}\xi_1\xi_2 + \zeta_1 \tag{9.25}$$

where $\xi_1\xi_2$ is the interaction term; η_1, ξ_1 and ξ_2 are deviated from their means; and ζ_1 is independent of ξ_1 and ξ_2 with $E(\zeta_1)$ equal to zero. Suppose that the following measurement model holds:

$$\begin{aligned}
y_1 &= \eta_1 \\
x_1 &= \xi_1 + \delta_1 \\
x_2 &= \lambda_{21}\xi_1 + \delta_2 \\
x_3 &= \xi_2 + \delta_3 \\
x_4 &= \lambda_{42}\xi_2 + \delta_4
\end{aligned} \tag{9.26}$$

where δ_i ($i = 1$ to 4) has an expected value of zero, is independent of ζ_1, ξ_1 and ξ_2, and δ_i is independent of δ_j for $i \neq j$. To develop measures of $\xi_1\xi_2$, we form the product indicators of x_1x_3, x_2x_3, x_1x_4, and x_2x_4. (Combining other pairs of indicators such as x_1x_2 or x_3^2 would not provide measures of $\xi_1\xi_2$.) The measurement equations for these new indicators are

$$\begin{aligned}
x_1x_3 &= \xi_1\xi_2 + \xi_1\delta_3 + \xi_2\delta_1 + \delta_1\delta_3 \\
x_2x_3 &= \lambda_{21}\xi_1\xi_2 + \lambda_{21}\xi_1\delta_3 + \xi_2\delta_2 + \delta_2\delta_3 \\
x_1x_4 &= \lambda_{42}\xi_1\xi_2 + \xi_1\delta_4 + \lambda_{42}\xi_2\delta_1 + \delta_1\delta_4 \\
x_2x_4 &= \lambda_{21}\lambda_{42}\xi_1\xi_2 + \lambda_{21}\xi_1\delta_4 + \lambda_{42}\xi_2\delta_2 + \delta_2\delta_4
\end{aligned} \tag{9.27}$$

The list of latent variables is expanded. The 15 latent exogenous variables and error terms for equations (9.26) and (9.27) are ξ_1, ξ_2, $\xi_1\xi_2$, δ_1, δ_2, δ_3, δ_4, $\xi_1\delta_3$, $\xi_1\delta_4$, $\xi_2\delta_1$, $\xi_2\delta_2$, $\delta_1\delta_3$, $\delta_2\delta_3$, $\delta_1\delta_4$, and $\delta_2\delta_4$. The λ_{21} and λ_{42} coefficients are the only unknown parameters in (9.27). Kenny and Judd (1984) show that if ξ_1, ξ_2, δ_1, δ_2, δ_3, δ_4, and ζ_1 are normally distributed with means of zero, then all 15 of the latent exogenous variables (except ξ_1

and ξ_2) are uncorrelated and have the following variances:

$$
\begin{aligned}
&\mathrm{VAR}(\xi_1) = \phi_{11}, && \mathrm{VAR}(\xi_2\delta_1) = \phi_{22}^2[\mathrm{VAR}(\delta_1)]^2 \\
&\mathrm{VAR}(\xi_2) = \phi_{22}, && \mathrm{VAR}(\xi_2\delta_2) = \phi_{22}^2[\mathrm{VAR}(\delta_2)]^2 \\
&\mathrm{VAR}(\xi_1\xi_2) = \phi_{11}^2\phi_{22}^2 + \phi_{12}^2, && \mathrm{VAR}(\delta_1\delta_3) = \mathrm{VAR}(\delta_1)\mathrm{VAR}(\delta_3) \\
&\mathrm{VAR}(\delta_1), \quad \mathrm{VAR}(\delta_3), && \mathrm{VAR}(\delta_1\delta_4) = \mathrm{VAR}(\delta_1)\mathrm{VAR}(\delta_4) \qquad (9.28) \\
&\mathrm{VAR}(\delta_2), \quad \mathrm{VAR}(\delta_4), && \mathrm{VAR}(\delta_2\delta_3) = \mathrm{VAR}(\delta_2)\mathrm{VAR}(\delta_3) \\
&\mathrm{VAR}(\xi_1\delta_3) = \phi_{11}^2[\mathrm{VAR}(\delta_3)]^2, && \mathrm{VAR}(\delta_2\delta_4) = \mathrm{VAR}(\delta_2)\mathrm{VAR}(\delta_4) \\
&\mathrm{VAR}(\xi_1\delta_4) = \phi_{11}^2[\mathrm{VAR}(\delta_4)]^2
\end{aligned}
$$

The variances of the product terms are functions of the variances of the errors (δ_i) and ξ_i variables. Thus the inclusion of the interaction term $\xi_1\xi_2$ leads to only one new unconstrained parameter to estimate, γ_{13}. With the imposition of the linear and nonlinear restrictions in equations (9.25) to (9.28), this model is identified.

There are at least three obstacles to implementing this method: (1) it requires nonlinear constraints on parameters [see (9.23), (9.24), (9.27), and (9.28)], (2) the assumption of a multinormal distribution for the product indicators is not tenable, and (3) nonnormality of the latent variables or disturbances for the nonproduct variables could lead to incorrect constraints. Point 1 is not a serious barrier. The nonlinear constraints can be directly implemented in programs such as COSAN (McDonald 1978; Fraser 1980) or LINCS (Schoenberg 1987). Alternatively, equality constraints and phantom variables can indirectly create these nonlinear restrictions (see Wong and Long 1987 or Hayduk 1987). The violation of the multinormality assumption results from forming the products of normally distributed variables (e.g., x_1^2, x_2^2, x_1x_3) that do not have a normal distribution. The ML estimator is consistent even when the observed variables are nonnormal, but the chi-square and coefficient tests of statistical significance may be invalid. To the extent that the elements of **S** no longer have the covariances required for GLS (see Chapter 4), the GLS-based significance tests also are not justified, though the GLS estimator remains consistent. Later in this chapter I introduce a weighted least squares fitting function that leads to an asymptotically valid chi-square estimator and standard errors for nonnormally distributed observed variables. Thus the nonnormality of the product variables can be treated.

There remains a third obstacle. Kenny and Judd (1984) derived the nonlinear constraints by assuming that the nonproduct latent and disturbance variables are normally distributed. We can test the normality of the nonproduct latent and disturbance variables by testing for the normality of the nonproduct observed variables. This follows since a normal distribution in the nonproduct variables is implied by the normality of the latent and disturbance variables of which it is composed. I describe tests of normality later in this chapter. Nonnormality of the nonproduct observed variables means that the researcher needs to investigate alternative distributional assumptions and to determine if new nonlinear restrictions are required.

Corrected covariance matrices is a second method of treating quadratic or interactive latent variables (Bohrnstedt and Marwell 1978; Busemeyer and Jones 1983; Heise 1986). To illustrate it, I return to equation (9.25):

$$\eta_1 = \gamma_{11}\xi_1 + \gamma_{12}\xi_2 + \gamma_{13}\xi_1\xi_2 + \zeta_1$$

It is easy to show that

$$\boldsymbol{\Gamma}' = \boldsymbol{\Phi}^{-1}\boldsymbol{\Sigma}_{\xi\eta_1} \tag{9.29}$$

where $\boldsymbol{\Gamma} = [\gamma_{11} \quad \gamma_{12} \quad \gamma_{13}]$, $\boldsymbol{\Phi}$, and $\boldsymbol{\Sigma}_{\xi\eta_1}$ are the population covariance matrices of $\boldsymbol{\xi}$ $(= [\xi_1 \quad \xi_2 \quad \xi_1\xi_2])$ with $\boldsymbol{\xi}$ and of $\boldsymbol{\xi}$ with η_1, respectively. When $y_1 = \eta_1$, $x_1 = \xi_1$, and $x_2 = \xi_2$, then $x_1x_2 = \xi_1\xi_2$,

$$y_1 = \gamma_{11}x_1 + \gamma_{12}x_2 + \gamma_{13}x_1x_2 + \zeta_1 \tag{9.30}$$

and,

$$\boldsymbol{\Gamma}' = \boldsymbol{\Sigma}_{xx}^{-1}\boldsymbol{\Sigma}_{xy_1} \tag{9.31}$$

where $\boldsymbol{\Sigma}_{xx}$ and $\boldsymbol{\Sigma}_{xy_1}$ are the population covariance matrices of $\mathbf{x}$ $(= [x_1 \quad x_2 \quad x_1x_2])$ with $\mathbf{x}$ and of $\mathbf{x}$ with y_1. But if x_1 or x_2 contains error, $\boldsymbol{\Gamma}' \neq \boldsymbol{\Sigma}_{xx}^{-1}\boldsymbol{\Sigma}_{xy_1}$ (see Chapter 5). If we desire $\boldsymbol{\Gamma}'$, we need $\boldsymbol{\Phi}^{-1}$ and $\boldsymbol{\Sigma}_{\xi\eta_1}$. The basic idea of the corrected covariance matrix method is to adjust $\boldsymbol{\Sigma}_{xx}$ and $\boldsymbol{\Sigma}_{xy_1}$ so that they equal $\boldsymbol{\Phi}$ and $\boldsymbol{\Sigma}_{\xi\eta_1}$. Then $\boldsymbol{\Gamma}'$ can be recovered using (9.29).

Busemeyer and Jones's (1983) formulation of the corrected covariance matrix method assumes (1) that there is a single indicator for each nonproduct latent variable with $x_1 = \xi_1 + \delta_1$ and $x_2 = \xi_2 + \delta_2$, where $E(\delta_i) = 0$, (2) that δ_1, δ_2, and ζ_1 are distributed independently of ξ_1 and ξ_2 and of each other, (3) that ξ_1, ξ_2, δ_1, δ_2, and ζ_1 are normally distributed, (4) that ξ_1, ξ_2, x_1, and x_2 are in deviation form, and (5) that $E(\zeta_1) = 0$. Under these conditions the relations between Σ_{xx} and Φ and Σ_{xy_1} and $\Sigma_{\xi\eta_1}$ are

$$\Sigma_{xx} = \Phi + \Theta_\delta \tag{9.32}$$

$$\Sigma_{xy_1} = \Sigma_{\xi\eta_1} \tag{9.33}$$

We can form Φ as $(\Sigma_{xx} - \Theta_\delta)$, whereas no adjustment is needed for Σ_{xy_1}. With this we can write Γ' as

$$\Gamma' = (\Sigma_{xx} - \Theta_\delta)^{-1}\Sigma_{xy_1} \tag{9.34}$$

Without knowing Θ_δ, the covariance matrix for the errors, we can progress no further. Assuming that we have the reliability of x_1 and x_2, the results of Bohrnstedt and Marwell (1978) provide the reliability of x_1x_2. With these we can construct Θ_δ and then form Γ' from (9.34). In practice, we do not have the population covariance matrices but substitute consistent sample estimators of them. For instance, the sample counterpart to equation (9.34) is

$$\hat{\Gamma}' = \left(\mathbf{S}_{xx} - \hat{\Theta}_\delta\right)^{-1}\mathbf{S}_{xy_1} \tag{9.35}$$

where $\mathbf{S}_{xx}$, $\hat{\Theta}_\delta$, and $\mathbf{S}_{xy_1}$ are consistent estimators of Σ_{xx}, Θ_δ, and Σ_{xy_1}. Thus the corrected covariance matrix method adjusts the sample covariance matrices of the observed variables so that they are consistent estimators of the corresponding population covariance matrices that can then be used to estimate the parameters of the latent variable model.

The major limitations of this procedure are (1) it allows only a single indicator per latent variable, (2) the error variances of the nonproduct indicators or independent estimates of them must be known, (3) tests of statistical significance for the overall model or individual parameter estimates are not available, and (4) violation of the normality and independence assumptions for the nonproduct latent variables and nonproduct disturbances could lead to incorrect adjustments to the covariance matrices. The first three points make this method more restrictive than the product indicators method. Early Monte Carlo simulation work by Heise (1986) suggests that the corrected covariance matrix method provides erratic parameter estimates when the reliability of the nonproduct variables are moderate to low.

The two preceding methods treat quadratic and interaction terms in the latent variable model. It also is possible for such terms to occur in the measurement model (e.g., $x_1 = \lambda_{11}\xi_1 + \lambda_{12}\xi_2 + \lambda_{13}\xi_1\xi_2 + \delta_1$). McDonald

(1967a, 1967b), Etezadi-Amoli and McDonald (1983), and Mooijaart and Bentler (1986) discuss methods for these situations.

INSTRUMENTAL-VARIABLE (IV) ESTIMATORS

A disadvantage of the ML, GLS, and ULS estimators is that they require iterative procedures to minimize a fitting function. As such they can be computationally costly. The choice of starting values affects the number of iterations (see Chapter Four, Appendix 4C) such that starting values close to the final estimates require fewer iterations. Instrumental Variable (IV) estimators are noniterative and provide consistent estimators of parameters. (See Bentler 1982 for an alternative noniterative, consistent estimator.) Being noniterative, IV estimators are less costly than the ML, GLS, or ULS estimators. In addition to their value in and of themselves, the IV estimators often provide good starting values for the iterative procedures. For instance, Jöreskog and Sörbom's (1986b) LISREL program uses an IV estimator for its starting values.

In this section I discuss IV estimators. To define an IV, I begin with a simple regression example:

$$y_1 = \gamma_{11} x_1 + \zeta_1 \tag{9.36}$$

where $E(\zeta_1) = 0$ and y_1 and x_1 are in deviation form. Usually we assume that $\mathrm{COV}(x_1, \zeta_1)$ is zero. Suppose that ζ_1 correlates with x_1. This could be due to many things such as a y_1 directly influencing x_1, omitted third variables correlated with x_1 and with a direct effect on y_1, or measurement error in x_1. Whatever the cause of $\mathrm{COV}(x_1, \zeta_1) \neq 0$, the consequence is that the usual OLS estimator of γ_{11} is inconsistent.

The IV estimator of γ_{11} requires an "instrumental" variable, say, z_1. To be an instrument, z_1 must satisfy the following conditions:

$$\mathrm{COV}(z_1, \zeta_1) = 0 \tag{9.37}$$

$$\mathrm{COV}(z_1, x_1) \neq 0 \tag{9.38}$$

So the instrument must be uncorrelated with the disturbance but correlated with the variable for which it is an instrument. Taking the covariance of z_1 with both sides of equation (9.36) leads to

$$\mathrm{COV}(y_1, z_1) = \gamma_{11}\,\mathrm{COV}(x_1, z_1) \tag{9.39}$$

$$\gamma_{11} = \frac{\mathrm{COV}(x_1, z_1)}{\mathrm{COV}(y_1, z_1)} \tag{9.40}$$

Substituting the sample covariances for the population covariances in (9.40) forms the IV estimator of γ_{11}:

$$\hat{\gamma}_{11} = \frac{s_{x_1 z_1}}{s_{y_1 z_1}} \tag{9.41}$$

The asymptotic variance of $\hat{\gamma}_{11}$ is

$$\left(\frac{\psi_{11}}{N}\right)\left(\frac{\mathrm{VAR}(z_1)}{[\mathrm{COV}(z_1, x_1)]^2}\right) \tag{9.42}$$

We can estimate ϕ_{11} and ψ_{11} with

$$\hat{\phi}_{11} = s_{x_1 x_1} \tag{9.43}$$

$$\hat{\psi}_{11} = s_{y_1 y_1} + \hat{\gamma}_{11}^2 \hat{\phi}_{11} - 2\,\hat{\gamma}_{11}\, s_{x_1 y_1} \tag{9.44}$$

For regression/econometric models (i.e., $\mathbf{y} = \mathbf{B}\mathbf{y} + \boldsymbol{\Gamma}\mathbf{x} + \boldsymbol{\zeta}$), the IV and the related two-stage least squares (2SLS) estimators of $\mathbf{B}$ and $\boldsymbol{\Gamma}$ and their asymptotic standard errors are well known (see, e.g., Johnston 1984; Fox 1984). The IV estimator, however, is less known in confirmatory factor analysis. Over two decades ago Madansky (1964) noted that IV techniques could estimate factor analysis models. More recent work by Hägglund (1982) has led to an IV estimator (called FABIN3) that provides the starting values of the factor loadings in LISREL VI. I illustrate the principles of the IV estimator in factor analysis with the air quality example presented in Chapter 7. Recall that this model has a single latent variable of general air quality (ξ_1) with four indicators: rated overall air quality (x_1), air clarity (x_2), air color (x_3), and air odor (x_4).

The measurement model is

$$x_1 = \xi_1 + \delta_1 \tag{9.45}$$

$$x_2 = \lambda_{21}\xi_1 + \delta_2 \tag{9.46}$$

$$x_3 = \lambda_{31}\xi_1 + \delta_3 \tag{9.47}$$

$$x_4 = \lambda_{41}\xi_1 + \delta_4 \tag{9.48}$$

where x_i ($i = 1$ to 4) and ξ_1 are in deviation form and I assume the original model where $\mathrm{COV}(\delta_i, \delta_j) = 0$ for $i \neq j$ and $\mathrm{COV}(\xi_1, \delta_i) = 0$ for all i. Equation (9.45) implies that ξ_1 equals ($x_1 - \delta_1$). Substituting this quantity for ξ_1 in (9.46) leads to

$$x_2 = \lambda_{21}x_1 + \delta_2 - \lambda_{21}\delta_1$$
$$= \lambda_{21}x_1 + u \tag{9.49}$$

where $u = \delta_2 - \lambda_{21}\delta_1$. It is tempting to use the ordinary least squares (OLS) estimator of λ_{21} in (9.49) since both x_1 and x_2 are observed variables, as is true in most regression equations. The OLS estimator for λ_{21} is s_{21}/s_{11}. For this to be consistent, however, x_1 must be uncorrelated with u. Since u contains δ_1, the error of measurement for x_1, this assumption is violated in nearly all cases, and the OLS estimator is inconsistent. To bypass this problem, start by placing x_3 and x_4 in a column vector and postmultiply both sides of (9.49) by the transpose of this vector:

$$x_2[x_3 \quad x_4] = \lambda_{21}x_1[x_3 \quad x_4] + u[x_3 \quad x_4] \tag{9.50}$$

Next take expected values of both sides of (9.50):

$$E(x_2[x_3 \quad x_4]) = \lambda_{21}E(x_1[x_3 \quad x_4]) + E(u[x_3 \quad x_4]) \tag{9.51}$$

$$[\sigma_{23} \quad \sigma_{24}] = \lambda_{21}[\sigma_{13} \quad \sigma_{14}] \tag{9.52}$$

where σ_{ij} is the population covariance of i and j. The last term in (9.51) is zero because x_3 and x_4 are uncorrelated with u ($u = \delta_2 - \lambda_{21}\delta_1$) in the population. The x_3 and x_4 variables act as instrumental variables for the x_2 equation. With some algebraic manipulation of (9.52), we can solve for λ_{21}. Substituting consistent estimators of σ_{ij} (e.g., s_{ij}) leads to

$$\hat{\lambda}_{21} = \left([s_{13} \quad s_{14}]\begin{bmatrix} s_{13} \\ s_{14} \end{bmatrix}\right)^{-1}[s_{13} \quad s_{14}]\begin{bmatrix} s_{23} \\ s_{24} \end{bmatrix} \tag{9.53}$$

Though (9.53) provides a consistent estimator of λ_{21}, there remains one problem with it. Hägglund (1982) argues that it is not an efficient estimator, and he proposes a weighted one that is asymptotically more efficient:

$$\hat{\lambda}_{21} = \left[[s_{13} \quad s_{14}]\begin{bmatrix} s_{33} & s_{34} \\ s_{34} & s_{44} \end{bmatrix}^{-1}\begin{bmatrix} s_{13} \\ s_{14} \end{bmatrix}\right]^{-1}\left[[s_{13} \quad s_{14}]\begin{bmatrix} s_{33} & s_{34} \\ s_{34} & s_{44} \end{bmatrix}^{-1}\begin{bmatrix} s_{23} \\ s_{24} \end{bmatrix}\right] \tag{9.54}$$

This weighted estimator assumes that the fourth-order cumulants of the x's and u are zero (see Hägglund 1982). Substituting the sample covariances (see equation 7.82) into (9.54) results in

$$\hat{\lambda}_{21} = 1.703 \tag{9.55}$$

To obtain $\hat{\lambda}_{31}$, the same series of steps as that followed for $\hat{\lambda}_{21}$ is repeated, except that x_2 and x_4 are the new instrumental variables. The equation for $\hat{\lambda}_{31}$ is (9.54), after replacing all subscripts of 3 to 2 and changing $[s_{23} \quad s_{24}]'$ to $[s_{32} \quad s_{34}]'$ to reflect the new dependent variable and the new instruments. A similar process creates an equation for $\hat{\lambda}_{41}$. Using these formulas, the factor loadings are

$$\hat{\Lambda}_x = \begin{bmatrix} 1.000 \\ 1.703 \\ 1.683 \\ 0.726 \end{bmatrix} \tag{9.56}$$

These match the initial estimates for the factor loadings provided by LISREL VI.

The instrumental variable estimator can be generalized to more complicated factor analysis problems. To do so, partition the observed variable vector $\mathbf{x}$ into three parts:

$$\mathbf{x} = \begin{bmatrix} \mathbf{x}_A \\ x_B \\ \mathbf{x}_C \end{bmatrix} \tag{9.57}$$

The vector $\mathbf{x}_A$ is $n \times 1$ and contains the observed variables that provide a scale for the latent variables, x_B is one of the x_i's not included in $\mathbf{x}_A$, and $\mathbf{x}_C$ contains all the remaining x_i variables eliminating $\mathbf{x}_A$ and x_B. The measurement model is

$$\mathbf{x} = \begin{bmatrix} \mathbf{x}_A \\ x_B \\ \mathbf{x}_C \end{bmatrix} = \begin{bmatrix} \mathbf{I}_n \\ \boldsymbol{\lambda}'_B \\ \boldsymbol{\Lambda}_C \end{bmatrix} \boldsymbol{\xi} + \boldsymbol{\delta} \tag{9.58}$$

where $\mathbf{I}_n$ is an $n \times n$ identity matrix for the observed variables that provide scales for $\boldsymbol{\xi}$, $\boldsymbol{\lambda}'_B$ is an $1 \times n$ vector containing the coefficients for the ξ variables that influence x_B, and $\boldsymbol{\Lambda}_C$ is a $(q - n - 1) \times n$ matrix of the remaining coefficients of factor loadings for $\mathbf{x}_C$. In the first part of the air quality example, x_1 was $\mathbf{x}_A$, x_2 was x_B, and x_3 and x_4 were $\mathbf{x}_C$. The vector $\mathbf{x}_C$ contains the instrumental variables for an equation. The estimator

(FABIN3) for λ'_B is

$$\hat{\lambda}'_B = \left(\mathbf{S}_{AC}\mathbf{S}_{CC}^{-1}\mathbf{S}_{CA}\right)^{-1}\mathbf{S}_{AC}\mathbf{S}_{CC}^{-1}\mathbf{S}_{CB} \tag{9.59}$$

where $\mathbf{S}_{ij}$ are the sample covariance matrices of i and j. The variables falling into x_B and $\mathbf{x}_C$ rotate depending on which coefficients are estimated.

Following the preceding steps leads to $\hat{\Lambda}_x$. The estimators for $\hat{\Theta}_\delta$ and $\hat{\Phi}$ remain to be described. Hägglund (1982, 213–214) shows that for a given $\hat{\Lambda}_x$, the value of $\hat{\Theta}_\delta$ that minimizes the discrepancy between the sample covariance matrix $\mathbf{S}$ and the implied covariance matrix $\Sigma(\hat{\theta})$ according to least squares principles is

$$\operatorname{diag}\left(\hat{\Theta}_\delta\right) = (\mathbf{I} - \mathbf{D} * \mathbf{D})^{-1}\mathbf{g} \tag{9.60}$$

where $\mathbf{D}$ equals $\hat{\Lambda}_x(\hat{\Lambda}'_x\hat{\Lambda}_x)^{-1}\hat{\Lambda}'_x$, the "$*$" means elementwise multiplication, and $\mathbf{g}$ is a $q \times 1$ column vector formed from the diagonal elements of $(\mathbf{S} - \mathbf{DSD})$. The off-diagonal elements of $\hat{\Theta}_\delta$ are zero. Using $\hat{\Theta}_\delta$, from (9.60) he then shows that $\hat{\Phi}$ equals

$$\hat{\Phi} = \left(\hat{\Lambda}'_x\hat{\Lambda}_x\right)^{-1}\hat{\Lambda}'_x\left(\mathbf{S} - \hat{\Theta}_\delta\right)\hat{\Lambda}_x\left(\hat{\Lambda}'_x\hat{\Lambda}_x\right)^{-1} \tag{9.61}$$

Finally, the implied covariance matrix, $\Sigma(\hat{\theta})$ is

$$\Sigma(\hat{\theta}) = \hat{\Lambda}_x\hat{\Phi}\hat{\Lambda}'_x + \hat{\Theta}_\delta \tag{9.62}$$

Based on these formulas, the remaining parameter estimates for the air quality model are

$$\operatorname{diag}\left(\hat{\Theta}_\delta\right) = [0.062 \quad 0.379 \quad 0.135 \quad 0.126] \tag{9.63}$$

$$\hat{\Phi} = 0.270 \tag{9.64}$$

$$\Sigma(\hat{\theta}) = \begin{bmatrix} 0.331 & & & \\ 0.459 & 1.160 & & \\ 0.454 & 0.772 & 0.898 & \\ 0.196 & 0.333 & 0.330 & 0.268 \end{bmatrix} \tag{9.65}$$

The estimates in (9.63) and (9.64) match the initial estimates provided by LISREL VI.

Hägglund (1982) also provides a means to obtain asymptotic standard errors for $\hat{\Lambda}_x$. For each $\hat{\lambda}'_B$ [see (9.59)], the asymptotic covariance matrix is

$$\text{ACOV}(\hat{\lambda}'_B) = (N^{-1})\sigma_{uu}(\Sigma_{AC}\Sigma_{CC}^{-1}\Sigma_{CA})^{-1} \tag{9.66}$$

where Σ_{AC}, Σ_{CC}, and Σ_{CA} are the population covariance matrices for the variables in A and C, C and C, and C and A, respectively. Estimators of Σ_{ij} matrices are the corresponding elements in $\Sigma(\hat{\theta})$ [see (9.62)]. The σ_{uu} estimator is (see Hägglund, 1982, 217)

$$\hat{\sigma}_{uu} = \hat{\sigma}_{BB} - 2\hat{\lambda}'_B\hat{\Sigma}_{AB} + \hat{\lambda}'_B\hat{\Sigma}_{AA}\hat{\lambda}_B \tag{9.67}$$

Applying (9.66) to each row of the factor loadings (other than the rows of fixed elements) and taking the square roots of the diagonal elements leads to the estimated asymptotic standard errors.

The coefficients with their standard errors (in parentheses) for the air quality example are

$$\hat{\Lambda}_x = \begin{bmatrix} 1.00 \\ (-) \\ 1.703 \\ (0.204) \\ 1.683 \\ (0.163) \\ 0.726 \\ (0.108) \end{bmatrix} \tag{9.68}$$

Hägglund (1982) does not suggest a means of calculating standard errors for the other estimates in the model.

Jöreskog and Sörbom's (1986b) IV or 2SLS estimator for the general structural equation model combines the IV or 2SLS estimator for confirmatory factor analysis with that for the observed variable econometric model. The first step is to estimate a measurement model which merges $\mathbf{y}$ and $\mathbf{x}$ into a single CFA:

$$\begin{bmatrix} \mathbf{y} \\ \mathbf{x} \end{bmatrix} = \begin{bmatrix} \Lambda_y & 0 \\ 0 & \Lambda_x \end{bmatrix} \begin{bmatrix} \boldsymbol{\eta} \\ \boldsymbol{\xi} \end{bmatrix} + \begin{bmatrix} \boldsymbol{\epsilon} \\ \boldsymbol{\delta} \end{bmatrix} \tag{9.69}$$

The estimator (9.59) for factor analysis provides estimates of the factor loadings. Equation (9.60) gives estimates of the error covariance matrix, and equation (9.61) leads to estimates of the covariance matrix for $\boldsymbol{\eta}$ and $\boldsymbol{\xi}$. Once we have the covariance matrix for the latent variables ($\boldsymbol{\eta}$ and $\boldsymbol{\xi}$), we can apply standard IV or 2SLS procedures, treating the latent variable covariance matrix as if it were the covariance matrix for observed variables. This leads to estimates of $\mathbf{B}$ and $\boldsymbol{\Gamma}$. An estimate of $\boldsymbol{\Phi}$ is part of the covariance matrix for $\boldsymbol{\eta}$ and $\boldsymbol{\xi}$. Finally, an estimate of $\boldsymbol{\Psi}$ is

$$\hat{\boldsymbol{\Psi}} = (\mathbf{I} - \hat{\mathbf{B}})\hat{\boldsymbol{\Sigma}}_{\eta\eta}(\mathbf{I} - \hat{\mathbf{B}})' - \hat{\boldsymbol{\Gamma}}\hat{\boldsymbol{\Phi}}\hat{\boldsymbol{\Gamma}}' \tag{9.70}$$

Hägglund's (1982) estimators assume that $\boldsymbol{\Theta}_\delta$ and $\boldsymbol{\Theta}_\epsilon$ are diagonal matrices. When this assumption fails, his estimators will not be consistent. This occurs because some "instruments" are correlated with the composite error term u [see, e.g., (9.49)]. A simple solution is to remove from $\mathbf{x}_C$ any x's that are correlated with u and to proceed as usual. This requires a sufficient number of instruments in $\mathbf{x}_C$.

A second aspect of the IV method is that it is a limited information estimator in that it treats a single equation at a time and does not take into account the complete information on all other structural equations in the model. The ML and GLS estimators, for instance, solve for all parameters in all parts of the model simultaneously. Limited information estimators, like the IV or 2SLS ones, have disadvantages as well as advantages. A disadvantage is that information in other parts of the model is ignored, even though it might improve estimator efficiency. The full-information estimators such as ML incorporate this information and can thereby reduce the variance of estimators. But this is a two-edged sword. If there is specification error in the system (e.g., omitted variables or correlated errors that are not included), this can affect many or all of the parameter estimates for full-information procedures. For the IV and 2SLS methods, the specification error is likely to have more isolated influences since not all of the model is considered at once.

Another consideration for the IV and 2SLS estimators is that asymptotic standard errors are available only for a subset of the model parameters. In addition we know little about the small or moderate sample behavior of the IV and 2SLS estimators for factor analysis or in structural equations with latent variables.

DISTRIBUTIONAL ASSUMPTIONS

The F_{ML} fitting function derives from the multinormal distribution of the observed variables (see Chapter 4, Appendices 4A and 4B). The F_{ML} or

Table 9.1 Properties of ML and GLS Estimators with and without Multinormal Observed Variables

	Properties of ML and GLS Estimators			
Observed Variable Distribution	Consistency	Asymptotic Efficiency	ACOV($\hat{\boldsymbol{\theta}}$)	Chi-square Estimator
Multinormal	yes	yes	correct	correct
No Kurtosis	yes	yes	correct	correct
Elliptical	yes	yes	incorrect	incorrect
"Arbitrary"	yes	no	incorrect	incorrect

F_{GLS} functions also are justified when the distribution of the observed variables has no excess kurtosis (see Chapter 4). Under either of these conditions and with a valid model, the $\hat{\boldsymbol{\theta}}$ from F_{ML} or F_{GLS} is a consistent and asymptotic efficient estimator and $(N-1)F_{\text{ML}}$ or $(N-1)F_{\text{GLS}}$ at their respective $\hat{\boldsymbol{\theta}}$'s have asymptotic chi-square distributions suitable for tests of overall model fit. We also have asymptotic covariance matrices for $\hat{\boldsymbol{\theta}}$ from F_{ML} or F_{GLS} that enable significance tests of individual or multiple parameters.[1]

How dependent are these properties on multinormality or no excess kurtosis? The purposes of this section are (1) to examine the consequences of violating these assumptions, (2) to present tests to detect these violations, and (3) to propose estimators that are robust for nonnormal data.

Consequences

Table 9.1 summarizes the consequences of the observed variables' distribution for several properties of the ML and GLS estimators. Proofs of these properties and a description of the necessary regularity conditions are in Browne (1982, 1984).

An important characteristic of $\hat{\boldsymbol{\theta}}$, whether from F_{ML}, F_{GLS}, or even F_{ULS}, is that it remains consistent. So, as the sample size grows larger, $\hat{\boldsymbol{\theta}}$ converges to $\boldsymbol{\theta}$ even for nonnormal distributions.

[1]The assumptions of homoscedasticity and no autocorrelation for disturbances or errors across observations also are made. In contrast to regression/econometric models (e.g., Johnston 1984), little work is available on the violations of these assumptions in the general structural equation model. Arminger and Schoenberg (1987), however, show that a consistent estimator of the asymptotic covariance matrix of $\hat{\boldsymbol{\theta}}$ can be formed when heteroscedasticity is present.

The first row of Table 9.1 shows that for multinormal variables, the ML and the GLS estimators also are asymptotically efficient and the usual asymptotic covariance matrix [ACOV($\hat{\boldsymbol{\theta}}$)] and chi-square estimators [$(N - 1)F_{\mathrm{ML}}$ or $(N - 1)F_{\mathrm{GLS}}$] apply. A little less restrictive than multinormality is when the ACOV(s_{ij}, s_{gh}) equals $N^{-1}(\sigma_{ig}\sigma_{jh} + \sigma_{ih}\sigma_{jg})$. This occurs when the kurtoses of the marginal and multivariate distributions are the same as those for a multinormal distribution. That is, the observed variables have no excess kurtosis. As stated previously, the ML and GLS estimators retain all four properties.[2]

When the observed variables have excessive kurtosis (e.g., "fatter" or "thinner" tails than the normal distribution), only the consistency of the estimator is guaranteed. But we have some special cases where additional properties may hold. One is for elliptical distributions (see row 3 of Table 9.1). I will describe elliptical distributions later, but in brief they are distributions without skewness and they have the same degree of kurtosis for each observed variable. For elliptical distributions, the $\hat{\boldsymbol{\theta}}$ that minimizes F_{ML} is asymptotically efficient.[3] However, the asymptotic covariance matrix, standard errors, and chi-square estimator based on F_{ML} are incorrect. Without adjustments, tests of significance may be in error.

The $\hat{\boldsymbol{\theta}}$ from F_{ML} or F_{GLS} lack even asymptotic efficiency when the observed variables have arbitrary distributions. But here, too, there are exceptions. For example, in Chapter 4, I treated alternative assumptions for **x** that still lead to desirable asymptotic properties in an observed variable model when **x** is nonnormal. Recent work has begun to determine a broader range of conditions when the ML test statistics apply to other nonnormal observed variables (e.g., Shapiro 1987; Anderson and Amemiya 1985). In a factor analysis model, $x_i = \lambda_{i1}\xi_1 + \delta_i$, $i = 1$ to 4, where $\lambda_{11} = 1$, ξ_1 and δ_i's are mutually independent, and each distributed as a χ^2 with one degree of freedom, Satorra and Bentler (1987) show that $(N - 1)F_{\mathrm{ML}}$ at $\hat{\boldsymbol{\theta}}$ is asymptotically distributed as a chi-square and that the usual ACOV($\hat{\boldsymbol{\theta}}$) is asymptotically correct for all but $\hat{\boldsymbol{\Theta}}_\delta$ and $\hat{\phi}_{11}$. Thus the usual F_{ML}-derived test statistics (or a subset of them) have applicability under a broader range of conditions than initially thought.

Most simulation examinations of the robustness of the F_{ML} or F_{GLS} estimators and test statistics have examined nonnormality created by categorizing continuous variables. A few have examined nonnormal continuous variables. Typically, the parameter estimates do not appear biased by

[2]The asymptotic efficiency property refers to asymptotic efficiency within the class of GLS estimators but not necessarily for all possible estimators (Browne 1984, 68).

[3]This result is for models which are invariant with respect to constant scaling factors (Browne 1982, 1984).

nonnormality, but the chi-square estimates and estimates of $\mathrm{ACOV}(\hat{\boldsymbol{\theta}})$ are sometimes off (e.g., Browne 1984, 77–81; Tanaka and Bentler 1984). These results are largely in agreement with the analytic asymptotic results.

In sum, violation of multinormality does not affect the consistency of the F_{ML}, F_{GLS}, or F_{ULS} estimators of $\boldsymbol{\theta}$. But excessive kurtosis usually eliminates asymptotic efficiency and makes the estimated asymptotic covariance matrix and the chi-square estimator potentially inaccurate. Thus the presence of nonnormal observed variables can affect tests of statistical significance. Before I examine how to correct for nonnormal distributions, I will explain methods of detecting them.

Detection

Tests of normality or excessive kurtosis are helpful in assessing the appropriateness of the ML or GLS estimators. Often all the observed variables should be examined. In cases such as in some regression-econometric models, the normality of the disturbances are of concern. Another application is in latent variable models with interaction or quadratic terms. Here the nonproduct observed variables should be tested for normality, but not the product terms.

The purpose of this section is to present procedures by which to evaluate the univariate or multivariate skewness, kurtosis, or normality of random variables. If these tests reveal problems, then it is advisable to turn to the weighted least squares or elliptical estimators, which I will present after the tests.

The moments around the mean of a distribution reveal departures from normality. For a random variable X with a population mean of μ_1, the rth moment about the mean is defined as

$$\mu_r = E(X - \mu_1)^r, \qquad \text{for } r > 1 \tag{9.71}$$

The univariate normal distribution is completely characterized by its mean (μ_1) and its variance (μ_2). Its higher-order moments are either zero or can be written as functions of its mean or variance. Thus departures from normality are detectable by examining the higher-order moments. The standardized third and fourth moments are

$$\text{Standardized third moment} = \frac{\mu_3}{(\mu_2)^{3/2}} \tag{9.72}$$

$$\text{Standardized fourth moment} = \frac{\mu_4}{\mu_2^2} \tag{9.73}$$

If a distribution is normal, its standardized third moment is 0 and its standardized fourth moment is 3.

Figure 9.3 illustrates distributions with skewness or kurtosis that deviate from that of a normal distribution. Part (a) displays a normal distribution (B) with two skewed distributions (A and C). The distribution to the left (A) has a long tail extending to the left and has *negative skewness*. Its standardized third moment, equation (9.72), is less than zero. The right-hand distribution (C) in Figure 9.3(a) has a tail that extends to the right and has *positive skewness* so that the standardized third moment, equation (9.72), is greater than zero.

Figure 9.3(b) illustrates kurtosis. Distribution B is a normal curve. Distribution A has *positive kurtosis* and has thinner tails than the normal (B). Its standardized fourth moment is greater than three. In contrast, the distribution C of part (b) has fatter tails than the normal (B) and has *negative kurtosis*. Its standardized fourth moment, equation (9.73), is less than three.

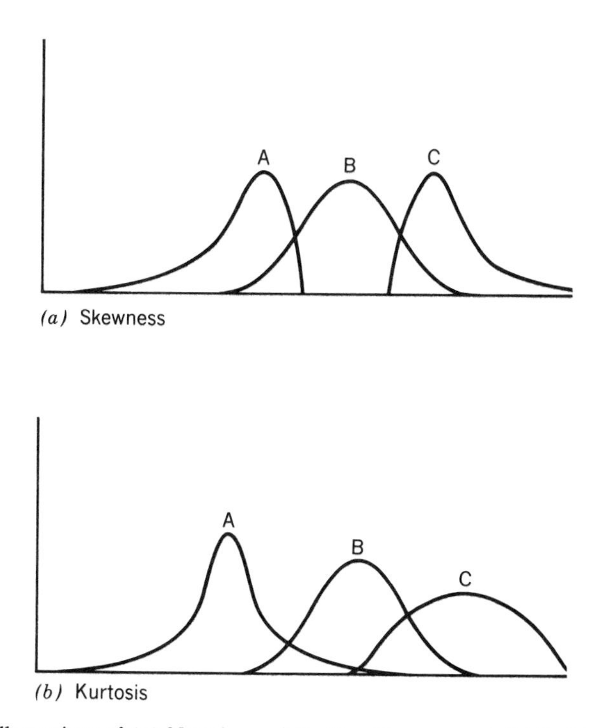

Figure 9.3 Illustrations of (a) Negative and Positive Skewness and (b) Positive and Negative Kurtosis

The sample counterparts of (9.72) and (9.73) are

$$\sqrt{b_1} = \frac{m_3}{(m_2)^{3/2}} \tag{9.74}$$

$$b_2 = \frac{m_4}{(m_2)^2} \tag{9.75}$$

where m_r is the rth sample moment defined as

$$m_r = \frac{\sum (X - \bar{X})^r}{N} \tag{9.76}$$

To illustrate these measures, I use two indicators of industrialization in developing countries taken from the political democracy and industrialization data of Chapter 8. I include the untransformed variables of GNP per capita (x_1) and energy consumption per capita (x_2) as well as the logarithmic transformation of these variables ($\ln x_1$ and $\ln x_2$). (The transformed variables were used in the earlier analyses.) The measures of skewness ($\sqrt{b_1}$) and kurtosis (b_2) are

	x_1	$\ln x_1$	x_2	$\ln x_2$
$\sqrt{b_1}$	1.761	0.259	3.087	−0.353
b_2	6.138	2.307	14.638	2.495
$b_2 - 3$	3.138	−0.693	11.638	−0.505

The last row ($b_2 - 3$) is calculated so that a kurtosis equal to the normal is zero, a kurtosis less than the normal is negative, and a kurtosis higher than the normal is positive.

It is evident that the untransformed variables deviate further from normality than the transformed ones. The large positive values of $\sqrt{b_1}$ for x_1 and x_2 show a positive skew for these variables that is greatly reduced with the logarithmic transformation. Indeed, logarithmic energy consumption per capita ($\ln x_2$) has a slightly negative skew. Large, positive kurtosis is present for x_1 and x_2 which changes to a slightly negative kurtosis for $\ln x_1$ and $\ln x_2$.

Sampling fluctuations might explain the deviations found, so it is necessary to test the statistical significance of the discrepancies. Many tests are available. I summarize one set recommended by D'Agostino (1986) in Table 9.2. The top left-hand side of the table provides the steps to form $Z_{\sqrt{b_1}}$

Table 9.2 Test Statistics for Univariate Skewness, $H_0: \sqrt{\beta_1} = 0$, or Kurtosis, $H_0: \beta_2 - 3 = 0$

Skewness	Kurtosis
$H_0: \sqrt{\beta_1} = 0$	$H_0: \beta_2 - 3 = 0$
Compute:	*Compute:*
(1) $\sqrt{b_1}$	(1) b_2
(2) $a_1 = \sqrt{b_1}\left[\frac{(N+1)(N+3)}{6(N-2)}\right]^{1/2}$	(2) $E(b_2) = \frac{3(N-1)}{N+1}$
(3) $a_2 = \frac{3(N^2+27N-70)(N+1)(N+3)}{(N-2)(N+5)(N+7)(N+9)}$	(3) $\mathrm{var}(b_2) = \frac{24N(N-2)(N-3)}{(N+1)^2(N+3)(N+5)}$
(4) $a_3 = -1 + [2(a_2-1)]^{1/2}$	(4) $c_1 = \frac{b_2 - E(b_2)}{[\mathrm{var}(b_2)]^{1/2}}$
(5) $a_4 = \left[\frac{1}{\log a_3}\right]^{1/2}$	(5) $c_2 = \frac{6(N^2-5N+2)}{(N+7)(N+9)}\left[\frac{6(N+3)(N+5)}{N(N-2)(N-3)}\right]^{1/2}$
(6) $a_5 = \left[\frac{2}{a_3-1}\right]^{1/2}$	(6) $c_3 = 6 + \frac{8}{c_2}\left[\frac{2}{c_2} + \left(1 + \frac{4}{c_2^2}\right)^{1/2}\right]$
if N ≥ 8:	*if N ≥ 20:*
$Z_{\sqrt{b_1}} = a_4 \log\left\{\frac{a_1}{a_5} + \left[\left(\frac{a_1}{a_5}\right)^2 + 1\right]^{1/2}\right\}$	$Z_{b_2} = \frac{1 - \left(\frac{2}{9c_3}\right) - \left[(1-2/c_3)/\{1 + c_1[2/(c_3-4)]^{1/2}\}\right]^{1/3}}{\left(\frac{2}{9c_3}\right)^{1/2}}$
if N ≥ 150:	*if N ≥ 1000:*
$Z_{\sqrt{b_1}} = a_1$	$Z_{b_2} = c_1$
$Z_{\sqrt{b_1}}$ approx. $N(0,1)$	Z_{b_2} approx. $N(0,1)$

that approximates a standardized normal variable for a test of skewness (H_0: $\sqrt{\beta_1} = 0$). The $Z_{\sqrt{b_1}}$ formula works well for $N \geq 8$. If $N \geq 150$, $Z_{\sqrt{b_1}}$ can be set to a_1. The right-hand side presents the steps to form an approximately $N(0, 1)$ test statistic Z_{b_2} for kurtosis (H_0: $\beta_2 - 3 = 0$). This works best when $N \geq 20$. The simpler calculation of $Z_{(b_2-3)} = c_1$ is adequate only for very large samples ($N \geq 1000$). An "omnibus" test statistic of the joint hypothesis of skewness and kurtosis equal to that of a normal distribution (i.e., H_0: $\sqrt{\beta_1} = 0$ and $\beta_2 - 3 = 0$) is

$$K^2 = Z_{\sqrt{b_1}}^2 + Z_{b_2}^2 \tag{9.77}$$

The K^2 statistic approximates a chi-square distribution with 2 df. D'Agostino (1986, 391) recommends $N \geq 100$ for this test.

I apply these procedures to the per capita GNP and energy consumption variables:

	x_1	$\ln x_1$	x_2	$\ln x_2$
$Z_{\sqrt{b_1}}$	4.98	0.98	6.84	−1.32
Z_{b_2}	3.28	−1.54	5.29	−0.88
K^2	35.6	3.32	74.7	2.52

At any conventional significance level I reject the hypotheses of zero skewness and a kurtosis of 3 for x_1 and x_2 for all test statistics, but not for the $\ln x_1$ and the $\ln x_2$. Since $N < 100$, the K^2 test should be interpreted cautiously, but its results are consistent with the other ones. As in all hypothesis tests, these do not "prove" that the $\ln x_1$ and $\ln x_2$ have normal distributions. Rather, they show that their skewnesses and kurtoses are not statistically different from that of normal distributions, though the same is not true of x_1 and x_2. Thus normal distributions for the transformed variables may be reasonable approximations, though the untransformed variables appear nonnormal.

We usually are concerned about *multinormality* or multivariate kurtosis or skewness, while the preceding tests are *univariate* ones. These have merit in that they pinpoint variables that sharply deviate from a normal distribution. Furthermore, if the observed variables have a multinormal distribution, then the marginal distributions of each observed variable should have the kurtosis and skew of a normal variable. If any fail to, the multivariate distribution cannot be multinormal. One possible check for multinormality or for excessive kurtosis or skew is to test the marginal distribution of each observed variable with the preceding tests. Since multiple tests are performed, a Bonferroni adjustment to the overall level of significance can be made by using $\alpha/(p + q)$ for the probability level of each test, where α is the overall level of significance (D'Agostino 1986, 409). Thus for ten observed variables and the 0.05 overall level of significance, each test would be made at the 0.005 level. A limitation of this procedure is that it is possible for variables to have normal marginal distributions but not to be multinormally distributed.

Mardia (1970, 1974, 1985) proposed *multivariate* tests of skewness and kurtosis that can address this problem. His sample measure of skewness, $b_{1,p}$ is

$$b_{1,p} = \left(\frac{1}{N^2}\right) \sum_{i=1}^{N} \sum_{j=1}^{N} \left\{(\mathbf{z}_i - \bar{\mathbf{z}})' \mathbf{S}^{-1} (\mathbf{z}_j - \bar{\mathbf{z}})\right\}^3 \tag{9.78}$$

where $\mathbf{z}_i$ is a column vector of the values for all variables for the ith observation, $\mathbf{z}_j$ is the same for the jth observation, and $\bar{\mathbf{z}}$ is the corresponding column vector of sample means. The sample measure of multivariate kurtosis is

$$b_{2,p} = \frac{1}{N} \sum_{i=1}^{N} \left\{(\mathbf{z}_i - \bar{\mathbf{z}})' \mathbf{S}^{-1} (\mathbf{z}_i - \bar{\mathbf{z}})\right\}^2 \tag{9.79}$$

Mardia provides several test statistics based on $b_{1,p}$ and $b_{2,p}$. Table 9.3 presents $W(b_{1,p})$ and $W(b_{2,p})$ that asymptotically have standardized normal distributions and compare favorably with other possible test statistics (Mardia 1974; Mardia and Foster 1983; Mardia 1985). An omnibus test for the joint hypothesis of no multivariate skew or excess kurtosis is

$$K^2 = W(b_{1,p})^2 + W(b_{2,p})^2 \tag{9.80}$$

which approximates a chi-square distribution with 2 df.

The eight indicators of political democracy used in a CFA in Chapter 7 have $W(b_{1,p}) = 3.99$ ($p < 0.001$), $W(b_{2,p}) = 0.17$ ($p = 0.86$), and $K^2 = 15.99$ ($p < 0.001$). The results suggest a nonnormal multivariate distribution.[4] The five subjective measures of line length (see Chapter 7) have $W(b_{1,p}) = 4.45$ ($p < 0.001$), $W(b_{2,p}) = 2.29$ ($p = 0.02$), and $K^2 = 25.04$ ($p < 0.001$). Little univariate skewness was evident but significant negative univariate kurtosis was present for each variable. These results raise ques-

[4]I consider the impact of the ordinal nature of these indicators later in this chapter.

Table 9.3 Test Statistics for Multivariate Skewness, H_0: $\beta_{1,p} = 0$ or Kurtosis, H_0: $\beta_{2,p} = p(p+2)$, where p is the Number of Variables

Skewness	Kurtosis
H_0: $\beta_{1,p} = 0$	H_0: $\beta_{2,p} = p(p+2)$
Compute:	*Compute:*
(1) $b_{1,p}$	(1) $b_{2,p}$
(2) $W(b_{1,p}) = [12p(p+1)(p+2)]^{-1/2} \times \left\{ \left[27Np^2(p+1)^2(p+2)^2 b_{1,p} \right]^{1/3} - 3p(p+1)(p+2) + 4 \right\}$	(2) $E(b_{2,p}) = \dfrac{(N-1)p(p+2)}{N+1}$
	(3) $\operatorname{var}(b_{2,p}) = 8p(p+2)N^{-1}$
	(4) $\operatorname{stnd}(b_{2,p}) = \dfrac{b_{2,p} - E(b_{2,p})}{\left[\operatorname{var}(b_{2,p})\right]^{1/2}}$
	(5) $f_1 = 6 + \left[8p(p+2)(p+8)^{-2}\right]^{1/2} N^{1/2} \left\{ \left[\left(\frac{1}{2}\right)p(p+2)\right]^{1/2} \times (p+8)^{-1}N^{1/2} + \left[1 + \frac{1}{2}p(p+2)(p+8)^{-2}N\right]^{1/2} \right\}$
	(6) $W(b_{2,p}) = 3\left(\dfrac{f_1}{2}\right)^{1/2} \left[1 - \left(\dfrac{2}{9f_1}\right) - \left(\dfrac{1 - 2/f_1}{1 + \operatorname{stnd}(b_{2,p})\left[2/(f_1 - 4)\right]^{1/2}} \right)^{1/3} \right]$

Sources: For skewness, Mardia (1985, 219); for Kurtosis, Mardia and Foster (1983).

tions about the accuracy of the test statistics from F_{ML} or F_{GLS} applied to these variables, an issue that I will return to later.

Correction

When nonnormality or excessive kurtosis threatens the validity of the ML or GLS significance tests, we have several possible corrections. One is to find transformations of the variables that lead them to better approximate multinormality or that remove the excess kurtosis. With successful transformations we can apply F_{GLS} or F_{ML} to the transformed data and proceed as usual. Second, Browne (1982, 1984) provides adjustments to the usual test statistics and standard errors so that modified significance tests from F_{ML}, F_{GLS}, and even F_{ULS} results are asymptotically correct. (See also Arminger and Schoenberg 1987 on robust chi-square tests and asymptotic standard errors.) Third, it may be possible to employ bootstrap resampling procedures to form nonparametric significance tests (Bollen and Stine 1987). However, unless we have special models or limit the number of iterations, the bootstrap method coupled with fully iterative solutions may be too time-consuming to be practical. Also neither the second nor the third solution corrects the lack of asymptotic efficiency of $\hat{\boldsymbol{\theta}}$ from F_{MLS}, F_{GLS}, or F_{ULS}.

The fourth approach is to employ an alternative estimator that allows for nonnormality and is asymptotically efficient. The weighted least squares (WLS) estimator is one such method. The purpose of this section is to provide an overview of the WLS estimator and a special case of it, the elliptical estimator.

Weighted Least Squares Estimators

The WLS fitting function is

$$F_{WLS} = [\mathbf{s} - \boldsymbol{\sigma}(\boldsymbol{\theta})]'\mathbf{W}^{-1}[\mathbf{s} - \boldsymbol{\sigma}(\boldsymbol{\theta})] \tag{9.81}$$

where $\mathbf{s}$ is a vector of $\frac{1}{2}(p+q)(p+q+1)$ elements obtained by placing the nonduplicated elements of $\mathbf{S}$ in a vector, $\boldsymbol{\sigma}(\boldsymbol{\theta})$ is the corresponding same-order vector of $\boldsymbol{\Sigma}(\boldsymbol{\theta})$, and $\boldsymbol{\theta}$ is the $t \times 1$ vector of free parameters. The $\mathbf{W}^{-1}$ is $\frac{1}{2}(p+q)(p+q+1) \times \frac{1}{2}(p+q)(p+q+1)$ positive-definite weight matrix.

Values of $\boldsymbol{\theta}$ are selected so as to minimize the weighted sum of squared deviations of $\mathbf{s}$ from $\boldsymbol{\sigma}(\boldsymbol{\theta})$. This is analogous to weighted least squares regression analysis where a weighted, squared difference between the observed and predicted dependent variable is minimized by the selection of regression coefficients. Here the observed and predicted values are covariances rather than individual values.

As with F_{ML}, F_{GLS}, and F_{ULS}, the $\hat{\boldsymbol{\theta}}$ from F_{WLS} is a consistent estimator of $\boldsymbol{\theta}$ when $\boldsymbol{\Sigma} = \boldsymbol{\Sigma}(\boldsymbol{\theta})$ under very general conditions. A remarkable result of Browne (1982, 1984) is that if $\mathbf{W}$ is chosen to equal or to be a consistent estimator of the asymptotic covariance matrix of $\mathbf{s}$ with $\mathbf{s}$, then the $\hat{\boldsymbol{\theta}}$ from F_{WLS} is asymptotically efficient within the class of functions that fall under (9.81). Furthermore a correct ACOV($\hat{\boldsymbol{\theta}}$) and chi-square test are available. The chi-square estimator is $(N-1)F_{\text{WLS}}$ at $\hat{\boldsymbol{\theta}}$. The asymptotic covariance matrix is

$$N^{-1}\left[\left(\frac{\partial \boldsymbol{\sigma}(\boldsymbol{\theta})}{\partial \boldsymbol{\theta}}\right)\boldsymbol{\Sigma}_{ss}^{-1}\left(\frac{\partial \boldsymbol{\sigma}(\boldsymbol{\theta})}{\partial \boldsymbol{\theta}}\right)'\right]^{-1} \tag{9.82}$$

where $\boldsymbol{\Sigma}_{ss}$ is the covariance matrix of the limiting distribution of $\sqrt{N}\mathbf{s}$.

Again an analogy to weighted least squares regression is evident. Many weight matrices could be selected without destroying the consistency of the weighted least squares regression estimator. The optimal choice, however, is the covariance matrix of the disturbance terms, whereas in (9.81) the optimal choice is the covariance matrix of the sample covariances.

In general, the asymptotic covariance of s_{ij} with s_{gh} is

$$\text{ACOV}(s_{ij}, s_{gh}) = N^{-1}(\sigma_{ijgh} - \sigma_{ij}\sigma_{gh}) \tag{9.83}$$

where σ_{ijgh} is, $E(X_i - \mu_i)(X_j - \mu_j)(X_g - \mu_g)(X_h - \mu_h)$, the fourth-order moment around the mean, and σ_{ij} and σ_{gh} are the population covariances of X_i with X_j and X_g with X_h, respectively. Note that (9.83) holds under general conditions, without specifying a particular distribution as long as all eighth-order moments of the observed variables' distribution are finite. The WLS fitting function with the optimal $\mathbf{W}$ is sometimes called the *arbitrary distribution function* (ADF), or the *distribution-free estimator*, because of this property. Each of the elements of $\mathbf{W}$ in F_{WLS} is an estimate of (9.83). A sample estimator of fourth-order element, σ_{ijgh}, is

$$s_{ijgh}^{*} = \frac{1}{N}\sum_{t=1}^{N}\left(Z_{it} - \bar{Z}_i\right)\left(Z_{jt} - \bar{Z}_j\right)\left(Z_{gt} - \bar{Z}_g\right)\left(Z_{ht} - \bar{Z}_h\right) \tag{9.84}$$

and the σ_{ij} and σ_{gh} estimators are

$$s_{ij}^* = \frac{1}{N} \sum_{t=1}^{N} (Z_{it} - \bar{Z}_i)(Z_{jt} - \bar{Z}_j) \tag{9.85a}$$

$$s_{gh}^* = \frac{1}{N} \sum_{t=1}^{N} (Z_{gt} - \bar{Z}_g)(Z_{ht} - \bar{Z}_h) \tag{9.85b}$$

The s_{ijgh}^* and s_{ij}^* and s_{gh}^* are marked with asterisks to indicate that division is by N rather than by $(N - 1)$, which is typical for the unbiased sample covariances.[5]

I illustrate the F_{WLS} fitting function in its ADF form with the eight indicators of political democracy that failed the univariate and multinormality tests in the last section. Table 9.4 lists a subset of the estimates, standard errors, and the chi-square value from F_{GLS} on the left-hand side and those from F_{WLS} on the right-hand side. The biggest difference in the parameter estimates is that most of the ones from F_{GLS} are larger than the ADF ones from F_{WLS}, whereas the standard errors from F_{WLS} tend to be less than those from F_{GLS}. The largest discrepancy is the bigger chi-square estimate for F_{WLS}.

Specializations of F_{WLS}

If the observed variables have a multinormal distribution, then σ_{ijgh} equals $\sigma_{ij}\sigma_{gh} + \sigma_{ig}\sigma_{jh} + \sigma_{ih}\sigma_{jg}$, and equation (9.83) simplifies to

$$\text{ACOV}(s_{ij}, s_{gh}) = N^{-1}(\sigma_{ig}\sigma_{jh} + \sigma_{ih}\sigma_{jg}) \tag{9.86}$$

This expression only involves products of covariances rather than the fourth-order moment as in (9.82). The optimal weight matrix $\mathbf{W}$ would consist of products of covariances in the form of (9.86), and Browne

[5]Browne (1984, 72) provides unbiased estimators for $(\sigma_{ijgh} - \sigma_{ij}\sigma_{gh})$ of

$$\frac{N(N-1)}{(N-2)(N-3)}(s_{ijgh}^* - s_{ij}^* s_{gh}^*)$$
$$- \left[\frac{N}{(N-2)(N-3)}\right]\left(s_{ig}^* s_{jh}^* + s_{ih}^* s_{jg}^* - \left(\frac{2}{N-1}\right)(s_{ij}^* s_{gh}^*)\right).$$

For large samples, this adjustment should not make too great a difference. The $\mathbf{W}$ using these as elements may not be positive-definite, though the chances of this occurring diminish as N becomes substantially larger than $\frac{1}{2}(p + q)(p + q + 1)$.

Table 9.4 Comparison of F_{GLS} and F_{WLS} Estimates for Eight Indicators of Political Democracy, 1960 and 1965

	Estimate (standard error)	
Parameter	F_{GLS}	F_{WLS}
λ_{11} $(=\lambda_{52})$	1.00	1.00
	(—)	(—)
λ_{21} $(=\lambda_{62})$	1.12	1.11
	(0.14)	(0.11)
λ_{31} $(=\lambda_{72})$	1.22	1.05
	(0.13)	(0.09)
λ_{41} $(=\lambda_{82})$	1.21	1.16
	(0.13)	(0.09)
VAR(ϵ_1)	1.64	1.30
	(0.42)	(0.34)
VAR(ϵ_2)	7.10	7.12
	(1.38)	(1.11)
VAR(ϵ_3)	3.80	3.53
	(0.95)	(0.78)
VAR(ϵ_4)	3.13	2.72
	(0.73)	(0.63)
VAR(ϵ_5)	2.47	2.29
	(0.53)	(0.57)
VAR(ϵ_6)	3.91	3.77
	(0.88)	(0.68)
VAR(ϵ_7)	3.15	2.60
	(0.69)	(0.44)
VAR(ϵ_8)	3.01	3.45
	(0.71)	(0.73)
χ^2	14.31	25.53
df	16	16
prob	0.58	0.06

(1974, 209–210) shows that the F_{WLS} function (9.81) equals

$$\tfrac{1}{2}\operatorname{tr}\{[\mathbf{S} - \boldsymbol{\Sigma}(\boldsymbol{\theta})]\mathbf{V}^{-1}\}^2 \tag{9.87}$$

where $\mathbf{V}^{-1}$ is a $(p + q) \times (p + q)$ weight matrix. (I use $\mathbf{V}$ in place of $\mathbf{W}$ since the dimensions of these weight matrices differ.) Equation (9.87) is the general form of the GLS estimator introduced in chapter four. With suitable choices of $\mathbf{V}$ ($= \mathbf{S}$, $\hat{\boldsymbol{\Sigma}}$, or $\mathbf{I}$) we get F_{GLS}, F_{MLS}, or F_{ULS}. Thus these three fitting functions are specializations of the more general F_{WLS} function of (9.81).

Elliptical Estimator

Another specialization of F_{WLS} follows when the observed variables have a multivariate elliptical distribution. The general density form for an elliptical distribution is

$$C|\mathbf{V}|^{-1/2}h\left[(\mathbf{z}-\boldsymbol{\mu})'\mathbf{V}^{-1}(\mathbf{z}-\boldsymbol{\mu})\right] \tag{9.88}$$

where C is a constant, h is a nonnegative function, $\mathbf{z}$ is the vector of observed variables with means of $\boldsymbol{\mu}$, and $\mathbf{V}$ is a positive definite matrix. The elliptical distribution includes a number of other distributions as special cases. For instance, if C is $(2\pi)^{(-1/2)(p+q)}$, $\mathbf{V}$ is $\boldsymbol{\Sigma}$, and the function $h(.)$ is $\exp[(-\frac{1}{2})(\mathbf{z}-\boldsymbol{\mu})'\boldsymbol{\Sigma}^{-1}(\mathbf{z}-\boldsymbol{\mu})]$, then (9.88) becomes

$$(2\pi)^{(-1/2)(p+q)}|\boldsymbol{\Sigma}|^{-1/2}\exp\left[\left(-\tfrac{1}{2}\right)(\mathbf{z}-\boldsymbol{\mu})'\boldsymbol{\Sigma}^{-1}(\mathbf{z}-\boldsymbol{\mu})\right] \tag{9.89}$$

which is the familiar density for multinormal variables. Other examples of elliptical distributions include the multivariate t-distribution and the contaminated normal distribution. Elliptical distributions have zero skewness but can have kurtosis that deviates from the multinormal. They can have positive or negative kurtosis as long as all the variables have a common degree of kurtosis of

$$K = \frac{\sigma_{iiii}}{3\sigma_{ii}^2} - 1 \tag{9.90}$$

where K is the common kurtosis parameter, σ_{iiii} is the fourth-order moment around the mean, and σ_{ii} is the second-order moment around the mean (i.e., the variance). This means that univariate kurtosis measures for each variable should be equal within sampling error. [Berkane and Bentler 1987 describe a significance test for this hypothesis.] Also K is zero if the distribution is multinormal.

To explain the fitting function for variables with elliptical distributions, consider the weighted least squares function in (9.81). For an elliptical distribution the fourth-order moment equals:

$$\sigma_{ijgh} = (K+1)(\sigma_{ij}\sigma_{gh} + \sigma_{ig}\sigma_{jh} + \sigma_{ih}\sigma_{jg}) \tag{9.91}$$

Combining (9.83) with (9.91) leads to

$$\text{ACOV}(s_{ij}, s_{gh}) = N^{-1}\left[(K+1)(\sigma_{ig}\sigma_{jh} + \sigma_{ih}\sigma_{jg}) + K(\sigma_{ij}\sigma_{gh})\right] \tag{9.92}$$

It follows that the optimal weight matrix for F_{WLS} when the variables have an elliptical distribution are functions of N, the covariances, and the common kurtosis parameter, K.

Browne (1984, 74–75) shows that for elliptical distributions with optimal weight matrices the F_{WLS} of (9.81) may be written as

$$F_E = \tfrac{1}{2}(K+1)^{-1}\text{tr}\left[(\mathbf{S} - \boldsymbol{\Sigma}(\boldsymbol{\theta}))\mathbf{V}^{-1}\right]^2 - C_1\left[\text{tr}(\mathbf{S} - \boldsymbol{\Sigma}(\boldsymbol{\theta}))\mathbf{V}^{-1}\right]^2 \tag{9.93}$$

where C_1 is a constant that equals $K/[4(K+1)^2 + 2(p+q)K(K+1)]$, $\mathbf{V}$ is the weight matrix, and F_E stands for the elliptical fitting function. Equation (9.93) is similar to the general GLS fitting function [see (9.87)]. In fact, if the distributions have no kurtosis (i.e., $K = 0$), C_1 is zero, $(K+1)^{-1}$ is one, and the fitting function is the same as equation (9.87). Thus the F_{GLS}, F_{ULS}, and F_{ML} also are special cases of (9.93) where the kurtosis parameter, K, is zero. Selecting $\mathbf{V}$ to be a consistent estimator of $\boldsymbol{\Sigma}$ leads to an asymptotic efficient estimator of $\boldsymbol{\theta}$ with $(N-1)\ F_E$ at $\hat{\boldsymbol{\theta}}$ asymptotically distributed as a chi-square variate. The two most common selections for $\mathbf{V}$ are $\mathbf{S}$ or $\hat{\boldsymbol{\Sigma}}$.

To implement the elliptical fitting function requires an estimate of the kurtosis parameter, K. One possibility is

$$\hat{K}_1 = \frac{b_{2,(p+q)} - (p+q)(p+q+2)}{(p+q)(p+q+2)} \tag{9.94}$$

where $b_{2,(p+q)}$ is Mardia kurtosis statistic for $(p+q)$ observed variables [see (9.79)]. Another is to average the b_2 univariate kurtosis estimates (9.75) for each variable as

$$\hat{K}_2 = [3(p+q)]^{-1}\sum_{i=1}^{p+q} b_{2i} \tag{9.95}$$

Equation (9.95) follows since if the variables have a multivariate elliptical distribution, the kurtosis values should be the same. Bentler (1985, 50–51) provides several other estimators of K, but the two just given should be adequate for most problems.

None of the empirical examples in the previous chapters had significant and equal kurtosis combined with no skewness for the observed variables. However, the line length example from Chapter 7 did show significant multivariate kurtosis with near-equal univariate kurtosis statistics (about -1.1). I use this to illustrate the elliptical estimator. The five judges' assessments of the length of 60 lines are labeled x_1 to x_5. Table 9.5 reports the estimates of a CFA of x_1 to x_5, all loading on one factor with no correlated errors of measurement. It includes the estimates from F_{GLS} and

Table 9.5 Generalized Least Squares (GLS) and Elliptical Generalized Least Squares (EGLS) for Five Estimates (x_1 to x_5) of Line Length $N = 60$

	Estimate (standard error)	
Parameter	F_{GLS}	F_E
λ_{11}	3.48	3.48
	(0.11)	(0.11)
λ_{21}	1.09	1.09
	(0.05)	(0.05)
λ_{31}	1.87	1.87
	(0.06)	(0.07)
λ_{41}	1.13	1.13
	(0.04)	(0.04)
λ_{51}	1.00	1.00
	(—)	(—)
ϕ_{11}	1.15	1.16
	(0.22)	(0.24)
VAR(δ_1)	0.22	0.22
	(0.06)	(0.07)
VAR(δ_2)	0.09	0.09
	(0.02)	(0.02)
VAR(δ_3)	0.16	0.16
	(0.03)	(0.04)
VAR(δ_4)	0.02	0.02
	(0.01)	(0.01)
VAR(δ_5)	0.04	0.04
	(0.01)	(0.01)
χ^2	3.26	2.85
df	5	5
prob	0.66	0.72

from F_E with $\mathbf{V} = \mathbf{S}$. The estimates from F_{ML} are essentially the same and are not shown. The estimates from F_{GLS} and F_E are extremely close. The chi-square estimate is slightly lower for F_E (= 2.85) than for F_{GLS} (= 3.26), whereas the standard errors tend to be slightly higher for F_E. Overall, for this example the elliptical estimator makes little difference even though the variables show considerable kurtosis.

Summary of F_{WLS}

The WLS (arbitrary distribution function) estimator has several advantages as well as drawbacks. Chief among its advantages is that it makes minimal assumptions about the distribution of the observed variables. Provided the weight matrix is chosen so that its elements contain consistent estimators of the asymptotic covariances of s_{ij} with s_{gh}, it is an efficient estimator which provides an asymptotic covariance matrix of $\hat{\boldsymbol{\theta}}$ and a chi-square test statistic. The F_{WLS} applies also to the analysis of sample correlation matrices as long as the weight matrix reflects the asymptotic covariance of the correlations r_{ij} with r_{gh}. This is not the same weight matrix as that for s_{ij} with s_{gh} (see de Leeuw 1983).

One disadvantage of F_{WLS} is computational. In its asymptotic distribution-free form, it requires the inversions of a $\frac{1}{2}(p+q)(p+q+1)$ square matrix. With a moderate size problem of 10 observed variables, $\mathbf{W}$ is a 55×55 matrix to invert. For models with many more variables, F_{WLS} becomes less feasible.[6] If we can assume that the distribution has only minor kurtosis or is elliptical, then we can turn to other estimators that are less demanding, computationally. The F_{ML}, F_{GLS}, F_{ULS}, and F_E fitting functions are simpler, special cases of F_{WLS}.

Also a concern with F_{WLS} is that its sample size requirements for convergence seem larger than those for the F_{ML}, F_{GLS}, or F_{ULS}. Not much is known about F_{WLS}'s small or moderate sample properties. So the results of models estimated with F_{WLS} in less than large samples require cautious interpretation. The political democracy CFA ($N = 75$) is an example, though no serious problems were obvious in the results.

Finally, it is not clear that F_{WLS} or F_E outperforms F_{ML}, F_{GLS}, or F_{ULS} in the cases where only moderate nonnormality is present. The problem is knowing when the nonnormality is severe enough to require F_{WLS}. Until more is learned about the robustness of the ML or GLS estimators, the prudent strategy is to compare the results of these estimators to those of WLS when nonnormality or significant kurtosis is a problem.

[6] Bentler and Dijkstra (1985) suggest "linearized estimators" and tests based on one-step improvements over initial consistent estimators to lessen the computations.

CATEGORICAL OBSERVED VARIABLES

Throughout the book I have assumed that the observed and latent variables are continuous. Strictly speaking, this assumption is always violated for the observed variables due to the limits of measuring instruments. For instance, if a ruler measures only to the nearest quarter of an inch, we will group measures of 1.30 in., 1.21 in., 1.33 in., and 1.19 in. into a single category of $1\frac{1}{4}$ in. The scale jumps in one-quarter inch intervals, whereas the latent length variable has more detailed calibrations. In many applications this categorical measure of length would cause little difficulty since the observed variable still assumes a large number of categories. In other situations where we require fine gradations, this crudeness of scale would not be tolerable. More coarsely categorized measures are common in the social sciences. Examples include education grouped into elementary, high school, and college education or income set into four or five ranks. Likert attitude scales with five or seven scale points are another frequent type of observed variables in social science research. In these and other examples the crudeness of the measure is considerable, and the question is the degree to which this produces misleading findings.

In this section I examine (1) the model assumptions violated with categorical indicators, (2) the consequences of the violations, and (3) some corrective procedures. There is a vast literature on categorical variables (see, e.g., Lazarsfeld and Henry 1968; Goodman 1972; Bishop, Fienberg, and Holland 1975; Amemiya 1981; Maddala 1983). I restrict the discussion to continuous latent variables with dichotomous or ordinal indicators and to simple extensions of the structural equation model.

Assumptions Violated and Consequences

Consider $\mathbf{y}^*$, a $p \times 1$ vector of continuous indicators of $\boldsymbol{\eta}$ which conforms to the usual measurement model:

$$\mathbf{y}^* = \boldsymbol{\Lambda}_y \boldsymbol{\eta} + \boldsymbol{\epsilon} \tag{9.96}$$

where $E(\boldsymbol{\epsilon}) = 0$, $\boldsymbol{\epsilon}$ is uncorrelated with $\boldsymbol{\eta}$, and $\boldsymbol{\eta}$ is in deviation form. We do not observe $\mathbf{y}^*$; instead, we have $\mathbf{y}$ with some, or possibly all, indicators in $\mathbf{y}$, categorical versions of $\mathbf{y}^*$. The y_1^*, for example, might be normally distributed, whereas y_1, is a four-category ordinal variable. If so, $\mathbf{y} \neq \mathbf{y}^*$ for at least some rows, and

$$\mathbf{y} \neq \boldsymbol{\Lambda}_y \boldsymbol{\eta} + \boldsymbol{\epsilon} \tag{9.97}$$

Thus one consequence is that the measurement model for $\mathbf{y}^*$ does not hold for $\mathbf{y}$. I refer to $\mathbf{y}^*$ as the latent *continuous* indicators, whereas $\mathbf{y}$ contains the observed indicators, some of which are ordinal versions of the corresponding $\mathbf{y}^*$ variables. Analogous arguments hold for $\mathbf{x}$ and $\mathbf{x}^*$.

A second consequence is that the distribution of the ordinal variables generally differs from that for the latent continuous indicators. The $\text{ACOV}(s_{ij}, s_{gh})$ is not likely to equal $\text{ACOV}(s_{ij}^*, s_{gh}^*)$, where s_{ij} and s_{gh} are elements of the covariance matrix of $\mathbf{y}$ and $\mathbf{x}$ and s_{ij}^* and s_{gh}^* are the corresponding elements for the covariance matrix of $\mathbf{y}^*$ and $\mathbf{x}^*$. Even when $\mathbf{y}^*$ and $\mathbf{x}^*$ are multinormal, the ordinal versions, $\mathbf{y}$ and $\mathbf{x}$, can be highly nonnormal. We can, of course, apply the ADF estimator (see last section) to correct this problem. However, ordinal variables also can create heteroscedastic disturbances or errors (Goldberger, 1964, 248–250) and the ADF estimator assumes homoscedasticity.

A more serious consequence of ordinal variables is the violation of the covariance structure hypothesis. Assume that for $\boldsymbol{\Sigma}^*$, the population covariance matrix of $\mathbf{y}^*$ and $\mathbf{x}^*$, that $\boldsymbol{\Sigma}^* = \boldsymbol{\Sigma}(\boldsymbol{\theta})$. In general, $\boldsymbol{\Sigma}$, the population covariance matrix of the categorical $\mathbf{y}$ and $\mathbf{x}$, does not equal $\boldsymbol{\Sigma}^*$ and $\boldsymbol{\Sigma} \neq \boldsymbol{\Sigma}(\boldsymbol{\theta})$. So the covariance structure hypothesis holds for the continuous latent indicators, but not necessarily for the ordinal observed indicators. Suppose that $\mathbf{S}$ is a consistent estimator of $\boldsymbol{\Sigma}$ and $\mathbf{S}^*$ is a consistent estimator of $\boldsymbol{\Sigma}^*$. The parameter estimator based on $\mathbf{S}$ and any one of the fitting functions (e.g., F_{ML}) is likely to be an inconsistent estimator of $\boldsymbol{\theta}$, the true parameter vector. For instance, in a simple regression equation where all variables are standardized to a mean of zero and a variance of one, the consistent estimator of the regression coefficient is the sample correlation, r^*, between y^* and x^*. The population counterpart to r^* is ρ^*. If y and x are the standardized ordinal versions of y^* and x^*, the regression coefficient is r, the sample correlation of y with x. Assuming that $\text{plim}(r)$ exists, it usually does not equal ρ^*, the true parameter. In the more general case the magnitude of the inconsistency depends on the relation of $\boldsymbol{\Sigma}$ to $\boldsymbol{\Sigma}^*$.

Robustness Studies

How robust are the model estimates and tests of statistical significance to these problems? Some Monte Carlo simulation work gives a rough idea of robustness. One set of studies compares correlations for ordinal variables to their continuous counterparts ($\boldsymbol{\Sigma}$ to $\boldsymbol{\Sigma}^*$). This provides information on the extent to which the covariance structure hypothesis is violated. Correlations are compared since the metric of the ordinal variables is usually arbitrary.

Work by Wylie (1976), Martin (1978), Bollen and Barb (1981), Olsson, Drasgow, and Dorans (1982), and others suggests that the Pearson correlation coefficients between categorized measures are generally less than the correlation of the corresponding continuous variables. The greatest attenuation occurs with few categories (e.g., < 5) for either categorical variable and with opposite skews for the categorical variables. As the number of categories increases and the marginal distributions become similar, the differences in correlations lessens and Σ comes closer to Σ^*.

This research is limited to analyzing correlations. Only a few robustness studies have directly addressed structural equation models with categorical variables. Olsson (1979a) analyzed a one-factor model with either six or 12 variables. The number of categories ranged from two to nine, the skewness of the observed variables was varied, and the factor loadings ranged from high to low. He found that the worst fit as measured by a "pseudo-χ^2" variate occurred with opposite skewed variables and with high factor loadings. The chi-square did not seem to be greatly affected by the number of categories. However, the fewer the scale points and the greater the differences in skewness, the more attenuation in factor loadings resulted.

Johnson and Creech (1983) simulated a structural equation model with three latent variables and with two normal indicators per latent variable. Both coefficients of their latent variable model were 0.6 or 0.4, whereas all lambdas were 0.6 or 0.9. Collapsed variables were grouped into two, three, four, five, 10, 20, or 36 categories and assigned consecutive integers. The cases were classified to approximate "normal" or uniform marginal distributions for the categorized measures. They found that correlated measurement errors were created in the model with categorized variables. These were most likely when the lambdas were high (0.9), the number of categories was small, and the distribution was uniform. The correlated errors were substantially less when the lambdas were 0.6. The analysis of the categorized indicators showed a tendency to underestimate the coefficients of the latent variable model, but by relatively small amounts under all conditions with most differences in standardized coefficients smaller (in absolute value) than 0.04. The largest differences tended to occur for two, three, or four categories. They did not report the chi-square or estimates of the lambdas for the categorized models.

Babakus, Ferguson, and Jöreskog (1987) examined a one-factor model with four normally distributed indicators. The continuous y^*'s were collapsed into five categories each with marginal distributions either U-shaped or with skews of 0, 0.5, or 1.5. They found that analyzing the Pearson correlation matrix generally led to underestimates of factor loadings and overestimates of their standard errors. Too large a chi-square value resulted from excessive skewness, but it was about right when the marginals were bell-shaped.

Boomsma (1983) analyzed the robustness of the ML estimator to nonnormal, discrete distributions. The categories for the observed variables ranged from two to five, the skewnesses were varied over several values, and he looked at several different models. An important difference of Boomsma's analysis compared to the previous ones is that the structural equation model holds for Σ [i.e., $\Sigma = \Sigma(\boldsymbol{\theta})$], the covariance matrix of the discrete variables $\mathbf{y}$ and $\mathbf{x}$, and it does not match Σ^* [i.e., $\Sigma^* \neq \Sigma(\boldsymbol{\theta})$], the covariance matrix of $\mathbf{y}^*$ and $\mathbf{x}^*$. Boomsma (1983) found no bias in the parameter estimates but inaccurate asymptotic standard errors. The chi-square showed only a small effect of categorization and low degrees of skewness. With high skewness, the chi-square tended to be too big.

Muthén and Kaplan (1985) compared ML, GLS, and ADF (WLS) estimators of factor analysis for nonnormal categorical variables. They specified a structural model that holds true for both the categorical variables [i.e., $\Sigma = \Sigma(\boldsymbol{\theta})$] and the continuous variables [i.e., $\Sigma^* = \Sigma(\boldsymbol{\theta})$]. They found that ML and GLS chi-square tests and estimated standard errors were quite robust, except when the observed variables had large skewnesses or kurtoses. Browne's ADF estimator performed well even in these cases. Muthén and Kaplan (1985) suggest that if variables have skewnesses and kurtoses from -1.0 to $+1.0$, not much distortion will occur if we use F_{ML} and F_{GLS}.

I illustrate some of the preceding findings with the perceived line length example. Recall that five judges estimated the length of 60 lines drawn on separate index cards. I treated the estimates of length from the judges as five indicators of perceived length. The ML estimates for these data are listed in the first column of Table 9.6. I then created three categories—short, medium, and long—in which to classify each judge's estimates. For each variable, the range was calculated, and it was divided into three equal intervals. Two types of categorical measures were assembled. The first scored each category by consecutive integers (1, 2, and 3). The second coded the midpoint value for each category and for each judge, and assigned the midpoints as the code for the categories. The covariance matrices of the collapsed variables were then analyzed with the ML fitting function. These estimates are in the last two columns of Table 9.6. A comparison of the "Rank" column with the first column shows the loss of the original metric for the lengths. The fact that the first judge's estimates are in different units (metric units) than the other four is no longer evident. However, the midpoint scoring of categories leads to estimates of λ_{ij} that are close to the ones for the original variables and that reflect the original units.

The midpoint categorical estimate for ϕ_{11} is a bit lower than the original one (0.94 vs. 1.15). The error variance estimates for the midpoint analysis generally are much larger than the original ones. For judge 1, for example, the error variance is about 0.8 with the midpoint ranked data, though it is about 0.2 with the original data. This suggests a tendency for categorical

Table 9.6 Comparison of Estimates from ML for Line Length Example for Original Measures from Five Judges, Three-Category Rank Measures, and Three-Category Midpoint Measures ($N = 60$)

	Estimate (standard error)		
Parameter	Original Variables	Rank (3 categories)	Midpoints (3 categories)
λ_{11}	3.48	1.00	3.65
	(0.11)	(0.06)	(0.22)
λ_{21}	1.09	0.99	1.23
	(0.05)	(0.06)	(0.07)
λ_{31}	1.87	0.98	1.78
	(0.07)	(0.07)	(0.13)
λ_{41}	1.13	1.07	1.10
	(0.04)	(0.06)	(0.06)
λ_{51}	1.00	1.00	1.00
	(—)	(—)	(—)
ϕ_{11}	1.15	0.62	0.94
	(0.22)	(0.13)	(0.20)
VAR(δ_1)	0.23	0.04	0.81
	(0.06)	(0.01)	(0.18)
VAR(δ_2)	0.09	0.04	0.10
	(0.02)	(0.01)	(0.02)
VAR(δ_3)	0.16	0.11	0.55
	(0.03)	(0.02)	(0.11)
VAR(δ_4)	0.02	0.01	0.01
	(0.01)	(0.01)	(0.01)
VAR(δ_5)	0.05	0.09	0.13
	(0.01)	(0.02)	(0.03)
χ^2	4.0	21.2	21.2
df	5	5	5
prob	0.549	< 0.001	< 0.001

measures to lead to larger error variances than their continuous counterparts. Generally, the standard errors are larger for the midpoint estimates than for the original data.

An additional consequence of categorizing is correlated measurement errors. The Lagrangian multiplier (LM) (see Chapter 7) for the original variables shows no strong indication of correlated errors. In contrast, for the midpoint categorical data, very large first-order LM statistics appear for the off-diagonal elements of the error covariance matrix. This finding is consistent with the simulation work by Johnson and Creech (1983) who found correlated errors being created in models with categorical measures.

Finally, the chi-square estimate of 21.2 with 5 df is considerably larger than the chi-square of 4.0 for the original variables. At first this seems to contradict Olsson's (1979) work which found that the number of scale points has little effect on chi-square estimates. This is particularly true since the marginal skewnesses of the original and categorized variables are relatively low (max $\sqrt{b_1} \approx 0.15$).

However, Olsson (1979a) took little account of the influence of kurtosis on the chi-square estimates. As shown earlier, the kurtoses of the original variables are lower than a normal distribution. Categorization increased the kurtoses of these variables. The normalized Mardia coefficient of kurtosis is 17.2 for the rank and midpoint variables, whereas it is 2.3 for the original measures. It appears that by increasing the kurtosis of categorical variables, the chi-square estimate can be increased relative to the original measures. To explore this possibility, I created a new set of ranks designed to have less negative kurtosis (marginals roughly 20%, 60%, 20%). This model has a chi-square estimate of 6.1 that is not significant even at the 0.25 level. It also is far closer to the original model's chi-square estimate of 4.06. Thus it does appear to be the kurtosis of the variables rather than the number of scale points that leads to the higher chi-square estimates.

Summary of Robustness to Categorization

Analysis of the robustness of structural equation techniques of categorical variables is at an early stage of development. The evidence to date suggests several consequences of treating ordinal indicators as if they were continuous. First, excessive kurtosis and skewness in ordinal versions of normal latent indicators adversely affect the chi-square and z-tests of statistical significance from F_{ML} or F_{GLS} results. Second, the chi-square estimate seems to be more strongly influenced by the kurtosis and skewness of the ordinal variables than by their number of categories. Third, when the covariance structure hypothesis holds for the latent continuous indicators $[\Sigma^* = \Sigma(\theta)]$ but not for the ordinal variables $[\Sigma \neq \Sigma(\theta)]$, the standardized coefficient estimates appear to be attenuated.[7] The attenuation is inversely related to the number of categories, being greatest with two or three categories and much less with seven or more. Fourth, having few categories with large initial factor loadings makes correlated measurement errors more

[7]This pattern was not evident in the *unstandardized* loadings for the midpoint scoring of the line length example where a systematic bias in the factor loadings was absent.

likely for the model with ordinal indicators. Finally, these findings must be regarded as preliminary until researchers determine whether they generalize to a more diverse set of models under a broader range of conditions. For instance, we know little about robustness when the latent continuous indicators are nonnormal.

Corrective Procedures

The last section described several consequences of using ordinal indicators in place of latent continuous indicators:

$$\begin{aligned} \mathbf{y} &\neq \mathbf{\Lambda}_y \boldsymbol{\eta} + \boldsymbol{\epsilon} \\ \mathbf{x} &\neq \mathbf{\Lambda}_x \boldsymbol{\xi} + \boldsymbol{\delta} \end{aligned} \tag{9.98}$$

$$\mathbf{\Sigma} \neq \mathbf{\Sigma}(\boldsymbol{\theta}) \tag{9.99}$$

$$\operatorname{ACOV}(s_{ij}, s_{gh}) \neq \operatorname{ACOV}(s_{ij}^*, s_{gh}^*) \tag{9.100}$$

Any corrective procedures should address each of these problems. Starting with (9.98), we see that the linear model relating $\mathbf{y}$ ($\mathbf{x}$) to $\boldsymbol{\eta}$ ($\boldsymbol{\xi}$) is no longer appropriate. A nonlinear function is required to relate the observed ordinal indicators ($\mathbf{y}$ and $\mathbf{x}$) to the latent continuous ones ($\mathbf{y}^*$ and $\mathbf{x}^*$). An illustration for an ordinal indicator y_1 is

$$y_1 = \begin{cases} 1, & \text{if } y_1^* \leq a_1 \\ 2, & \text{if } a_1 < y_1^* \leq a_2 \\ \vdots & \vdots \\ c-1, & \text{if } a_{c-2} < y_1^* \leq a_{c-1} \\ c, & \text{if } a_{c-1} < y_1^* \end{cases} \tag{9.101}$$

where c is the number of categories for y_1, a_i ($i = 1, 2, \ldots, c-1$) is the category threshold, and y_1^* is the latent continuous indicator that determines the values of y_1 as it crosses different thresholds. Figure 9.4 illustrates the threshold model for an ordinal variable y_1 that has three categories and two thresholds with y_1^*, the latent continuous indicator. If y_1^* is less than a_1, y_1 is in category one, for $a_1 < y_1^* \leq a_2$, y_1 is in category two, and if y_1^* exceeds a_2, y_1 is in category three. A similar nonlinear relation holds for the other ordinal y's and x's. If y_i or x_j are continuous, the simple relation is $y_i = y_i^*$ or $x_j = x_j^*$, and these variables do not require threshold models.

For ordinal (including dichotomous) indicators we must determine the thresholds. We can estimate them if we know y_i^*'s and x_j^*'s distributions

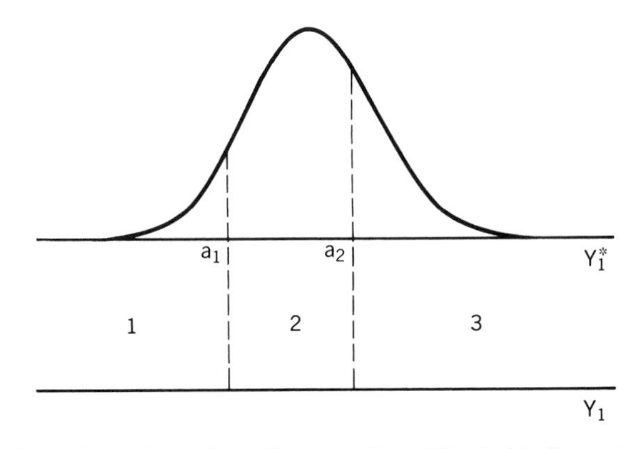

Figure 9.4 Illustration of a Three Category, Two Threshold Y_1 and an Underlying Y_1^* Variable

and the sample proportion of cases in each category of y_i and x_j. The most common assumption is that y^* and x^* are multinormally distributed. If so, the marginal distributions of these variables are normal. Since the scales of $\mathbf{y}^*$ and $\mathbf{x}^*$ that correspond to ordinal variables are arbitrary, we can standardize each to a mean of zero and a variance of one. An estimate of the threshold is

$$a_i = \Phi^{-1}\left(\sum_{k=1}^{i} \frac{N_k}{N}\right), \qquad i = 1, 2, \ldots, c-1 \tag{9.102}$$

where $\Phi^{-1}(.)$ is the inverse of the standardized normal distribution function, N_k is the number of cases in the kth category, and c is the total number of categories for y or x.

As an illustration, consider a categorized version of the 1960 free press measure (e.g., y_1) used in the model of industrialization and political democracy for developing societies.[8] The marginal frequencies, proportions, and cumulative proportions of free press ($N = 75$) are

Category	**1**	**2**	**3**	**4**	**5**	**6**	**7**	**8**
Frequency	8	13	5	13	5	22	4	5
Proportion	0.1067	0.1733	0.0667	0.1733	0.0667	0.2933	0.0533	0.0667
Cumulative Proportion	0.1067	0.2800	0.3467	0.5200	0.5867	0.8800	0.9333	1.000

[8]I collapsed the original variable into the eight categories shown here.

Referring to a table for a standardized normal variable, we see that these cumulative proportions correspond to the following thresholds:

Threshold	a_1	a_2	a_3	a_4	a_5	a_6	a_7
Estimate	−1.24	−0.58	−0.39	0.05	0.22	1.17	1.50

If the latent continuous free press indicator, y_1^*, is -0.53, y_1 is in category three; a y_1^* of 0.79 means y_1 of six; and so on. We can estimate the thresholds for all ordinal indicators in the same fashion. The marginal distributions of the ordinal variables can have kurtosis or skew since we do not assume that the ordinal variables are normal, even though we do assume normality for the latent continuous indicators. Of course, if the latent continuous indicators have nonnormal distributions, then the thresholds would be different.

In sum, to correct the problem that $\mathbf{y} \neq \mathbf{\Lambda}_y \boldsymbol{\eta} + \boldsymbol{\epsilon}$ and $\mathbf{x} \neq \mathbf{\Lambda}_x \boldsymbol{\xi} + \boldsymbol{\delta}$ when some y's and some x's are ordinal, we add a threshold model to the usual measurement model. The threshold model relates the ordinal y's and x's to their latent continuous counterparts of y^*'s and x^*'s. The more familiar measurement model holds for $\mathbf{y}^*$ and $\mathbf{x}^*$ so that:

$$\mathbf{y}^* = \mathbf{\Lambda}_y \boldsymbol{\eta} + \boldsymbol{\epsilon} \tag{9.103}$$

$$\mathbf{x}^* = \mathbf{\Lambda}_x \boldsymbol{\xi} + \boldsymbol{\delta} \tag{9.104}$$

The second consequence of ordinal indicators is that the covariance structure hypothesis usually does not hold for the covariance matrix of the observed variables [i.e., $\mathbf{\Sigma} \neq \mathbf{\Sigma}(\boldsymbol{\theta})$].[9] However, the model does hold for the latent continuous indicators so that $\mathbf{\Sigma}^* = \mathbf{\Sigma}(\boldsymbol{\theta})$, where $\mathbf{\Sigma}^*$ is the covariance matrix of $\mathbf{y}^*$ and $\mathbf{x}^*$. The need, then, is to estimate $\mathbf{\Sigma}^*$ to take the place of an estimate of $\mathbf{\Sigma}$ in the analysis.

To estimate $\mathbf{\Sigma}^*$ from the ordinal indicators, we must assume a distribution for the latent continuous indicators. As with the threshold model, the typical assumption is that $\mathbf{y}^*$ and $\mathbf{x}^*$ are multinormal. We can estimate a correlation for every pair of variables in $\mathbf{y}^*$ and $\mathbf{x}^*$. If both variables are continuous, then the Pearson correlation is applicable. If both variables are ordinal, then the correlation between the underlying continuous indicators is called a *polychoric* correlation. A special case of the polychoric correlation for two dichotomous indicators is the *tetrachoric* correlation. Finally,

[9] I say usually because it is possible to have a covariance structure hypothesis that holds for both the ordinal and continuous indicators (see Muthén and Kaplan 1985).

when one variable is ordinal and the other is continuous, a *polyserial* correlation results.

The ML estimation of the polychoric correlation between two ordinal indicators illustrates the principles involved in estimating the correlations between the latent continuous indicators. For a $c \times d$ table of two ordinal variables, say, x and y, the log likelihood is (Olsson 1979b)

$$\ln L = A + \sum_{i=1}^{c} \sum_{j=1}^{d} N_{ij} \ln(\pi_{ij}) \tag{9.105}$$

where c and d are the number of categories for the first and second ordinal indicators, A is an irrelevant constant, and N_{ij} is the frequency of observations in the ith and jth categories. The thresholds for x are a_i, $i = 0, 1, \ldots, c$, and the thresholds for y are b_j, $j = 0, 1, \ldots, d$, where $a_0 = b_0 = -\infty$ and $a_c = b_d = +\infty$. Also

$$\begin{aligned} \pi_{ij} = {} & \Phi_2(a_i, b_j) - \Phi_2(a_{i-1} b_j) \\ & - \Phi_2(a_i, b_{j-1}) + \Phi_2(a_{i-1}, b_{j-1}) \end{aligned} \tag{9.106}$$

where $\Phi_2(.,.)$ is the bivariate normal distribution function with correlation ρ.

Following ML principles, we can take the partial derivatives on $\ln L$ in (9.105) with respect to $\rho, a_1, \ldots, a_{c-1}, b_1, \ldots, b_{d-1}$. Simultaneously solving for ρ and all the thresholds requires numerical methods and is computationally intensive since it involves numerical integration. Olsson (1979b, 447) provides the necessary first derivatives.

An alternative two-step method is to first estimate the thresholds a_i and b_j from the univariate marginal distributions, as was done above. Then conditional upon these threshold values, solve for ρ. Olsson (1979b) suggests that this two-step method performs nearly as well as the more complicated simultaneous estimation procedure. ML estimation of tetrachoric correlations are just a special case of the polychoric ones just described. Olsson, Drasgow, and Dorans (1982) describe the ML estimation of polyserial correlations.

Table 9.7 provides the Pearson correlations (upper diagonal) and polychoric correlations for the eight indicators of political democracy example. Without exception, the polychoric correlations are larger than the corresponding Pearson correlations. The polychoric correlation matrix is a consistent estimator of $\boldsymbol{\Sigma}^*$, with which we can test the hypothesis that $\boldsymbol{\Sigma}^* = \boldsymbol{\Sigma}(\boldsymbol{\theta})$.

Table 9.7 Pearson Correlations (Upper Diagonal) and Polychoric Correlations (Lower Diagonal) between Indicators of Political Democracy (x_1 to x_8)

	x_1	x_2	x_3	x_4	x_5	x_6	x_7	x_8
x_1		0.585	0.677	0.686	0.733	0.635	0.638	0.670
x_2	0.658		0.440	0.707	0.547	0.727	0.564	0.646
x_3	0.748	0.511		0.563	0.560	0.408	0.613	0.512
x_4	0.739	0.760	0.670		0.659	0.655	0.670	0.726
x_5	0.771	0.618	0.627	0.707		0.561	0.662	0.646
x_6	0.723	0.834	0.501	0.737	0.661		0.575	0.733
x_7	0.722	0.659	0.709	0.769	0.716	0.721		0.677
x_8	0.710	0.689	0.591	0.792	0.691	0.801	0.787	

Note: Number of categories for ordinal indicators are 8, 5, 4, 5, 9, 5, 4, and 5, respectively.

Analysis of $\hat{\Sigma}^*$ leads to consistent estimators of $\boldsymbol{\theta}$ with any of the fitting functions I have presented. However, the standard errors, z-tests, chi-square tests, and other significance tests are not correct for F_{ML}, F_{GLS}, or F_{ULS}. A better choice is F_{WLS}:

$$F_{\mathrm{WLS}} = [\hat{\boldsymbol{\rho}} - \boldsymbol{\sigma}(\boldsymbol{\theta})]'\mathbf{W}^{-1}[\hat{\boldsymbol{\rho}} - \boldsymbol{\sigma}(\boldsymbol{\theta})] \tag{9.107}$$

where $\hat{\boldsymbol{\rho}}$ is a $\frac{1}{2}(p+q)(p+q+1) \times 1$ vector containing the polychoric, polyserial, and Pearson correlation coefficients for the nonredundant correlations between all pairs of $\mathbf{y}^*$ and $\mathbf{x}^*$ variables, $\boldsymbol{\sigma}(\boldsymbol{\theta})$ is the corresponding same dimension vector for the implied covariance matrix, and $\mathbf{W}$ is a consistent estimator of the asymptotic covariance matrix of $\hat{\boldsymbol{\rho}}$. See Muthén (1984) for an estimator of $\mathbf{W}$.

Table 9.8 lists the estimates and their standard errors (s.e.) for Λ_x for the political democracy CFA panel model with four indicators in 1960 and four in 1965. The first column lists the F_{WLS} estimates and asymptotic s.e. from the analysis of the polychoric correlations, with the weight matrix as provided in Muthén's (1987) LISCOMP program. The second column shows the results of applying F_{ML} to the polychoric correlations. Since F_{ML} uses the wrong weight matrix, we expect the s.e.'s and chi-square estimates to be inaccurate, even though the estimator of $\boldsymbol{\theta}$ is consistent. Finally, the third column provides the results of applying F_{ML} to the Pearson correlation matrix of the ordinal indicators. Here even consistency is not guaranteed.

Comparing the factor loadings, we find that those from F_{WLS} and F_{ML} for the polychoric correlations are the closest. Since both estimators are

Table 9.8 Factor Loading Estimates, Standard Errors, and Chi-Square Estimate from F_{WLS} and F_{ML} Applied to Polychoric Correlations and F_{ML} Applied to Pearson Correlations, Eight Indicators of Political Democracy

	Estimate (standard error)		
Parameter	F_{WLS} (polychoric)	F_{ML} (polychoric)	F_{ML} (Pearson's correlation)
λ_{11}	1.00^c	1.00^c	1.00^c
	(—)	(—)	(—)
λ_{21}	0.89	0.90	0.87
	(0.06)	(0.11)	(0.13)
λ_{31}	0.94	0.89	0.83
	(0.06)	(0.11)	(0.13)
λ_{41}	1.01	1.04	1.02
	(0.05)	(0.10)	(0.12)
λ_{52}	1.00^c	1.00^c	1.00^c
	(—)	(—)	(—)
λ_{62}	1.07	1.03	0.94
	(0.07)	(0.13)	(0.14)
λ_{72}	1.08	1.10	1.02
	(0.07)	(0.12)	(0.14)
λ_{82}	1.08	1.09	1.06
	(0.08)	(0.12)	(0.14)
χ^2	16.2	22.5	14.4
df	13	13	13
p-value	0.24	0.05	0.35

Note: ^{c} = constrained parameter.

consistent when analyzing the polychoric correlations, this is not surprising. The standard errors are smallest for the $\hat{\lambda}_{ij}$ from F_{WLS}. For this example, F_{ML} results in inflated asymptotic s.e. when we analyze the polychoric correlations.

The last column with F_{ML} applied to the Pearson correlations of the ordinal indicators has factor loadings that are lower than the first column (except for λ_{41}). This matches the simulation work that has found factor loading attenuation when analyzing the Pearson correlations of ordinal measures. But the attenuation is surprisingly small for most loadings. The asymptotic s.e. in the third column are incorrect and are much larger than those in the first column. The chi-square estimates for columns (1) and (3) are relatively close. That for F_{ML} of the polychoric correlations is much higher. These results agree with the simulation work of Babakus, Ferguson, and Jöreskog (1987). However, this example must be viewed with caution because of the small sample size ($N = 75$). Large sample properties for the estimators are known, but moderate to small sample properties have not been studied.

We can extend this approach to noncontinuous indicators in several directions. One is to include censored or truncated indicators. Consider a *censored* variable first:

$$y_1 = y_1^*, \qquad \text{if } y_1^* \geq L \tag{9.108}$$

$$y_1 = L, \qquad \text{if } y_1^* < L \tag{9.109}$$

The y_1 indicator is censored from below so that any y_1^*'s less than a cutoff of L are coded as L, whereas above L, y_1 equals y_1^*. Analogous types of equations hold for indicators censored from above or censored from above and below. An example of a censored from below variable is y_1^* equal to the propensity to purchase an automobile and y_1 is the purchase price. In any given year a large proportion of the sample does not buy a car, and hence y_1 is zero even though y_1^* is not identical for these nonbuyers. Indicators *truncated* from above or below have similar equations, except that rather than y_1 taking a particular value, y_1 is not observed if y_1^* is above or below some cutoff value.

For censored or truncated variables, we can follow a strategy analogous to that for ordinal variables: construct a threshold model, estimate the correlation (covariance) matrix $\hat{\Sigma}^*$, find the inverse of the asymptotic covariance matrix of the elements of $\hat{\Sigma}^*$, and use this as the weight matrix in F_{WLS}. See Muthén (1987) and Jöreskog and Sörbom (1986b) for further discussion.

Another generalization of models with noncontinuous indicators is to estimate means and intercepts. Muthén's (1984, 1987) LISCOMP model does this with two equations:

$$\boldsymbol{\eta} = \boldsymbol{\alpha} + \mathbf{B}\boldsymbol{\eta} + \boldsymbol{\Gamma}\mathbf{x} + \boldsymbol{\zeta} \tag{9.110}$$

$$\mathbf{y}^* = \boldsymbol{\upsilon} + \boldsymbol{\Lambda}\boldsymbol{\eta} + \boldsymbol{\epsilon} \tag{9.111}$$

where $\boldsymbol{\alpha}$ and $\boldsymbol{\upsilon}$ are intercept terms and the other variables and assumptions are the same as previously. The exception is that we assume that $\mathbf{x} = \boldsymbol{\xi}$ for the exogenous variables and Muthén assumes a multinormal distribution of

$\mathbf{y}^*$ *conditional on* $\mathbf{x}$. We need not assume a distribution for $\mathbf{x}$ so that it can consist of dummy variables or other nonnormally distributed variables.

For this model, the implied covariance matrix of $\mathbf{y}^*$ conditional upon $\mathbf{x}$ is

$$(\Sigma^*|\mathbf{x}) = \Lambda(\mathbf{I} - \mathbf{B})^{-1}\Psi(\mathbf{I} - \mathbf{B})'^{-1}\Lambda' + \Theta \tag{9.112}$$

and the mean of $\mathbf{y}^*$ given $\mathbf{x}$ is

$$E(\mathbf{y}^*|\mathbf{x}) = \boldsymbol{\upsilon} + \Lambda(\mathbf{I} - \mathbf{B})^{-1}\boldsymbol{\alpha} + \Lambda(\mathbf{I} - \mathbf{B})^{-1}\Gamma\mathbf{x} \tag{9.113}$$

where Ψ and Θ are the covariance matrices of $\boldsymbol{\zeta}$ and $\boldsymbol{\epsilon}$ respectively. In cases where $\mathbf{x}$ is absent, the conditioning on $\mathbf{x}$ has no meaning, and the last term of (9.113) drops out.

Muthén (1984, 987) proposed a three-stage procedure whereby these equations are estimated with limited information methods and the F_{WLS} fitting function. The method is similar to those described earlier for the ordinal indicators, with the addition of the intercept terms and $\mathbf{x}$ variables. Finally, Muthén further generalizes this model so that it applies to multigroup analyses.

At the time of this writing, the most general software to fit models with multiple noncontinuous indicators are Muthén's (1987) LISCOMP program and Jöreskog and Sörbom's (1986a) PRELIS program. The PRELIS program produces an estimate of Σ^* and the weight matrix for F_{WLS} to use in conjunction with LISREL VII.

SUMMARY

In this chapter I presented extensions of the general model of Chapter 8. I demonstrated an alternative notation that can expand the applications of structural equation models so that earlier restrictions such as not allowing indicators to directly affect one another are no longer needed. I also showed that means and intercepts, equality and inequality constraints, and interaction-quadratic terms can be included in models. A new generation of more general elliptical and ADF estimators was presented. Finally, the robustness of techniques to categorical data and new procedures that explicitly consider such indicators were reviewed.

Structural equation models remain an extremely active area of social science research. This chapter showed some of the areas that are likely to be developed further. In particular, it is probable that more work will appear on interaction and quadratic terms since current solutions are not fully satisfactory. Similarly, further innovations in the treatment of categorical data are likely, perhaps resulting in less restrictive assumptions than multinormality for the underlying variables and additional ways to reduce the computational burden of these procedures. Ways to reduce computation for WLS (ADF) estimators and the development of new limited information estimators are two other desirable goals.

APPENDIX 9A LISREL PROGRAM FOR MODEL IN FIGURE 9.1(c)

Programming models in the notation of equations (9.1) to (9.5) with software such as LISREL or EQS is straightforward. With EQS the user specifies an equation for each endogenous variable and includes all latent or observed explanatory variables on the right-hand side of the equation. In LISREL use only the LAMBDA-Y, BETA, and PSI matrices. Below I list the LISREL program to estimate the model shown in Figure 9.1 (c).

```
OBJEC. & SUBJ. SES EXAMPLE IN FIGURE 9.1 (c)
DA NI=5 NO=432 MA=CM
KM
*
1.000
0.292   1.000
0.282    .184   1.000
 .166    .383    .386   1.000
 .231    .277    .431    .537   1.000
SD
*
21.277   2.198   0.640   0.670   0.627
LA
*
'X2' 'X1' 'Y2' 'Y1' 'Y3'
SE
'Y1' 'Y2' 'Y3' 'X1' 'X2'
MO NY=5 NE=6 PS=FI TE=ZE BE=FU,FI
LE
*
'OBJINC' 'OBJOCC' 'LATSES' 'SUBJINC' 'SUBJOCC' 'SUBJGEN'
FR BE 3 1 BE 3 2 BE 4 1 BE 4 3 BE 5 2 BE 5 3
FR PS 1 1 PS 2 1 PS 2 2 PS 3 3 PS 4 4 PS 5 5 PS 6 6
ST 1 BE 6 3 LY 1 4 LY 2 5 LY 3 6 LY 4 1 LY 5 2
ST 1.063 BE 4 3
ST .776 BE 5 3
ST .078 BE 3 1
```

```
ST .004 BE 3 2
ST 4.831 PS 1 1
ST 452.711 PS 2 2
ST 13.656 PS 2 1
ST .165 PS 3 3
ST .209 PS 4 4
ST .282 PS 5 5
ST .181 PS 6 6
OU SE TV RS MI TO NS
```

APPENDIX A

Matrix Algebra Review

This appendix presents some of the basic definitions and properties of matrices. Many of the matrices in the appendix are named the same as the matrices that appear in the book's notation. I do this so that readers may become more familiar with the structural equation notation at the same time they are reviewing matrix algebra. Those wishing a more detailed treatment of matrices should refer to Lunneborg and Abbott (1983), Searle (1982), Graybill (1983), or Hadley (1961).

SCALARS, VECTORS, AND MATRICES

A basic distinction is that between scalars, vectors, and matrices. A single element, value, or quantity is referred to as a *scalar*. For instance, the covariance between x_1 and x_2, $\text{COV}(x_1, x_2)$, is a scalar, as is the number 5 and the regression coefficient β_{21}.

When two or more scalars are written in a row or a column, they form a row or column *vector*. The *order* of a row vector is $1 \times c$, where the 1 indicates one row and c is the number of columns (elements); the order of a column vector is $r \times 1$. Three examples of vectors are

$$\boldsymbol{\xi} = \begin{bmatrix} \xi_1 \\ \xi_2 \\ \xi_3 \end{bmatrix}, \qquad \mathbf{a}' = \begin{bmatrix} 1 & 4 & 5 & 2 \end{bmatrix}, \qquad \boldsymbol{\Gamma}' = \begin{bmatrix} \gamma_{11} & \gamma_{21} \end{bmatrix}$$

Boldface print shows that a symbol (e.g., $\boldsymbol{\xi}$) refers to a vector (or matrix) rather than a scalar. In the preceding examples $\boldsymbol{\xi}$ is a 3×1 column vector and $\mathbf{a}'$ and $\boldsymbol{\Gamma}'$ are row vectors of orders 1×4 and 1×2, respectively. The prime superscript distinguishes row vectors (e.g., $\mathbf{a}'$) from column vectors (e.g., $\boldsymbol{\xi}$). The prime stands for the transpose operator, which I will define later.

A *matrix* is a group of elements arranged into rows and columns. Matrices also are represented by **boldface** symbols. The order of a matrix is indicated by $r \times c$, where r is the number of rows and c is the number of columns. The order $r \times c$ also is the dimension of the matrix. As examples, consider

$$\mathbf{S} = \begin{bmatrix} s_{11} & s_{12} \\ s_{21} & s_{22} \end{bmatrix}, \qquad \boldsymbol{\Gamma} = \begin{bmatrix} 0 & \gamma_{12} \\ \gamma_{21} & 0 \\ 0 & 0 \end{bmatrix}$$

The dimension of $\mathbf{S}$ is 2×2, whereas $\boldsymbol{\Gamma}$ is a 3×2 matrix.

Vectors and scalars are special cases of matrices. A row vector is a $1 \times c$ matrix, a column vector is an $r \times 1$ matrix, and a scalar is a 1×1 matrix. Two matrices, say $\mathbf{S}$ and $\mathbf{T}$, are equal if they are of the same order and if every element s_{ij} of $\mathbf{S}$ equals the corresponding element t_{ij} of $\mathbf{T}$ for all i's and j's. For instance, here $\mathbf{S}$ and $\mathbf{T}$ are equal, but $\mathbf{S}$ and $\mathbf{T}^*$ are not:

$$\mathbf{S} = \begin{bmatrix} s_{11} & 0 \\ s_{21} & s_{22} \end{bmatrix}, \qquad \mathbf{T} = \begin{bmatrix} s_{11} & 0 \\ s_{21} & s_{22} \end{bmatrix}$$

$$\mathbf{S} = \mathbf{T}$$

$$\mathbf{S} = \begin{bmatrix} s_{11} & 0 \\ s_{21} & s_{22} \end{bmatrix}, \qquad \mathbf{T}^* = \begin{bmatrix} s_{11} & 0 \\ 0 & s_{22} \end{bmatrix}$$

$$\mathbf{S} \neq \mathbf{T}^*, \qquad \text{for } s_{21} \neq 0$$

MATRIX OPERATIONS

To add two or more matrices, they must be of the same dimension or order. The resulting matrix is of the same order, with each element equal to the sum of the corresponding elements in each of the matrices added together. For instance,

$$\mathbf{B} = \begin{bmatrix} \beta_{11} & \beta_{12} \\ \beta_{21} & \beta_{22} \end{bmatrix}, \qquad \mathbf{I} = \begin{bmatrix} 1 & 0 \\ 0 & 1 \end{bmatrix}$$

$$\mathbf{B} + \mathbf{I} = \begin{bmatrix} \beta_{11} + 1 & \beta_{12} \\ \beta_{21} & \beta_{22} + 1 \end{bmatrix}$$

Two useful properties of matrix addition, for any matrices **S**, **T**, and **U** of the same order, are as follows:

1. $\mathbf{S} + \mathbf{T} = \mathbf{T} + \mathbf{S}$.
2. $(\mathbf{S} + \mathbf{T}) + \mathbf{U} = \mathbf{S} + (\mathbf{T} + \mathbf{U})$.

Two matrices can be multiplied only when the number of columns in the first matrix equals the number of rows in the second matrix. If this condition is met the matrices are said to be *conformable* with respect to multiplication. If the first matrix has order $a \times b$ and the second matrix has order $b \times c$, then the resulting product of these two matrices is a matrix of order $a \times c$.

The *ij* element of the product matrix is derived by multiplying the elements of the *i*th row of the first matrix by the corresponding elements of the *j*th column of the second matrix and summing all terms. Consider two matrices **S** and **T** of orders $a \times b$ and $b \times c$, respectively:

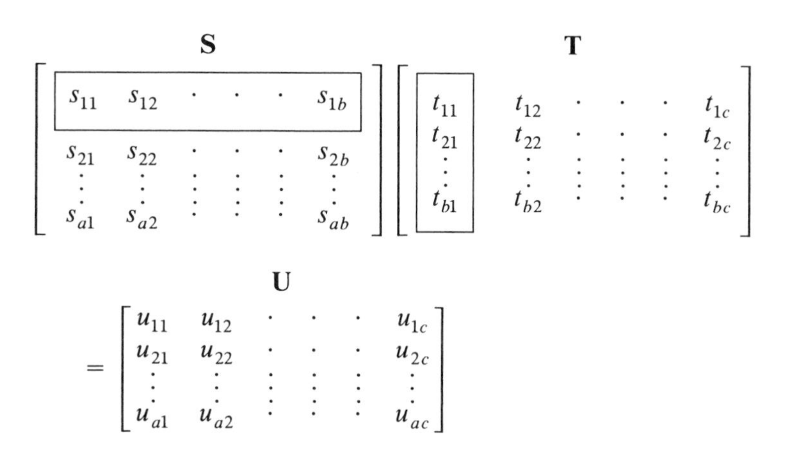

To form the matrix product of **S** times **T**, labeled **U**, start with the first row of **S**, which is enclosed in a box. Each element in this row is multiplied by the corresponding element in the first column of **T**, which is also enclosed in a box. The sum of these b products equals u_{11} of **U**. In other words $u_{11} = s_{11}t_{11} + s_{12}t_{21} + \cdots + s_{1b}t_{b1}$. This may be stated more generally as

$$u_{ij} = \sum_{k=1}^{b} s_{ik}t_{kj}$$

for each element of **U**.

For example,

$$\mathbf{B} = \underset{3 \times 3}{\begin{bmatrix} 0 & \beta_{12} & 0 \\ \beta_{21} & 0 & \beta_{23} \\ 0 & 0 & 0 \end{bmatrix}}, \qquad \boldsymbol{\eta} = \underset{3 \times 1}{\begin{bmatrix} \eta_1 \\ \eta_2 \\ \eta_3 \end{bmatrix}}$$

$$\mathbf{B}\boldsymbol{\eta} = \underset{3 \times 1}{\begin{bmatrix} \beta_{12}\eta_2 \\ \beta_{21}\eta_1 + \beta_{23}\eta_3 \\ 0 \end{bmatrix}}$$

(The term $\mathbf{B}\boldsymbol{\eta}$ appears in the latent variable model.)

Some properties of matrix multiplication for any matrices **S**, **T**, and **U** that are conformable are as follows:

1. $\mathbf{ST} \neq \mathbf{TS}$ (except in special cases).
2. $(\mathbf{ST})\mathbf{U} = \mathbf{S}(\mathbf{TU})$.
3. $\mathbf{S}(\mathbf{T} + \mathbf{U}) = \mathbf{ST} + \mathbf{SU}$.
4. $c(\mathbf{S} + \mathbf{T}) = c\mathbf{S} + c\mathbf{T}$ (where c is a scalar).

These properties come into play at many points throughout the book. The order in which matrices are multiplied is important. For this reason *premultiplication* and *postmultiplication* by a matrix are distinguished. For instance, if $\mathbf{U} = \mathbf{ST}$, we can say that **U** results from the premultiplication of **T** by **S** or the postmultiplication of **S** by **T**.

The *transpose* of a matrix interchanges its rows and columns. The transpose of a matrix is indicated by a prime (′) symbol following the matrix. As an example, consider Γ and Γ':

$$\Gamma = \underset{3 \times 2}{\begin{bmatrix} \gamma_{11} & 0 \\ \gamma_{21} & 0 \\ 0 & \gamma_{32} \end{bmatrix}}, \qquad \Gamma' = \underset{2 \times 3}{\begin{bmatrix} \gamma_{11} & \gamma_{21} & 0 \\ 0 & 0 & \gamma_{32} \end{bmatrix}}$$

The first row of Γ is the first column of Γ', the second row is the second column, and the third row is the third column. The order of Γ is 3×2, whereas the order of Γ' is 2×3. The transpose of an $a \times b$ order matrix leads to a $b \times a$ matrix.

Some useful properties of the transpose operator are listed below:

1. $(\mathbf{S}')' = \mathbf{S}$.
2. $(\mathbf{S} + \mathbf{T})' = \mathbf{S}' + \mathbf{T}'$ (where $\mathbf{S}$ and $\mathbf{T}$ have the same order).
3. $(\mathbf{ST})' = \mathbf{T}'\mathbf{S}'$ (where matrices are conformable for multiplication).
4. $(\mathbf{STU})' = \mathbf{U}'\mathbf{T}'\mathbf{S}'$ (where matrices are conformable for multiplication).

Some additional matrix types and matrix operations are important for square matrices. A *square* matrix is a matrix that has the same number of rows and columns. An example of a square matrix is

$$\Gamma = \begin{bmatrix} \gamma_{11} & 0 & \gamma_{13} \\ 0 & \gamma_{22} & 0 \\ 0 & \gamma_{32} & \gamma_{33} \end{bmatrix}$$

The dimension of the Γ matrix is 3×3.

The *trace* is defined for a square matrix. It is the sum of the elements on the main diagonal. For an $n \times n$ matrix $\mathbf{S}$,

$$\operatorname{tr}(\mathbf{S}) = \sum_{i=1}^{n} s_{ii}$$

Properties of the trace include:

1. $\operatorname{tr}(\mathbf{S}) = \operatorname{tr}(\mathbf{S}')$.
2. $\operatorname{tr}(\mathbf{ST}) = \operatorname{tr}(\mathbf{TS})$ (if $\mathbf{T}$ and $\mathbf{S}$ conform for multiplication).
3. $\operatorname{tr}(\mathbf{S} + \mathbf{T}) = \operatorname{tr}(\mathbf{S}) + \operatorname{tr}(\mathbf{T})$ (if $\mathbf{S}$ and $\mathbf{T}$ conform for addition).

The trace appears in the fitting functions and indices of goodness of fit for many structural equation techniques.

If all the elements above (or below) the main diagonal of a square matrix are zero, the matrix is *triangular*. For instance, the $\mathbf{B}$ matrix for recursive models (which I discuss in Chapter 4) may be written as a triangular matrix. The $\mathbf{B}$ matrix contains the coefficients of the effects of the endogenous latent variables on one another. To illustrate, one such $\mathbf{B}$ matrix is

$$\mathbf{B} = \begin{bmatrix} 0 & 0 & 0 \\ \beta_{21} & 0 & 0 \\ \beta_{31} & \beta_{32} & 0 \end{bmatrix}$$

Note that in this case the main diagonal elements are zero. However, triangular matrices may have nonzero entries in the main diagonal.

A *diagonal* matrix is a square matrix that has some nonzero elements along the main diagonal and zeros elsewhere. For instance, $\mathbf{\Theta}_\delta$, the covariance matrix (see Chapter 2), of the errors of measurement for the $\mathbf{x}$ variables, commonly is assumed to be diagonal. For δ_1, δ_2, and δ_3, the population covariance matrix $\mathbf{\Theta}_\delta$ might look as follows:

$$\mathbf{\Theta}_\delta = \begin{bmatrix} \text{VAR}(\delta_1) & 0 & 0 \\ 0 & \text{VAR}(\delta_2) & 0 \\ 0 & 0 & \text{VAR}(\delta_3) \end{bmatrix}$$

The zeros above and below the main diagonal represent the assumption that the errors of measurement for different variables are uncorrelated.

A *symmetric* matrix is a square matrix that equals its transpose (e.g., $\mathbf{S} = \mathbf{S}'$). The typical correlation and covariance matrices are symmetric since the *ij* element equals the *ji* element. For instance,

$$\mathbf{\Sigma} = \begin{bmatrix} \text{VAR}(x_1) & \text{COV}(x_1, x_2) & \text{COV}(x_1, x_3) \\ \text{COV}(x_2, x_1) & \text{VAR}(x_2) & \text{COV}(x_2, x_3) \\ \text{COV}(x_3, x_1) & \text{COV}(x_3, x_2) & \text{VAR}(x_3) \end{bmatrix}$$

For all the variables, the covariance of x_i and x_j equals the covariance of x_j and x_i. Sometimes symmetric matrices, such as $\mathbf{\Sigma}$, are written with blanks above the main diagonal because these terms are redundant.

An *identity* matrix, I, is a square matrix with ones down the main diagonal and zeros elsewhere. The 3×3 identity matrix is

$$\mathbf{I} = \begin{bmatrix} 1 & 0 & 0 \\ 0 & 1 & 0 \\ 0 & 0 & 1 \end{bmatrix}$$

Properties of the identity matrix, I, include the following:

1. $\mathbf{IS} = \mathbf{SI} = \mathbf{S}$ (for any $\mathbf{I}$ and $\mathbf{S}$ conformable for multiplication).
2. $\mathbf{I} = \mathbf{I}'$.

A vector that consists of all ones is a *unit* vector:

$$\mathbf{1} = \begin{bmatrix} 1 \\ 1 \\ 1 \end{bmatrix} \qquad \mathbf{1}' = [1 \quad 1 \quad 1]$$

Unit vector products have some interesting properties. If you premultiply a matrix by a conformable unit vector, the result is a row vector whose elements are the column sums of the matrix. For example,

$$\mathbf{1}'\mathbf{T} = [1 \quad 1 \quad 1]\begin{bmatrix} 4 & -1 \\ 0 & 1 \\ 2 & 2 \end{bmatrix} = [6 \quad 2]$$

Postmultiplying a matrix by a conforming unit vector leads to a column vector of row sums:

$$\mathbf{T1} = \begin{bmatrix} 4 & -1 \\ 0 & 1 \\ 2 & 2 \end{bmatrix}\begin{bmatrix} 1 \\ 1 \end{bmatrix} = \begin{bmatrix} 3 \\ 1 \\ 4 \end{bmatrix}$$

Finally, if we both premultiply and postmultiply a matrix by conforming unit vectors, a scalar that equals the sum of all the matrix elements results:

$$\mathbf{1}'\mathbf{T1} = [1 \quad 1 \quad 1]\begin{bmatrix} 4 & -1 \\ 0 & 1 \\ 2 & 2 \end{bmatrix}\begin{bmatrix} 1 \\ 1 \end{bmatrix} = [8]$$

Using unit vectors and some of the other matrix properties, we can compute the covariance matrix. Consider $\mathbf{X}$ an $N \times p$ matrix of N observations for p variables. The $1 \times p$ row vector of means for $\mathbf{X}$ is formed as

$$\left(\frac{1}{N}\right)\mathbf{1}'\mathbf{X}$$

The deviation form of $\mathbf{X}$ requires subtracting from $\mathbf{X}$ a matrix whose columns consist of $N \times 1$ vectors of the means for the corresponding variables in $\mathbf{X}$. So every element in the first column equals the mean of the first variable in $\mathbf{X}$, every element in the second column equals the mean of the second column of **X**, and so on. This matrix of means is

$$\mathbf{1}\left(\frac{1}{N}\right)\mathbf{1}'\mathbf{X}$$

which, when subtracted from **X**, forms deviation from the mean scores:

$$\mathbf{X} - \mathbf{1}\left(\frac{1}{N}\right)\mathbf{1}'\mathbf{X}$$

If the preceding deviation score matrix is represented by **Z**, then the $p \times p$ unbiased sample covariance matrix estimator **S** is

$$\mathbf{S} = \left(\frac{1}{(N-1)}\right)\mathbf{Z}'\mathbf{Z}$$

A numerical example illustrates these calculations:

$$\mathbf{X} = \begin{bmatrix} 2 & 3 & 1 \\ -1 & 1 & 1 \\ 0 & 4 & 2 \\ -1 & 0 & 0 \end{bmatrix}$$

$$\left(\frac{1}{N}\right)\mathbf{1}'\mathbf{X} = \left(\frac{1}{4}\right)[1 \quad 1 \quad 1 \quad 1]\begin{bmatrix} 2 & 3 & 1 \\ -1 & 1 & 1 \\ 0 & 4 & 2 \\ -1 & 0 & 0 \end{bmatrix} = [0 \quad 2 \quad 1]$$

$$\mathbf{1}\left(\frac{1}{N}\right)\mathbf{1}'\mathbf{X} = \begin{bmatrix} 1 \\ 1 \\ 1 \\ 1 \end{bmatrix}[0 \quad 2 \quad 1] = \begin{bmatrix} 0 & 2 & 1 \\ 0 & 2 & 1 \\ 0 & 2 & 1 \\ 0 & 2 & 1 \end{bmatrix}$$

$$\mathbf{Z} = \mathbf{X} - \mathbf{1}\left(\frac{1}{N}\right)\mathbf{1}'\mathbf{X} = \begin{bmatrix} 2 & 1 & 0 \\ -1 & -1 & 0 \\ 0 & 2 & 1 \\ -1 & -2 & -1 \end{bmatrix}$$

$$\mathbf{S} = \left(\frac{1}{(N-1)}\right)\mathbf{Z}'\mathbf{Z} = \begin{bmatrix} \operatorname{var}(x_1) & \operatorname{cov}(x_1, x_2) & \operatorname{cov}(x_1, x_3) \\ \operatorname{cov}(x_2, x_1) & \operatorname{var}(x_2) & \operatorname{cov}(x_2, x_3) \\ \operatorname{cov}(x_3, x_1) & \operatorname{cov}(x_3, x_2) & \operatorname{var}(x_3) \end{bmatrix}$$

$$= \frac{1}{3}\begin{bmatrix} 6 & 5 & 1 \\ 5 & 10 & 4 \\ 1 & 4 & 2 \end{bmatrix}$$

The last line is the unbiased sample covariance matrix and contains the variances of the variables in the main diagonal and the covariances of all pairs of variables in the off-diagonal elements. All covariance matrices are square and symmetric.

Suppose that I form a diagonal matrix from the main diagonal of **S**:

$$\mathbf{D} = \begin{bmatrix} \text{var}(x_1) & 0 & 0 \\ 0 & \text{var}(x_2) & 0 \\ 0 & 0 & \text{var}(x_3) \end{bmatrix}$$

where $\text{var}(x_i)$ represents the sample variance of x_i. The square root of **D**, represented as $\mathbf{D}^{1/2}$, has standard deviations in its diagonal:

$$\mathbf{D}^{1/2} = \begin{bmatrix} [\text{var}(x_1)]^{1/2} & 0 & 0 \\ 0 & [\text{var}(x_2)]^{1/2} & 0 \\ 0 & 0 & [\text{var}(x_3)]^{1/2} \end{bmatrix}$$

and $\mathbf{D}^{-1/2}$ has the inverse of the standard deviations in its main diagonal. If **S** is postmultiplied by $\mathbf{D}^{-1/2}$, the first column is divided by the standard deviation of x_1, the second column is divided by the standard deviation of x_2, and the third column is divided by the standard deviation of x_3:

$$\mathbf{SD}^{-1/2} = \begin{bmatrix} [\text{var}(x_1)]^{1/2} & \dfrac{\text{cov}(x_1, x_2)}{[\text{var}(x_2)]^{1/2}} & \dfrac{\text{cov}(x_1, x_3)}{[\text{var}(x_3)]^{1/2}} \\ \dfrac{\text{cov}(x_2, x_1)}{[\text{var}(x_1)]^{1/2}} & [\text{var}(x_2)]^{1/2} & \dfrac{\text{cov}(x_2, x_3)}{[\text{var}(x_3)]^{1/2}} \\ \dfrac{\text{cov}(x_3, x_1)}{[\text{var}(x_1)]^{1/2}} & \dfrac{\text{cov}(x_3, x_2)}{[\text{var}(x_2)]^{1/2}} & [\text{var}(x_3)]^{1/2} \end{bmatrix}$$

Premultiplying $\mathbf{SD}^{-1/2}$ by $\mathbf{D}^{-1/2}$ leads to

$$\mathbf{D}^{-1/2}\mathbf{SD}^{-1/2} = \begin{bmatrix} 1 & \dfrac{\text{cov}(x_1, x_2)}{[\text{var}(x_1)\text{var}(x_2)]^{1/2}} & \dfrac{\text{cov}(x_1, x_3)}{[\text{var}(x_1)\text{var}(x_3)]^{1/2}} \\ \dfrac{\text{cov}(x_2, x_1)}{[\text{var}(x_2)\text{var}(x_1)]^{1/2}} & 1 & \dfrac{\text{cov}(x_2, x_3)}{[\text{var}(x_2)\text{var}(x_3)]^{1/2}} \\ \dfrac{\text{cov}(x_3, x_1)}{[\text{var}(x_3)\text{var}(x_1)]^{1/2}} & \dfrac{\text{cov}(x_3, x_2)}{[\text{var}(x_3)\text{var}(x_2)]^{1/2}} & 1 \end{bmatrix}$$

The resulting matrix is the sample correlation matrix with the off-diagonal elements equal to the correlations of the x_i and x_j variables. These results generalize to any dimension matrix. If the covariance matrix $\mathbf{S}$ is pre- and postmultiplied by $\mathbf{D}^{-1/2}$, where $\mathbf{D}^{-1/2}$ is the diagonal matrix with the standard deviations of $\mathbf{x}$ in its diagonal, then the resulting matrix is the sample correlation matrix.

The nonnegative integer power of square matrices occurs in the decomposition of effects in path analysis. It is defined as the number of times a matrix is multiplied by itself. For instance,

$$\begin{bmatrix} 0 & \beta_{12} \\ \beta_{21} & 0 \end{bmatrix}^2 = \begin{bmatrix} 0 & \beta_{12} \\ \beta_{21} & 0 \end{bmatrix}\begin{bmatrix} 0 & \beta_{12} \\ \beta_{21} & 0 \end{bmatrix} = \begin{bmatrix} \beta_{12}\beta_{21} & 0 \\ 0 & \beta_{21}\beta_{12} \end{bmatrix}$$

For any square matrix, $\mathbf{S}$, a scalar quantity exists called the *determinant* of $\mathbf{S}$, represented as $|\mathbf{S}|$ or $\det \mathbf{S}$. In the case of a 2×2 matrix the determinant is

$$\begin{vmatrix} s_{11} & s_{12} \\ s_{21} & s_{22} \end{vmatrix} = s_{11}s_{22} - s_{12}s_{21}$$

If $\mathbf{S}$ is a 3×3 matrix, the determinant is

$$\begin{vmatrix} s_{11} & s_{12} & s_{13} \\ s_{21} & s_{22} & s_{23} \\ s_{31} & s_{32} & s_{33} \end{vmatrix} = s_{11}s_{22}s_{33} - s_{12}s_{21}s_{33} + s_{12}s_{23}s_{31} - s_{13}s_{22}s_{31}$$

$$+ s_{13}s_{21}s_{32} - s_{11}s_{23}s_{32}$$

As the order of $\mathbf{S}$ increases, the formula for the determinant becomes more complicated. There is a general rule for calculating determinants for a square matrix of any order. To explain this rule, the concepts of a *minor* and a *cofactor* need to be defined. The minor of a matrix is the determinant of the matrix obtained when the ith row and jth column of a matrix are removed. Consider the following matrix $\mathbf{S}$:

$$\mathbf{S} = \begin{bmatrix} s_{11} & s_{12} & s_{13} \\ s_{21} & s_{22} & s_{23} \\ s_{31} & s_{32} & s_{33} \end{bmatrix}$$

The minor with respect to s_{11}, represented as $|\mathbf{S}_{11}|$, is

$$|\mathbf{S}_{11}| = \begin{vmatrix} s_{22} & s_{23} \\ s_{32} & s_{33} \end{vmatrix} = s_{22}s_{33} - s_{23}s_{32}$$

The minor of s_{22} is

$$|\mathbf{S}_{22}| = \begin{vmatrix} s_{11} & s_{13} \\ s_{31} & s_{33} \end{vmatrix} = s_{11}s_{33} - s_{13}s_{31}$$

The cofactor of the element s_{ij} is defined as $(-1)^{i+j}$ times the minor of s_{ij},

$$C_{ij} = (-1)^{i+j}|\mathbf{S}_{ij}|$$

The cofactors of each element of matrix **S** placed in the appropriate ijth location creates a new matrix:

$$\begin{bmatrix} +|\mathbf{S}_{11}| & -|\mathbf{S}_{12}| & +|\mathbf{S}_{13}| \\ -|\mathbf{S}_{21}| & +|\mathbf{S}_{22}| & -|\mathbf{S}_{23}| \\ +|\mathbf{S}_{31}| & -|\mathbf{S}_{32}| & +|\mathbf{S}_{33}| \end{bmatrix}$$

The determinant of a matrix can be found by multiplying each element in any given row (column) in the **S** matrix by the corresponding cofactor in the preceding matrix. This is then summed over all elements in the row (column). For example, if we do this for the first row of **S**, we obtain

$$\begin{aligned} s_{11}|\mathbf{S}_{11}| - s_{12}|\mathbf{S}_{12}| + s_{13}|\mathbf{S}_{13}| &= s_{11}(s_{22}s_{33} - s_{23}s_{32}) \\ &\quad - s_{12}(s_{21}s_{33} - s_{23}s_{31}) \\ &\quad + s_{13}(s_{21}s_{32} - s_{22}s_{31}) \\ &= s_{11}s_{22}s_{33} - s_{11}s_{23}s_{32} \\ &\quad - s_{12}s_{21}s_{33} + s_{12}s_{23}s_{31} \\ &\quad + s_{13}s_{21}s_{32} - s_{13}s_{22}s_{31} \\ &= s_{11}s_{22}s_{33} - s_{12}s_{21}s_{33} \\ &\quad + s_{12}s_{23}s_{31} - s_{13}s_{22}s_{31} \\ &\quad + s_{13}s_{21}s_{32} - s_{11}s_{23}s_{32} \end{aligned}$$

Note that this formula is identical to the earlier formula for the determinant of a 3×3 matrix. Slightly different arrangements of terms occur depending on the row or column selected for expansion. However, regardless of which formula is chosen, the determinant will be the same.

Useful properties of the determinant for square and conformable $\mathbf{S}$ and $\mathbf{T}$ and a scalar c include:

1. $|\mathbf{S}'| = |\mathbf{S}|$.
2. If $\mathbf{S} = c\mathbf{T}$, then $|\mathbf{S}| = c^q|\mathbf{T}|$ (where q is the order of $\mathbf{S}$).
3. If $\mathbf{S}$ has two identical rows (or columns), $|\mathbf{S}| = 0$.
4. $|\mathbf{ST}| = |\mathbf{S}||\mathbf{T}|$.

The determinant appears in the fitting functions for the estimators of structural equations. It also is useful in finding the *rank* and *inverse* of matrices.

The *inverse* of a square matrix $\mathbf{S}$ is that matrix $\mathbf{S}^{-1}$ that, when $\mathbf{S}$ is pre- or postmultiplied by $\mathbf{S}^{-1}$, produces the identity matrix, $\mathbf{I}$:

$$\mathbf{SS}^{-1} = \mathbf{S}^{-1}\mathbf{S} = \mathbf{I}$$

The inverse of a matrix is calculated from the *adjoint* and the determinant of a matrix. The adjoint of a matrix is the transpose of the matrix of cofactors defined earlier. Using the 3×3 $\mathbf{S}$ matrix, the adjoint of $\mathbf{S}$ is

$$\operatorname{adj} \mathbf{S} = \begin{bmatrix} +|\mathbf{S}_{11}| & -|\mathbf{S}_{21}| & +|\mathbf{S}_{31}| \\ -|\mathbf{S}_{12}| & +|\mathbf{S}_{22}| & -|\mathbf{S}_{32}| \\ +|\mathbf{S}_{13}| & -|\mathbf{S}_{23}| & +|\mathbf{S}_{33}| \end{bmatrix}$$

The inverse matrix is

$$\mathbf{S}^{-1} = \frac{1}{|\mathbf{S}|}(\operatorname{adj} \mathbf{S})$$

To illustrate the calculation of the inverse, consider the simple case of a two-variable covariance matrix:

$$\mathbf{S} = \begin{bmatrix} 20 & 10 \\ 10 & 20 \end{bmatrix}$$

$$|\mathbf{S}| = (20)(20) - (10)(10)$$
$$= 400 - 100 = 300$$

$$\text{Matrix of cofactors of } \mathbf{S} = \begin{bmatrix} 20 & -10 \\ -10 & 20 \end{bmatrix}$$

$$\operatorname{adj} \mathbf{S} = \begin{bmatrix} 20 & -10 \\ -10 & 20 \end{bmatrix}$$

(The adjoint of a symmetric matrix of cofactors equals the matrix of cofactors.)

$$\mathbf{S}^{-1} = \frac{1}{300}\begin{bmatrix} 20 & -10 \\ -10 & 20 \end{bmatrix}$$

Multiplying $\mathbf{S}^{-1}$ by $\mathbf{S}$ yields a 2×2 identity matrix.

Note that the inverse $\mathbf{S}^{-1}$ does not exist if $|\mathbf{S}| = 0$. If a matrix has a zero determinant, it is called a *singular* matrix.

Two properties of inverses for conformable square matrices $\mathbf{S}$, $\mathbf{T}$, and $\mathbf{U}$ are the following:

1. $(\mathbf{S}')^{-1} = (\mathbf{S}^{-1})'$.
2. $(\mathbf{ST})^{-1} = \mathbf{T}^{-1}\mathbf{S}^{-1}$; $(\mathbf{STU})^{-1} = \mathbf{U}^{-1}\mathbf{T}^{-1}\mathbf{S}^{-1}$.

In manipulating the latent variable equations, we sometimes need to take inverses. In addition the inverse appears in explanations of the fitting functions and in several other topics.

Another important property of a matrix is its *rank*. The *rank* of a matrix, $\mathbf{S}$, is the maximum number of independent columns or rows of $\mathbf{S}$, where $\mathbf{S}$ is any $a \times b$ order matrix. Another way to define the rank is as the order of the largest square submatrix of $\mathbf{S}$ whose determinant is nonzero.

The properties of ranks for matrices $\mathbf{S}$ and $\mathbf{T}$ include the following:

1. $\text{rank}(\mathbf{S}) \leq \min(a, b)$, where a is the number of rows, b is the number of columns.
2. $\text{rank}(\mathbf{ST}) \leq \min[\text{rank}(\mathbf{S}), \text{rank}(\mathbf{T})]$.

The rank of matrices appears in the discussion of identification of models in Chapter 4.

Eigenvalues and eigenvectors are important characteristics of square matrices. If a vector $\mathbf{u} \neq 0$, a scalar e, and an $n \times n$ matrix $\mathbf{S}$ exist such that

$$\mathbf{Su} = e\mathbf{u}$$

then $\mathbf{u}$ is an eigenvector and e is an eigenvalue of $\mathbf{S}$. The eigenvectors and eigenvalues are sometimes referred to as latent vectors and latent values, or characteristic vectors and characteristic roots. (Often such an equation is represented as $\mathbf{Ax} = \lambda\mathbf{x}$. I depart from this practice so that λ is not confused with the factor loadings which use the same symbol.)

The preceding equation may be rewritten as

$$\mathbf{Su} - e\mathbf{u} = 0$$
$$(\mathbf{S} - e\mathbf{I})\mathbf{u} = 0$$

Only if $(\mathbf{S} - e\mathbf{I})$ is singular, does a nontrivial solution[1] for this equation exist. If $(\mathbf{S} - e\mathbf{I})$ is singular, then

$$|\mathbf{S} - e\mathbf{I}| = 0$$

Solving this equation for e provides the eigenvalues.

To illustrate, suppose that $\mathbf{S}$ is a 2×2 correlation matrix:

$$\mathbf{S} = \begin{bmatrix} 1.00 & 0.50 \\ 0.50 & 1.00 \end{bmatrix}$$

The $(\mathbf{S} - e\mathbf{I})$ matrix is

$$\mathbf{S} - e\mathbf{I} = \begin{bmatrix} 1.00 - e & 0.50 \\ 0.50 & 1.00 - e \end{bmatrix}$$

The determinant is

$$|\mathbf{S} - e\mathbf{I}| = (1.00 - e)^2 - 0.25$$
$$= e^2 - 2e + 0.75$$

The two solutions for e, 1.5 and 0.5, are the eigenvalues for this 2×2 correlation matrix. Each eigenvalue, e, has a set of eigenvectors, $\mathbf{u}$, associated with it. For example, the e of 1.5 leads to the following:

$$(\mathbf{S} - e\mathbf{I})\mathbf{u} = 0$$

$$\begin{bmatrix} 1.00 - e & 0.50 \\ 0.50 & 1.00 - e \end{bmatrix} \begin{bmatrix} u_1 \\ u_2 \end{bmatrix} = 0$$

$$\begin{bmatrix} -0.50 & 0.50 \\ 0.50 & -0.50 \end{bmatrix} \begin{bmatrix} u_1 \\ u_2 \end{bmatrix} = 0$$

$$-0.50u_1 + 0.50u_2 = 0$$
$$0.50u_1 - 0.50u_2 = 0$$

[1]A trivial solution for e would exist if $\mathbf{u} = 0$. As specified here, I assume that $\mathbf{u}$ is a nonzero vector.

From this you can see that $u_1 = u_2$ and that an infinite set of values would work as the eigenvector for the eigenvalue of 1.5.

Though the eigenvalues for the preceding example and for all real symmetric matrices are real numbers, this need not be true for nonsymmetric matrices. When the eigenvalue is a complex number, say $z = a + ib$, where a and b are real constants and $i = \sqrt{-1}$, we commonly refer to the modulus or norm of z which is $(a^2 + b^2)^{1/2}$.

Some useful properties of the eigenvalues for a symmetric or nonsymmetric square matrix **S** are

1. A $b \times b$ matrix **S** has b eigenvalues (some may take the same value).
2. The product of all eigenvalues for **S** equals $|\mathbf{S}|$.
3. The number of nonzero eigenvalues of **S** equals the rank of **S**.
4. The sum of the eigenvalues of **S** equals the tr **S**.

Eigenvalue and eigenvectors play a large role in traditional factor analyses. In this book they are useful in the decomposition of effects in path analysis discussed in Chapter 8.

Quadratic forms are represented by

$$\underset{(1 \times b)(b \times b)(b \times 1)}{\mathbf{x}'\mathbf{S}\mathbf{x}}$$

which equals

$$\sum_i x_i^2 s_{ii} + \sum_j \sum_{>i} x_i x_j (s_{ij} + s_{ji})$$

Usually **S** for a quadratic form is a symmetric matrix so that

$$\mathbf{x}'\mathbf{S}\mathbf{x} = \sum_i x_i^2 s_{ii} + 2\sum_j \sum_{>i} x_i x_j s_{ij}$$

Quadratic forms result in a scalar. **S** is positive-definite if $\mathbf{x}'\mathbf{S}\mathbf{x}$ is positive for all nonzero **x** vectors. If this quadratic form is nonnegative for all nonzero **x**, then **S** is positive-semidefinite. The eigenvalues of a positive definite matrix are all positive. If **S** is positive-definite, then **S** is nonsingular. The eigenvalues of a positive semidefinite matrix are positive or zero. Negative-definite and negative-semidefinite have analogous definitions and properties.

Occasionally, structural equation models analyzed with the LISREL program (Jöreskog and Sörbom 1984) may report that a matrix is not positive-definite.

For instance, suppose that we analyze the following sample covariance matrix **S**:

$$\begin{bmatrix} 7 & 3 & 4 \\ 3 & 2 & 1 \\ 4 & 1 & 3 \end{bmatrix}$$

S is not positive-definite, since $\mathbf{x}'\mathbf{S}\mathbf{x}$ is zero for some $\mathbf{x} \neq 0$ (e.g., $\mathbf{x}' = [1 \ -1 \ -1]$). Indeed, **S** is singular ($|\mathbf{S}| = 0$), and singular matrices are not positive-definite.

Consider the following three matrices:

$$\begin{bmatrix} 2 & 3 \\ 3 & 1 \end{bmatrix} \quad \begin{bmatrix} -2 & 1 \\ 1 & 1.5 \end{bmatrix} \quad \begin{bmatrix} 0 & 0 \\ 0 & 2 \end{bmatrix}$$

Assume that these are estimates of the covariance matrix of the disturbances from two equations. None is positive-definite. The failure of the first two to be positive-definite indicates a problem. In the first case the covariance ($= 3$) and variances (2 and 1) imply an impossible correlation value ($= 3/\sqrt{2}$). The middle matrix has an impossible negative disturbance variance ($= -2$). Whether the nonpositive definite nature of the last matrix is troublesome depends on whether the variance of the first disturbance should be zero. Identity relations (e.g., $\eta_1 = \eta_2 + \eta_3$) or when measurement error is absent (e.g., $x_1 = \xi_1$) are two situations where zero disturbance variances make sense. However, when zero is not a plausible value, then the analyst must determine the source of this improbable value.

The vec operator is the operation of forming a vector from a matrix by stacking each column of a matrix one under the other. For instance:

$$\operatorname{vec} \mathbf{B} = \operatorname{vec}\begin{bmatrix} 0 & \beta_{12} \\ \beta_{21} & 1 \end{bmatrix} = \begin{bmatrix} 0 \\ \beta_{21} \\ \beta_{12} \\ 1 \end{bmatrix}$$

The vec operator appears in Chapter 8.

A *Kronecker product* (or a *direct product*) of two matrices, $\mathbf{S}(p \times q)$ and $\mathbf{T}(m \times n)$, is defined as

$$\mathbf{S} \otimes \mathbf{T} = \begin{bmatrix} s_{11}\mathbf{T} & \cdots & s_{1q}\mathbf{T} \\ \vdots & & \\ s_{p1}\mathbf{T} & \cdots & s_{pq}\mathbf{T} \end{bmatrix}$$

Each element of the left matrix, $\mathbf{S}$, is multiplied by $\mathbf{T}$ to form a submatrix $s_{ij}\mathbf{T}$. All of these submatrices combined result in a $pm \times qn$ matrix. An example is:

$$[\gamma_{11} \quad \gamma_{12}] \otimes \begin{bmatrix} 1 & \beta_{12} \\ \beta_{21} & 1 \end{bmatrix} = \begin{bmatrix} \gamma_{11} & \gamma_{11}\beta_{12} & \gamma_{12} & \gamma_{12}\beta_{12} \\ \gamma_{11}\beta_{21} & \gamma_{11} & \gamma_{12}\beta_{21} & \gamma_{12} \end{bmatrix}$$

Kronecker's products appear in the formulas for the asymptotic standard errors of indirect effects in Chapter 8.

APPENDIX B

Asymptotic Distribution Theory

Asymptotic theory describes the behavior of random variables (or constants) as the sample size increases toward infinity. Its appeal is due to several factors. One is that it is sometimes extremely difficult, or even impossible, to establish the properties of estimators in finite samples, whereas the "large sample" or asymptotic properties may be more easily known. *Consistency* is one such asymptotic property of great importance. Although we may accept an estimator that is biased in small samples, we are hesitant to use one that is an inconsistent estimator—that is, one that does not converge on the population parameter even when the sample size is arbitrarily large.

Asymptotic properties are salient for the estimators of this book. One example is the maximum likelihood (ML) estimator which I describe in Chapter 4. It has largely unknown small sample properties but well-established and desirable asymptotic properties. The same is true for the generalized least squares and the weighted least squares estimators of Chapters 4 and 9, respectively.

In this section I provide a brief introduction to two aspects of asymptotic theory: *convergence in probability* and *convergence in distribution*. I refer the reader to White (1984), Theil (1971), and Rao (1973) for further details on asymptotic theory.

CONVERGENCE IN PROBABILITY

Consider a sequence of random variables $\hat{\theta}_1, \hat{\theta}_2, \hat{\theta}_3, \ldots, \hat{\theta}_N, \ldots,$ where the subscript of $\hat{\theta}$ refers to the sample size from which $\hat{\theta}$ comes. For instance,

$\hat{\theta}_2$ means that $\hat{\theta}$ is from a sample of size 2, and $\hat{\theta}_N$ refers to $\hat{\theta}$ from a sample of N observations. The $\hat{\theta}$ can be an estimator (e.g., $\hat{\theta} = \overline{X}$, or $\hat{\theta} = \hat{\beta}$) or any other random variable. The random variable $\hat{\theta}_N$ converges in probability to a constant θ if

$$\lim_{N \to \infty} P\left[|\hat{\theta}_N - \theta| < \delta\right] = 1, \qquad \text{for any } \delta > 0 \tag{B.1}$$

The $P[\cdot]$ refers to the probability of the expression within brackets being true, δ is a positive, arbitrarily small number, and the limit of $P[\cdot]$ is taken as N goes to infinity. This condition states that the absolute value of the deviation of $\hat{\theta}_N$ from θ becomes less than any positive δ as N grows larger. In the limit as $N \to \infty$, this condition is satisfied with a probability of one. Thus, as the sample size grows larger, $\hat{\theta}_N$ converges in probability to θ. A briefer notation for (B.1) is

$$\operatorname*{plim}_{N \to \infty} \hat{\theta}_N = \theta \tag{B.2}$$

The "plim" stands for the "probability limit of." Sometimes the subscript on $\hat{\theta}_N$ is dropped and is implicit, as is the fact that the limit is taken as $N \to \infty$, so that an expression equivalent to (B.2) is $\operatorname{plim} \hat{\theta} = \theta$. If (B.2) is true, then $\hat{\theta}_N$ is a *consistent* estimator of θ. A simple illustration is for X, a random normal variable with $\hat{\theta}_N = \overline{X}_N$, and $\theta = \mu$ (the population mean). For any sample size, the $E(\overline{X}_N)$ is μ. The variance of $\overline{X}_N$ is σ^2/N, where σ^2 is the variance of the random variable X. As N goes to infinity, the variance of $\overline{X}$ goes to zero and $\overline{X}_N$ converges in probability to μ or $\operatorname{plim} \overline{X} = \mu$. Thus $\overline{X}_N$ is a consistent estimator of μ.

Several useful properties of plims are listed next. The X and Y refer to any random variables, including estimators (such as $\hat{\theta}$ represented earlier). These variables need not be independent. The c represents a constant.

$$\begin{aligned}
\operatorname{plim}(c) &= c \\
\operatorname{plim}(X + Y) &= \operatorname{plim}(X) + \operatorname{plim}(Y) \\
\operatorname{plim}(cX) &= c \operatorname{plim}(X) \\
\operatorname{plim}(XY) &= \operatorname{plim}(X)\operatorname{plim}(Y) \\
\operatorname{plim}(Y^{-1}) &= [\operatorname{plim}(Y)]^{-1} \\
\operatorname{plim}(XY^{-1}) &= \operatorname{plim} X[\operatorname{plim}(Y)]^{-1}
\end{aligned} \tag{B.3}$$

Analogous properties hold for random or constant matrices. For example, $\text{plim}(\mathbf{cX} + \mathbf{Y}) = \mathbf{c}\,\text{plim}(\mathbf{X}) + \text{plim}(\mathbf{Y})$, where $\mathbf{c}$ is a constant vector, $\mathbf{X}$ and $\mathbf{Y}$ are matrices of random variables, and the vector and matrices conform for multiplication and addition.

CONVERGENCE IN DISTRIBUTIONS

Corresponding to the sequence of random variables $\hat{\theta}_1, \hat{\theta}_2, \ldots, \hat{\theta}_N, \ldots,$ is a sequence of distribution functions, $F_1(\ ), F_2(\ ), \ldots, F_N(\ ), \ldots$. If $F_N(\ )$ converges to a distribution function $F(\ )$ as N goes to infinity, then $F(\ )$ is the limiting distribution of $\hat{\theta}_N$. In other words, $\hat{\theta}_N$ *converges in distribution* to $F(\ )$ as $N \to \infty$. When $\text{plim}\,\hat{\theta}_N$ equals a constant, then $F(\ )$ is a *degenerate distribution* since it converges on a single value. However, it is often possible to study the distribution of $\hat{\theta}_N$ as it approaches a degenerate distribution. Studies of these asymptotic or limiting distributions are useful in situations where the finite sample distributions are unknown or are difficult to derive. With large samples the asymptotic distribution can be a reasonable approximation for the distribution of a random variable or an estimator.

Suppose that $\hat{\theta}_N$ is an estimator of a parameter θ and that $\text{plim}\,\hat{\theta}_N$ equals θ. Since $\hat{\theta}_N$ converges to a single value θ as N grows large, it has a degenerate distribution. However, if I multiply $(\hat{\theta}_N - \theta)$ by $\sqrt{N}$, this distribution can be analyzed and used to understand the large sample behavior of the $\hat{\theta}_N$. If, for instance, $\sqrt{N}(\hat{\theta}_N - \theta)$ begins to look like a normal distribution with a mean of zero and a variance of V, I represent this as

$$\sqrt{N}\left(\hat{\theta}_N - \theta\right) \xrightarrow{D} N(0, V) \tag{B.4}$$

where $\xrightarrow{D}$ means "converges in distribution" and $N(0, V)$ indicates a normal distribution with a mean of zero and a variance of V. In this case $\hat{\theta}_N$ is distributed as an asymptotically normal variable with a mean of θ and asymptotic variance of V/N:

$$\hat{\theta}_N \sim AN\left(\theta, \frac{V}{N}\right) \tag{B.5}$$

where AN abbreviates asymptotically normal. I sometimes use the shorthand of $\text{AVAR}(\hat{\theta}_N)$ to refer to N^{-1} times the variance of the limiting distribution of $\hat{\theta}_N$ and I represent the sample estimate of AVAR($\hat{\theta}_N$) by avar($\hat{\theta}_N$).

A simple illustration of these ideas is the sample mean, $\bar{X}_N$. If X has a mean of μ and a variance of σ^2, the central limit theorem states that the limiting distribution of $\sqrt{N}(\bar{X}_N - \mu)$ is normal with a mean of zero and a variance of σ^2, regardless of the distribution of X:

$$\sqrt{N}\left(\bar{X}_N - \mu\right) \xrightarrow{D} N(0, \sigma^2) \tag{B.6}$$

Equation (B.6) is the same as the general expression in equation (B.4), where $\hat{\theta}_N = \bar{X}_N$, $\theta = \mu$, and $V = \sigma^2$. Analogous to equations (B.5), $\bar{X}_N \sim AN(\mu, \sigma^2/N)$, and the AVAR($\bar{X}_N$) is σ^2/N. The sample estimate of AVAR($\bar{X}$) is avar($\bar{X}_N$) $= s^2/N$, where s^2 is the unbiased estimator of σ^2.

Another important property is *asymptotic efficiency*. It refers to the estimator within the class of consistent estimators that has the lowest asymptotic variance.

Much of the preceding discussion is easily generalized to vectors (or matrices) of parameter estimators (or other random variables). For instance, for a vector $\hat{\boldsymbol{\theta}}_N$, we can have

$$\sqrt{N}\left(\hat{\boldsymbol{\theta}}_N - \boldsymbol{\theta}\right) \xrightarrow{D} N(\mathbf{O}, \mathbf{V}) \tag{B.7}$$

where now $\mathbf{V}$ is the covariance matrix of the limiting distribution of $\sqrt{N}(\hat{\boldsymbol{\theta}}_N - \boldsymbol{\theta})$. I use ACOV($\hat{\boldsymbol{\theta}}_N$) to refer to the asymptotic covariance matrix, $N^{-1}\mathbf{V}$, and acov($\hat{\boldsymbol{\theta}}_N$) is the sample estimate of ACOV($\hat{\boldsymbol{\theta}}_N$). To indicate specific asymptotic covariances between two random variables, I use ACOV($\hat{\theta}_1$, $\hat{\theta}_2$), where $\hat{\theta}_1$ and $\hat{\theta}_2$ are two estimators (or more generally two random variables) with the sample estimate shown as acov($\hat{\theta}_1$, $\hat{\theta}_2$). Note that here and elsewhere in the book I omit the subscript of N to simplify the notation so that ACOV($\hat{\theta}_1$, $\hat{\theta}_2$) is the same as ACOV($\hat{\theta}_{N1}$, $\hat{\theta}_{N2}$).

Sometimes asymptotic unbiasedness is confused with consistency. *Asymptotic unbiased* means that as N grows large, the expected value of $\hat{\theta}_N$ is θ. More formally,

$$\lim_{N \to \infty} E\left(\hat{\theta}_N\right) = \theta \tag{B.8}$$

A consistent estimator often is asymptotically unbiased but this is not always the case (Srinivasan 1970, 538–539).

Also, some authors define the asymptotic variance as

$$N^{-1} \lim_{N \to \infty} E\left[\hat{\theta}_N - \lim_{N \to \infty} E\left(\hat{\theta}_N\right)\right]^2 \tag{B.9}$$

This need not equal N^{-1} times the variance of the limiting distribution (i.e., $\text{AVAR}(\hat{\theta}_N)$ as defined earlier). Similarly, the limits (as $N \to \infty$) of the covariance matrix of an estimator, $\hat{\boldsymbol{\theta}}_N$, can differ from the covariance matrix of the limiting distribution of the estimator. With few exceptions, my use of the terms asymptotic variance or asymptotic covariance refer to N^{-1} times the variance or covariance of the limiting distribution.

In Chapters 4, 5, 8, and 9 I make the most use of the asymptotic theory reviewed in this appendix.

References

Afifi, A. A., and R. M. Elashoff (1966). Missing observations in multivariate statistics I: Review of the literature. *Journal of the American Statistical Association*, **61**:595–604.

Aigner, D. J. (1974). MSE dominance of least squares with errors of observations. *Journal of Econometrics*, **2**:365–72.

Aigner, D. J., and A. S. Goldberger, eds. (1977). *Latent Variables in Socio-economic Models*. Amsterdam: North-Holland.

Aigner, D. J., C. Hsiao, A. Kapteyn, and T. Wansbeek (1984). Latent variable models in econometrics. In Z. Griliches and M. D. Intriligator, eds., *Handbook of Econometrics*, Vol. 2. Amsterdam: North-Holland, pp. 1321–1393.

Akaike, H. (1974). A new look at the statistical model identification. *IEEE Transactions on Automatic Control*, **AC-19**:716–723.

Allison, P. D. (1987). Estimation of linear models with incomplete data. In C. C. Clogg, ed., *Sociological Methodology 1987*. Washington, D.C.: American Sociological Association, pp. 71–103.

Althauser, R. P., and T. A. Heberlein (1970). A causal assessment of validity and the multitrait-multimethod matrix. In E. Borgatta, ed., *Sociological Methodology 1970*. San Francisco: Jossey-Bass, pp. 151–169.

Alwin, D. F., and R. M. Hauser (1975). The decomposition of effects in path analysis. *American Sociological Review*, **40**:37–47.

Alwin, D. F., and D. J. Jackson (1979). Measurement models for response errors in surveys: Issues and applications. In K. F. Schuessler, ed., *Sociological Methodology 1980*. San Francisco: Jossey-Bass, pp. 68–119.

Amemiya, T. (1981). Qualitative response models: A survey. *Journal of Economic Literature*, **19**:483–536.

Anderson, J., and D. W. Gerbing (1984). The effects of sampling error on convergence, improper solutions and goodness-of-fit indices for maximum likelihood confirmatory factor analysis. *Psychometrika*, **49**:155–173.

Anderson, T. W. (1958). *An Introduction to Multivariate Statistical Analysis*. New York: Wiley.

Anderson, T. W., and Y. Amemiya (1985). The asymptotic normal distribution of estimators in factor analysis under general conditions. Technical Report No. 12, Stanford University, Palo Alto, California.

Anderson, T. W., and H. Rubin (1956). Statistical inference in factor analysis. *Proceedings of the Third Berkeley Symposium for Mathematical Statistics Problems*, **5**:111–150.

Arminger, G., and R. Schoenberg (1987). Construction of general tests for misspecification with applications to covariance structure models. Paper presented at the 1987 Convention of the American Sociological Association. Chicago, Illinois.

Babakus, E., C. E. Ferguson, Jr., and K. G. Jöreskog (1987). The sensitivity of confirmatory maximum likelihood factor analysis to violations of measurement scale and distributional assumptions. *Journal of Marketing Research*, **24**:222–228.

Bagozzi, R. P., ed. (1982). Special Issue on Causal Modeling. *Journal of Marketing*, **19**:403–584.

Bagozzi, R. P. (1980). *Causal Models in Marketing*. New York: Wiley.

Baker, L. A., and D. W. Fulker (1983). Incomplete covariance matrices and LISREL. *Data Analyst*, **1**:3–5.

Bard, Y. (1974). *Nonlinear Parameter Estimation*. New York: Academic Press.

Bearden, W. O., S. Sharma, and J. E. Teel, (1982). Sample size effects on chi-square and other statistics used in evaluating causal models. *Journal of Marketing Research*, **19**:425–430.

Bekker, P. A., and D. S. G. Pollock (1986). Identification of linear stochastic models with covariance restrictions. *Journal of Econometrics*, **31**:179–208.

Belsley, D. A., E. Kuh, and R. E. Welsh (1980). *Regression Diagnostics: Identifying Influential Data and Sources of Collinearity*. New York: Wiley.

Ben-Israel, A., and T. N. Greville (1974). *Generalized Inverses: Theory and Applications*. New York: Wiley.

Bentler, P. M. (1986a). Structural modeling and Psychometrika: An historical perspective on growth and achievements. *Psychometrika*, **51**:35–51.

Bentler, P. M. (1986b). Lagrange multiplier and Wald tests for EQS and EQS/PC. Unpublished manuscript. Los Angeles: BMDP Statistical Software.

Bentler, P. M. (1985). Theory and implementation of EQS: A structural equations program. Los Angeles: BMDP Statistical Software.

Bentler, P. M. (1983). Simultaneous equation systems as moment structure models. *Journal of Econometrics*, **22**:13–42.

Bentler, P. M. (1982). Confirmatory factor analysis via noniterative estimation: A fast, inexpensive method. *Journal of Marketing Research*, **19**:417–424.

Bentler, P. M. (1980). Multivariate analysis with latent variables: Causal modeling. *Annual Review of Psychology*, **31**:419–456.

Bentler, P. M., and D. G. Bonett (1980). Significance tests and goodness-of-fit in the analysis of covariance structures. *Psychological Bulletin*, **88**:588–600.

Bentler, P. M., and T. Dijkstra (1985). Efficient estimation via linearization in structural models. In P. R. Krishnaiah, ed., *Multivariate Analysis VI*. Amsterdam: North-Holland, pp. 9–42.

Bentler, P. M., and E. H. Freeman (1983). Tests for stability in linear structural equation systems. *Psychometrika*, **48**:143–145.

Bentler, P. M., and S. Y. Lee (1983). Covariance structures under polynomial constraints: Applications to correlation and alpha-type structural models. *Journal of Educational Statistics*, **8**:207–222.

Bentler, P. M., and D. G. Weeks (1980). Multivariate analysis with latent variables. In P. R. Krishnaiah and L. Kanal, eds., *Handbook of Statistics*. Vol. 2. Amsterdam: North-Holland, pp. 747–771.

Berkane, M., and P. M. Bentler (1987). Distribution of kurtoses, with estimators and tests of homogeneity of kurtosis. *Statistics and Probability Letters*, **5**:201–207.

Bielby, W. T. (1986). Arbitrary metrics in multiple-indicator models of latent variables. *Sociological Methods and Research*, **15**:3–23.

Bielby, W. T., and R. M. Hauser (1977). Structural equation models. *Annual Review of Sociology*, **3**:137–161.

Bishop, Y. M., T. S. Fienberg, and P. Holland (1975). *Discrete Multivariate Analysis*. Cambridge, MA: MIT Press.

Blalock, H. M. (1979). Measurement and conceptualization problems: The major obstacle to integrating theory and research. *American Sociological Review*, **44**:881–894.

Blalock, H. M., ed. (1971). *Causal Models in the Social Sciences*. Chicago: Aldine-Atherton.

Blalock, H. M. (1971). Causal models involving unmeasured variables in stimulus-response situations. In H. M. Blalock, ed., *Causal Models in the Social Sciences*. Chicago: Aldine-Atherton, pp. 335–347.

Blalock, H. M. (1967). Causal inferences, closed populations, and measures of association. *American Political Science Review*, **61**:130–136.

Blalock, H. M. (1964). *Causal Inferences in Nonexperimental Research*. Chapel Hill: University of North Carolina.

Blalock, H. M. (1963). Making causal inferences for unmeasured variables from correlations among indicators. *American Journal of Sociology*, **69**:53–62.

Blalock, H. M. (1961). Correlation and causality: The multivariate case. *Social Forces*, **39**:246–251.

Bock, R. D., and R. E. Borgman (1966). Analysis of covariance structures. *Psychometrika*, **31**:507–534.

Bohrnstedt, G. W. (1969). Observations on the measurement of change. In E. F. Borgatta, ed., *Sociological Methodology 1969*. San Francisco: Jossey-Bass, pp. 113–133.

Bohrnstedt, G. W., and G. Marwell (1978). The reliability of products of two random variables. In K. F. Schuessler, ed., *Sociological Methodology 1978*. San Francisco: Jossey-Bass, pp. 254–273.

Bollen, K. A. (1988). A new incremental fit index for general structural equation models. A paper presented at 1988 Southern Sociological Society Meetings. Nashville, Tennessee.

Bollen, K. A. (1987a). Outliers and improper solutions: A confirmatory factor analysis example. *Sociological Methods and Research*, **15**:375–384.

Bollen, K. A. (1987b). Total, direct, and indirect effects in structural equation models. In C. C. Clogg, ed., *Sociological Methodology 1987*, Washington, D.C.: American Sociological Association, pp. 37–69.

Bollen, K. A. (1986). Sample size and Bentler and Bonett's nonnormed fit index. *Psychometrika*, **51**:375–377.

Bollen, K. A. (1984). Multiple indicators: Internal consistency or no necessary relationship? *Quality and Quantity*, **18**:377–385.

Bollen, K. A. (1982). A confirmatory factor analysis of subjective air quality. *Evaluation Review*, **6**:521–535.

Bollen, K. A. (1980). Issues in the comparative measurement of political democracy. *American Sociological Review*, **45**:370–390.

Bollen, K. A. (1979). Political democracy and the timing of development. *American Sociological Review*, **44**:572–587.

Bollen, K. A., and K. H. Barb (1981). Pearson's *R* and coarsely categorized measures. *American Sociological Review*, **46**:232–239.

Bollen, K. A., and R. W. Jackman (1985). Regression diagnostics: An expository treatment of outliers and influential cases. *Sociological Methods and Research*, **13**:510–542.

Bollen, K. A., and K. G. Jöreskog (1985). Uniqueness does not imply identification: A note on confirmatory factor analysis. *Sociological Methods and Research*, **14**:155–163.

Bollen, K. A., and J. Liang (1988). Some properties of Hoelter's CN. *Sociological Research and Methods*, **16**:492–503.

Bollen, K. A., and R. C. Schwing (1987). Air pollution-mortality models. A demonstration of the effects of random measurement error. *Quality and Quantity*, **21**:37–48.

Bollen, K. A., and R. Stine (1987). Bootstrapping structural equation models: variability of indirect effects and goodness of fit measures. Unpublished manuscript.

Boomsma, A. (1983). *On the Robustness of LISREL (Maximum Likelihood Estimation) against Small Sample Size and Nonnormality*. Amsterdam: Sociometric Research Foundation.

Boomsma, A. (1982). The robustness of LISREL against small sample sizes in factor analysis models. In K. G. Jöreskog and H. Wold, eds., *Systems under Indirect Observation, Part I*. Amsterdam: North-Holland, pp. 149–173.

Boudon, R. (1965). A method of linear causal analysis: Dependence analysis. *American Sociological Review*, **30**:365–373.

Brown, C. Hendricks (1983). Asymptotic comparison of missing data procedures for estimating factor loadings. *Psychometrika*, **48**:269–291.

Browne, M. W. (1984). Asymptotic distribution free methods in analysis of covariance structures. *British Journal of Mathematical and Statistical Psychology*, **37**:62–83.

Browne, M. W. (1982). Covariance structures. In D. M. Hawkins, ed., *Topics in Multivariate Analysis*. Cambridge: Cambridge University Press, pp. 72–141.

Browne, M. W. (1974). Generalized least-squares estimators in the analysis of covariance structures. *South African Statistical Journal*, **8**:1–24.

Buse, A. (1982). The likelihood ratio, Wald, and Lagrange multiplier tests: An expository note. *The American Statistician*, **36**:153–157.

Busemeyer, J. R., and L. E. Jones (1983). Analysis of multiplicative combination rules when the causal variables are measured with error. *Psychological Bulletin*, **93**:549–562.

Byron, R. P. (1972). Testing for misspecification in econometric systems using full information. *International Economic Review*, **13**:745–756.

Campbell, D. T. (1963). From description to experimentation: Interpreting trends from quasi-experiments. In C. W. Harris, ed., *Problems in Measuring Change*. Madison: University of Wisconsin Press, pp. 212–242.

Campbell, D. T., and D. W. Fiske (1959). Convergent and discriminant validation by the multitrait-multimethod matrix. *Psychological Bulletin*, **56**:81–105.

Campbell, D. T., and J. C. Stanley (1966). *Experimental and Quasi-Experimental Designs for Research*. Boston: Houghton Mifflin.

Carmines, E. G., and McIver, J. P. (1981). Analyzing models with unobserved variables: Analysis of covariance structures. In G. W. Bohrnstedt and E. F. Borgatta, eds., *Social Measurement: Current Issues*. Beverly Hills, CA: Sage, pp. 65–115.

Cartwright, N. (1983). *How the Laws of Physics Lie*. Oxford: Clarendon Press.

Central Intelligence Agency (1981). *Patterns of International Terrorism: 1980. A Research Paper*. Washington, D.C.: Central Intelligence Agency.

Cermak, G. W. (1983). An experimental test of two models of attribute integration. General Motors Research Publication, GMR-4386. Warren, Michigan.

Cermak G. W., and K. A. Bollen (1984). Observer consistency in judging extent of cloud cover. *Atmospheric Environment*, **17**:2107–2110.

Cermak, G. W., and K. A. Bollen (1982). The relationship between judged air quality: Results of a survey. *Journal of the Air Pollution Control Association*, **32**:86–88.

Cliff, N. (1983). Some cautions concerning the application of causal modeling methods. *Multivariate Behavioral Research*, **18**:115–126.

Cook, R. D., and S. Weisberg (1982). Criticism and influence analysis in regression. In S. Leinhardt, ed., *Sociological Methodology 1982*. San Francisco: Jossey-Bass, pp. 313–362.

Cook, T. D., and D. T. Campbell (1976). The design and conduct of quasi-experiments and true experiments in field settings. In M. D. Dunnette, ed., *Handbook of Industrial and Organizational Psychology*. New York: Rand McNally, pp. 223–326.

Costner, H. L. (1971). Utilizing causal models to discover flaws in experiments. *Sociometry*, **34**:398–410.

Costner, H. L., and R. Schoenberg (1973). Diagnosing indicator ills in multiple indicator models. In A. S. Goldberger and O. D. Duncan, eds., *Structural Equation Models in the Social Sciences*. New York: Seminar Press, pp. 167–199.

Cronbach, L. J. (1951). Coefficient alpha and the internal structure of tests. *Psychometrika*, **16**:297–334.

Cudeck, R., and M. W. Browne (1983). Cross-validation of covariance structures. *Multivariate Behavioral Research*, **18**:147–167.

D'Agostino, R. B. (1986). Tests for the normal distribution. In R. B. D'Agostino and M. A. Stephens, eds., *Goodness-of-Fit Techniques*. New York: Marcel Dekker, pp. 367–419.

Daniel, C., and F. S. Wood (1980). *Fitting Equations to Data*. New York: Wiley.

de Leeuw, J. (1983). Models and methods for the analysis of correlation coefficients. *Journal of Econometrics*, **22**:113–138.

de Leeuw, J., W. J. Keller, and T. Wansbeek, eds. (1983). Interfaces between econometrics and psychometrics. *Annals of Applied Econometrics*, supplement to *Journal of Econometrics*, **22**:1–243.

Dempster, A. P. (1971). An overview of multivariate data analysis. *Journal of Multivariate Analysis*, **1**:316–346.

Dhrymes, P. (1978). *Introductory Econometrics*. New York: Springer-Verlag.

Dijkstra, T. K. (1981). *Latent Variables in Linear Stochastic Models*. Amsterdam: Sociometric Research Foundation.

Duncan, O. D. (1975). *Introduction to Structural Equation Models*. New York: Academic Press.

Duncan, O. D. (1969). Some linear models for two-wave, two-variable panel analysis. *Psychological Bulletin*, **72**:177–182.

Duncan, O. D. (1966). Path analysis: Sociological examples. *American Journal of Sociology*, **72**:1–16.

Duncan, O. D., A. O. Haller, and A. Portes (1968). Peer influences on aspirations: A reinterpretation. *American Journal of Sociology*, **74**:119–137.

Dunlap, J. W., and E. E. Cureton (1930). On the analysis of causation. *Journal of Educational Psychology*, **21**:657–680.

Engle, R. F. (1984). Wald, likelihood ratio, and Lagrange Multiplier tests in econometrics. In A. Griliches, and M. D. Intriligator, eds., *Handbook of Econometrics*, Amsterdam: North-Holland, pp. 776–826.

Englehart, M. D. (1936). The technique of path coefficients. *Psychometrika*, **1**:287–293.

Etezadi-Amoli, J., and R. P. McDonald (1983). A second generation nonlinear factor analysis. *Psychometrika*, **48**:315–342.

Ferber, R., and W. Z. Hirsch (1982). *Social Experimentation and Economic Policy*. Cambridge: Cambridge University Press.

Fisher, F. M. (1970). A correspondence principle for simultaneous equation models. *Econometrica*, **38**:73–92.

Fisher, F. M. (1966). *The Identification Problem in Econometrics*. New York: McGraw-Hill.

Folmer, H. (1981). Measurement of the effects of regional policy instruments by means of linear structural equation models and panel data. *Environment and Planning* A, **13**:1435–1448.

Fox, J. (1984). *Linear Statistical Models and Related Methods*. New York: Wiley.

Fox, J. (1980). Effect analysis in structural equation models. *Sociological Methods and Research*, **9**:3–28.

Fox, K. A. (1958). *Econometric Analysis for Public Policy*. Ames: Iowa State University Press.

Fraser, C. (1980). *COSAN User's Guide*. Toronto: The Ontario Institute for Studies in Education.

Freedman, D. (1986). As others see us: A case study in path analysis. *Journal of Education Statistics*, **12**:101–128.

Fuller, W. A., and M. A. Hidiroglou (1978). Regression estimation after correcting for attenuation. *Journal of the American Statistical Association*, **73**:99–105.

Gerbing, D. W., and J. C. Anderson (1985). The effects of sampling error and model characteristics on parameter estimation for maximum likelihood confirmatory factor analysis. *Multivariate Behavioral Research*, **20**:255–271.

Gerbing, D. W., and J. C. Anderson (1984). On the meaning of within-factor correlated measurement errors. *Journal of Consumer Research*, **11**:572–580.

Gibbons, D. I., and G. C. McDonald (1980). Examining regression relationships between air pollution and mortality. General Motors Research Publication, GMR-3278. Warren, Michigan.

Glymour, C., R. Scheines, P. Spirtes, and K. Kelly (1987). *Discovering Causal Structure*. Orlando, FL: Academic Press.

Goldberg, S. (1958). *Introduction to Difference Equations*. New York: Wiley.

Goldberger, A. S. (1972). Structural equation methods in the social sciences. *Econometrica*, **40**:979–1001.

Goldberger, A. S. (1964). *Econometric Theory*. New York: Wiley.

Goldberger, A. S., and O. D. Duncan, eds. (1973). *Structural Equation Models in the Social Sciences*. New York: Academic Press.

Goldfeld, S. M., and R. E. Quandt (1972). *Nonlinear Methods in Econometrics*. Amsterdam: North-Holland.

Goodman, L. A. (1972). A general model for the analysis of surveys. *American Journal of Sociology*, **77**:1035–1086.

Graff, J. (1979). Verallgemeinertes LISREL-Modell. Unpublished manuscript. Mannheim, Germany.

Graff, J., and P. Schmidt (1982). A general model for decomposition of effects. In K. G. Jöreskog and H. Wold, eds., *Systems under Indirect Observation*. Amsterdam: North-Holland, pp. 131–148.

Granger, C. W. J. (1969). Investigating causal relations by econometric models and cross-spectral methods. *Econometrica*, **37**:424–438.

Graybill, F. A. (1983). *Matrices with Applications in Statistics*. 2d ed. Belmont, CA: Wadsworth.

Griliches, Z. (1957). Specification bias in estimates of production functions. *Journal of Farm Economics*, **39**:8–20.

Gruvaeus, G. T., and K. G. Jöreskog (1970). A computer program for minimizing a function of several variables. Research Bulletin of Educational Testing Service. Princeton, New Jersey.

Haavelmo, T. (1953). Methods of measuring the marginal propensity to consume. In W. C. Hoods and T. C. Koopmans, eds., *Studies in Econometric Methods*. New York: Wiley, pp. 75–91.

Hadley, G. (1961). *Linear Algebra*. Reading, MA: Addison-Wesley.

Hägglund, G. (1982). Factor analysis by instrumental variable methods. *Psychometrika*, **47**:209–222.

Haitovsky, Y. (1968). Missing data in regression analysis. *Royal Statistical Society*, series B, **30**:67–82.

Harman, H. H. (1976). *Modern Factor Analysis*. Chicago: University of Chicago Press.

Hauser, R. M. (1973). Disaggregating a social-psychological model of educational attainment. In A. S. Goldberger and O. D. Duncan, eds., *Structural Equation Models in the Social Sciences*. New York: Academic Press, pp. 255–284.

Hayduk, L. A. (1987). *Structural Equation Modeling with LISREL*. Baltimore: Johns Hopkins University.

Haynam, G. E., Z. Govindarajulu, and F. C. Leone (1973). Tables of the cumulative chi-square distribution. In H. L. Harter and D. B. Owen, eds., *Selected Tables in Mathematical Statistics*. Providence, RI: American Mathematical Society, pp. 1–78.

Heckman, J. T. (1979). Sample selection bias as a specification error. *Econometrica*, **45**:153–161.

Heise, D. R. (1986). Estimating nonlinear models correcting for measurement error. *Sociological Methods and Research*, **14**:447–472.

Heise, D. R. (1970). Causal inferences from panel data. In E. F. Borgatta and G. W. Bohrnstedt, eds., *Sociological Methodology 1970*. San Francisco: Jossey-Bass, pp. 3–27.

Heise, D. R. (1969). Separating reliability and stability in test-retest correlation. *American Sociological Review*, **34**:93–101.

Herting, J. R., and H. L. Costner (1985). Respecification in multiple indicator models. In H. M. Blalock, ed., *Causal Models in the Social Sciences*. 2d ed. New York: Aldine, pp. 321–393.

Hill, K. Q. (1982). Retest reliability for trust in government and governmental responsiveness measures: A research note. *Political Methodology*, **8**:33–46.

Hoelter, J. W. (1983). The analysis of covariance structures: Goodness-of-fit indices. *Sociological Methods and Research*, **11**:325–344.

Holland, P. W. (1986). Statistics and causal inferences. *Journal of the American Statistical Association*, **81**:945–960.

Howe, W. G. (1955). Some contributions to factor analysis. Report No. ORNL-1919. Oak Ridge National Laboratory, Oak Ridge, Tennessee.

Huba, G. J., and P. M. Bentler (1983). Test of a drug use causal model using asymptotically distribution free methods. *Journal of Drug Education*, **13**:3–13.

Hume, D. (1977 [1739]). *A Treatise on Human Nature*. New York: Dutton.

Jeffreys, H. (1983). *Scientific Inference*. Cambridge: Cambridge University Press.

Jennrich, R. I., and D. T. Thayer (1973). A note on Lawley's formulas for standard errors in maximum likelihood factor analysis. *Psychometrika*, **38**:571–580.

Johnson, D. R., and J. C. Creech (1983). Ordinal measures in multiple indicators models: A simulation study of categorization error. *American Sociological Review*, **48**:398–407.

Johnston, J. (1984). *Econometric Methods*. New York: McGraw-Hill.

Jöreskog, K. G. (1981). Analysis of covariance structures. *Scandanavian Journal of Statistics*, **8**:65–92.

Jöreskog, K. G. (1979a). Basic ideas of factor and component analysis. In K. G. Jöreskog and D. Sörbom, *Advances in Factor Analysis and Structural Equation Models*. Cambridge, MA: Abt, pp. 5–20.

Jöreskog, K. G. (1979b). A general approach to confirmatory maximum likelihood factor analysis with addendum. In K. G. Jöreskog and D. Sörbom, *Advances in Factor Analysis and Structural Equation Models*. Cambridge, MA: Abt, pp. 21–43.

Jöreskog, K. G. (1979c). Analyzing psychological data by structural analysis of covariance matrices. In K. G. Jöreskog and D. Sörbom, *Advances in Factor Analysis and Structural Equation Models*. Cambridge, MA: Abt, pp. 45–100.

Jöreskog, K. G. (1979d). Statistical estimation of structural models in longitudinal-developmental investigation. In J. R. Nesselrode and P. B. Baltes, Eds., *Longitudinal Research in the Study of Behavior and Development*. New York: Academic Press, pp. 303–352.

Jöreskog, K. G. (1978). Structural analysis of covariance and correlation matrices. *Psychometrika*, **43**:443–477.

Jöreskog, K. G. (1977). Structural equation models in the social sciences: specification, estimation and testing. In P. R. Krishnaiah, ed., *Applications of Statistics*. Amsterdam: North-Holland, pp. 265–287.

Jöreskog, K. G. (1974). Analyzing psychological data by structural analysis of covariance matrices. In R. C. Atkinson, D. H. Krantz, R. D. Luce, and P. Suppes, eds., *Contemporary Developments in Mathematical Psychology*. San Francisco: Freeman, pp. 1–56.

Jöreskog, K. G. (1973). A general method for estimating a linear structural equation system. In A. S. Goldberger and O. D. Duncan, eds., *Structural Equation Models in the Social Sciences*. New York: Academic Press, pp. 85–112.

Jöreskog, K. G. (1971). Statistical analysis of sets of congeneric tests. *Psychometrika*, **36**:109–133.

Jöreskog, K. G. (1970). A general method for analysis of covariance structures. *Biometrika*, **57**:239–251.

Jöreskog, K. G. (1969). A general approach to confirmatory maximum likelihood factor analysis. *Psychometrika*, **34**:183–202.

Jöreskog, K. G. (1967). Some contributions to maximum likelihood factor analysis. *Psychometrika*, **32**:443–482.

Jöreskog, K. G., and A. S. Goldberger (1975). Estimation of a model with multiple indicators and multiple causes of a single latent variable. *Journal of the American Statistical Association*, **70**:631–639.

Jöreskog, K. G., and A. S. Goldberger (1972). Factor analysis by generalized least squares. *Psychometrika*, **37**:243–260.

Jöreskog, K. G., and D. Sörbom (1986a). *PRELIS: A preprocessor for LISREL*. Mooresville, IN: Scientific Software, Inc.

Jöreskog, K. G., and D. Sörbom (1986b). *LISREL VI: Analysis of Linear Structural Relationships by Maximum Likelihood and Least Square Methods*. Mooresville, IN: Scientific Software, Inc.

Jöreskog, K. G., and D. Sörbom (1988). *LISREL 7: A Guide to the Program and Applications*. Chicago: SPSS, Inc.

Jöreskog, K. G., and H. Wold (1982). *Systems under Indirect Observation, Part I and Part II*. Amsterdam: North-Holland.

Judge, G. G., W. E. Griffiths, R. C. Hill, and T. Lee (1980). *The Theory and Practice of Econometrics*. New York: Wiley.

Keesling, J. W. (1972). Maximum Likelihood Approaches to Causal Analysis. Ph.D. dissertation. Department of Education: University of Chicago.

Kendall, M. G., and A. Stuart (1963). *The Advanced Theory of Statistics, Vol. 1: Distribution and Theory*. London: Griffin.

Kendall, M. G., and A. Stuart (1979). *The Advanced Theory of Statistics, Vol. 3: Inference and Relationship*. London: Griffin.

Kennedy, W. J., and J. E. Gentle (1980). *Statistical Computing*. New York: Marcel Dekker.

Kenny, D. A. (1979). *Correlation and Causality*. New York: Wiley.

Kenny, D. A., and C. M. Judd (1984). Estimating the non-linear and interactive effects of latent variables. *Psychological Bulletin*, **96**:201–210.

Kim, J., and J. Curry (1977). The treatment of missing data in multivariate analysis. *Sociological Methods and Research*, **6**:215–240.

Kluegel, J. R., R. Singleton, Jr., and C. E. Starnes (1977). Subjective class identification: A multiple indicator approach. *American Sociological Review*, **42**:599–611.

Kmenta, J. (1971). *Elements of Econometrics*. New York: Macmillan.

Lave, L. B., and E. P. Seskin (1977). *Air Pollution and Human Health*. Baltimore: Johns Hopkins.

Lave, L. B., and E. P. Seskin (1970). Air pollution and human health. *Science*, **169**:723–733.

Lawley, D. N. (1940). The estimation of factor loadings by the method of maximum likelihood. *Proceedings of the Royal Society of Edinburgh*, **60**:64–82.

Lawley, D. N., and A. E. Maxwell (1971). *Factor Analysis as a Statistical Method*. London: Butterworth.

Lazarsfeld, P. F. (1959). Latent structure analysis. In S. Koch, ed., *Psychology: A Study of Science*. Vol. 3. New York: McGraw-Hill, pp. 476–543.

Lazarsfeld, P. F., and N. W. Henry (1968). *Latent Structure Analysis*. Boston: Houghton Mifflin.

Lee, Sik-Yum (1986). Estimation for structural equation models with missing data. *Psychometrika*, **51**:93–99.

Lee, S. Y. (1980). Estimation of covariance structure models with parameters subject to functional restraints. *Psychometrika*, **45**:309–324.

Leinhardt, S., and S. S. Wasserman (1978). Exploratory data analysis: An Introduction to selected models. In K. F. Schuessler, ed., *Sociological Methodology 1979*. San Francisco: Jossey-Bass, pp. 311–365.

Levi, M. D. (1973). Errors in the variables bias in the presence of correctly measured variables. *Econometrica*, **41**:985–986.

Lieberson, S. (1985). *Making It Count*. Berkeley: University of California Press.

Ling, R. (1983). Review of correlation and causation by Kenny. *Journal of the American Statistical Association*, **77**:489–491.

Lord, F. M., and M. R. Novick (1968). *Statistical Theories of Mental Test Scores*. Reading, MA: Addison-Wesley.

Lunneborg, C. E., and R. D. Abbott (1983). *Elementary Multivariate Analysis for the Behavioral Sciences: Applications of Basic Structure*. Amsterdam: North-Holland.

MacCallum, R. (1986). Specification searches in covariance structure modeling. *Psychological Bulletin*, **100**:107–120.

Mackie, J. L. (1974). *The Cement of the Universe: A Study of Causation*. Oxford: Oxford University Press.

Madansky, A. (1964). Instrumental variables in factor analysis. *Psychometrika*, **29**:105–113.

Maddala, G. S. (1983). *Limited-Dependent and Qualitative Variables in Econometrics*. Cambridge: Cambridge University Press.

Mardia, K. V. (1985). Mardia's test of multinormality. In S. Kotz and N. L. Johnson, eds., *Encyclopedia of Statistical Sciences*. Vol. 5. New York: Wiley, pp. 217–221.

Mardia, K. V. (1974). Applications of some measures of multivariate skewness and kurtosis in testing normality and robustness studies. *Sankhya*, B, **36**:115–128.

Mardia, K. V. (1970). Measures of multivariate skewness and kurtosis with applications. *Biometrika*, **57**:519–530.

Mardia, K. V., and K. Foster (1983). Omnibus tests of multinormality based on skewness and kurtosis. *Communication in Statistics*, **12**:207–221.

Marsh, H. W., and D. Hocevar (1985). Application of confirmatory factor analysis to the study of self-concept: First and higher-order factor models and their invariance across groups. *Psychological Bulletin*, **97**:562–582.

Martin, W. S. (1978). Effects of scaling on the correlation coefficient: Additional considerations. *Journal of Marketing Research*, **15**:304–308.

Matsueda, R. L., and W. T. Bielby (1986). Statistical power in covariance structure models. In N. B. Tuma, ed., *Sociological Methodology 1986*. Washington, D.C.: American Sociological Association, pp. 120–158.

McArdle, J. J., and McDonald, R. P. (1984). Some algebraic properties of the reticular action model for moment structures. *British Journal of Mathematical and Statistical Psychology*, **37**:234–251.

McCallum, B. T. (1972). Relative asymptotic bias from errors of omission and measurement. *Econometrica*, **40**:757–758.

McDonald, J. A., and D. A. Clelland (1984). Textile workers and union sentiment. *Social Forces*, **63**:502–521.

McDonald, R. P. (1982). A note on the investigation of local and global identifiability. *Psychometrika*, **47**:101–103.

McDonald, R. P. (1980). A simple comprehensive model for the analysis of covariance structures: Some remarks on applications. *British Journal of Mathematical and Statistical Psychology*, **33**:161–83.

McDonald, R. P. (1978). A simple comprehensive model for the analysis of covariance structures. *British Journal of Mathematical and Statistical Psychology*, **31**:59–72.

McDonald, R. P. (1967a). Numerical methods for polynomial models in non-linear factor analysis. *Psychometrika*, **32**:77–112.

McDonald, R. P. (1967b). Nonlinear factor analysis. *Psychometric Monograph*, No. 15.

McDonald, R. P., and E. J. A. Burr (1967). A comparison of four methods of constructing factor scores. *Psychometrika*, **32**:381–401.

McDonald, R. P., and W. R. Krane (1979). A Monte Carlo study of local identifiability and degrees of freedom in the asymptotic likelihood ratio test. *British Journal of Mathematical and Statistical Psychology*, **32**:121–131.

McFatter, R. M. (1979). The use of structural equation models in interpreting regression equations including suppressor and enhancer variables. *Applied Psychological Measurement*, **3**:123–135.

Miller, A. D. (1971). Logic of causal analysis: From experimental to nonexperimental designs. In H. M. Blalock, ed., *Causal Models in the Social Sciences*. Chicago: Aldine-Atherton, pp. 273–294.

Mooijaart, A., and P. M. Bentler (1986). Random polynomial factor analysis. In E. Diday et al., eds., *Data Analysis and Informatics*. Amsterdam: Elsevier Science, pp. 241–250.

Moran, P. A. P. (1961). Path coefficients reconsidered. *Australian Journal of Statistics*, **3**:87–93.

Mulaik, S. (1972). *The Foundations of Factor Analysis*. New York: McGraw-Hill.

Muthén, B. (1987). *LISCOMP: Analysis of Linear Structural Equations with a Comprehensive Measurement Model*. Mooresville, IN: Scientific Software, Inc.

Muthén, B. (1984). A general structural equation model with dichotomous, ordered categorical and continuous latent variable indicators. *Psychometrika*, **49**:115–132.

Muthén, B. (1983). Latent variable structural equation modeling with categorical data. *Journal of Econometrics*, **22**:43–65.

Muthén, B. (1982). Some categorical response models with continuous latent variables. In K. G. Jöreskog and H. Wold, eds., *Systems under Indirect Observation*. Amsterdam: North-Holland, pp. 65–79.

Muthén, B. (1979). A structural probit model with latent variables. *Journal of the American Statistical Association*, **74**:807–811.

Muthén, B. and K. G. Jöreskog (1983). Selectivity problems in quasi-experimental studies. *Evaluation Review*, **7**:139–174.

Muthén, B., and D. Kaplan (1985). A comparison of some methodologies for the factor analysis of non-normal Likert variables. *British Journal of Mathematical and Statistical Psychology*, **38**:171–189.

Nagel, E. (1965). Types of causal explanations in science. In D. Lerner, ed., *Cause and Effect*. New York: Free Press, pp. 11–26.

Namboordiri, N. K., L. F. Carter, and H. M. Blalock (1975). *Applied Multivariate Analysis and Experimental Designs*. New York: McGraw-Hill.

Olsson, U. (1979a). On the robustness of factor analysis against crude classification of the observations. *Multivariate Behavioral Research*, **14**:485–500.

Olsson, U. (1979b). Maximum likelihood estimation of the polychoric correlation coefficient. *Psychometrika*, **44**:443–460.

Olsson, U., F. Drasgow, and N. J. Dorans (1982). The polyserial correlation coefficient. *Psychometrika*, **47**:337–347.

Pearson, K. (1901). On lines and planes of closest fit to systems of points in space. *Philosophical Magazine*, **6**:559–572.

Pedhazur, E. J. (1982). *Multiple Regression in Behavioral Research*. New York: Holt, Rinehart and Winston.

Popper, K. R. (1968). *The Logic of Scientific Discovery*. New York: Harper.

Rao, C. R. (1973). *Linear Statistical Inference and Its Applications*. 2d ed. New York: Wiley.

Reisenzein, R. (1986). A structural equation analysis of Weiner's attribution-affect model of helping behavior. *Journal of Personality and Social Psychology*, **50**:1123–1133.

Rindskopf, D. (1984a). Structural equation models: Empirical identification, Heywood cases and related problems. *Sociological Methods and Research*, **13**:109–119.

Rindskopf, D. (1984b). Using phantom and imaginary latent variables to parameterize constraints in linear structural models. *Psychometrika*, **49**:37–47.

Rindskopf, D. (1983). Parameterizing inequality constraints on unique variances in linear structural models. *Psychometrika*, **48**:73–83.

Robinson, P. M. (1974). Identification, estimation, and large sample theory for regressions containing unobservable variables. *International Economic Review*, **15**:680–692.

Rosenthal, R. (1966). *Experimenter Effects in Behavioral Research*. New York: Appleton-Century Crofts.

Rothenberg, T. J. (1971). Identification in parametric models. *Econometrica*, **39**:577–591.

Rubin, D. B. (1976). Inference and missing data. *Biometrika*, **63**:581–592.

Russell, B. (1912–1913). On the notion of cause. *Proceedings of the Aristotelian Society*, **13**:1–26.

Sachs, L. (1982). *Applied Statistics: A Handbook of Techniques*. Translated by Zenon Reynarowych. New York: Springer-Verlag.

Saris, W. E., W. M. de Pijper, and J. Mulder (1978). Optimal procedures for estimation of factor scores. *Sociological Methods and Research*, **7**:85–106.

Saris, W. E., W. M. Pijper, and P. Zegwaart (1979). Detection of specification errors in linear structural equation models. In K. F. Schuessler, ed., *Sociological Methodology 1979*. San Francisco: Jossey-Bass, pp. 151–171.

Saris, W. E., A. Satorra, and D. Sörbom (1987). The detection and correction of specification errors in structural equation models. In C. C. Clogg, ed., *Sociological Methodology 1987*. Washington, D.C.: American Sociological Association, pp. 105–130.

Saris, W. E., J. den Ronden, and A. Satorra (forthcoming). Testing structural equation models. In P. F. Cultance and J. R. Ecob, eds., *Structural Modeling*. Cambridge: Cambridge University Press.

Saris, W. E., and L. H. Stronkhorst (1984). *Causal Modelling in Non-Experimental Research*. Amsterdam: Sociometric Research.

Satorra, A., and P. M. Bentler (1986). Robustness properties of ML statistics in covariance structure analysis. Unpublished manuscript.

Satorra, A., and W. E. Saris (1985). Power of the likelihood ratio test in covariance structure analysis. *Psychometrika*, **50**:83–90.

Schneider, B. (1970). Relationships between various criteria of leadership in small groups. *Journal of Social Psychology*, **82**:253–261.

Schoenberg, R. (1987). *LINCS: Linear Covariance Structure Analysis Users Guide*. Kensington, MD: RJS Software.

Schwarz, G. (1978). Estimating the dimensions of a model. *Annals of Statistics*, **6**:461–464.

Schwertman, N. C., and D. M. Allen (1979). Smoothing an indefinite variance-covariance matrix. *Journal of Statistical Computing and Simulation*, **9**:183–194.

Searle, S. R. (1982). *Matrix Algebra Useful for Statistics*. New York: Wiley.

Shapiro, A. (1987). Robustness properties of the MDF analysis of moment structures. *South African Statistical Journal*, **21**:39–62.

Simon, H. A. (1971 [1954]). Spurious correlation: A causal interpretation. In H. M. Blalock, ed., *Causal Models in the Social Sciences*. Chicago: Aldine-Atherton, pp. 5–17.

Simon, J. L. (1969). The effect of income on fertility. *Population Studies*, **23**:327–341.

Simon, J. L. (1968). The effect of income on the suicide rate: A paradox resolved. *American Journal of Sociology*, **74**:302–303.

Sims, C. A. (1972). Money, income, and causality. *American Economic Review*, **62**:540–552.

Sobel, M. (1986). Some new results on indirect effects and their standard errors in covariance structure models. In N. B. Tuma, ed., *Sociological Methodology 1986*. Washington, D.C.: American Sociological Association, pp. 159–186.

Sobel, M. (1982). Asymptotic confidence intervals for indirect effects in structural equation models. In S. Leinhardt, ed., *Sociological Methodology 1982*. San Francisco: Jossey-Bass, pp. 290–312.

Sobel, M., and G. W. Bohrnstedt (1985). Use of null models in evaluating the fit of covariance structure models. In N. B. Tuma, ed., *Sociological Methodology 1985*. San Francisco: Jossey-Bass, pp. 152–178.

Sörbom, D. (1975). Detection of correlated errors in longitudinal data. *British Journal of Mathematical and Statistical Psychology*, **28**:138–151.

Spearman, C. (1904). General intelligence, objectively determined and measured. *American Journal of Psychology*, **15**:201–293.

Srinivasan, T. N. (1970). Approximations to finite sample moments of estimators whose exact sampling distributions are unknown. *Econometrica*, **38**:533–541.

Srole, L. (1956). Social integration and certain corollaries: An exploratory study. *American Sociological Review*, **21**:709–716.

Stapelton, D. C. (1977). Analyzing political participation data with a MIMIC model. In K. F. Schuessler, ed., *Sociological Methodology 1978*. San Francisco: Jossey-Bass, pp. 52–74.

Steiger, J. H., A. Shapiro, and M. W. Browne (1985). On the multivariate asymptotic distribution of sequential chi-square statistics. *Psychometrika*, **50**:253–264.

Stolzenberg, R. M. (1979). The measurement and decomposition of causal effects in nonlinear and nonadditive models. In K. Schuessler, ed., *Sociological Methodology 1980*. San Francisco: Jossey-Bass, pp. 459–488.

Suppes, P. (1970). *A Probabilistic Theory of Causality*. Amsterdam: North-Holland.

Tanaka, J. S., and P. M. Bentler (1984). Quasi-likelihood estimation in asymptotically efficient covariance structure models. *Proceedings of the Social Statistics Session of the American Statistical Association*, pp. 658–662.

Tanaka, J. S., and G. J. Huba (1985). A fit index for covariance structure models under arbitrary GLS estimation. *British Journal of Mathematical and Statistical Psychology*, **38**:197–201.

Theil, H. (1957). Specification errors and the estimation of economic relationships. *Review of the International Statistical Institute*, **25**:41–51.

Thurstone, L. L. (1947). *Multiple-Factor Analysis*. Chicago: University of Chicago Press.

Timm, N. H. (1970). The estimation of variance-covariance and correlation matrices from incomplete data. *Psychometrika*, **35**:417–437.

Tucker, L. R., and C. Lewis (1973). A reliability coefficient for maximum likelihood factor analysis. *Psychometrika*, **38**:1–10.

Tukey, J. W. (1954). Causation, regression and path analysis. In O. K. Kempthorne, T. A. Bancroft, J. W. Gowen and J. L. Lush, eds., *Statistics and Mathematics in Biology*. Ames: Iowa State University Press, pp. 35–66.

Turner, M. E., and C. D. Stevens (1959). The regression analysis of causal paths. *Biometrics*, **15**:236–258.

Valentine, T. J. (1980). Hypothesis tests and confidence intervals for mean elasticities calculated from linear regression equations. *Economic Letters*, **4**:363–367.

Wald, A. (1950). A note on the identification of economic relations. In T. C. Koopmans, ed., *Statistical Inference in Dynamic Economic Models*. New York: Wiley, pp. 238–244.

Wallace, W. A. (1972). *Causality and Scientific Explanation*. Vol. 1. Ann Arbor: University of Michigan Press.

Weiss, D. J., and M. L. Davison (1981). Test theory and methods. *Annual Review of Psychology*, **32**:629–658.

Werts, C. E., and R. L. Linn (1970). Path analysis: Psychological examples. *Psychological Bulletin*, **74**:193–212.

Werts, C. E., D. A. Rock, and J. Grandy. (1979). Confirmatory factor analysis applications: Missing data problems and comparisons of path models between populations. *Multivariate Behavioral Research*, **14**:199–213.

Wheaton, B., B. Muthén, D. F. Alwin, and G. F. Summers (1977). Assessing reliability and stability in panel models. In D. R. Heise, ed., *Sociological Methodology 1977*. San Francisco: Jossey-Bass, pp. 84–136.

White, Halbert (1984). *Asymptotic Theory for Econometricians*. Orlando, FL: Academic Press.

Wickens, M. R. (1972). A note on the use of proxy variables. *Econometrica*, **40**:759–761.

Wiley, D. E. (1973). The identification problem for structural equation models with unmeasured variables. In A. S. Goldberger and O. D. Duncan, eds., *Structural Equation Models in the Social Sciences*. New York: Academic Press, pp. 69–83.

Wold, H. (1956). Causal inferences from observational data. *Journal of the Royal Statistical Society*, A, **119**:28–60.

Wong, S. K., and J. S. Long (1987). Parameterizing nonlinear constraints in models with latent variables. Technical Report. Washington State University.

Wonnacott, R. J., and T. H. Wonnacott (1979). *Econometrics*. New York: Wiley.

Wright, S. (1960). Path coefficients and path regressions: Alternative or complementary concepts? *Biometrics*, **16**:189–202.

Wright, S. (1954). The interpretation of multivariate systems. In O. K. Kempthorne, T. A. Bancroft, J. W. Gowen and J. L. Lush, eds., *Statistics and Mathematics in Biology*. Ames: Iowa State University Press, pp. 11–33.

Wright, S. (1934). The method of path coefficients. *Annals of Mathematical Statistics*, **5**:161–215.

Wright, S. (1921). Correlation and Causation. *Journal of Agricultural Research*, **20**:557–585.

Wright, S. (1918). On the Nature of Size Factors. *Genetics*, **3**:367–374.

Wylie, P. B. (1976). Effects of coarse grouping and skewed marginal distributions on the Pearson product moment correlation coefficient. *Educational and Psychological Measurement*, **36**:1–7.

Zellner, A. (1984). Causality and econometrics. In A. Zellner, *Basic Issues in Econometrics*. Chicago: University of Chicago Press, pp. 35–74.

Index

HOLLANDER and WOLFE • Nonparametric Statistical Methods
IMAN and CONOVER • Modern Business Statistics
JESSEN • Statistical Survey Techniques
JOHNSON • Multivariate Statistical Simulation
JOHNSON and KOTZ • Distributions in Statistics
Discrete Distributions
Continuous Univariate Distributions—1
Continuous Univariate Distributions—2
Continuous Multivariate Distributions
JUDGE, HILL, GRIFFITHS, LÜTKEPOHL and LEE • Introduction to the Theory and Practice of Econometrics, *Second Edition*
JUDGE, GRIFFITHS, HILL, LÜTKEPOHL and LEE • The Theory and Practice of Econometrics, *Second Edition*
KALBFLEISCH and PRENTICE • The Statistical Analysis of Failure Time Data
KISH • Statistical Design for Research
KISH • Survey Sampling
KUH, NEESE, and HOLLINGER • Structural Sensitivity in Econometric Models
KEENEY and RAIFFA • Decisions with Multiple Objectives
LAWLESS • Statistical Models and Methods for Lifetime Data
LEAMER • Specification Searches: Ad Hoc Inference with Nonexperimental Data
LEBART, MORINEAU, and WARWICK • Multivariate Descriptive Statistical Analysis: Correspondence Analysis and Related Techniques for Large Matrices
LINHART and ZUCCHINI • Model Selection
LITTLE and RUBIN • Statistical Analysis with Missing Data
McNEIL • Interactive Data Analysis
MAGNUS and NEUDECKER • Matrix Differential Calculus with Applications in Statistics and Econometrics
MAINDONALD • Statistical Computation
MALLOWS • Design, Data, and Analysis by Some Friends of Cuthbert Daniel
MANN, SCHAFER and SINGPURWALLA • Methods for Statistical Analysis of Reliability and Life Data
MARTZ and WALLER • Bayesian Reliability Analysis
MASON, GUNST, and HESS • Statistical Design and Analysis of Experiments with Applications to Engineering and Science
MIKÉ and STANLEY • Statistics in Medical Research: Methods and Issues with Applications in Cancer Research
MILLER • Beyond ANOVA, Basics of Applied Statistics
MILLER • Survival Analysis
MILLER, EFRON, BROWN, and MOSES • Biostatistics Casebook
MONTGOMERY and PECK • Introduction to Linear Regression Analysis
NELSON • Applied Life Data Analysis
OSBORNE • Finite Algorithms in Optimization and Data Analysis
OTNES and ENOCHSON • Applied Time Series Analysis: Volume I, Basic Techniques
OTNES and ENOCHSON • Digital Time Series Analysis
PANKRATZ • Forecasting with Univariate Box-Jenkins Models: Concepts and Cases
PLATEK, RAO, SARNDAL and SINGH • Small Area Statistics: An International Symposium
POLLOCK • The Algebra of Econometrics
RAO and MITRA • Generalized Inverse of Matrices and Its Applications
RÉNYI • A Diary on Information Theory
RIPLEY • Spatial Statistics
RIPLEY • Stochastic Simulation
ROSS • Introduction to Probability and Statistics for Engineers and Scientists
ROUSSEEUW and LEROY • Robust Regression and Outlier Detection
RUBIN • Multiple Imputation for Nonresponse in Surveys